MINNESOTA STUDIES IN THE PHILOSOPHY OF SCIENCE

Minnesota Studies in the PHILOSOPHY OF SCIENCE

HERBERT FEIGL AND GROVER MAXWELL, GENERAL EDITORS

VOLUME VI

Induction, Probability, and Confirmation

EDITED BY

GROVER MAXWELL AND ROBERT M. ANDERSON, JR.

FOR THE MINNESOTA CENTER FOR PHILOSOPHY OF SCIENCE

UNIVERSITY OF MINNESOTA PRESS, MINNEAPOLIS

PRINTED IN THE UNITED STATES OF AMERICA

PUBLISHED IN CANADA BY BURNS & MACEACHERN LIMITED, DON MILLS, ONTARIO

Library of Congress Catalog Card Number: 74-83133
ISBN 0-8166-0736-2

Preface

Much of the inspiration for this, the sixth volume of the *Minnesota Studies in the Philosophy of Science* was provided by a conference on confirmation theory held at the Minnesota Center for Philosophy of Science in 1968. In fact, the contributors to this volume, with a couple of important exceptions, are those who participated in the conference. However, most of the papers in the volume were written quite recently, several years after the conference, and the others have been extensively augmented and revised. No special significance should be attached to the order of the essays in the volume; very roughly, the book begins with articles treating general and fundamental problems of confirmation theory and its foundations, and these are followed by topics of a more specific or a more specialized nature.

We are sincerely grateful to the contributors for their patience and cooperation, but most of all for their excellent contributions, to Caroline Cohen, the Center secretary, for her usual indispensable assistance, to Demetria C. Q. Leong for her assistance with the index, and to the University of Minnesota Press.

December 1974

Grover Maxwell
Robert M. Anderson, Jr.

Contents

Contents

MINNESOTA STUDIES IN THE PHILOSOPHY OF SCIENCE

WESLEY C. SALMON

Confirmation and Relevance

Item: One of the earliest surprises to emerge from Carnap's precise and systematic study of confirmation was the untenability of the initially plausible Wittgenstein confirmation function c†. Carnap's objection rested on the fact that c† precludes "learning from experience" because it fails to incorporate suitable relevance relations. Carnap's alternative confirmation function c* was offered as a distinct improvement because it does sustain the desired relevance relations.[1]

Item: On somewhat similar grounds, it has been argued that the heuristic appeal of the concept of partial entailment, which is often used to explain the basic idea behind inductive logic, rests upon a confusion of relevance with nonrelevance relations. Once this confusion is cleared up, it seems, the apparent value of the analogy between full and partial entailment vanishes.[2]

Item: In a careful discussion, based upon his detailed analysis of relevance, Carnap showed convincingly that Hempel's classic conditions of adequacy for any explication of the concept of confirmation are vitiated by another confusion of relevance with nonrelevance relations.[3]

Item: A famous controversy, in which Popper charges that Carnap's theory of confirmation contains a logical inconsistency, revolves around the same issue. As a result of this controversy, Carnap acknowledged in the preface to the second edition of *Logical Foundations of Probability*

AUTHOR'S NOTE: I wish to express gratitude to the National Science Foundation for support of research on inductive logic and probability. Some of the ideas in this paper were discussed nontechnically in my article "Confirmation," *Scientific American*, 228, 5 (May 1973), 75–83.

[1] Rudolf Carnap, *Logical Foundations of Probability* (Chicago: University of Chicago Press, 1950), sec. 110A.

[2] Wesley C. Salmon, "Partial Entailment as a Basis for Inductive Logic," in Nicholas Rescher, ed., *Essays in Honor of Carl G. Hempel* (Dordrecht: Reidel, 1969), and "Carnap's Inductive Logic," *Journal of Philosophy*, 64 (1967), 725–39.

[3] Carnap, *Logical Foundations*, secs. 86–88.

that the first edition had been unclear with regard to this very distinction between relevance and nonrelevance concepts.[4]

Item: A new account of statistical explanation, based upon relations of relevance, has recently been proposed as an improvement over Hempel's well-known account, which is based upon relations of high degree of confirmation.[5]

Item: The problem of whether inductive logic can embody rules of acceptance — i.e., whether there are such things as inductive inferences in the usual sense — has been a source of deep concern to inductive logicians since the publication of Carnap's *Logical Foundations of Probability* (1950). Risto Hilpinen has proposed a rule of inductive inference which, he claims, avoids the "lottery paradox," thus overcoming the chief obstacle to the admission of rules of acceptance. Hilpinen's rule achieves this feat by incorporating a suitable combination of relevance and high confirmation requirements.[6]

The foregoing enumeration of issues is designed to show the crucial importance of the relations between confirmation and relevance. Yet, in spite of the fact that many important technical results have been available at least since the publication of Carnap's *Logical Foundations of Probability* in 1950, it seems to me that their consequences for the concept of confirmation have not been widely acknowledged or appreciated. In the first three sections of this paper, I shall summarize some of the most important facts, mainly to make them available in a single concise and relatively nontechnical presentation. All the material of these sections is taken from the published literature, much of it from the latter chapters of Carnap's book. In section 4, in response to questions raised by Adolf Grünbaum,[7] I shall apply some of these considerations to the Duhemian problems, where, to the best of my knowledge, they have not previously been brought to bear. In section 5, I shall attempt to pinpoint the source of some apparent difficulties with relevance concepts, and in the final section, I shall try to draw some morals from these results. All in all, I believe that many of the facts are shocking and counterintuitive

[4] Karl R. Popper, *The Logic of Scientific Discovery* (New York: Basic Books, 1959), appendix 9; Carnap, *Logical Foundations*, preface to the 2nd ed., 1962.

[5] Wesley C. Salmon, *Statistical Explanation and Statistical Relevance* (Pittsburgh: University of Pittsburgh Press, 1971).

[6] Risto Hilpinen, *Rules of Acceptance and Inductive Logic*, in *Acta Philosophica Fennica*, XXII (Amsterdam: North-Holland, 1968).

[7] In private conversation. At the same time he reported that much of his stimulus for raising these questions resulted from discussions with Professor Laurens Laudan.

and that they have considerable bearing upon current ideas about confirmation.

1. Carnap and Hempel

As Carnap pointed out in *Logical Foundations of Probability*, the concept of confirmation is radically ambiguous. If we say, for example, that the special theory of relativity has been confirmed by experimental evidence, we might have either of two quite distinct meanings in mind. On the one hand, we may intend to say that the special theory has become an accepted part of scientific knowledge and that it is very nearly certain in the light of its supporting evidence. If we admit that scientific hypotheses can have numerical degrees of confirmation, the sentence, on this construal, says that the degree of confirmation of the special theory on the available evidence is high. On the other hand, the same sentence might be used to make a very different statement. It may be taken to mean that some particular evidence — e.g., observations on the lifetimes of mesons — renders the special theory more acceptable or better founded than it was in the absence of this evidence. If numerical degrees of confirmation are again admitted, this latter construal of the sentence amounts to the claim that the special theory has a higher degree of confirmation on the basis of the new evidence than it had on the basis of the previous evidence alone.

The discrepancy between these two meanings is made obvious by the fact that a hypothesis *h*, which has a rather low degree of confirmation on prior evidence *e*, may have its degree of confirmation raised by an item of positive evidence *i* without attaining a high degree of confirmation on the augmented body of evidence *e.i*. In other words, a hypothesis may be confirmed (in the second sense) without being confirmed (in the first sense). Of course, we may believe that hypotheses can achieve high degrees of confirmation by the accumulation of many positive instances, but that is an entirely different matter. It is initially conceivable that a hypothesis with a low degree of confirmation might have its degree of confirmation increased repeatedly by positive instances, but in such a way that the confirmation approaches ¼ (say) rather than 1. Thus, it may be possible for hypotheses to be repeatedly confirmed (in the second sense) without ever getting confirmed (in the first sense). It can also work the other way. A hypothesis *h* that already has a high degree of confirmation on evidence e may still have a high degree of confirmation on evidence $e.i.$, even though the addition of evidence i does not raise the degree of confirmation of h. In this case, h is confirmed (in the first sense) without being confirmed (in the second sense) on the basis of additional evidence i.

If we continue to speak in terms of numerical degrees of confirmation, as I shall do throughout this paper, we can formulate the distinction between these two senses of the term "confirm" clearly and concisely. For uniformity of formulation, let us assume some background evidence e (which may, upon occasion, be the tautological evidence t) as well as some additional evidence i on the basis of which degrees of confirmation are to be assessed. We can define "confirm" in the first (absolute; non-relevance) sense as follows:

D1. Hypothesis h is confirmed (in the absolute sense) by evidence i in the presence of background evidence $e =_{df} c(h,e.i) > b$, where b is some chosen number, presumably close to 1.

This concept is absolute in that it makes no reference to the degree of confirmation of h on any other body of evidence.[8] The second (relevance) sense of "confirm" can be defined as follows:

D2. Hypothesis h is confirmed (in the relevance sense) by evidence i in the presence of background evidence $e =_{df} c(h,e.i) > c(h,e)$.

This is a relevance concept because it embodies a relation of change in degree of confirmation. Indeed, Carnap's main discussion of this distinction follows his technical treatment of relevance relations, and the second concept of confirmation is explicitly recognized as being identical with the concept of positive relevance.[9]

It is in this context that Carnap offers a detailed critical discussion of Hempel's criteria of adequacy for an explication of confirmation.[10] Car-

[8] The term "absolute probability" is sometimes used to refer to probabilities that are not relative or conditional. E.g., Carnap's null confirmation $c_0(h)$ is an absolute probability, as contrasted with $c(h,e)$ in which the degree of confirmation of h is relative to, or conditional upon, e. The distinction I am making between the concepts defined in D1 and D2 is quite different. It is a distinction between two different types of confirmation, where one is a conditional probability and the other is a relevance relation defined in terms of conditional probabilities. In this paper, I shall not use the concept of absolute probability at all; in place of null confirmation I shall always use the confirmation $c(h,t)$ on tautological evidence, which is equivalent to the null confirmation, but which is a conditional or relative probability.

[9] Carnap, *Logical Foundations*, sec. 86.

[10] These conditions of adequacy are presented in Carl G. Hempel, "Studies in the

nap shows conclusively, I believe, that Hempel has conflated the two concepts of confirmation, with the result that he has adopted an indefensible set of conditions of adequacy. As Carnap says, he is dealing with two explicanda, not with a single unambiguous one. The incipient confusion is signaled by his characterization of

> the quest for general objective criteria determining (A) whether, and — if possible — even (B) to what degree, a hypothesis *h* may be said to be corroborated by a given body of evidence e. . . . The two parts of this . . . problem can be related in somewhat more precise terms as follows:
> (A) To give precise definitions of the two nonquantitative relational concepts of confirmation and disconfirmation; *i.e.* to define the meaning of the phrases 'e confirms *h*' and 'e disconfirms *h*'. (When e neither confirms nor disconfirms *h*, we shall say that e is neutral, or irrelevant, with respect to *h*.)
> (B) (1) To lay down criteria defining a metrical concept "degree of confirmation of *h* with respect to e," whose values are real numbers . . .[11]

The parenthetical remark under (A) makes it particularly clear that a relevance concept of confirmation is involved there, while a nonrelevance concept of confirmation is obviously involved in (B).

The difficulties start to show up when Hempel begins laying down conditions of adequacy for the concept of confirmation (A) (as opposed to degree of confirmation (B)). According to the very first condition "entailment is a special case of confirmation." This condition states:

H-8.1 Entailment Condition. Any sentence which is entailed by an observation report is confirmed by it.[12]

If we are concerned with the absolute concept of confirmation, this condition is impeccable, for $c(h,e) = 1$ if e entails *h*. It is not acceptable, however, as a criterion of adequacy for a relevance concept of confirmation. For suppose our hypothesis *h* is "$(\exists x)Fx$" while evidence e is "*Fa*" and evidence *i* is "*Fb*." In this case, *i* entails *h*, but *i* does not confirm *h* in the relevance sense, for $c(h,e.i) = 1 = c(h,e)$.

Carnap offers further arguments to show that the following condition has a similar defect:

Logic of Confirmation," *Mind*, 54 (1945), 1–26, 97–121. Reprinted, with a 1964 postscript, in Carl G. Hempel, *Aspects of Scientific Explanation* (New York: Free Press, 1965). Page references in succeeding notes will be to the reprinted version.

[11] Hempel, *Aspects of Scientific Explanation*, p. 6. Hempel's capital letters "*H*" and "*E*" have been changed to lowercase for uniformity with Carnap's notation.

[12] *Ibid.*, p. 31. Following Carnap, an "H" is attached to the numbers of Hempel's conditions.

H-8.3 Consistency Condition. Every logically consistent observation report is logically compatible with the class of all the hypotheses which it confirms.[13]

This condition, like the entailment condition, is suitable for the absolute concept of confirmation, but not for the relevance concept. For, although no two incompatible hypotheses can have high degrees of confirmation on the same body of evidence, an observation report can be positively relevant to a number of different and incompatible hypotheses, provided that none of them has too high a prior degree of confirmation on the background evidence e. This happens typically when a given observation is compatible with a number of incompatible hypotheses — when, for example, a given bit of quantitative data fits several possible curves.

The remaining condition Hempel wished to adopt is as follows:

H-8.2 Consequence Condition. If an observation report confirms every one of a class k of sentences, then it also confirms any sentence which is a logical consequence of k.[14]

It will suffice to look at two conditions that follow from it:[15]

H-8.21 Special Consequence Condition. If an observation report confirms a hypothesis h, then it also confirms every consequence of h.

H-8.22 Equivalence Condition. If an observation report confirms a hypothesis h, then it also confirms every hypothesis that is logically equivalent with h.

The equivalence condition must hold for both concepts of confirmation. Within the formal calculus of probability (which Carnap's concept of degree of confirmation satisfies) we can show that, if h is equivalent to h', then $c(h,e) = c(h',e)$, for any evidence e whatever. Thus, if h has a high degree of confirmation on $e.i$, h' does also. Likewise, if i increases the degree of confirmation of h, it will also increase the degree of confirmation of h'.

The special consequence condition is easily shown to be valid for the nonrelevance concept of confirmation. If h entails k, then $c(k,e) \geqq c(h,e)$; hence, if $c(h,e.i) > b$, then $c(k,e.i) > b$. But here, I think, our intuitions mislead us most seductively. It turns out, as Carnap has shown with great clarity, that the special consequence condition fails for the

[13] *Ibid.*, p. 33.
[14] *Ibid.*, p. 31.
[15] *Ibid.*

relevance concept of confirmation. It is entirely possible for *i* to be positively relevant to *h* without being positively relevant to some logical consequence *k*. We shall return in section 3 to a more detailed discussion of this fact.

The net result of the confusion of the two different concepts is that obviously correct statements about confirmation relations of one type are laid down as conditions of adequacy for explications of concepts of the other type, where, upon examination, they turn out to be clearly unsatisfactory. Carnap showed how the entailment condition could be modified to make it acceptable as a condition of adequacy.[16] As long as we are dealing with the relevance concept of confirmation, it looks as if the consistency condition should simply be abandoned. The equivalence condition appears to be acceptable as it stands. The special consequence condition, surprisingly enough, cannot be retained.

Hempel tried to lay down conditions for a nonquantitative concept of confirmation, and we have seen some of the troubles he encountered. After careful investigation of this problem, Carnap came to the conclusion that it is best to establish a quantitative concept of degree of confirmation and then to make the definition of the two nonquantitative concepts dependent upon it, as we have done in D1 and D2 above.[17] Given a quantitative concept and the two definitions, there is no need for conditions of adequacy like those advanced by Hempel. The nonquantitative concepts of confirmation are fully determined by those definitions, but we may, if we wish, see what general conditions such as H-8.1, H-8.2, H-8.21, H-8.22, H-8.3 are satisfied. In a 1964 postcript to the earlier article, Hempel expresses general agreement with this approach of Carnap.[18] Yet, he does so with such equanimity that I wonder whether he, as well as many others, recognize the profound and far-reaching consequences of the fact that the relevance concept of confirmation fails to satisfy the special consequence condition and other closely related conditions (which will be discussed in section 3).

2. Carnap and Popper

Once the fundamental ambiguity of the term "confirm" has been pointed out, we might suppose that reasonably well-informed authors

[16] Carnap, *Logical Foundations*, p. 473.
[17] *Ibid.*, p. 467.
[18] Hempel, *Aspects of Scientific Explanation*, p. 50.

could easily avoid confusing the two senses. Ironically, even Carnap himself did not remain entirely free from this fault. In the preface to the second edition of *Logical Foundations of Probability* (1962), he acknowledges that the first edition was not completely unambiguous. In the new preface, he attempts to straighten out the difficulty.

In the first edition, Carnap had distinguished a triplet of confirmation concepts:[19]

1. Classificatory — e confirms h.
2. Comparative — e confirms h more than e′ confirms h'.
3. Quantitative — the degree of confirmation of h on e is u

In the second edition, he sees the need for two such triplets of concepts.[20] For this purpose, he begins by distinguishing what he calls "concepts of firmness" and "concepts of increase of firmness." The concept of confirmation we defined above in D1, which was called an "absolute" concept, falls under the heading "concepts of firmness." The concept of confirmation we defined in D2, and called a "relevance" concept, falls under the heading "concepts of increase of firmness." Under each of these headings, Carnap sets out a triplet of classificatory, comparative, and quantitative concepts:

I. Three Concepts of Firmness

I-1. h is firm on the basis of e.	$c(h,e) > b$, where b is some fixed number.
I-2. h is firmer on e than is h' on e′.	$c(h,e) > c(h',e')$.
I-3. The degree of firmness of h on e is u.	$c(h,e) = u$.

To deal with the concepts of increase of firmness, Carnap introduces a simple relevance measure $D(h,i) =_{df} c(h,i) - c(h,t)$. This is a measure of what might be called "initial relevance," where the tautological evidence t serves as background evidence. The second triplet of concepts is given as follows:

II. Three Concepts of Increase of Firmness

II-1. h is made firmer by i. $D(h,i) > 0$.

II-2. The increase in firmness of h due to i is greater than the increase of firmness of h' due to i'. $D(h,i) > D(h',i')$.

[19] Carnap, *Logical Foundations*, sec. 8.

[20] Carnap, *Logical Foundations*, 2nd ed., pp. xv–xvi.

II-3. The amount of increase of firmness of h due to i is w.

$$D(h,i) = w.$$

Given the foregoing arrays of concepts, any temptation we might have had to identify the absolute (nonrelevance) concept of confirmation with the original classificatory concept, and to identify the relevance concept of confirmation with the original comparative concept, while distinguishing both from the original quantitative concept of degree of confirmation, can be seen to be quite mistaken. What we defined above in D1 as the absolute concept of confirmation is clearly seen to coincide with Carnap's new classificatory concept I-1, while our relevance concept of confirmation defined in D2 obviously coincides with Carnap's other new classificatory concept II-1. Carnap's familiar concept of degree of confirmation (probability$_1$) is obviously his quantitative concept of firmness I-3, while his new quantitative concept II-3 coincides with the concept of degree of relevance. Although we shall not have much occasion to deal with the comparative concepts, it is perhaps worth mentioning that the new comparative concept II-2 has an important use. When we compare the strengths of different tests of the same hypothesis we frequently have occasion to say that a test that yields evidence i is better than a test that yields evidence i'; sometimes, at least, this means that i is more relevant to h than is i' — i.e., the finding i would increase the degree of confirmation of h by a greater amount than would the finding i'.

It is useful to have a more general measure of relevance than the measure D of initial relevance. We therefore define the relevance of evidence i to hypothesis h on (in the presence of) background evidence e as follows:[21]

D3. $R(i,h,e) = c(h,e.i) - c(h,e)$.

Then we can say:

D4. i is positively relevant to h on e $=_{df} R(i,h,e) > 0$.
i is negatively relevant to h on e $=_{df} R(i,h,e) < 0$.
i is irrelevant to h on e $=_{df} R(i,h,e) = 0$.

[21] Carnap introduces the simple initial relevance measure D for temporary heuristic purposes in the preface to the second edition. In ch. 6 he discusses both our relevance measure R and Keynes's relevance quotient $c(h,e.i)/c(h,e)$, but for his own technical purposes he adopts a more complicated relevance measure $r(i,h,e)$. For purposes of this paper, I prefer the simpler and more intuitive measure R, which serves as well as Carnap's measure r in the present context. Since this measure differs from that used by Carnap in ch. 6 of *Logical Foundations*, I use a capital "R" to distinguish it from Carnap's lowercase symbol.

Hence, the classificatory concept of confirmation in the relevance sense can be defined by the condition $R(i,h,e) > 0$. Using these relevance concepts, we can define a set of concepts of confirmation in the relevance sense as follows:

D5. *i* confirms *h* given e $=_{df}$ *i* is positively relevant to *h* on e.
i disconfirms *h* given e $=_{df}$ *i* is negatively relevant to *h* on e.
i neither confirms nor disconfirms *h* given e $=_{df}$ *i* is irrelevant to *h* on e.

Having delineated his two triplets of concepts, Carnap then acknowledges that his original triplet in the first edition was a mixture of concepts of the two types; in particular, he had adopted the classificatory concept of increase of firmness II-1, along with the comparative and quantitative concepts of firmness I-2 and I-3. This fact, plus some misleading informal remarks about these concepts, led to serious confusion.[22]

As Carnap further acknowledges, Popper called attention to these difficulties, but he fell victim to them as well.[23] By equivocating on the admittedly ambiguous concept of confirmation, he claimed to have derived a contradiction within Carnap's formal theory of probability$_1$. He offers the following example:

> Consider the next throw with a homogeneous die. Let x be the statement 'six will turn up'; let y be its negation, that is to say, let $y = \bar{x}$; and let z be the information 'an even number will turn up'.
>
> We have the following absolute probabilities:
>
> $p(x) = 1/6$; $p(y) = 5/6$; $p(z) = 1/2$.
>
> Moreover, we have the following relative probabilities:
>
> $p(x,z) = 1/3$; $p(y,z) = 2/3$.
>
> We see that x is supported by the information z, for z raises the probability of x from $1/6$ to $2/6 = 1/3$. We also see that y is undermined by z, for z lowers the probability of y by the same amount from $5/6$ to $4/6 = 2/3$. Nevertheless, we have $p(x,z) < p(y,z)$.[24]

From this example, Popper quite correctly draws the conclusion that there are statements x, y, and z such that z confirms x, z disconfirms y, and y has a higher degree of confirmation on z than x has. As Popper points out quite clearly, this result would be logically inconsistent *if* we were to take the term "confirm" in its nonrelevance sense. It would be self-contradictory to say,

[22] Carnap, *Logical Foundations*, 2nd ed., pp. xvii–xix.
[23] *Ibid.*, p. xix, fn.
[24] Popper, *The Logic of Scientific Discovery*, p. 390.

> The degree of confirmation of x on z is high, the degree of confirmation of y on z is not high, and the degree of confirmation of y on z is higher than the degree of confirmation of x on z.

The example, of course, justifies no such statement; there is no way to pick the number *b* employed by Carnap in his definition of confirmation in the (firmness) sense I-1 according to which we could say that the degree of confirmation of x on z is greater than *b* and the degree of confirmation of y on z is not greater than *b*. The proper formulation of the situation is:

> The evidence z is positively relevant to x, the evidence z is negatively relevant to y, and the degree of confirmation of x on z is less than the degree of confirmation of y on z.

There is nothing even slightly paradoxical about this statement.

Popper's example shows clearly the danger of equivocating between the two concepts of confirmation, but it certainly does not show any inherent defect in Carnap's system of inductive logic, for this system contains both degree of confirmation c(*h*,e) and degree of relevance r(*i*,*h*,e). The latter is clearly and unambiguously defined in terms of the former, and there are no grounds for confusing them.[25] The example shows, however, the importance of exercising great care in our use of English language expressions in talking about these exact concepts.

3. The Vagaries of Relevance

It can be soundly urged, I believe, that the verb "to confirm" is used more frequently in its relevance sense than in the absolute sense. When we say that a test confirmed a hypothesis, we would normally be taken to mean that the result was positively relevant to the hypothesis. When we say that positive instances are confirming instances, it seems that we are characterizing confirming evidence as evidence that is positively relevant to the hypothesis in question. If we say that several investigators independently confirmed some hypothesis, it would seem sensible to understand that each of them had found positively relevant evidence. There is no need to belabor this point. Let us simply assert that the term "confirm" is often used in its relevance sense, and we wish to investigate some of the properties of this concept. In other words, let us agree for now to use the term "confirm" solely in its relevance sense

[25] Carnap, *Logical Foundations*, sec. 67.

(unless some explicit qualification indicates otherwise), and see what we will be committed to.

It would be easy to suppose that, once we are clear on the two senses of the term *confirm* and once we resolve to use it in only one of these senses in a given context, it would be a simple matter to tidy up our usage enough to avoid such nasty equivocations as we have already discussed. This is not the case, I fear. For, as Carnap has shown by means of simple examples and elegant arguments, the relevance concept of confirmation has some highly counterintuitive properties.

Suppose, for instance, that two scientists are interested in the same hypothesis *h*, and they go off to their separate laboratories to perform tests of that hypothesis. The tests yield two positive results, *i* and *j*. Each of the evidence statements is positively relevant to *h* in the presence of common background information e. Each scientist happily reports his positive finding to the other. Can they now safely conclude that the net result of both tests is a confirmation of *h*? The answer, amazingly, is no! As Carnap has shown, two separate items of evidence can each be positively relevant to a hypothesis, while their conjunction is negative to that very same hypothesis. He offers the following example:[26]

Example 1. Let the prior evidence e contain the following information. Ten chess players participate in a chess tournament in New York City; some of them are local people, some from out of town; some are junior players, some are seniors; some are men (M), some are women (W). Their distribution is known to be as follows:

	Local players	Out-of-towners
Juniors	M, W, W	M, M
Seniors	M, M	W, W, W

Table 1

Furthermore, the evidence e is supposed to be such that on its basis each of the ten players has an equal chance of becoming the winner, hence 1/10 . . . It is assumed that in each case [of evidence that certain players have been eliminated] the remaining players have equal chances of winning.

Let *h* be the hypothesis that a man wins. Let *i* be the evidence that a local player wins; let *j* be the evidence that a junior wins. Using the equiprobability information embodied in the background evidence e, we can read the following values directly from table 1:

[26] *Ibid.*, pp. 382–83. Somewhat paraphrased for brevity.

$$\begin{array}{llll} c(h,e) = 1/2 & c(h,e.i) = 3/5 & R(i,h,e) = 1/10 \\ & c(h,e.j) = 3/5 & R(j,h,e) = 1/10 \\ & c(h,e.i.j) = 1/3 & R(i.j,h,e) = -1/6 \end{array}$$

Thus, *i* and *j* are each positively relevant to *h*, while the conjunction *i.j* is negatively relevant to *h*. In other words, *i* confirms *h* and *j* confirms *h* but *i.j* disconfirms *h*.

The setup of example 1 can be used to show that a given piece of evidence may confirm each of two hypotheses individually, while that same evidence disconfirms their conjunction.[27]

Example 2. Let the evidence e be the same as in example 1. Let *h* be the hypothesis that a local player wins; let *k* be the hypothesis that a junior wins. Let *i* be evidence stating that a man wins. The following values can be read directly from table 1:

$$\begin{array}{lll} c(h,e) = 1/2 & c(h,e.i) = 3/5 & R(i,h,e) = 1/10 \\ c(k,e) = 1/2 & c(k,e.i) = 3/5 & R(i,k,e) = 1/10 \\ c(h.k,e) = 3/10 & c(h.k,e.i) = 1/5 & R(i,h.k,e) = -1/10 \end{array}$$

Thus, *i* confirms *h* and *i* confirms *k*, but *i* disconfirms *h.k*.

In the light of this possibility it might transpire that a scientist has evidence that supports the hypothesis that there is gravitational attraction between any pair of bodies when at least one is of astronomical dimensions and the hypothesis of gravitational attraction between bodies when both are of terrestrial dimensions, but which disconfirms the law of universal gravitation! In the next section we shall see that this possibility has interesting philosophical consequences.

A further use of the same situation enables us to show that evidence can be positive to each of two hypotheses, and nevertheless negative to their disjunction.[28]

Example 3. Let the evidence e be the same as in example 1. Let *h* be the hypothesis that an out-of-towner wins; let *k* be the hypothesis that a senior wins. Let *i* be evidence stating that a woman wins. The following values can be read directly from table 1:

$$\begin{array}{lll} c(h,e) = 1/2 & c(h,e.i) = 3/5 & R(i,h,e) = 1/10 \\ c(k,e) = 1/2 & c(k,e.i) = 3/5 & R(i,k,e) = 1/10 \\ c(h \vee k,e) = 7/10 & c(h \vee k,e.i) = 3/5 & R(i,h \vee k,e) = -1/10 \end{array}$$

Thus, *i* confirms *h* and *i* confirms *k*, but *i* nevertheless disconfirms *h* v *k*.

[27] *Ibid.*, pp. 394–95.
[28] *Ibid.*, p. 384.

Imagine the following situation:[29] a medical researcher finds evidence confirming the hypothesis that Jones is suffering from viral pneumonia and also confirming the hypothesis that Jones is suffering from bacterial pneumonia — yet this very same evidence disconfirms the hypothesis that Jones has pneumonia! It is difficult to entertain such a state of affairs, even as an abstract possibility.

These three examples illustrate a few members of a large series of severely counterintuitive situations that can be realized:

i. Each of two evidence statements may confirm a hypothesis, while their conjunction disconfirms it. (Example 1.)

ii. Each of two evidence statements may confirm a hypothesis, while their disjunction disconfirms it. (Example 2a, Carnap, *Logical Foundations*, p. 384.)

iii. A piece of evidence may confirm each of two hypotheses, while it disconfirms their conjunction. (Example 2.)

iv. A piece of evidence may confirm each of two hypotheses, while it disconfirms their disjunction. (Example 3.)

This list may be continued by systematically interchanging positive relevance (confirmation) and negative relevance (disconfirmation) throughout the preceding statements. Moreover, a large number of similar possibilities obtain if irrelevance is combined with positive or negative relevance. Carnap presents a systematic inventory of all of these possible relevance situations.[30]

In section 1, we mentioned that Hempel's special consequence condition does not hold for the relevance concept of confirmation. This fact immediately becomes apparent upon examination of statement iv above. Since h entails $h \vee k$, and since i may confirm h while disconfirming $h \vee k$, we have an instance in which evidence confirms a statement but fails to confirm one of its logical consequences. Statement ii, incidentally, shows that the converse consequence condition, which Hempel discusses but does not adopt,[31] also fails for the relevance concept of confirmation. Since $h.k$ entails h, and since i may confirm h without confirming $h.k$, we have an instance in which evidence confirms a hypothesis without confirming at least one statement from which that hypothesis follows. The failure of the special consequence condition and the converse consequence condition appears very mild when compared with the much

[29] This example is adapted from *ibid.*, p. 367.
[30] *Ibid.*, secs. 69, 71.

stronger results i–iv, and analogous ones. While one might, without feeling too queasy, give up the special consequence condition — the converse consequence condition being unsatisfactory on the basis of much more immediate and obvious considerations — it is much harder to swallow possibilities like i–iv without severe indigestion.

4. Duhem and Relevance

According to a crude account of scientific method, the testing of a hypothesis consists in deducing observational consequences and seeing whether or not the facts are as predicted. If the prediction is fulfilled we have a positive instance; if the prediction is false the result is a negative instance. There is a basic asymmetry between verification and falsification. If, from hypothesis h, an observational consequence o can be deduced, then the occurrence of a fact o' that is incompatible with o (o' entails $\sim o$) enables us to infer the falsity of h by good old modus tollens. If, however, we find the derived observational prediction fulfilled, we still cannot deduce the truth of h, for to do so would involve the fallacy of affirming the consequent.

There are many grounds for charging that the foregoing account is a gross oversimplification. One of the most familiar, which was emphasized by Duhem, points out that hypotheses are seldom, if ever, tested in isolation; instead, auxiliary hypotheses a are normally required as additional premises to make it possible to deduce observational consequences from the hypothesis h that is being tested. Hence, evidence o' (incompatible with o) does not entail the falsity of h, but only the falsity of the conjunction $h.a$. There are no deductive grounds on which we can say that h rather than a is the member of the conjunction that has been falsified by the negative outcome. To whatever extent auxiliary hypotheses are invoked in the deduction of the observational consequence, to that extent the alleged deductive asymmetry of verification and falsification is untenable.

At this point, a clear warning should be flashed, for we recall the strange things that happen when conjunctions of hypotheses are considered. Example 2 of the previous section showed that evidence that disconfirms a conjunction $h.a$ can nevertheless separately confirm each of the conjuncts. Is it possible that a *negative* result of a test of the hy-

[31] Hempel, *Aspects of Scientific Explanation*, pp. 32–33.

pothesis h, in which auxiliary hypotheses a were also involved, could result in the *confirmation* of the hypothesis of interest h *and* in a *confirmation* of the auxiliary hypotheses a as well?

It might be objected, at this point, that in the Duhemian situation o' is not merely negatively relevant to $h.a$; rather,

(1) o' entails $\sim(h.a)$.

This objection, though not quite accurate, raises a crucial question. Its inaccuracy lies in the fact that h and a together do not normally entail o'; in the usual situation some initial conditions are required along with the main hypothesis and the auxiliary hypotheses. If this is the case, condition (1) does not obtain. We can deal with this trivial objection to (1), however, by saying that, since the initial conditions are established by observation, they are to be taken as part of the background evidence e which figures in all of our previous investigations of relevance relations. Thus, we can assert that, in the presence of background evidence e, o can be derived from $h.a$. This allows us to reinstate condition (1).

Unfortunately, condition (1) is of no help to us. Consider the following situation:

Example 4. The evidence e contains the same equiprobability assumptions as the evidence in example 1, except for the fact that the distribution of players is as indicated in the following table:

	Local players	Out-of-towners
Juniors	W	M, M
Seniors	M, M	M, W, W, W, W

Table 2

Let h be the hypothesis that a local player wins; let k be the hypothesis that a junior wins. Let i be evidence stating that a man wins. In this case, condition (1) is satisfied; the evidence i is logically incompatible with the conjunction $h.k$. The following values can be read directly from the table:

$$c(h,e) = 0.3 \qquad c(h,e.i) = 0.4 \qquad R(i,h,e) = 0.1$$
$$c(k,e) = 0.3 \qquad c(k,e.i) = 0.4 \qquad R(i,k,e) = 0.1$$
$$c(h.k,e) = 0.1 \qquad c(h.k,e.i) = 0 \qquad R(i,h.k,e) = -0.1$$

This example shows that evidence i, even though it conclusively refutes the conjunction $h.k$, nevertheless confirms both h and k taken individually.

Here is the situation. Scientist Smith comes home late at night after a hard day at the lab. "How did your work go today, dear?" asks his wife.

"Well, you know the Smith hypothesis, h_s, on which I have staked my entire scientific reputation? And you know the experiment I was running today to test my hypothesis? Well, the result was negative."

"Oh, dear, what a shame! Now you have to give up your hypothesis and watch your entire reputation go down the drain!"

"Not at all. In order to carry out the test, I had to make use of some auxiliary hypotheses."

"Oh, what a relief — saved by Duhem! Your hypothesis wasn't refuted after all," says the philosophical Mrs. Smith.

"Better than that," Smith continues, "I actually confirmed the Smith hypothesis."

"Why that's wonderful, dear," replies Mrs. Smith, "you found you could reject an auxiliary hypothesis, and that in so doing, you could establish the fact that the test actually confirmed your hypothesis? How ingenious!"

"No," Smith continues, "it's even better. I found I had confirmed the auxiliary hypotheses as well!"

This is the Duhemian thesis reinforced with a vengeance. Not only does a negative test result fail to refute the test hypothesis conclusively — it may end up confirming both the test hypothesis and the auxiliary hypotheses as well.

It is very tempting to suppose that much of the difficulty might be averted if only we could have sufficient confidence in our auxiliary hypotheses. If a medical researcher has a hypothesis about a disease which entails the presence of a certain microorganism in the blood of our favorite victim Jones, it would be outrageous for him to call into question the laws of optics as applied to microscopes as a way of explaining failure to find the bacterium. If the auxiliary hypotheses are well enough established beforehand, we seem to know where to lay the blame when our observational predictions go wrong. The question is how to establish the auxiliary hypotheses in the first place, for if the Duhemian is right, no hypotheses are ever tested in isolation. To test *any* hypothesis, according to this view, it is necessary to utilize auxiliary hypotheses; consequently, to establish our auxiliary hypotheses *a* for use in the tests of *h*, we would need some other auxiliary hypotheses *a′* to carry out the tests of *a*. A vicious regress threatens.

A more contextual approach might be tried.[32] Suppose that *a* has been used repeatedly as an auxiliary hypothesis in the *successful* testing of other hypotheses *j*, *k*, *l*, etc. Suppose, that is, that the conjunctions *j.a*, *k.a*, *l.a*, etc., have been tested and repeatedly confirmed — i.e., all test results have been positively relevant instances. Can we say that *a* has been highly confirmed as a result of all of these successes? Initially, we might have been tempted to draw the affirmative conclusion, but by now we know better. Examples similar to those of the previous section can easily be found to show that evidence positively relevant to a conjunction of two hypotheses can nevertheless be negative to each conjunct.[33] It is therefore logically possible for each confirmation of a conjunction containing *a* to constitute a disconfirmation of *a* — and indeed a disconfirmation of the other conjunct as well in each such case.

There is one important restriction that applies to the case in which new observational evidence refutes a conjunction of two hypotheses, namely, hypotheses that are incompatible on evidence *e.i* can have, at most, probabilities that add to one. If *e.i* entails $\sim(h.k)$

$$c(h,e.i) + c(k,e.i) \leq 1.$$

Since we are interested in the case in which *i* is positively relevant to both *h* and *k*, these hypotheses must also satisfy the condition

$$c(h,e) + c(k,e) < 1.$$

We have here, incidentally, what remains of the asymmetry between confirmation and refutation. If evidence *i* refutes the conjunction *h.k*, that fact places an upper limit on the sum of the probabilities of *h* and *k* relative to *e.i*. If, however, evidence *i* confirms a conjunction *h.k* while disconfirming each of the conjuncts, there is no lower bound (other than zero) on the sum of their degrees of confirmation on *i*.

In this connection, let us recall our ingenious scientist Smith, who turned a refuting test result into a positive confirmation of both his pet hypothesis h_S and his auxiliary hypotheses *a*. We see that he must have been working with a test hypothesis or auxiliaries (or both) which had rather low probabilities. We might well question the legitimacy of using hypotheses with degrees of confirmation appreciably less than one as

[32] This paper was motivated by Grünbaum's questions concerning this approach. See his "Falsifiability and Rationality" to be published in a volume edited by Joseph J. Kockelmans (proceedings of an international conference held at Pennsylvania State University).

[33] Carnap, *Logical Foundations*, pp. 394–95, 3b, is just such an example.

auxiliary hypotheses. If Smith's auxiliaries a had decent degrees of confirmation, his own hypothesis h_s must have been quite improbable. His clever wife might have made some choice remarks about his staking an entire reputation on so improbable a hypothesis. But I should not get carried away with dramatic license. If we eliminate all the unnecessary remarks about staking his reputation on h_s, and regard it rather as a hypothesis he finds interesting, then its initial improbability may be no ground for objection. Perhaps every interesting general scientific hypothesis starts its career with a very low prior probability. Knowing, as we do, that a positively relevant instance may disconfirm both our test hypothesis and our auxiliaries, while a negative instance may confirm them both, there remains a serious, and as yet unanswered, question how any hypothesis ever can become either reasonably well confirmed or reasonably conclusively disconfirmed (in the absolute sense). It obviously is still an open question how we could ever get any well-confirmed hypotheses to serve as auxiliaries for the purpose of testing other hypotheses.

Suppose, nevertheless, that we have a hypothesis *h* to test and some auxiliaries *a* that will be employed in conducting the test and that somehow we have ascertained that *a* has a higher prior confirmation than *h* on the initial evidence e:

$$c(a,e) > c(h,e).$$

Suppose, further, that as the result of the test we obtain negative evidence o′ which refutes the conjunction *h.a*, but which confirms both *h* and *a*. Thus, o′ entails $\sim(h.a)$ and

$$c(h,e.o') > c(h,e) \qquad c(a,e.o') > c(a,e).$$

We have already seen that this can happen (example 4). But now we ask the further question, is it possible that the posterior confirmation of *h* is greater than the posterior confirmation of *a*? In other words, can the negative evidence o′ confirm both conjuncts and do so in a way that reverses the relation between *h* and *a*? A simple example will show that the answer is affirmative.

Example 5. The Department of History and Philosophy of Science at Polly Tech had two openings, one in history of science and the other in philosophy of science. Among the 1000 applicants for the position in history, 100 were women. Among the 2000 applicants for the position in philosophy, 100 were women. Let h be the hypothesis that the history job was filled by a woman; let k be the hypothesis that the philosophy job was filled by a woman. Since both selections were made by the use of a fair lottery device belonging to the inductive logician in the department,

$$c(h,e) = .1$$
$$c(k,e) = .05$$
$$c(h,e) > c(k,e).$$

Let i be the evidence that the two new appointees were discovered engaging in heterosexual intercourse with each other in the office of the historian. It follows at once that

$$c(h.k,e.i) = 0$$
$$c(h,e.i) + c(k,e.i) = 1$$

i.e., one appointee was a woman and the other a man, but we do not know which is which. Since it is considerably more probable, let us assume, that the office used was that of the male celebrant, we assign the values

$$c(h,e.i) = .2 \qquad c(k,e.i) = .8$$

with the result that

$$c(h,e.i) < c(k,e.i).$$

This illustrates the possibility of a reversal of the comparative relation between the test hypothesis and auxiliaries as a result of refuting evidence. It shows that a's initial superiority to h is no assurance that it will still be so subsequent to the refuting evidence. If, prior to the negative test result, we had to choose between h and a, we would have preferred a, but after the negative outcome, h is preferable to a.

There is one significant constraint that must be fulfilled if this reversal is to occur in the stated circumstances. If our auxiliary hypotheses a are initially better confirmed than our test hypothesis h, and if the conjunction $h.a$ is refuted by evidence o' that is positively relevant to both h and a, and if the posterior confirmation of h is greater than the posterior confirmation of a, then the prior confirmation of a must have been less than 1/2. For,

$$c(h,e.o') + c(a,e.o') \leqq 1$$

and

$$c(h,e.o') > c(a,e.o').$$

Hence,

$$c(a,e.o') < 1/2.$$

Moreover,

$$c(a,e) < c(a,e.o').$$

Therefore,

$$c(a,e) < 1/2.$$

It follows that if a is initially more probable than h and also initially more probable than its own negation $\sim a$, then it is impossible for a refuting instance o' which confirms both h and a to render h more probable than a. Perhaps that is some comfort. If our auxiliaries are more probable than not, and if they are better established before the test than our test hypothesis h, then a refuting test outcome which confirms both h and a cannot make h preferable to a.

But this is not really the tough case. The most serious problem is whether a refutation of the conjunction $h.a$ can render h more probable than a by being positively relevant to h and negatively relevant to a, even when a is initially *much* more highly confirmed than h. You will readily conclude that this is possible; after all of the weird outcomes we have discussed, this situation seems quite prosaic. Consider the following example:

Example 6. Let

e = Brown is an adult American male
h = Brown is a Roman Catholic
k = Brown is married

and suppose the following degrees of confirmation to obtain:

$$c(h,e) = .3$$
$$c(k,e) = .8$$
$$c(h.k,e) = .2.$$

Let i be the information that Brown is a priest — that is, an ordained clergyman of the Roman Catholic, Episcopal, or Eastern Orthodox church. Clearly, i refutes the conjunction $h.k$, so

$$c(h.k,e.i) = 0.$$

Since the overwhelming majority of priests in America are Roman Catholic, let us assume that

$$c(h,e.i) = .9$$

and since some, but not all, non–Roman Catholic priests marry, let

$$c(k,e.i) = .05.$$

We see that i is strongly relevant to both h and k; in particular, it is positively relevant to h and negatively relevant to k. Moreover, while k has a much higher degree of confirmation than h relative to the prior evidence e, h has a much higher degree of confirmation than k on the posterior evidence $e.i$. Thus, the refuting evidence serves to reverse the preferability relation between h and k.

It might be helpful to look at this situation diagrammatically and to think of it in terms of class ratios or frequencies. Since class ratios satisfy the mathematical calculus of probabilities, they provide a useful device for establishing the possibility of certain probability relations. With our background evidence e let us associate a reference class A, and with our hypotheses h and k let us associate two overlapping subclasses B and C respectively. With our additional evidence i let us associate a further subclass D of A. More precisely, let

$$e = x \in A,\ h = x \in B,\ k = x \in C,\ i = x \in D.$$

Since we are interested in the case in which the prior confirmation of k is high and the prior confirmation of h is somewhat lower, we want most of A to be included in C and somewhat less of A to be included in B. Moreover, since our hypotheses h and k should not be mutually exclusive on the prior evidence e alone, B and C must overlap. However, neither B nor C can be a subset of the other; they must be mutually exclusive within D, since h and k are mutually incompatible on additional evidence i. Moreover, because we are not considering cases in which $e.i$ entails either h or k alone, the intersections of D with B and C must both be nonnull. We incorporate all of these features in Figure 1. In order to achieve the desired result — that is, to have the posterior confirmation of h greater than the posterior confirmation of k — it is only necessary to draw D so that its intersection with B is larger than its intersection with C. This is obviously an unproblematic condition to fulfill. Indeed, there is no difficulty in arranging it so that the proportion of D occupied by its intersection with B is larger than the proportion of A occupied by its intersection with C. When this condition obtains, we can not only say that the evidence i has made the posterior confirmation of h greater than the posterior confirmation of k (thus

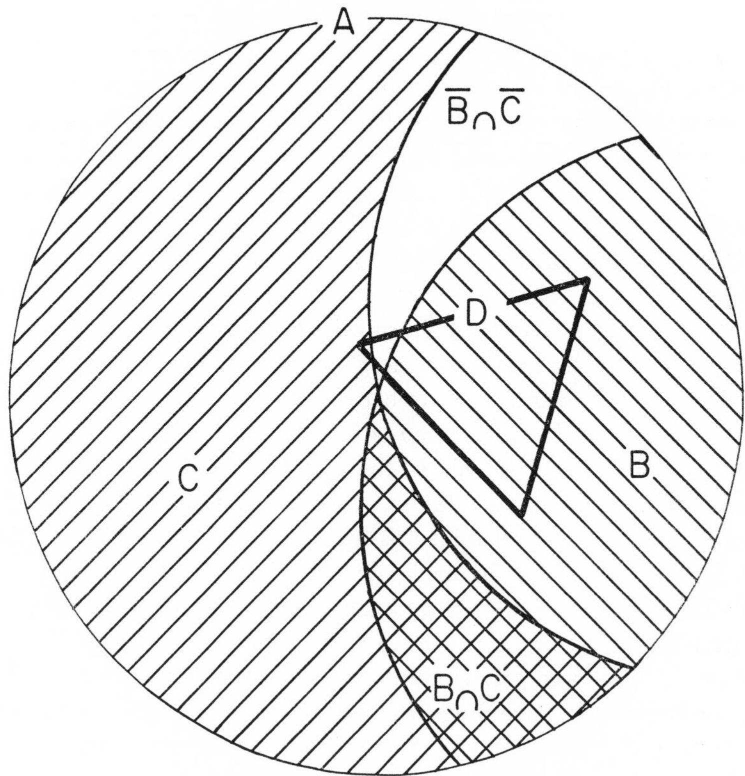

Figure 1

reversing their preferability standing), but also that the posterior confirmation of *h* is greater than the prior confirmation of *k*. Translated into the Duhemian situation, this means that not only can the refuting evidence o′ disconfirm the initially highly probable auxiliary hypotheses a, but it can also confirm the test hypothesis *h* to the extent that its posterior confirmation makes it more certain than were the auxiliaries before the evidence o′ turned up. This set of relationships is easily seen to be satisfied by example 6 if we let

A = American men	*B* = Roman Catholics
C = married men	*D* = priests.

It is evident from Figure 1 that C can be extremely probable relative to A without presenting any serious obstacle to the foregoing outcome, provided *D* is very much smaller than A. There seems little ground for assurance that a refuting experimental result will generally leave the auxiliaries intact, rather than drastically disconfirming them and radically elevating the degree of confirmation of the test hypothesis. This state of affairs can apparently arise with distressing ease, and there are no evident constraints that need to be fulfilled in order for it to happen. There seems to be no basis for confidence that it does not happen frequently in real scientific situations. If this is so, then whenever our auxiliaries have an initial degree of confirmation that falls ever so slightly short of certainty, we cannot legitimately infer with confidence that the test hypothesis *h*, rather than the auxiliaries, is disconfirmed by the refuting instance. Thus, no matter how high the initial probability of the auxiliaries *a* (provided it is somewhat less than certainty), it is still possible for a finding that entails the falsity of the conjunction *h.a* to constitute a confirmation for either the one or the other. We certainly cannot say that the negative finding disconfirms *h* rather than *a* on the ground that *a* is more probable initially than *h*.

A parallel remark can be made about an instance that confirms the conjunction *h.a*. Such an instance might disconfirm either conjunct, and we have no way of saying which. In view of these dismal facts, we may well repeat, with even greater emphasis, the question posed earlier: how can any hypothesis (including those we need as auxiliary hypotheses for the testing of other hypotheses) ever be reasonably well confirmed or disconfirmed (in the absolute sense)?

The (to my mind) shocking possibilities that we have been surveying arise as consequences of the fact that we have been construing "confirm" in the relevance sense. What happens, I believe, is that we acknowledge with Carnap and Hempel that the classificatory concept of confirmation is best understood, in most contexts, as a concept of positive relevance defined on the basis of some quantitative degree of confirmation function. Then, in contexts such as the discussion of the Duhemian thesis, we proceed to talk casually about confirmation, forgetting that our intuitive notions are apt to be very seriously misleading. The old ambiguity of the absolute and the relevance sense of confirmation infects our intuitions, with the result that all kinds of unwarranted suppositions insinuate themselves. We find it extraordinarily difficult to keep a firm mental grasp upon such strange possibilities as we have seen in this section and the previous one.

5. Analysis of the Anomalies

There is, of course, a striking contrast between the "hypotheses" and "evidence" involved in our contrived examples, on the one hand, and the genuine hypotheses and evidence to be found in actual scientific practice, on the other. This observation might easily lead to the charge that the foregoing discussion is not pertinent to the logic of actual scientific confirmation, as opposed to the theory of confirmation constructed by Carnap on highly artificial and oversimplified languages. This irrelevance is demonstrated by the fact, so the objection might continue, that the kinds of problems and difficulties we have been discussing simply do not arise when real scientists test serious scientific hypotheses.

This objection, it seems to me, is wide of the mark. I am prepared to grant that such weird possibilities as we discussed in previous sections do not arise in scientific practice; at least, I have no concrete cases from the current or past history of science to offer as examples of them. This is, however, a statement of the problem rather than a solution. Carnap has provided a number of examples that, on the surface at least, seem to make a shambles of confirmation; why do they not also make a shambles of science itself? There can be no question that, for example, one statement can confirm each of two other statements separately while at the same time disconfirming their disjunction or conjunction. If that sort of phenomenon never occurs in actual scientific testing, it must be because we know something more about our evidence and hypotheses than merely that the evidence confirms the hypotheses. The problem is to determine the additional factors in the actual situation that block the strange results we can construct in the artificial case. In this section, I shall try to give some indications of what seems to be involved.

The crux of the situation seems to be the fact that we have said very little when we have stated merely that a hypothesis h has been confirmed by evidence i. This statement means, of course, that i raises the degree of confirmation of h, but that *in itself* provides very little information. It is by virtue of this paucity of content that we can go on and say that this same evidence i confirms hypothesis k as well, without being justified in saying anything about the effect of i upon the disjunction or the conjunction of h with k.

This state of affairs seems strange to intuitions that have been thoroughly conditioned on the extensional relations of truth-functional logic.

Probabilities are not extensional in the same way. Given the truth-values of h and k we can immediately ascertain the truth-values of the disjunction and the conjunction. The degrees of confirmation ("probability values") of h and k do not, however, determine the degree of confirmation of either the disjunction or the conjunction. This fact follows immediately from the addition and multiplication rules of the probability calculus:

$$(2) \qquad c(h \vee k,e) = c(h,e) + c(k,e) - c(h.k,e)$$

$$(3) \qquad c(h.k,e) = c(h,e) \times c(k,h.e) = c(k,e) \times c(h,k.e).$$

To determine the probability of the disjunction, we need, in addition to the values of the probabilities of the disjuncts, the probability of the conjunction. The disjunctive probability is the sum of the probabilities of the two disjuncts if they are mutually incompatible in the presence of evidence e, in which case $c(h.k,e) = 0$.[34] The probability of the conjunction, in turn, depends upon the probability of one of the conjuncts alone and the conditional probability of the other conjunct given the first.[35] If

$$(4) \qquad c(k,h.e) = c(k,e)$$

the multiplication rule assumes the special form

$$(5) \qquad c(h.k,e) = c(h,e) \times c(k,e)$$

in which case the probability of the conjunction is simply the product of the probabilities of the two conjuncts. When condition (4) is fulfilled, h and k are said to be independent of one another.[36] Independence, as thus defined, is obviously a relevance concept, for (4) is equivalent to the statement that h is irrelevant to k, i.e, $R(h,k,e) = 0$.

We can now see why strange things happen with regard to confirmation in the relevance sense. If the hypotheses h and k are mutually exclusive in the presence of e (and a fortiori in the presence of $e.i$), then

[34] The condition $c(h.k,e) = 0$ is obviously sufficient to make the probability of the disjunction equal to the sum of the probabilities of the disjuncts, and this is a weaker condition than e entails $\sim(h.k)$. Since the difference between these conditions has no particular import for the discussion of this paper, I shall, in effect, ignore it.

[35] Because of the commutativity of conjunction, it does not matter whether the probability of h conditional only on e or the probability of k conditional only on e is taken. This is shown by the double equality in formula (3).

[36] Independence is a symmetric relation; if h is independent of k then k will be independent of h.

(6) $c(h \vee k,e) = c(h,e) + c(k,e)$
(7) $c(h \vee k,e.i) = c(h,e.i) + c(k,e.i)$

so that if

(8) $c(h,e.i) > c(h,e)$ and $c(k,e.i) > c(k,e)$

it follows immediately that

(9) $c(h \vee k,e.i) > c(h \vee k,e)$.

Hence, in this special case, if *i* confirms *h* and *i* confirms *k*, then *i* must confirm their disjunction. This results from the fact that the relation between *h* and *k* is the same in the presence of e.*i* as it is in the presence of e alone.[37]

If, however, *h* and *k* are not mutually exclusive on evidence e we must use the general formulas

(10) $c(h \vee k,e) = c(h,e) + c(k,e) - c(h.k,e)$
(11) $c(h \vee k,e.i) = c(h,e.i) + c(k,e.i) - c(h.k,e.i)$.

Now, if it should happen that the evidence *i* drastically alters the relevance of *h* to *k* in just the right way our apparently anomalous results can arise. For then, as we shall see in a moment by way of a concrete (fictitious) example, even though condition (8) obtains — i.e., *i* confirms *h* and *i* confirms *k* — condition (9) may fail. Thus, if

(12) $c(h.k,e.i) > c(h.k,e)$

it may happen that

(13) $c(h \vee k,e.i) < c(h \vee k,e)$

i.e., *i* disconfirms $h \vee k$. Let us see how this works.

Example 7. Suppose *h* says that poor old Jones has bacterial pneumonia, and *k* says that he has viral pneumonia. I am assuming that these are the only varieties of pneumonia, so that $h \vee k$ says simply that he has pneumonia. Let evidence e contain the results of a superficial diagnosis as well as standard medical background knowledge about the disease, on the basis of which we can establish degrees of confirmation for *h*, *k*, *h.k*, and $h \vee k$. Suppose, moreover, that the probability on e that Jones has both viral and bacterial pneumonia is quite low, that is, that people do not often get them both simultaneously. For the sake of definiteness, let us

[37] To secure this result it is not necessary that $c(h.k,e) = c(h.k,e.i) = 0$; it is sufficient to have $c(h.k,e) = c(h.k,e.i)$, though obviously this condition is not necessary either.

introduce some numerical values. Suppose that on the basis of the superficial diagnosis it is 98 percent certain that Jones has one or the other form of pneumonia, but the diagnosis leaves it entirely uncertain which type he has. Suppose, moreover, that on the basis of e there is only a 2 percent chance that he has both. We have the following values:

$$c(h,e) = .50 \qquad c(k,e) = .50$$
$$c(h \vee k,e) = .98 \qquad c(h.k,e) = .02.$$

These values satisfy the addition formula (2). Suppose, now, that there is a certain test which indicates quite reliably those rare cases in which the subject has both forms of pneumonia. Let *i* be the statement that this test was administered to Jones with a positive result, and let this result make it 89 percent certain that Jones has both types. Assume, moreover, that the test rarely yields a positive result if the patient has only one form of pneumonia (i.e., when the positive result occurs for a patient who does not have both types, he usually has neither type). In particular, let

$$c(h,e.i) = .90, \quad c(k,e.i) = .90, \quad c(h.k,e.i) = .89$$

from which it follows that

$$c(h \vee k,e.i) = .91 < c(h \vee k,e) = .98.$$

The test result *i* thus confirms the hypothesis that Jones has bacterial pneumonia and the hypothesis that Jones has viral pneumonia, but it disconfirms the hypothesis that Jones has pneumonia!

It achieves this feat by greatly increasing the probability that he has both. This increase brings about a sort of clustering together of cases of viral and bacterial pneumonia, concomitantly decreasing the proportion

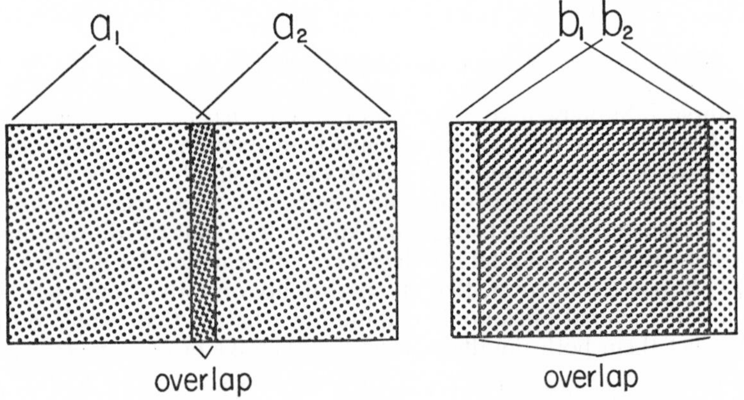

Figure 2

of people with only one of the two types. The effect is easily seen diagrammatically in Figure 2. Even though the rectangles in 2-b are larger than those in 2-a, those in 2-b cover a smaller total area on account of their very much greater degree of overlap. Taking the rectangles to represent the number of cases of each type, we see graphically how the probability of each type of pneumonia can increase simultaneously with a decrease in the overall probability of pneumonia. The evidence i has significantly altered the relevance relation between h and k. Using the multiplication formula (3), we can establish that

$$c(k,h.e) = .04 \qquad c(k,h.e.i) \cong .99$$
$$R(h,k,e) = -0.46 \qquad R(h,k,e.i) \cong .09$$

In the presence of e alone, h is negatively relevant to k; in the presence of i as well, h becomes positively relevant to k. There is nothing outlandish in such changes of relevance in the light of additional evidence. This case thus exemplifies condition (13) by satisfying condition (12).

A similar analysis enables us to understand how an item of evidence can confirm each of two hypotheses, while disconfirming — indeed, even while conclusively refuting — their conjunction. If hypotheses h and k are independent of each other in the presence of evidence $e.i$ and also in the presence of e alone, the following relations obtain:

$$(14) \qquad c(h.k,e) = c(h,e) \times c(k,e)$$
$$(15) \qquad c(h.k,e.i) = c(h,e.i) \times c(k,e.i)$$

so that if

$$(16) \qquad c(h,e.i) > c(h,e) \text{ and } c(k,e.i) > c(k,e)$$

it follows immediately that

$$(17) \qquad c(h.k,e.i) > c(h.k,e).$$

Hence, in this special case, if i confirms h and i confirms k, then i must confirm $h.k$.

A different situation obtains if h and k are not independent on both e and $e.i$; in that case we must use the general formulas

$$(18) \qquad c(h.k,e) = c(h,e) \times c(k,h.e)$$
$$(19) \qquad c(h.k,e.i) = c(h,e.i) \times c(k,h.e.i).$$

Even given that condition (16) still obtains, so that

$$(20) \qquad c(k,e.i) > c(k,e)$$

it is still possible that

(21) $c(k,h.e.i) < c(k,h.e)$[38]

which makes it possible, in turn, that

(22) $c(h.k,e.i) < c(h.k,e)$.

Since, according to (20) and (21),

(23) $c(k,e.i) - c(k,e) = R(i,k,e) > 0$

(24) $c(k,h.e.i) - c(k,h.e) = R(i,k,h.e) < 0$

the possibility of i confirming each of two hypotheses while disconfirming their conjunction depends upon the ability of h to make a difference in the relevance of i to k. We said above, however, that the occurrence of the strange confirmation phenomena depends upon the possibility of a change in the relevance of the hypotheses to one another in the light of new evidence. These characterizations are, however, equivalent to one another, for the change in relevance of i to k brought about by h is equal to the change in relevance of h to k brought about by i, that is,

(25) $R(h,k,e) - R(h,k,e.i) = R(i,k,e) - R(i,k,h.e)$.[39]

We can therefore still maintain that the apparently anomalous confirma-

[38] To establish the compatibility of (20) and (21), perhaps a simple example, in addition to the one about to be given in the text, will be helpful. Let

e = X is a man. i = X is American.

h = X is very wealthy. k = X vacations in America.

Under this interpretation, relation (20) asserts: It is more probable that an American man vacations in America than it is that a man (regardless of nationality) vacations in America. Under the same interpretation, relation (21) asserts: It is less probable that a very wealthy American man will vacation in America than it is that a very wealthy man (regardless of nationality) will vacation in America. The interpretation of formula (20) seems like an obviously true statement; the interpretation of (21) seems likely to be true owing to the apparent tendency of the very wealthy to vacation abroad. There is, in any case, no contradiction in assuming that every very wealthy American man vacations on the French Riviera, while every very wealthy man from any other country vacations in America.

[39] This equality can easily be shown by writing out the relevance terms according to their definitions as follows:

$R(h,k,e) =_{df} c(k,h.e) - c(k,e)$

$R(h,k,e.i) =_{df} c(k.h,e.i) - c(k,e.i)$

$R(h,k,e) - R(h,k,e.i) = c(k,h.e) - c(k,e) - c(k,h.e.i) + c(k,e.i)$ (*).

$R(i,k,e) =_{df} c(k,e.i) - c(k,e)$

$R(i,k,h.e) =_{df} c(k,h.e.i) - c(k,h.e)$

$R(i,k,e) - R(i,k,h.e) = c(k,e.i) - c(k,e) - c(k,h.e.i) + c(k,h.e)$ (**).

The right-hand sides of equations (*) and (**) obviously differ only in the arrangement of terms.

tion situation arises from the ability of new evidence to change relations of relevance between the hypotheses, as was suggested by our initial examination of the general addition and multiplication rules (2) and (3).

Let us illustrate the conjunctive situation with another concrete (though fictitious) example.

Example 8. Suppose that the evidence e tells us that two radioactive atoms A and B decay, each ejecting a particle, and that the probability in each case is 0.7 that it is an alpha particle, 0.2 that it is a negative electron, and 0.1 that it is a positive electron (positron). Assume that the two emissions are independent of one another. Let h be the statement that atom A emits a negative electron; let k be the statement that atom B emits a negative electron. We have the following probabilities:

$$c(h,e) = .2, \quad c(k,e) = .2, \quad c(h.k,e) = .04.$$

Let i be the observation that the two particles approach one another and annihilate upon meeting. Since this occurrence requires the presence of one positive and one negative electron, i entails $\sim(h.k)$. At the same time, since a negative electron must have been present, and since it is just as probable that it was emitted by atom A as atom B, we have

$$c(h,e.i) = .5 \quad \text{and} \quad c(k,e.i) = .5.$$

Hence, evidence i, which refutes the conjunction of the two hypotheses, confirms each one of them.[40]

This occurs because the evidence i makes the hypotheses h and k, which were independent of one another on evidence e alone, into mutually exclusive and exhaustive alternatives, i.e.,

$$c(k,h.e) - c(k,e) = R(h,k,e) = 0$$
$$c(k,h.e.i) - c(k,e.i) = R(h,k,e.i) = -.5.$$

Hypotheses that were totally irrelevant to each other in the absence of evidence i become very strongly relevant in the presence of i. Again, there is nothing especially astonishing about such a change in relevance as a result of new evidence.

Since, as we have seen, all the trouble seems to arise out of a change in the relevance of one hypothesis to the other as a result of new evidence, the most immediate suggestion might be to choose hypotheses h and k whose mutual relevance relations will not change in the light of

[40] This case constitutes a counterexample to the Hempel consistency condition H-8.3 discussed in sec. 2 above.

the new evidence *i*. We have noted that this constancy of relevance is guaranteed if we begin with hypotheses that are mutually exclusive on evidence *e*; they remain mutually exclusive on any augmented evidence *e.i*. But when we use the conjunction of a test hypothesis *h* with auxiliary hypotheses *a* in order to attempt a test of *h*, we certainly do not want *h* and *a* to be mutually exclusive — that is, we do not want to be in the position of knowing that we must reject our test hypothesis *h* if we are prepared to accept the auxiliaries *a*, even without the addition of any new evidence *i*. It would be more reasonable to insist that the auxiliary hypotheses *a* should themselves be neutral (irrelevant, independent) to the test hypothesis *h*. If that condition is satisfied, we can accept the auxiliary hypotheses *a* and still keep an entirely open mind regarding *h*. We cannot, however, demand that *h* and *a* remain irrelevant to one another after the new evidence *i* has been obtained. The interesting test situation is that in which, given *e*, *h.a* entails some observational consequence *o*. If a result *o′* occurs which is incompatible with *o*, then our hypotheses *h* and *a*, which may have been independent in the presence of *e* alone, are mutually exclusive in the light of the new evidence *o′*. Thus, the very design of that kind of test requires hypotheses whose mutual relevance relations are bound to change in the face of new evidence. Several of our examples (5 — positions at Polly Tech; 6 — celibacy among priests; 8 — electron-positron annihilation) show exactly what can happen when new evidence renders independent hypotheses mutually incompatible.

6. Conclusions

The crude hypothetico-deductive account of scientific inference, according to which hypotheses are confirmed by deducing observational consequences which are then verified by observation, is widely recognized nowadays as an oversimplification (even leaving aside the Duhemian objections). One can hardly improve upon Russell's classic example. From the hypothesis, "Pigs have wings," in conjunction with the observed initial condition, "Pigs are good to eat," we can deduce the consequence, "Some winged things are good to eat." Upon observing that such winged creatures as ducks and turkeys are good to eat, we have a hypothetico-deductive confirmation of the hypothesis, "Pigs have wings."[41] I am inclined to agree with a wide variety of authors who hold that something akin to a Bayesian schema must be involved in the confirmation of scientific hypotheses. If this is correct, it is entirely possible to have positive hypothetico-deductive test results that do not confirm the hypothesis (i.e., that do not add anything to its degree of confirmation on prior evidence). To emphasize this point, Reichenbach aptly described the crude hypothetico-deductive inference as an instance of "the fallacy of incomplete schematization."[42] Recognition of the basic inadequacy of the hypothetico-deductive schema does no violence to the logic of science; it only shows that the methods of science are more complex than this oversimplified schema.

Acknowledging the fact that positive hypothetico-deductive instances may not be confirming instances, I have been discussing the logic of confirmation—that is, I have been investigating the conclusions that can be drawn from the knowledge that this or that evidence confirms this or that hypothesis. By and large, we have found this logic to be poverty-ridden. Although evidence i confirms hypotheses h and k, we have found that we cannot infer that i confirms $h.k$. Evidence i may in fact confirm $h.k$, but to draw that conclusion from the given premises would be another instance of the fallacy of incomplete schematization. Indeed, our investigations have revealed exactly what is missing in the inference. In addition to knowing that i is positively relevant to h and positively relevant to k, we must know what bearing i has on the relevance of h to k. If this is known quantitatively, and if the degrees of relevance of i to h and to k are also known quantitatively, we can ascertain the relevance of i to $h.k$ and to $h \vee k$. Without this quantitative knowledge, we cannot say much of anything. The moral is simple: even if we base our qualitative concept of confirmation (in the relevance sense) upon a quantitative concept of degree of confirmation, the resulting qualitative concept is not very serviceable. It is too crude a concept, and it doesn't carry enough information to be useful. In order to make any substantial headway in understanding the logic of evidential support of scientific hypotheses, we must be prepared to work with at least crude estimates of quantitative values of degree of confirmation

[41] Bertrand Russell, "Dewey's New 'Logic,'" in Paul Arthur Schilpp, ed., *The Philosophy of John Dewey* (New York: Tudor, 1939), p. 149.

[42] Hans Reichenbach, *The Theory of Probability* (Berkeley and Los Angeles: University of California Press, 1949), p. 96.

and degree of relevance. Then, in contexts such as the discussion of the Duhemian problem, we must bring the more sophisticated concepts to bear if we hope to achieve greater clarity and avoid logical fallacies. In detailing the shortcomings of the qualitative concept of confirmation, we have, in a way, shown that this sort of confirmation theory is a shambles, but we have done no more violence to the logic of science than to show that its embodies more powerful concepts.

If we are willing, as Carnap has done, to regard degree of confirmation (in the nonrelevance sense) as a probability — that is, as a numerical functor that satisfies the probability calculus — then we can bring the structure of the quantitative probability concept to bear on problems of confirmation. With this apparatus, which gives us the power of Bayes's theorem, we can aspire to a much fuller understanding of relations of confirmation (in both the absolute and the relevance senses).

We can also provide an answer to many who have claimed that confirmation is not a probability concept. Confirmation in the relevance sense is admittedly not a probability; as we have insisted, it is not to be identified with high probability. A quantitative concept of degree of relevance can nevertheless be defined in terms of a concept of degree of confirmation. Degree of relevance, as thus defined, is not a probability; it obviously can take on negative values, which degree of probability cannot do. It is a probability concept, however, in the sense that it is explicitly defined in terms of degree of confirmation which is construed as a probability concept. Thus, even though degree of confirmation in the relevance sense cannot be construed as degree of probability, this fact is no basis for concluding that the concept of probability is an inadequate or inappropriate tool for studying the logic of evidential relations between scientific hypotheses and observational evidence. Moreover, it provides no basis whatever for rejecting the notion that high probabilities as well as high content are what we want our scientific hypotheses eventually to achieve on the basis of experimental testing.

RICHARD C. JEFFREY

Carnap's Empiricism

"Our knowledge is of matters of fact and of relations of ideas. Logic — inductive and deductive — concerns relations of ideas. As to our factual knowledge, some of it we have direct, through the senses, and the rest we have from the direct part, through logic." This is a rough sketch of Carnap's empiricism. Let me now try to fill it in, to get a fair likeness.

1. The Confirmational Net

For at least his last thirty years, this much of Carnap's view was constant: Inductive logic ought to be representable by a network, the nodes of which are the sentences (or, indifferently, the propositions expressed by the sentences) of a formalized, interpreted language. The nodes of the net are connected by arrows, each of which has a number associated with it, viz., the degree of confirmation of the sentence at the arrow's head, given the sentence at the arrow's tail. Thus, the configuration $\mathrm{e} \underset{x}{\rightarrow} h$ means that $\mathrm{c}(h,\mathrm{e}) = x$, where c is the confirmation function we shall eventually adopt, and e and h are sentences in the explicitly structured language we shall eventually adopt, either in actual practice or as a "rational reconstruction" of our wayward linguistic practice.

One's beliefs at time t ought to be representable by a credence function, cr_t, which assigns to each node h a numerical value $cr_t(h)$ in the interval from 0 to 1. Carnap's paradigm was the case in which there is a sentence e_t which expresses the part of our knowledge which we obtain at time t directly, through the senses: the experiential component. If we define

$$\mathrm{e}(t) = \mathrm{e}_1 \cdot \mathrm{e}_2 \cdot \ . \ . \ . \ \cdot \mathrm{e}_t$$

AUTHOR'S NOTE: I wish to express gratitude to the National Science Foundation for support of my research on probability and scientific method. This paper forms a pair with Jeffrey (1973), which it partly overlaps.

then one's entire credence function at time t ought to be determined by the formula

$$(1)\qquad cr_t(h) = c(h,e(t)).$$

In other words, one's degree of belief in any sentence h ought to be the number associated with the arrow to h from the sentence $e(t)$. One might imagine one's degrees of credence in the various nodes of the net at time t to be represented by luminosities: Where $cr_t(h) = 1$, node h shines with a certain maximum brightness; where $cr_t(h) = 0$, node h is quite dark; and credences between 0 and 1 are represented by intermediate luminosities. In these terms, $e(t)$ might be regarded as the *origin* of all the light in the net at time t. The number associated with the arrow from $e(t)$ to h indicates the portion of the full luminosity of $e(t)$ which the net transmits to h. As t increases, some parts of the net increase in brightness while others dim.

There is a point to the image of the twinkling, luminous net. One can see that different nodes have different brightnesses at time t, even though the luminosities represent someone else's beliefs. Dropping the metaphor, this "seeing" is a matter of knowing the fellow's *preference ranking* at time t: If he would rather have h be true than have h' be true, then his preference at time t is for h over h', and we write hP_th'. If the relation P_t has certain properties (Bolker, 1967) there will exist an expected utility function eu_t where

$$(2)\qquad hP_th' \textit{ iff } eu_t(h) > eu_t(h')$$

from which one can derive the credence function cr_t by the relation

$$(3)\qquad cr_t(h) = \frac{eu_t(-h) - eu_t(h \vee -h)}{eu_t(-h) - eu_t(h)} \quad \textit{if } hP_t{-h} \textit{ or } {-h}P_th.$$

(In case the condition of applicability of (3) fails, use the method of section 7.3 in Jeffrey, 1965.) It is (3) which determines the *actual* credence function at time t of the man whose preferences at that time are given by the relation P_t. It is (1) which gives the *rational* credence function at time t of the man whose actual observational evidence up to time t is $e(t)$. In using the symbol 'cr_t' for both functions, I am assuming that the actual credence function is as it ought to be, given what the man has observed.

Now let us make a slight alteration in the image of the net, prompted by the fact that logically equivalent sentences are confirmationally indistinguishable. We write '[*h*]' for the logical equivalence class to which *h* belongs, i.e., for the set of sentences which are logically equivalent to *h*. Imagine that all nodes (= sentences) in [*h*] are gathered together into a supernode. All arrows between nodes in the same supernode have the number 1 associated with them, and all arrows from one supernode to another will have the same number associated with them. Now the nodes of *the reduced net* will be the supernodes of the original net, and we shall have the configuration $[e] \xrightarrow{x} [h]$ in the reduced net if and only if we had $e \xrightarrow[x]{} h$ in the original net. From now on, "net" will mean *reduced net.*

The structure of the (reduced) net is static. (By "structure" I mean not only the configuration of arrows, but also the numbers associated with the arrows.) In the cases which Carnap has discussed in his publications, the play of light from the nodes is to be explained by (a) the trajectory of the points [e(1)], [e(2)], . . . , at which experience pumps light into the net, and (b) the static structure of conductive arrows through which the light of experience is propagated throughout the net. (The number associated with an arrow indicates its conductivity.) In this way, the experiential element in knowledge, (a), is neatly separated from the inductive element, (b), which Carnap viewed as purely logical.

2. In What Sense Is c "Logical"?

What did Carnap mean by his repeated assertions that his c-functions are purely logical?

I think he had two things in mind, the first of which is unassailably true, but perhaps a disappointment: If the net is to be static, we ought to identify our chosen c-function in such a way that, once e and *h* are identified either ostensively (as in quotation-mark names) or by structural descriptions, it is a matter of calculation to determine the value which c assigns to the pair (*h*,e) — no empirical research should be needed. In Carnap's terms, the values which c assigns to such pairs ought to be determined by the semantical rules of the metalanguage in which such sentences as 'c^*('Pa_1', 'Pa_2') = 2/3' appear. If 'M' is the name of that metalanguage, all such sentences should be M-determinate, e.g., the one just given is M-true.

Thus the first thing Carnap had in mind, in speaking of 'c' as a *logical functor*, was a requirement about the form of definition of c, e.g., it would be all right to define it by using a computing machine program which, given descriptions of h and e, eventually prints out the value of c(h,e); but it would not do to define c by some stipulation which, when we seek the value of 'c('P_1a', 'P_2a')', instructs us to discover the limiting relative frequency of P_1's among P_2's, even though (as luck would have it) the computing machine program mentioned above always gives the very answers to which the unacceptable sort of stipulation would lead us to by way of empirical research. The *function* c would be the same in both of these cases, being describable by a certain set of ordered triples. But in the first case, the functor which has that function as its extension is logical, while in the second case the functor is factual.

Carnap thought this point important, and he argued for it by inviting us to imagine the trajectory [e(t)], [e($t-1$)], . . . , traced back to its source, in a primal state of innocence in which our experience is null: $[e_0] = [p \vee -p]$. The structure of the net is constant, throughout the trajectory, so that the numbers associated with the arrows are the same at time 0 as at all successive times. But at time 0 one had no experience, and therefore the values of cr_0 and, with them, the values of c, must be independent of experience: they are logical, not factual. Of course, this argument will have no weight with someone who rejects the picture of the static net. Its point is rather to show that once you accept that picture you are committed to a definition of c which makes it a matter of computation, not experience, to determine the numbers c(h,e).

This aspect of Carnap's claim that 'c' must be a logical functor is then a rather trivial truth about Carnap's program of representing inductive logic by means of a static confirmational net. Read in this way, the claim is disappointing because it leaves us with the question, "Granted that the form of definition ought to be 'logical,' how are we to identify the *right* c-function, among the infinity of functions you have already considered, and the infinity that you have not yet considered?"

Here, Carnap's answer is available in his paper on inductive intuition (Carnap, 1968) and in parts of his paper entitled "A Basic System of Inductive Logic" (Carnap and Jeffrey, 1971). I would simply add that if the program is carried on and succeeds in the sense that we eventually find a language as well as a confirmational net on that language, which conforms to the clear cases of our inductive practice and resolves the unclear ones in a way that convinces people, then in the finding we shall have sharpened and developed and tuned and fixed our inductive intuitions far beyond anything we have now, and the resulting system will of course be describable by the net we shall have found. The corresponding c-function (or small family of c-functions, all of which seem to do the job about equally well) will then be uniquely "logical" in the sense of according with our inductive logical intuitions, as they will then exist.

I conclude that in its first aspect Carnap's claim that 'c' is logical is a clear, analytic truth about the character of his program for developing inductive logic; and in its second aspect his claim has the same truth-value as the statement that in time Carnap's program will be successfully carried out. (If and only if the truth-value is *t*, we shall eventually have highly developed inductive intuitions which are describable by way of a c-function.)

3. Experience and Confirmation: Correspondence with Carnap

In 1957 I finished my doctoral dissertation and sent a summary to Carnap, who expressed interest in my treatment of probability kinematics, viz., an alternative to the use of the trajectory $[e_t]$ $(t = 1, 2, \ldots)$. (See Jeffrey, 1965, chapter 11.) His own, very different ideas on the subject can be seen in the following correspondence.

From Carnap, July 17, 1957

To the first point: *Your rejection of conditionalization.* I believe that this idea of yours has a certain relationship to an old idea of mine; I never worked it out because I was never able to find a satisfactory solution. . . .

My idea was as follows. The c-method (i.e., the rule: if e represents all observational results of A then it is reasonable for A, to believe in the hypothesis h to the degree c(h,e)) is an oversimplification because it proceeds as if A knew certain observational results with certainty, while in fact we can never have absolute certainty with respect to a factual sentence. This simplification is useful to a certain extent, like many other schematizations in science. But I thought that it would be worthwhile to search for a method which takes into consideration the uncertainty of the evidence. . . . Thus I considered the following problem: The person A should represent his evidence (at a given time point) not simply as a list or conjunction of observational sentences, but rather as a list of such sentences together with a number attached to each sentence, where the number is to indicate the subjective certainty of the sentence on the basis of the observational experience. Thus the new form of evidence E may [be as follows]:

$$E = \{e_1, b_1; \ldots ; e_i, b_i; \ldots \}, \text{ where } 0 < b_i < 1.$$

Let e_1 be "the ball just drawn is red." If A could look at the ball for some time with good illumination and under otherwise normal circumstances, then b_1 is close to 1. If on the other hand A could see the ball only during a short moment and the illumination was not good or the air was misty or otherwise the circumstances were not favorable for an observation, then b_1 may be rather low. The important point is that b_1 represents the certainty of e_1 merely on the basis of the observational experience, without regard to any inductive relation to earlier observations. My problem now was to find a belief function cr(h,E) which is to determine the rational degree of belief (credibility) for any sentence h of a given simple language L, for which we have a c-function. My guess was that cr(h,E) somehow was to be defined on the basis of the numbers b_i and the values of $c(h,—)$ with respect to the e_i or combinations of such. Many years ago I explained this problem to Helmer and Hempel (in connection with their papers with Oppenheim ca. 1945) and later, when I was in Princeton, to Kemeny and Putnam. As far as I know, none of us has made any serious efforts to find a solution. I was discouraged from continuing the search for a solution by the fact that very soon I found some difficulties and did not know how to overcome them. One of them arises even before sentences outside of E are considered. The difficulty lies in the question how to determine the value of $cr(e_1,E)$? The value cannot simply be taken as equal to b_1 because the value must also take into account the inductive influence on e_1 of the other sentences in E, and this influence is not taken into account in the number b_1. Suppose that $b_1 = 0.8$ (for example, because the illumination was not quite good). Even if we assume for the other sentences of E the simplest case, namely that every b_i (for $i \neq 1$) is so close to 1 that we can take it as approximately $= 1$, the difficulty still remains. Let E′ represent these other sentences of E. Let $c(e_1,E') = 0.9$ (E′ may for example describe a great number of balls drawn from the urn of which almost all are red). Then $cr(e_1,E)$ is neither simply $= b_1 = 0.8$, nor simply $= 0.9$. I would guess that it should be > 0.9, since E contains in comparison with E′ an additional observation about e_1 and this observation, although it has only $b_1 = 0.8$ nevertheless represents an additional favorable item of evidence for the truth of e_1.

When I read your letter I had first the impression that your a's were about the same as my b's, and that therefore your method was perhaps a solution of my problem. It is true that you say only: "Now suppose A changes his mind about e . . . : A's degree of belief in e changes from m(e) to a" without specifying the reasons for A's change of mind. I presumed that you had in mind a change motivated by an observation, or at least that this was one possible case of a rational change. But now I doubt that your a's are the same as my b's. For, if I understand you correctly, a_1 is the degree of belief in e_1 after the change, in other words, after the observation, hence, in my terminology $cr(e_1,E)$ which is different from b_1.

My main question now is this: does your theory still somehow supply the solution to my problem . . . ? If not, what is the rule of application which your theory would give to A for the determination of the new value a_1 for the sentence e_1? It seems to me that any practicable system of inductive logic must give to A the following: (1) rules for choosing a function m [viz, cr_0] as the initial belief function . . . and (2) rules for A to determine the rational change of his belief function step for step on the basis of new observations which he makes.

To Carnap, September 4, 1957

Let $c_i = cr(e_i, E)$. The difficulty you describe . . . is: to determine c_i as a function of $b_1, b_2, \ldots$. As you remark . . . it is your *c*'s rather than your *b*'s which correspond to my *a*'s, i.e., to the person's actual degrees of belief in the e's after his observations. Actually, the c's will not be identical with the *a*'s unless (i) *E* reports all the observational experience the person has ever had which is relevant to the e's, (ii) the person in question is rational, and (iii) there really is a function cr which uniquely describes *rational degree of credibility*.

Now the *a*'s are measurable by observing the believer's behavior, e.g. by offering to bet with him on the e's at various stakes, immediately after his observations, and noting which bets he accepts. But the *b*'s are high-level theoretical terms, relative to the *a*'s. I am inclined to turn your question around, and ask, not "How are the c's determined by the *b*'s?" but rather "How are the *b*'s determined by the c's (or, actually, by the a's)?"

This question is not factual; it asks for an explication, and might be rephrased: "How shall we define the *b*'s in such a way that they will adequately explicate that component of the rational observer's belief in the e's which corresponds to what he has seen with his own eyes?"

From Carnap, December 26, 1957

I am still thinking about the question which I raised in . . . my letter of July 17: which rule does your inductive logic give to the observer in order to determine when and how he should make a change according to your method? You are indeed right that the customary c-method has the disadvantage that it refers only to certain situations which can never be completely realized, namely cases where the observer knows the result of an observation with certainty. But the method gives at least a clear rule for these idealized situations:

(3) When you experience an observation o_i which is formulated by the sentence e_i, then add e_i to the prior evidence and take as rational degree of belief in any sentence h its c-value with respect to the total evidence.

You have so far not given any rule. You emphasize correctly that your a_i is behavioristically determinable. But this concerns only the *factual* question of the *actual* belief of A in e_i. But A desires to have a rule which tells him what is the *rational* degree of belief. One part of the rule should presumably be as follows:

(4) Make a change with respect to e_i if and only if you have made an observation whose result is formulated by e_i.

But then a rule of the following form must still be added:

(5) If the prior degree of belief in e_i is m_i, then take as its posterior value after the observation the value $m'_i =$ . . .

The new value m'_i might depend upon the clarity of the observation(or the feeling of certainty connected with it or something similar) and further on m_i. . . .

P.S. Now some comments on your letter of November 14. . . . You are right with your interpretation of Neurath's paradoxical denial of the comparison of sentences and facts; I criticized this already in "Wahrheit und Bewährung" (1935); partly translated in Feigl and Sellars, *Readings*, 1949. Your idea of clarifying the situation of observing and recording by imagining a machine doing this, seems to me very helpful. But I still think it might be useful to regard the deposits of the observational results in the memory organ of the machine as sentences (in the interior language of the machine). My coefficients b_i might then perhaps enter in this way: the machine deposits in its memory after each observation o_i not only the corresponding observation sentences e_i but also a number b_i which depends only on the circumstances of the observation (e.g., in the case of a visual observation, on the intensity of the illumination, the sharpness of the picture on the screen, etc.) but not on the results of earlier observations.

To Carnap, January 16, 1958

I think this special case [viz., what I called "conditionalization" and Carnap called "the customary c-method"] is an important one. In physics, for example, there are very high standards which an event must meet in order to count as an observation, and there is perhaps not much variation in degree of reliability among such events. At least, this is true if by *observation* one means observations of the positions of needles on meters or such events. But if one thinks of the observer as consisting of the human being together with a certain amount of apparatus, e.g. together with an oscilloscope, and thereupon speaks of "observing" the presence of harmonics superimposed on a sine wave of voltage (as engineers, at any rate, are prone to do), the situation is different. I am intrigued with this broader sense of "observation." It seems that in physics, one can dispense with it, and speak instead of observations in the narrower sense, to which we apply portions of physical theory in order to derive or induce the "observation" in the wider sense. But where the science is not so highly developed as physics is, it is difficult to dispense with "observations" in which a great deal of interpretation is mixed up; e.g. in psychology we might say we observe that someone is angry, or we might say, more cautiously, that we observe that the person flushes, trembles, and curses, and add that there are laws which say that when these symptoms are exhibited, the subject is usually angry. Empiricists have usually tried to isolate the empirical element, i.e., the observational element, by analyzing "observations" in the broad sense into observations in the narrow sense plus theory. But I'm not sure this is the best way. When we learn, as children, to use words like "angry," we do so in much the same way in which we learn to use words like "red" and "flush." So I am inclined to think that the difference between "angry" and "red" — as far as their relation to experience is concerned — is not that one is theoretical while the other is observational, but rather that one term can more reliably be attributed to a subject, on the basis of observation, than the other. But they differ in degree, not in kind. It is desirable, where possible, to be able to reduce observation reports to terms like "red," which are highly reliable; but I doubt if this is always possible, and I expect that if we study only those situations in which such reduction is possible, we are apt to get a distorted notion of meaning.

From Carnap, March 6, 1958

I am still inclined to take as primitive simple properties like "red" rather than complex ones like "angry." With respect to complex properties, great difficulties would arise for the question of dependencies; such dependencies would then have to be expressed by complicated meaning postulates. And for the inductive treatment of two families [where a family is a set of properties $P^i{}_1$, $P^i{}_2$, . . . , which are mutually exclusive and collectively exhaustive on logical grounds] it is quite essential that, e.g., $P^1{}_1$ be simpler than $(P^1{}_1 \cdot P^2{}_1) \vee (P^1{}_1 \cdot P^2{}_2)$, although both properties have the same logical width. If this condition of simplicity were not fulfilled, there would not be a good reason for the axiom of analogy A16 (Notes, p. 29) [viz., U.C.L.A. "Lecture Notes on Probability and Induction"].

4. Prospects for Carnap's Program

To some extent, I have exaggerated the futuristic aspects of Carnap's program for developing an account of induction based on a static confirmational net. *He* was quite prepared to consider the question, "What c-function is appropriate for a language in which the primitive predicates are organized into a finite number of finite families?" Such a language would be very far from the very expressive languages for which he hoped that suitable c-functions would eventually be discovered. He would consider such simple languages rather abstractly, seeking to define families of c-functions for them. The family of c-functions for such a language would be parametrized by a number of adjustable constants, whose values would be determined by the interpretations chosen for the primitive predicates. Examples of such parameters: the *logical widths* of the primitive predicates; the *logical distances* between them.

Carnap, of course, sided with Humpty Dumpty: "When *I* use a word, it means exactly what I choose it to mean — neither more nor less." Others (e.g., Quine and Alice) attacked this linguistic Jacobinism from the right: "The question is," said Alice, "whether you *can* make words mean so many different things." To this, Carnap would reply, with Humpty Dumpty: "The question is, which is to be master — that's all."

Consider the case in which a small child thinks there may be an elephant in the next room. He stands in the doorway and looks about, sees nothing out of the ordinary, but still thinks there *may* be an elephant there, e.g., behind the easy chair. Quine points out that there is no telling whether the child's ignorance of the sizes of elephants reveals lack of factual knowledge or lack of semantical understanding: it is only by an arbitrary stipulation that we can see it as a matter of definition (or, alternatively, as a matter of fact) that elephants are rather large, compared with easy chairs. But that did not trouble Carnap, who was quite prepared to make such stipulations, after comparing the merits of making them one way or another; he was a linguistic revisionist, and quite unabashed about it. He was not given pause by the fact that the distinction between primitive and defined terms is vacuous in connection with natural languages. He was prepared to impose and police such distinctions as part of the business of reconstructing language closer to the mind's desire. The instrument which he sought to forge was a unit consisting of a language and a c-function, fused. It was no part of his claim that either element now lies ready to hand, fully formed.

For my part, I subscribe to a form of right-wing deviationism called "subjectivism." Subjectivists, like Carnap, use relations like (3) to determine the actual credence function, cr_t, at time t. Thus, subjectivists can measure the luminosities of the various nodes. Using the relationship

$$c_t(h,e) = \frac{cr_t(h \cdot e)}{cr_t(e)} \text{ \textit{if} } cr_t\,(e) > 0$$

they then determine the number which is to be associated with the arrow from [e] to [h] *at time t*: subjectivists take a dynamic view of the net. So may Carnap, if he wishes, for $c_t(h,e)$ is simply $c(h,e \cdot e(t))$ if the subject is being completely rational. But subjectivists do not assume that there is an underlying, static net corresponding to the function c. Nor need they assume that the sentences in the nodes of the net belong to a rationally reconstructed language. One can indeed follow the sequence $c_t, c_{t+1}, \ldots$, some distance forward under favorable circumstances: perhaps the change from c_t to c_{t+1} is clearly to be explained by an observation which convinced the subject that e_{t+1} is true, so that we have

$$c_{t+1}(h,e) = c_t(h,e \cdot e_{t+1})$$

Perhaps, too, one can follow the path backward some distance — but not, on the subjectivistic view, back to an a priori c-function c_0 which determines an underlying static net of the Carnapian sort.

Over his last ten or fifteen years, Carnap kept reporting encouraging signs of movement: subjectivists like L. J. Savage were getting closer to his position. For his part, Savage explained the relative motion as originating in Carnap. Indeed, from a great enough distance, e.g., from the point of view of K. R. Popper, Savage and Carnap were Tweedledum and Tweedledee. Perhaps, indeed, the difference is largely temperamental: you are a subjectivist if, like Carnap, you take credence to be determined by (3) but, unlike Carnap, boggle at the idealization (or falsification) which is involved in the attempt to trace the sequence $c_t, c_{t-1}, \ldots$, all the way back to an underlying static net, corresponding to the a priori function $c_0 = c$. It is not that subjectivists, rejecting the a priori c-function, are better empiricists than Carnap. On the contrary, they lack the means to formulate empiricism as I did (for Carnap) at the beginning of this paper.

That formulation was designedly archaic. Carnap would have no truck with such psychologism ("relations of ideas") and was scrupulous in speaking rather of relations between sentences (or, later, between propositions or attributes, etc.). But the point of the archaism is to suggest that even after such Carnapian refinement, empiricism seems unworkably simplistic. Here I take the distinguishing mark of empiricism to be the insistence on isolating the experiential element in knowledge from the logical element — at least in principle — so that we can conceive of testing our beliefs by tracing their provenance in impressions or in *protokollsätze* or in sets of ordered pairs (e_i, b_i) where the number b_i indicates the degree of belief in e_i which should be warranted by the observation itself, apart from the confirmation or infirmation which e_i derives from its relationships to the rest of the observer's beliefs.

I have no proof that Carnap's program for fleshing out empiricism cannot work. I take the difficulties which Carnap discussed in our correspondence, fifteen years ago, to be symptoms of unworkability, but that was never Carnap's view. He was sure that a solution could be found, and had projected a section on the topic for *Studies in Inductive Logic and Probability* (1971) before he found time getting short. It may be that, had he lived longer, he would have solved the problem, and it may be that someone else will solve it yet. But my own uncertain sense of plausibility leads me off on a different tack.

REFERENCES

Bolker, E. (1967). "A Simultaneous Axiomatization of Utility and Subjective Probability," *Philosophy of Science*, vol. 34, pp. 333–40.

Carnap, R. (1968). "Inductive Logic and Inductive Intuition," in I. Lakatos, ed. *The Problem of Inductive Logic*. Amsterdam: North Holland.

Carnap, R., and R. Jeffrey, eds. (1971). *Studies in Inductive Logic and Probability*, vol. 1. Berkeley, Los Angeles, and London: University of California Press.

Jeffrey, R. (1965). *The Logic of Decision*. New York: McGraw-Hill.

Jeffrey, R. (1973). "Carnap's Inductive Logic," *Synthese*, vol. 25, pp. 299–306.

MARY HESSE

Bayesian Methods and the Initial Probabilities of Theories

I. Introduction

In the past, objections have been raised to the notion of a probabilistic confirmation theory on the grounds that it required assignment of initial probabilities to expressions of hypotheses and evidence and that these probability values are not directly verifiable as are probabilities in the classic theories of equipossible alternatives or in the statistical frequency theory. Such objections have often come from philosophers and statisticians of positivist bent[1] and are closely parallel to objections within science itself to introduction of theoretical concepts that are not directly verifiable. The deductivist analysis of science has, however, released us from the tyranny of the "directly verifiable," in showing how unobservables can be given a function in theories, and it prompts the suggestion with regard to a confirmation theory that initial probabilities might be construed as "theoretical terms" to be judged by how well they enable the theory to systematize and explicate inductive arguments that are

AUTHOR'S NOTE: This paper is an expanded version of my "Theories and the Transitivity of Confirmation," *Philosophy of Science*, 37 (1970), 50–63, which was in turn an expanded version of a paper delivered at the Colloquium on Confirmation held in the Minnesota Center for Philosophy of Science in July 1968. I am indebted to Professor Allan Birnbaum for helpful comments on a first draft of Section II. Since preparing the paper for publication in this volume, I have incorporated much of sections II–V into my book *The Structure of Scientific Inference* (London: Macmillan, 1974). These passages are reproduced here with the publisher's permission.

[1] For example, the frequentist statistician R. von Mises makes the quasi-operational demand, "A quantitative probability concept must be defined in terms of potentially unlimited sequences of observations and experiments." *Probability, Statistics, and Truth*, 2nd Eng. ed. (London: Allen & Unwin, 1957), p. viii, and the positivist philosopher H. Reichenbach claims for his account of probability that "there is no doubt that this finitization [of the frequency limit for finite sequences] will satisfy all requirements of the verifiability theory of meaning in its most rigorous form." *The Theory of Probability*, Eng. ed. (Berkeley and Los Angeles: University of California Press, 1949), p. 348.

actually used in scientific contexts. Just as the justification for assigning a numerical mass to the unobservable electron is found in the verifiable consequences of this assignment, so the justification for assigning a probability value to a hypothesis prior to evidence is to be found in the agreement or disagreement of the consequences of this assignment with the posterior probabilities of hypotheses which are implicit in our attitudes of acceptance or rejection of these hypotheses on the basis of evidence.

Not only is a more liberal climate in philosophy of science more congenial to the introduction of initial probabilities, but also developments within statistics itself have tended in the same direction. The subjective view of probability as "degree of belief" is interpreted in more objective fashion in the *personalist* analysis of "degree of *rational* belief," and the introduction of Bayesian methods has clarified use of inverse arguments from evidence and initial probabilities to the posterior probabilities of hypotheses.

Criticism of probabilistic methods in confirmation theory now comes on more detailed grounds, namely that our intuitive requirements for the confirmation of hypotheses by evidence do not satisfy the probability axioms.[2] Much confusion has arisen in this debate from failure to recognize that there are many different explicata closely connected in ordinary discourse with the concept of confirmation, but not all behaving in the same way. Among others, the notions of *acceptability* of hypotheses, *desirable risk* in action, *utilities, power, content, expectation of truth* or *success*, and *support* by evidence have all entered the discussion. All are initially as unclear as the notion of confirmation itself. But in this area it is impossible to solve all problems until we commit ourselves to a particular solution of some of them. I shall therefore concentrate here on the notion of confirmation or support of hypotheses by evidence, and begin by adopting the personalist interpretation of confirmation as

[2] Especially K. R. Popper, *Logic of Scientific Discovery* (London: Hutchinson, 1959), ch. 10 and appendix *vii, and *Conjectures and Refutations* (London: Routledge, 1963), chs. 1, 11, and pp. 388–91; and L. J. Cohen, *The Implications of Induction* (London: Methuen, 1970). For critical surveys of Popper's and Carnap's positions, see I. Lakatos, "Changes in the Problem of Inductive Logic," in I. Lakatos, ed., *The Problem of Inductive Logic* (Amsterdam: North Holland, 1968), p. 315; and Alex C. Michalos, *The Popper-Carnap Controversy* (The Hague: Nijhoff, 1971). For arguments on behalf of Bayesian methods, see W. C. Salmon, *The Foundations of Scientific Inference* (Pittsburgh: University of Pittsburgh Press, 1966), ch. 7.

degree of rational belief in hypotheses or predictions relative to evidence.

The first half of the paper (sections II–IV) rehearses the personalist's arguments from rational decision procedures for taking degree of rational belief to be measured by a probability function, and the objections to them. It is by now generally accepted that these arguments provide no grounds for particular assignments of initial probabilities, nor for the use of Bayes's theorem to obtain posterior distributions of belief from initial distributions together with observed evidence. I next investigate some suggested transformations of belief which are not Bayesian and also find these wanting. Finally in this critical analysis of Bayesian methods, the claims of personalists for "convergence of belief" with increasing evidence are investigated and found largely inapplicable to the conditions of confirmation in science.

The first half of the paper shows the inadequacy of relatively a priori approaches to Bayesian confirmation theory. The methods criticized are those in which the probability axioms are claimed to be derivable from rationality alone, and no constraints are put on the initial probabilities (except that some are required to be non-zero), since such constraints would not be derivable a priori. The second half of the paper (sections V and VI) attempts to develop a more modest Bayesian confirmation theory for science, in which intuitions concerning purely local inductive methods are used to suggest stronger constraints on the initial distribution of belief and on transformations from initial to posterior distributions. Some intuitive criteria of adequacy for a confirmation theory are found to be explicated immediately by a Bayesian theory; others, involving universal generalizations and theories, are more problematic, but are found to raise interesting general issues regarding the function of laws and theories.

II. Games against Nature

1. The Probability Axioms

The starting point of a personalist theory of confirmation is the assimilation of the scientist's problem in confronting hypotheses with evidence to *betting behavior,* and more generally to *decision-making under uncertainty,* or in a more perspicuous phrase, "games against nature." In order to bring to bear upon a man's subjective beliefs some objective norms to which all rational men should assent, the personalist proposes (following the original suggestion of F. P. Ramsey)[3] to interpret "degree of belief" in a prediction behaviorally in terms of the maximum betting odds a man would be prepared to accept on the occurrence of the predicted event, provided no considerations other than his belief influence his attitude to the bet (for example, being honest and having no money, and therefore being unwilling to take *any* risk of loss however strong his belief). Then it is held that a rational man will not allow his betting system to be such that an opponent who knows no more than he does can make a book against him which will result in his loss whatever the outcome of the bet (a *Dutch book*). A betting function is called *coherent* if it assigns real numbers in the closed interval (0,1) to the possible outcome of the bet in such a way that if betting odds are laid on the occurrence of the outcomes in the ratio of the numbers, a Dutch book cannot be made by the opponent. It can then be shown that a coherent betting function is a probability measure over the possible outcomes and conversely that any probability measure corresponds to some coherent betting function. Thus the personalist's case for a probabilistic confirmation theory rests on two proposed equivalences:

(A) Degree of rational belief $\equiv$ coherent betting behavior.

(B) Coherent betting function $\equiv$ probability measure.

We shall need first a convenient formulation of the probability axioms. We are concerned with expressions of hypotheses, evidence, and predictions satisfying the axioms of propositional logic. These propositions will be denoted by lowercase letters such as h, g, f, e, sometimes with suffixes, to indicate that some of the terms in a probability expression will generally be interpreted as hypotheses or laws (as h,g,f) and some as evidence or observable predictions (as e). All the axioms and theorems to be discussed will, however, be valid for any substitution of h-type letters for e-type letters, and conversely, unless something is specifically indicated to the contrary. Probability will be understood as a measure over a pair of propositions; thus, $p(h/e)$ is to be read as "the probability of h conditionally upon e," or "the probability of h given e." The *initial probability* of h will be written $p(h)$, but this is to be understood consistently with what has just been said as the probability of h given either the tautology or previous background evidence which does not enter into the current calculations.

[3] *Foundations of Mathematics*, ed. R. B. Braithwaite (London: Routledge, 1931), ch. 7.

The *probability axioms* are as follows:

(P.1) $0 \leq p(h/e) \leq 1$;

(P.2) *Equivalence*: If h and e are logically equivalent to h' and e' respectively, then $p(h/e) = p(h'/e')$;

(P.3) *Entailment*: If e logically implies h, then $p(h/e) = 1$;

(P.4) *Addition*: If e logically implies $\sim(h \,\&\, h')$, then $p(h \vee h'/e) = p(h/e) + p(h'/e)$;

(P.5) *Multiplication*: If e, e′ are consistent and not self-contradictory, $p(h \,\&\, e'/e) = p(e'/e)\, p(h/e' \,\&\, e)$.

Logical implication or entailment will be written $\rightarrow$.

If the requirement of coherence is strengthened by requiring that the odds are not such that a bettor must certainly either come out even or lose for every possible outcome (*strict coherence*), then it can be shown that the necessary and sufficient conditions are obtained by replacing axiom (P.3) by (P.3′):

(P.3′) If and only if e logically implies h, then $p(h/e) = 1$.

With this replacement the axioms define a *regular* probability measure.

Some immediate corollaries of the axioms will be useful.

(C.1) From the addition axiom it follows immediately that if on evidence e we have a set $S = \{h_1, h_2, \ldots, h_n\}$ of mutually exclusive and exhaustive hypotheses (of which $\{h, \sim h\}$ is the simplest special case), then their probabilities on e must sum to 1 for all e. For the disjunction $h_1 \vee h_2 \ldots \vee h_n$ must be true given e, and hence $p(h_1 \vee h_2 \ldots \vee h_n/e) = 1$, and by the addition axiom

$$p(h_1/e) + p(h_2/e) + \ldots + p(h_n/e) = 1.$$

(C.2) From entailment and addition, if e *falsifies* h, that is, entails that $\sim h$ is true, we have

$$p(h/e) = 1 - p(\sim h/e) = 0.$$

(C.3) If h_1 and h_2 are independent, that is, if the truth of one has no relevance to the truth or falsity of the other, then $p(h_1/h_2 \,\&\, e) = p(h_1/e)$, and $p(h_2/h_1 \,\&\, e) = p(h_2/e)$. The multiplication axiom then gives

$$p(h_1 \,\&\, h_2/e) = p(h_1/e)\, p(h_2/e)$$

which can be taken as the definition of the *probabilistic independence* of h_1 and h_2.

(C.4) The equivalence and multiplication axioms also yield the following result: if h_1 entails h_2, then the probability of h_1 cannot be greater than that of h_2, given any e. That is,

$$\text{if } h_1 \rightarrow h_2, \text{ then } p(h_1/e) \leq p(h_2/e).$$

(C.5) A rearrangement of the multiplication axiom itself is often wrongly called *Bayes's theorem* — wrongly, because, historically, Thomas Bayes proved a different expression from the multiplication axiom which should properly go by his name. Consider the multiplication axiom expressed as a relation between the initial probability of h and its *posterior probability* on evidence e, $p(h/e)$. Provided $p(e)$ is non-zero, we have

$$(1) \qquad p(h/e) = p(h \,\&\, e) \,/\, p(e) = p(h)\, p(e/h) \,/\, p(e).$$

This is what is often called Bayes's theorem, and we shall follow this historical misusage here.

In some statistical applications, $p(e/h)$ is called the *likelihood* of h on e. So we have informally

$$\text{Posterior probability of } h = \frac{(\text{Initial probability of } h) \text{ times } (\text{likelihood of } h \text{ on e})}{\text{Initial probability of e}}.$$

In deductive situations, where $h \rightarrow$ e, using axiom (3), this expression reduces to

$$\text{Posterior probability of } h = \frac{\text{Initial probability of } h}{\text{Initial probability of e}}.$$

(C.6) It is useful to notice as a corollary of Bayes's theorem that, provided neither $p(e/h)$ nor $p(e)$ is zero, if $p(h)$ is zero, so is $p(h/e)$ for every e, and by replacing h by $\sim h$ it also follows that if $p(h) = 1$, $p(\sim h) = 0$, and hence $p(h/e) = 1$ for all e.

(C.7) Consider now the expression $p(e)$. Taking S as the set of exhaustive and exclusive hypotheses as before, since $h_1 \vee h_2 \ldots \vee h_n$ is logically true for this set, we have the logical equivalence

$$e \equiv (h_1 \vee h_2 \ldots \vee h_n) \,\&\, e.$$

Therefore, by the equivalence axiom

$$p(e) = p((h_1 \vee h_2 \ldots \vee h_n) \,\&\, e).$$

And by the addition and multiplication axioms

$$(2) \qquad p(e) = \sum_{s=1}^{n} p(h_s) \; p(e/h_s)$$

or, the prior probability of e is the sum over all *h*'s of the product of initial probability and likelihood of each *h*.

(C.8) Bayes's theorem strictly speaking is obtained from (1) and (2):

$$p(h_r/e) = \frac{p(h_r)\, p(e/h_r)}{\sum_{s=1}^{n} p(h_s) \; p(e/h_s)}$$

for all $h_r (r = 1, 2, \ldots, n)$.

2. *Betting Rates*

With these specifications of coherence and probability, equivalence (B) for betting functions has been uncontroversially justified by the theorems of Kemeny and Shimony.[4] But the probability axioms place only very weak constraints upon the measures that can be assigned to propositions, since except in the circumstances mentioned in (P.3) and (P.3′), particular probability measures are not specified. In fact, if h_s is a set of mutually exclusive and exhaustive hypotheses, given e, then any set of measures $p(h_s/e)$ in the interval (0,1) such that

$$\sum_{s} p(h_s/e) = 1$$

is consistent with the axioms.

Justification of equivalence (A) is a much more difficult matter. The suggestion that degrees of belief should be explicated as betting behavior originally arose from a behaviorist or empiricist desire to unpack the apparently subjective notion of belief into terms of observable behavior. The highest odds a man is prepared to offer on an unknown outcome is taken to represent his subjective degree of belief in that outcome. Adding the objective requirement of coherence, we obtain the notion of *rational* betting behavior as an index of degrees of *rational* belief. "Rationality" is of course still a very weak notion here — a man is rational in this sense if he does not deliberately court loss; it is not necessary that

[4] J. G. Kemeny, "Fair Bets and Inductive Probabilities," *Journal of Symbolic Logic*, 20 (1955), 263; A. Shimony, "Coherence and the Axioms of Confirmation," *Journal of Symbolic Logic*, 20 (1955), 1.

he use any other criteria of comparison between his beliefs. He may assign a probability 5/6 to six coming up at the next throw or to the constitutional election of an anarchist government in Britain this year; so long as the rest of his beliefs cohere with these assignments he is rational.

The concept of coherent betting is clearly insufficient to determine useful probability assignments to hypotheses, and we shall later have to consider further constraints upon such assignments. Meanwhile, is it even *necessary* for a confirmation measure? Indeed is it at all applicable? We neglect for a moment the complication that a man might not be willing to back his actual beliefs by corresponding betting odds if he is unwilling to risk much loss under any circumstances. Someone who put an infinite value on his life would hardly ever cross any busy road, however small he believed the chances of accident to be. So we assume the stakes low enough to allow our bettor to contemplate the risk of loss and allow his bets to reflect his actual beliefs. Are his actual beliefs unreasonable if the corresponding betting rates are not coherent?

There are two ways in which an affirmative answer might be justified, one of which seems conclusive but is only dubiously relevant to "games against nature"; the other of which is more relevant but far from conclusive.

Suppose X and Y are tossing a die. X is of an optimistic turn of mind and offers even odds for each fall of the die. He might, of course, win on one or more tosses. But if the rules of the game are such that Y can propose any combination of bets on the same toss at the odds accepted by X, Y has only to place a sum m against each of the six possible falls of the die to be sure of winning an amount $(5/6 \times m/2) - (1/6 \times m/2) = m/3$ whatever the outcome. In the context of fair games the odds were intuitively unreasonable as well as incoherent.

But what has this situation to do with "games against nature"? Here we suppose the scientist asking himself what is his degree of belief in a hypothesis on the basis of evidence by considering how much he would bet on its truth. (We will suppose for the moment that the hypothesis is a singular prediction in order to avoid the difficulty that no bet on a universal hypothesis in an infinite domain can be conclusively settled in a finite time if the hypothesis is true.) Must his betting rates at modest stakes faithfully reflect his degrees of belief, and even if it is assumed that they must and that his degrees of belief are reasonable in an intuitive sense, must the rates be coherent? The argument from ideal games certainly does not prove either of these assumptions. The hypothesis in question is not the outcome of a closely specified situation which can be made to repeat itself in all relevant respects an indefinite number of times — it may in fact be a once-for-all prediction in conditions never again even approximately repeated. Nature is not an ingenious opponent who seeks out just those bets which will result in certain loss. And there is no clear relation between reasonable expectation of truth and the willingness to gamble on financial reward. The point does not need laboring that there is no close analogy between games and games against nature.

In a second form of the argument we concede most of the disanalogies referred to, but redescribe the gambling situation as one in which an addicted gambler plays many different games for short and long periods, picking his odds at random and playing against a variety of opponents, both subtle and disinterested. Then, by a generalization of the first argument, although nothing is certain, his *reasonable expectation* on coherent betting odds is to come out approximately even in the long run. This form of the argument is obviously not conclusive for coherence. It refers to expectations not certainties, and it may be intuitively reasonable on some occasions to assume expectations will not happen, and consequently to bet incoherently. We may be in a highly unexpected corner of the world, even in the long, but necessarily finite, run.

Further props can be knocked away from equivalence (A) by removing the restriction to singular predictions that was implicit in the notion of an "outcome." Many philosophers argue that scientific laws and theories are in their essence universal in scope over infinite domains. We shall later find reason to question this, or at least to question its relevance to the logic of confirmation, but if it is accepted, then degrees of belief in laws cannot be represented by betting rates in a game in which there is a fair chance of settling the bet. Nor could the rates in any case be strictly coherent. For, if we assume that a law can be falsified but never shown to be true, any bet upon its truth may be either lost or not settled, but it can never be won. The only odds on its truth a reasonable man would freely choose under these circumstances would surely be 0:1, for since in any case he cannot win anything, any positive sum he places on its truth may be lost, and at best he will simply not have to pay. A zero degree of belief would thereby be assigned to all laws alike, that is, the same degree that is assigned to statements known to be false. We might suggest forcing the scientist to adopt non-zero coherent odds however small in order to elicit his *relative* degrees of belief in different laws, but the concept of a forced bet only weakens still further the analogy between reasonable belief and betting behavior in ideal games.

3. *Decision-Making under Uncertainty*

The concept of coherent betting as a test for rational choice between hypotheses has been very considerably strengthened and generalized in statistical decision theory.[5] Initially this theory does not depend on inverse probability or the assignment of initial probabilities to propositions. It starts from the relatively uncontroversial assumption that a statistical hypothesis h of the form "If x is A, the probability that x is B is p" determines for h a *likelihood* p which is equal to the probability p of an A being a B if h is true. Suppose now we have a problem of choice between two hypotheses h_1 and h_2 with likelihoods p_1 and p_2 respectively, and we consider the possible outcomes of a test consisting of taking a sample of one individual which is A and observing whether it is B. We can set up a table of likelihoods for possible outcomes on the assumption first that h_1 is true and second that h_2 is true, as in Table 1. Before making the test we decide on a particular *decision strategy*, which is equivalent to a general rule for choosing one of the hypotheses, given a particular outcome of the test. In this case there are four strategies, as follows:

T_{11}: Choose h_1 on outcome B, and h_1 on outcome $\sim B$.
T_{12}: " h_1 " B, " h_2 " $\sim B$.
T_{21}: " h_2 " B, " h_1 " $\sim B$.
T_{22}: " h_2 " B, " h_2 " $\sim B$.

We can now set up a table for each T of the probability of making the *wrong* choice if h_1 is true and if h_2 is true (Table 2). By examining this table it can be seen that some of these strategies are "rational" in a sense similar to that adopted in coherent betting, and some are "irra-

[5] For introductory accounts, see N. Chernoff and L. E. Moses, *Elementary Decision Theory* (New York: John Wiley, 1959); R. B. Braithwaite, *Scientific Explanation* (Cambridge: At the University Press, 1953), ch. 7. See also R. D. Luce and H. Raiffa, *Games and Decisions* (New York: Wiley, 1957), and R. M. Thrall, C. H. Coombs, and R. L. Davis, *Decision Processes* (New York: John Wiley, 1954), especially J. Milnor, "Games against Nature," ch. 4.

	Outcomes	
	B	$\sim B$
h_1 true	p_1	$1-p_1$
h_2 true	p_2	$1-p_2$

Table 1

	T_{11}	T_{12}	T_{21}	T_{22}
h_1 true	0	$1-p_1$	p_1	1
h_2 true	1	p_2	$1-p_2$	0

Table 2

tional." Suppose $p_1 = 2/3$, $p_2 = 1/4$. Then, whatever hypothesis is true, T_{21} gives a greater probability of wrong choice than T_{12}, and T_{21} is said to be *dominated* by T_{12}. Any strategy which is dominated by some other strategy is said to be *inadmissible*. Strategies which are not dominated by any strategy are said to be *admissible*. In this example T_{11}, T_{12}, T_{22} are all admissible.

The very simple example just described can be generalized by the introduction of a continuum of admissible *randomized strategies*, defined by some probabilistic method of choosing between the strategies of the original admissible set. The example can also be generalized to cases of more than two hypotheses and more than one individual in the test sample, with corresponding multiplication of the number of possible outcomes and strategies. Further expansion of the concept of a decision theory involves introduction of *utilities* and *initial probabilities*. In most decision situations it is only in the artificial case of low stakes that choice of hypothesis should depend only upon minimizing the probability of false choice. In general we want to minimize the acceptable *risk* of false choice, whether this is measured in terms of financial, emotional, or some other undesirable loss. Thus our betting behavior would not generally reflect the full degree of our belief where the consequences of losing are in some way disastrous. Someone might not choose to enter into a marriage even though the probability of its success was somewhat greater than its failure. These considerations can be taken into account by defining a *utility function* over the hypotheses between which we are choosing in such a way that the utility of a hypothesis reflects the value to us of the hypothesis being true. Utility values are assumed additive for mutually exclusive hypotheses. In our example let the utilities, which are taken constant for each hypothesis, be represented by u_1, u_2 for h_1, h_2 respectively. Then if we are interested in somehow minimizing expected loss or utility, we shall be concerned with the products of u_1 or u_2 with the probabilities of false choice if h_1 or h_2 respectively are true, rather than with the probabilities of false choice themselves.

Initial probabilities may be introduced by noticing that if in calculating the expected loss of utility resulting from false choices of h_1 or h_2, we had had some means of knowing that their initial probabilities were unequal, say equal to v, $(1 - v)$, respectively, then the expected losses of utility would have been multiplied in this ratio. This consideration shows that although initial probabilities do not appear in the formulation of the decision problem, since likelihoods of hypotheses on evidence are the only probabilities admitted in this formulation, nevertheless there is a mathematical equivalence between the decision problem and calculations based on the assumption that the initial probability v is equal to 1/2.

Thus, if initial probabilities are admitted, the probabilities of false choice of Table 2 would be multiplied through in the first row by $w_1 = u_1 v$, and in the second row by $w_2 = u_2\ (1 - v)$. In fact only these products w_1, w_2 can affect subsequent choice of strategy and hypothesis. With the new representation of expected loss of utility exemplified by Table 3 we can now consider what criteria might be applied to selection among admissible strategies.

	T_{11}	T_{12}	T_{21}	T_{22}
h_1 true	0	$w_1(1 - p_1)$	$w_1 p_1$	w_1
h_2 true	w_2	$w_2 p_2$	$w_2(1 - p_2)$	0

Table 3

There is one type of admissible strategy with uniquely desirable properties. This is the so-called *minimax strategy*, which is obtained by finding the maximum expected loss in each strategy column, and picking that strategy or one of those strategies for which this maximum is least. The minimax strategy ensures that, independent of which hypothesis is true, the expected loss is smaller than the worst possible loss on any other strategy. In this sense the strategy is prudent, not to say timid. In general it leads to playing safe; if the decision is either to take action with risk or to take no action with no risk, it will choose inaction, and between hypotheses of equal prior probability and utility it will choose that which goes as little as possible beyond the evidence. Therefore in scientific contexts it is not often an interesting strategy. But it is not forced upon us by probabilistic decision methods, and it may be remarked parenthetically that the general criticism leveled, in particular by Popper, against probabilistic methods as always selecting the weakest hypothesis is therefore groundless.

The prudence of the minimax strategy is bought at the cost of the possibility of greater expected gain if the true hypothesis is chosen, and also of course by guarding against the possibility of greater expected loss if the false hypothesis is chosen. Introduction of a greater element of risk in exchange for higher potential winnings suggests averaging the expected losses on each strategy, and taking that strategy which minimizes the average instead of the maximum expected loss. In the special case of $w_1 = w_2$ this strategy is called the *maximum likelihood strategy* ("maximum" referring, with greater optimism than minimax adherents are likely to exhibit, to the possibility of *gain* rather than loss). The strategy is so-called after Fisher's method of hypothesis-choice, which asserts that the best hypothesis is that whose likelihood on the observed sample is a maximum. That is to say, in the above example, if the sample is B the best hypothesis is h_1, while if the sample is $\sim B$ the best hypothesis is h_2, giving T_{12} as the maximum likelihood strategy.

Fisher's method was developed in response to objections to inverse probability methods depending on initial probabilities of hypotheses. But, as we have seen, initial probabilities and also utilities can be introduced into a more general viewpoint, which yields Bayesian decision theory. When maximum likelihood is extended to become a method of maximizing expected gain of utility, the best strategy in this sense is called a *Bayes strategy*, and is obtained by minimizing the average value of the losses in each column of Table 3 and selecting the strategy that corresponds to this minimum loss or maximum gain. The name "Bayes" is appropriate because it can be shown that the Bayes strategy with parameters $w_1, w_2, \ldots, w_n$ for n hypotheses always selects that hypothesis which would be chosen on given evidence by calculating the posterior probabilities of all the hypotheses by Bayes's theorem and choosing that hypothesis whose posterior probability is maximum, provided the initial probabilities assumed in this calculation are in the ratios $w_1 : w_2 : \ldots : w_n$. Moreover, it can be shown that every admissible strategy is a Bayes strategy for some set of w's (all positive), and conversely that Bayes's strategies for positive w's are always admissible.

This brief account of the concepts of elementary decision theory is sufficient to indicate its generality as a theory of hypothesis-choice. It is tempting to conclude that if any theory of confirmation is going to be adequate for scientific inference, then Bayesian decision theory will be. The theory is certainly preferable to the analogy of confirmation with betting rates, since it removes some of the artificial restrictions under which that analogy describes "games against nature." It indicates how to take account of utilities if these are known. It is not, however, obvious how the notion of "utility" should be interpreted in relation to pure science, since at least monetary gains and losses do not seem to be directly involved. It might be held that all theories are equally valuable if true, so that utilities can be neglected. Or the notion of "epistemic utility" may be adopted, depending on the range or content of a hypothesis, thus incorporating the view that a scientific theory is more valuable the more powerful it is.[6]

With regard to the more basic objections leveled against the rationality of coherent betting rates, however, we are in no better case with admissible strategies. In the short run (and compared with the universal or near-universal pretensions of science all runs are short) inadmissible strategies may give more right answers than admissible ones. Suppose in the example above we adopt strategy T_{21}. Since we don't know whether we are in a typical corner of the universe, we don't know that in a comparatively small number of choices we are going to be faced with samples which will lead to greater loss than on, say, T_{12}. We may be presented with a series of B's and no not-B's, and yet h_2 may be true. Let us simplify by supposing that each sample is forgotten after it is observed, so that the choice of hypotheses starts afresh at each sampling. This enables us to consider the example as a sequence of independent single-sample decisions. Under these circumstances T_{21} will be right every time and T_{12} will be wrong. There are, however, two things that can be said in support of the irrationality of T_{21} in this case. First, if h_2 is true it becomes decreasingly probable with larger samples that we shall continue to sample not-B's and therefore that T_{21} will always choose correctly. And more telling than this long-run justification is the fact that *if* T_{21} is always successful for a given finite sample, there is always *some* admissible strategy which would have been equally successful for that sample, namely one which makes h_2 initially very probable. And in general, for any finite sample, there is always some set of initial prob-

[6] Suggested by C. G. Hempel, "Inductive Inconsistencies," *Aspects of Scientific Explanation* (New York: Free Press, 1965), p. 76, and developed by I. Levi, *Gambling with Truth* (New York: Knopf, 1967), ch. 5ff.

abilities which will choose the true hypothesis each time by Bayesian methods, if only we knew what that set is. But if a Bayes strategy with, say, $1 - v$ very near to zero is considered, it might be that no justification can be found for assuming h_2 so antecedently improbable, and a decision-maker who "irrationally" adopted T_{21} might prefer to describe his behavior in terms of a lucky break with an inadmissible strategy.

4. Bayesian Transformations

There is a further important respect in which the arguments for coherence and admissibility establish weaker conclusions than might at first sight appear. To see what this is, we need a more explicit formulation of the decision procedure. If only admissible strategies are adopted and utilities are assumed equal, it follows that any hypothesis-choice is practically equivalent to maximizing the posterior probability conditional upon evidence, where this posterior probability measure is defined over the set of all possible hypotheses. Admissibility requires only that this conditional measure satisfy the probability axioms.

Consider the position of the decision-maker at a given point of time, say t_0, when he has a certain amount of evidence, e_0, and a probability distribution over hypotheses. Suppose he adopts Bayes's strategy T. T tells him, for each possible combination of evidence statements he may *subsequently* observe to be true, which hypothesis he should choose, and it does this in accordance with the rule of taking the maximally probable hypothesis for each combination of evidence, which is equivalent to using Bayes's theorem. In the case of a bettor this means that he is prepared at time t_0 with coherent odds for any bet his opponent may suggest conditional upon whatever evidence may be observed. Under these circumstances the bettor cannot be presented with a Dutch book, and the decision-maker does not face the expectation of loss whatever the state of nature. It is very important to stress that all this refers only to *possibilities*, none of which are yet realized at t_0. It refers to what the decision-maker should be prepared to choose in hypothetical circumstances *before* more evidence is collected. It describes the static situation at t_0 in terms of a certain probability distribution satisfying the probability axioms.

The situation is, however, quite different when more evidence is collected at time t_1. Denote this new evidence by e_1. There is now a new decision situation in which a probability distribution $p_1(h/e_o \& e_1)$ is defined over all h consistent with $e_o \& e_1$. Admissibility demands only that p_1 be a probability measure; it does not demand that p_1 be related to p_o by the equations

$$p_1(h/e_o \& e_1) = p_o(h/e_o \& e_1) = \frac{p_o(h/e_o)\, p_o(e_1/h \& e_o)}{p_o(e_1/e_o)}$$

for all mutually consistent h, e_o, e_1. In other words, p_1 need not be related to the initial distribution of probabilities p_o in accordance with axiom (P.5). Only if it is so related, is the transformation *Bayesian*.[7]

Since the coherence requirement has nothing to say about what particular distribution of initial probabilities should be adopted, it cannot prevent this distribution from being changed with new evidence. Indeed a man may change his betting rates *without* new evidence by just changing his mood, or thinking more deeply, and yet remain coherent. Coherence does not merely leave the question of initial probabilities completely open; it also allows them to be changed for any reason or no reason, in ways that do not satisfy the Bayesian transformation. Consequently we have to envisage two different kinds of situation that may face the decision-maker. He may have to choose a set of hypotheses in various domains all at once at t_0, when his actual evidence is e_0. This corresponds, for example, to the scientist's situation in predicting the outcome of a complex of future experiments. In such a case arguments for admissibility indicate that he should adopt a Bayesian strategy, and this involves calculating conditional probabilities and maximizing. But on the other hand the decision-maker may be in a situation in which he is allowed to collect evidence sequentially. Then he has not only to select a particular Bayesian strategy T_o at t_0, but also to select strategies at t_1, t_2, . . . , as evidence e_1, e_2, . . . , comes in. These subsequent strategies may be the same as or different from T_o. The difference between the situations is that between prediction and learning. The learner, of course, also remains a predictor, because it is assumed that he is not merely soaking information in as time goes on, but learning from it to

[7] The non-derivability of Bayesian transformations from coherence was explicitly pointed out by J. M. Vickers, "Some Remarks on Coherence and Subjective Probability," *Philosophy of Science*, 32 (1965), 32, and I. Hacking, "Slightly More Realistic Personal Probability," *Philosophy of Science*, 34 (1967), 311. For a detailed discussion see D. H. Mellor, *The Matter of Chance* (Cambridge: At the University Press, 1971), ch. 2.

adapt to the future. And the learner has the predictor's problem of choice of initial probabilities, renewed on each occasion upon which he receives new evidence.

To conclude this brief survey, the best that can be said for a decision-theoretic formulation of confirmation theory is that it provides a sufficient framework in terms of which any choice-problem and subsequent decision *could* be represented, though possibly at the cost of highly implausible assumptions about utilities, initial probabilities, and posterior transformations. But clearly in itself this representation is vacuous as a method of induction; for unless there are some restrictions upon the initial distribution and some grounds for using a Bayes or some other posterior transformation rule on that distribution, the effect of evidence on the posterior distribution is quite arbitrary. Even if the overall probability distribution is required to be coherent or strictly coherent, that is, so long as it satisfies the probability axioms, a long run of black crows, for example, might be represented as good evidence for the statistical generalization "Most crows are white," if only the probability of this generalization was initially high enough or if Bayes's theorem is held to be irrelevant to the posterior transformation of probabilities. Rational decision procedures are not sufficient to capture any interesting *inductive* sense of rationality, for which further constraints must be found.

III. Non-Bayesian Transformations

The possibilities of assignment of initial probabilities and posterior transformations have been left wide open by the arguments of the previous section. The weakest possible response to this situation, which might be called *pure personalism*, is the view that the succession of probability distributions over hypotheses held by a believer is a purely *physical* or *causal* matter of belief-response to stimuli deriving from the evidence. If *no* normative constraints are put upon belief distributions other than coherence, this is equivalent to adopting an inductively arbitrary, and arbitrarily changing, initial probability distribution as a result of each new piece of evidence, and cannot in itself yield inductively or predictively useful beliefs except by accident. No serious inductive logician could be a pure personalist in just this sense, and in fact those who have emphasized the ultimately physical nature of the effect of evidence on belief have also placed other normative constraints upon the sequence of probability distributions.

An inductive theory of this type is suggested by Richard Jeffrey,[8] who makes a radical departure from Bayesian theory by rejecting the notion of incorrigible evidence. In orthodox Bayesian theory it is assumed that the posterior probabilities are conditionalized upon evidence which must be taken to have probability unity once it has been observed, that is to say, after observation of evidence e_r, the initial probability of e_r is necessarily replaced by the posterior probability $p(e_r/e_r) = 1$. But, Jeffrey suggests, observation is not always such a conclusive affair. We may not be quite sure whether we have observed e_r or not; we may be slightly more sure of its truth than before, but the light may have been bad, and we may wish to leave open the possibility that at some future time we may be led to withdraw e_r from the evidence under pressure of other evidence or at least to give it probability less than 1. A Bayesian might attempt to accommodate this idea by allowing withdrawals as well as accumulations of evidence in his sequence of posterior distributions, but he cannot deal in this way with "partly observed" evidence to which it is appropriate to give a probability value between 0 and 1. Or he might attempt to get beneath the uncertain evidence to something more conclusive which can be given probability 1 and can communicate a higher probability still short of 1 to the uncertain evidence. But any such attempt would be reminiscent of the search for more and more basic incorrigible observation statements, a program which has met conclusive refutation in recent philosophy of science.

Jeffrey's solution is a radical personalism in which not only is the initial distribution a physical rather than a rational bequest, but also every subsequent distribution shares this physical character. As we go around with our eyes more or less open, our degrees of belief in various propositions are physically caused to change. We usually cannot specify any evidence that can be asserted conclusively; thus the sequence of distributions is best not specified by conditional probabilities, but merely by suffixes indicating the order of the sequence. The distribution $p_r(h_s)$, for example, is the equivalent of the posterior distribution at t_r, and this distribution includes such probabilities as $p_r(e_j)$ with values less than 1, which are physically related to the uncertain and inexpressible results of

[8] R. Jeffrey, *The Logic of Decision* (New York: McGraw-Hill, 1965), chs. 11, 12. I have somewhat modified Jeffrey's notation in what follows.

observation. Observation-type propositions are distinguished from hypothesis-type propositions by being those propositions in which the changed probabilities physically *originate*. Changes are disseminated over the distribution for hypothesis-type propositions in such a way that coherence is preserved, according to the set of equations

$$p_r(h_s) = p_r(h_s/e_j)\, p_r(e_j) + p_r(h_s/{\sim}e_j)\, p_r({\sim}e_j),$$

where e_j is the only evidence-type proposition whose probability has changed, or according to a corresponding set of equations in the general case where several probabilities of evidence have changed. Jeffrey does require that in the sequence of distributions the conditional probabilities such as $p_r(h_s/e_j)$ remain invariant, so that for these probabilities the *r*-suffix can be dropped, and the above equation becomes

$$(3) \qquad p_r(h_s) = p(h_s/e_j)\, p_r(e_j) + p(h_s/{\sim}e_j)\, p_r({\sim}e_j).$$

Jeffrey seems to introduce this condition as just a device for distinguishing between those propositions which are directly affected by observation and those which are not.[9] If $p_r(h_s/e_j)$ does not satisfy the stability condition for some h_s, then this h_s must be included among those propositions in which changes of probability originate. It might be objected that if h_s is hypothesis-like, it is contrary to the spirit of the physical account (which is an attempt to retain a vestige of empiricism) to suppose that our belief in a *hypothesis* can be directly affected by observation in a way not directly related to the probabilities of evidence for that hypothesis. But even if it is accepted that there must be some rule for dissemination of probability-changes from evidence to hypotheses, why choose the invariance of conditional probability and not, for example, the likelihood $p_r(e_j/h_s)$ or some other function of the p_{r-1}-distribution?

Jeffrey's attempt to deal with the problem of uncertain evidence is courageous and ingenious.[10] Unfortunately it presupposes only coherence

[9] In private correspondence, which he kindly permits me to quote, Jeffrey tells me that he intends the condition just as one which defines an interesting situation, namely, that in which the observer judges that he made an observation directly relevant to e and to nothing else.

[10] I have proposed an alternative, Bayesian, method of accommodating uncertain evidence without assuming a stable observation language in "A Self-Correcting Observation Language," in B. van Rootselaar and J. F. Stahl, eds., *Logic, Methodology, and Philosophy of Science*, vol. 3 (Amsterdam: North Holland, 1968), 297, and in "Duhem, Quine, and a New Empiricism," *Knowledge and Necessity*, Royal Institute of Philosophy Lectures, vol. 3 (London: Macmillan, 1970), 191. After making my

constraints upon the sequence of distributions and could hardly form a sufficient basis of a confirmation theory. First, it should be noticed that in Jeffrey's theory, if (3) holds and $p_r(e_j) \neq p_{r-1}(e_j) \neq 1$, then the likelihoods of statistical hypotheses with respect to e must be assumed to change. For we have

$$p_r(e_j/h_s) = p_r(e_j)p(h_s/e_j) \;/\; p_r(h_s)$$

and by (3) this is not equal to

$$p_{r-1}(e_j/h_s) = p_{r-1}(e_j)p(h_s/e_j) \;/\; p_{r-1}(h_s).$$

Some personalists and most objectivists have regarded the likelihoods of statistical hypotheses as the most "objective" of the various probability functions entering Bayes's theorem.[11] Of course, Bayesians themselves replace $p_{r-1}(e_j/h_s)$ by $p_r(e_j/h_s \& e_j) = 1$ when e is said to have been observed. But for them there is at all times t_r a constant conditional likelihood $p(e_j/h_s)$ which is dependent only on the statistical hypothesis h_s and a random sample e_j, whether or not e_j has been observed at t_r. This constant likelihood cannot be expressed by Jeffrey, because all his probabilities, including likelihoods, are in effect conditional upon unexpressed (and inexpressible) evidence. The probability of e_j on h_s is well defined only if the truth of h_s is the only factor upon which e_j is conditional. Where the value of $p_r(e_j/h_s)$ in Jeffrey's theory differs from this "objective" likelihood, e_j must be construed as a sample conditional not only upon h_s but also upon some other unexpressed evidence, for example that only part of the domain of h_s is being sampled. It must be conceded to Jeffrey that such nonrandom sampling is nearer to practical scientific experimentation than the ideal situations of statistics, and indeed I shall accept this argument in the next section.

The most serious drawback to Jeffrey's proposal from the point of view of confirmation theory, however, is that it allows no control over confirmation relations. It is impossible to calculate, except conditionally at t_{r-1}, what the effect of e_j upon the probability of h_s will be when e_j has actually been observed at t_r. The conditional probability $p_{r-1}(h_s/e_j)$ bears no regular relation to $p_r(h_s)$, because the effect of actually

suggestion I discovered that essentially the same method is to be found in C. I. Lewis, *An Analysis of Knowledge and Valuation* (La Salle, Ill.: Open Court, 1946), p. 235f.

[11] This point has been made against Jeffrey by I. Levi, "Probability Kinematics," *British Journal for the Philosophy of Science*, 18 (1967), 197.

"observing" e_j has an unpredictable effect upon the whole distribution. But to have a confirmation theory at all we must be able to answer explicitly some general questions concerning the effect of evidence upon the probability of hypotheses. For example, we may wish to calculate, on our present belief distribution, what will be the effect on our belief in given hypotheses of the future observation of a certain piece of evidence. This might be done by means of Bayes's theorem or by some other posterior transformation so long as we have a rule for it which can be applied at t_{r-1} to calculate the effects of evidence at t_r. But Jeffrey's proposal provides no rule at all, neither Bayesian nor any other, and if there is *no* rule, no answer can *now* be given to questions of the type: Which experiment *will* most usefully test a given hypothesis, or Which hypothesis *will* be most acceptable on the basis of specified future evidence? Without some such rule, there can be no confirmation theory.

It does not yet follow, however, that the rule must be Bayesian. An attempt to provide an a priori argument for Bayesian transformations has been made by Abner Shimony,[12] by examining a purely analytic derivation of the probability axioms, where the derivation does not depend on the notion of coherent betting. The proof concerns only values of probability assumed to hold at a given time, but it might be adapted to the discussion of Bayesian transformations if its conditions were applicable to posterior probability at time t_1 in terms of initial probabilities at t_0, as well as to the conditional probabilities at t_0 which yield the multiplication axiom. The nontrivial assumptions involved in the proof assert a functional dependence of such expressions as $p(h \& e)$ upon $p(h/e)$ and $p(e)$, and of $p(h \vee h')$ upon $p(h)$ and $p(h')$, where h, h' are mutually exclusive. But Shimony concludes that such analytic assumptions, weak as they are, strain our intuitions about belief too far to be cogent a priori, even where axioms relative to a given evidence-set are concerned. A fortiori, it seems unlikely that such analytic arguments are capable of yielding Bayesian transformations from a belief distribution on one evidence-set to a belief distribution on another.

Shimony suggests that the justification of Bayesian methods must therefore rest on investigation of their consequences and the consequences of alternatives. I shall pursue the first type of investigation in sections V and VI below. As for the second, there is so far little to

[12] "Scientific Inference," in R. G. Colodny, ed., *Pittsburgh Studies in the Philosophy of Science*, vol. 4 (Pittsburgh: University of Pittsburgh Press, 1970), p. 79.

report, since few discussions of confirmation theory have explicitly abandoned Bayesian transformations and replaced them by an alternative rule. The most notable exception to this general neglect is Reichenbach's so-called "straight rule" of inference, and the related suggestions in Hempel and Oppenheim's early "Definition of 'Degree of Confirmation.'"[13] Reichenbach's rule is equivalent in the cases to which it is applicable to the maximum likelihood method, that is, it is the direction to expect the proportion of occurrences of a given property in the population to be the same as that in the observed sample, and this does not yield Bayesian transformations. The rule is not, however, generally applicable without further assumptions to probability distributions over all hypotheses, since it specifies only which the most likely hypothesis is on given evidence (in the technical sense of likelihood mentioned in C.5, Section II). It does not specify the distribution of probabilities on that evidence over competing hypotheses. To obtain this, as with maximum likelihood itself, *some* initial probability distribution must be assumed, even if this is a uniform distribution. The rule also suffers from the drawbacks of the maximum likelihood method for small samples, for example, it gives probability 1 to the next positive instance of a universal generalization on the basis of only one or a small number of observed positive instances.

Examples of restricted Bayesian methods are given by Suppes[14] in an analysis of various types of learning in behavioral psychology, where the constraints upon transformations necessary for learning are inferred in models simulating the actual behavior of organisms. In a simple case of choice between two possibilities in a sequence of trials with reinforcement for correct choice, the rule may be to guess on each occasion that possibility which turned out correct on the previous trial. This rule corresponds to "forgetting" all evidence before the immediately preceding trial, but if the probabilities of occurrence remain the same from trial to trial, and one possibility occurs more frequently than the other, eventually more guesses will be correct than not, and learning has been partially successful.

In general, the rule to forget all or most previous evidence, and use

[13] Reichenbach, *The Theory of Probability*, ch. 11; C. G. Hempel and P. Oppenheim, "Definition of 'Degree of Confirmation,'" *Philosophy of Science*, 12 (1945), 98.

[14] P. Suppes, "Concept Formation and Bayesian Decisions," in J. Hintikka and P. Suppes, eds., *Aspects of Inductive Logic* (Amsterdam: North Holland, 1966), p. 21, and "Probabilistic Inference and the Concept of Total Evidence," *ibid.*, p. 49.

the posterior distribution at t_0 as if it were the initial distribution in calculating the effect of subsequent evidence, is almost inescapable in any situation that is at all complex. It is certainly exemplified in the state of belief of the scientific community at any given time, which is far more dependent on theories that have recently proved successful than upon the detailed evidence which originally led to their acceptance and which may be forever lost in the back numbers of journals and the vaults of libraries. The rule to forget, however, is not Bayesian, for calculation of posterior probabilities by Bayes's theorem requires *total evidence* to be taken into account, and failing to take it into account may make the time-order of collection of evidence relevant to posterior distributions. If the total evidence of science since Galileo had been collected backwards, our beliefs calculated by the rule of forgetting would not now be the same as they are. (However, it must be remembered that the decision what to collect at any given time is itself dependent on the order in which evidence and theories have emerged, so this conclusion is not as paradoxical as it may look at first sight.) That Bayes's theorem cannot be a general requirement for historical development of scientific theories follows most obviously in cases of observation of new properties and invention of new concepts. If these have not previously been named in the language, then no hypotheses containing them can have explicitly been given any initial probability. Conversely, terms may disappear from the scientific language (for example, "phlogiston"), thus removing some hypotheses by giving them probability identically zero and causing non-Bayesian readjustment in the probabilities of others.

Much work remains to be done on the question of the relative efficiency of Bayes and other transformation rules in given sorts of circumstances. It may be assumed that which rules are most efficient will be a function both of the complexity of the learning mechanism (the size of its memory, for example), and also of the general characteristics of the world in which it is learning. With sufficiently complex learning devices one can envisage a "monitor" which checks the predictive success attained by a lower level probabilistic inductive device and changes the initial probabilities until better success rates are attained. Something like this process is suggested by Carnap for choosing a probability distribution out of his continuum of different distributions defined by a parameter λ in the real open interval $(0, \infty)$.[15] He shows that, if the universe is on the whole homogeneous, the distribution defined by $\lambda = 0$ will be expected to give greatest success in the long run, while if the universe is grossly heterogeneous, λ infinite will give greatest expected success. $\lambda = 0$ is in fact equivalent to the statistical method of maximum likelihood, and, in simple cases, to Reichenbach's straight rule, which is the direction to predict that the proportion of properties in the world will continue to be the proportion that has already been observed in samples. The success of this method clearly relies heavily on the assumption that the world is relatively homogeneous with respect to the samples, especially when these are small.

But of course the essence of the inductive problem is that we do not know a priori whether the universe is on the large scale homogeneous or heterogeneous, whether we are in a typical part of it, or how far the past success of any inductive method will be maintained in the future. If choice of pragmatically efficient initial distributions and transformation rules depends crucially on how well adapted the learning organism is to actual contingent conditions in the world, then it is unlikely that any detailed confirmation theory can be developed a priori, that is to say, independently of empirical investigations into how organisms (including ourselves) actually learn. It cannot be expected, therefore, that Bayesian methods will be useful except locally. But before going on to investigate how far they will carry us in this situation, I shall consider one further influential attempt to show that a priori conditions are sufficient for establishing a normative confirmation theory.

IV. Convergence of Opinion

Some personalist statisticians have attempted to provide a long-run justification of all types of scientific inference by adopting a single constraint on initial distributions, namely, that all hypotheses to be confirmed have a finite initial probability. Appeal is made to a convergence theorem for statistical hypotheses, according to which, under certain conditions that must be examined, the posterior probabilities of such hypotheses tend to 1 or 0 with increasing evidence for true or false hypotheses respectively, irrespective of the numerical values of the initial

[15] R. Carnap, *The Continuum of Inductive Methods* (Chicago: University of Chicago Press, 1951), secs. 18, 22.

probabilities of these hypotheses, provided only that the initial probabilities of true hypotheses are not less than some small fixed value. This theorem, if relevant to scientific inference in general, would be a very powerful aid to confirmation theory, for it would discharge us from discussing the details of the initial probability distribution if only this one condition upon it were satisfied.

Attempts to apply the convergence theorem in the context of confirmation theory are, however, subject to several kinds of misunderstanding and overoptimism. Thus, Savage claims to show rather generally how a person "typically becomes almost certain of the truth" when the amount of his experience increases indefinitely, and Edwards, Lindman, and Savage remark in a well-known expository article, "Although your initial opinion about future behavior of a coin may differ radically from your neighbor's, your opinion and his will ordinarily be so transformed by application of Bayes' theorem to the results of a long sequence of experimental flips as to become nearly indistinguishable."[16]

Assimilation of such results to scientific inference is certainly tempting, if only because it gives access to a large and familiar body of mathematical theory and technique. But personalist formulations of this kind, when applied to science, neglect the very stringent conditions on hypotheses and evidence under which the relevant convergence theorem is true and therefore omit to take account of the relevance of these conditions to typical cases of scientific inference. The second quotation also gives the misleading impression that Bayesian transformations are required for convergence of opinion. If this were the case it might be held to provide an argument for adoption of Bayesian transformations along the following lines. Inductive intuition suggests that in a satisfactory confirmation theory a true hypothesis should be one whose posterior probability tends to 1 after collection of a large amount of evidence. The convergence theorem says that this is the case under a number of necessary conditions. If these conditions included Bayesian transformations, it would be reasonable to adopt such transformations, since they would be necessary for explicating the inductive intuition. But, as we shall see, Bayesian transformations are *not* among the necessary conditions of the theorem. This indeed makes the theorem rather strong, since it is inde-

[16] L. J. Savage, *The Foundations of Statistics* (New York: John Wiley, 1954), p. 46; W. Edwards, H. Lindman, and L. J. Savage, "Bayesian Statistical Inference for Psychological Research," *Psychological Review*, 70 (1963), 197.

pendent of choice of nonzero initial probabilities at every stage of collection of evidence, and not just at the first stage. However, the remaining necessary conditions of the theorem do restrict its application in scientific contexts quite stringently. We shall now examine these conditions in detail.[17]

Consider a finite set S of N mutually exclusive and exhaustive hypotheses $\{h_s\}$. Let E_r be the conjunction of all evidence propositions $e_1, e_2, \ldots, e_r$ collected before time t_r. The convergence theorem proceeds as follows. By means of a law of large numbers it is first shown that, for h_1 true, the likelihood ratio $[p(E_r/h_1)]/[p(E_r/h_s)]$, for all $s \neq 1$, becomes greater than any fixed number η, however large, with second-order probability which approaches 1 as r increases. Then assuming for the moment the Bayesian transformation (which will be relaxed later), we have

$$(4) \qquad p(h_s/E_r) = \frac{p(h_s)\,p(E_r/h_s)}{\Sigma_s\, p(h_s)p(E_r/h_s)} \qquad (1 \leq s \leq N).$$

It follows at once that if $p(h_s) = 0$, no finite observation can make $p(h_s/E_r)$ nonzero, that is, if h_s is judged a priori to be impossible, no evidence can make it probable.

Suppose, however, that $p(h_s) > 0$ for all s. Consider the second-order probability P that $p(h_1/E_r) > \alpha$ for any fixed α as near to 1 as we please if h_1 is true. Using the result above for the likelihood ratios, it can be shown that, after a sufficiently large number of observations yielding E_r, P approaches 1 as r increases. It follows also that the posterior probabilities of all false hypotheses of the set S, given E_r, become as near to zero as we please with probability which approaches 1 as r increases.

The significance of the convergence theorem, more informally, is that, with sufficient evidence, the posterior probability of a true hypothesis is overwhelmingly likely to approach 1 and that of a false hypothesis to approach zero, *whatever the set of nonzero initial probabilities* when Bayesian transformations are assumed. The convergence of the second-order probability ensures not only that there is an E_r such that $p(h_1, E_r) > \alpha$ for fixed α, but also that it becomes progressively less probable that E_r is freak evidence which only temporarily favors a false hypothesis. Thus, the posterior probability of a true hypothesis in not only overwhelmingly likely to approach 1, but also to remain near 1 as evidence increases.

[17] See Savage, *The Foundations of Statistics*, pp. 46–50.

Conversely, if the posterior probability of a given hypothesis approaches 1 as evidence increases, there is only a small and decreasing second-order probability that it is not the true hypothesis. Moreover, if there are nonzero lower limits on the $p(h_s)$, lower limits of these second-order probabilities can be calculated. If, however, no lower limits on the $p(h_s)$ can be stated, nothing can be said about the speed of convergence, and in any finite time *any* given hypothesis of the set will have the highest probability for *some* distribution of initial probabilities. Thus for the theorem to be of any use in a finite time (that is, for it to be of any use at all), not only must the initial probabilities be nonzero, but they must also have some fixed finite lower limit, say ϵ, as near to zero as we please.

Under this condition, the requirement of Bayesian transformation can itself be relaxed. Denote probabilities at time t_r by p_r and assume fixed likelihoods, that is,

$$p_o(e_i/h_s) = p_1(e_i/h_s) =, \ldots, = p(e_i/h_s).$$

The limiting property of the likelihood ratios then remains as before, but we cannot now assume that $p_r(h_s) = p_{r-1}(h_s)$. The ratios (4), however, now yield

$$p_r(h_1/E_r) > a' p_r(h_1)$$

for any fixed a' near 1 and h_1 true, with second-order probability approaching 1 with increasing r. But since $p_r(h_1) \geqq \epsilon$, all r, this means $p_r(h_1/E_r) > a$ with a as near to 1 as we please, and we have the convergence theorem as before, but without the Bayesian assumption of constant initial probabilities. In other words, as long as the lower limit of $p(h_1)$ is fixed, however small it is, the posterior probability provided by sufficient evidence for the true hypothesis will approach maximal probability.

Now let us investigate the conditions of the theorem in more detail in relation to a set of scientific hypotheses. The conditions are as follows:

(i) $p(h_s) \geqq \epsilon > 0$, all s. The possibility of satisfying this condition for scientific hypotheses will have to be investigated later on; meanwhile we assume it satisfied.

(ii) No two hypotheses, h_s, h_s', yield the same value of $p(e_i/h_s)$ for all possible e_i. Or, if this condition is not satisfied by different hypotheses h_s, h_s', then only the disjunction of h_s and h_s' can be confirmed by any e_i.

(iii) Given a particular hypothesis h_s, the probability of making the test which yields e_i is independent of the particular order in which it and other evidence are observed. This will be called the *randomness assumption.* It is a substantial assumption in scientific contexts, because it is often known to be false and usually not known to be true; for it contemplates a situation in which we know nothing about the conditions which differentiate one observation from another. If, for example, the structure of the universe includes the true hypothesis "A proportion p of aircraft are jets," and we know that we are sampling aircraft, and *this is all we know,* then the probability that this aircraft is a jet is p on any occasion of observation. But the situations in which we wish to use the convergence theorem are unlikely to be so simple, because we are likely to want to confirm h_s on evidence which is known not to be random in this sense. This might be because we are able to sample only over a limited part of the spatial and temporal domain of h_s, or only over certain sorts of instances of h_s (the "observable" as opposed to the "theoretical" instances, for example). More obviously, there is something paradoxical about a requirement of random sampling in the context of scientific experiment, because it is universally accepted that science requires deliberately structured experiments, *not* random observation, and that this requirement is precisely what differentiates science from statistics. The difference of method must be regarded as a *consequence* of the nonapplicability of many statistical assumptions in the domain of science, not as its cause. Its cause lies rather in the inaccessibility of the whole domain to random sampling and to the desire to reach conclusions in as short a run as possible. That this is so is shown by the fact that where the random sampling assumptions seem reasonable or even inescapable, as in social surveys, star counts, genetics, and the like, statistical methods form an integral part of scientific procedure. Where these assumptions are not reasonable, convergence is inapplicable and structured experiment is required.

(iv) The e_i's are independent given h_s. That is, for all s,

$$p(e_1 \,\&\, e_2/h_s) = p(e_1/h_s)\, p(e_2/h_s).$$

I shall call this the *independence assumption.* Like the randomness assumption, it is not always valid in scientific contexts. It implies that the probability of observing a particular piece of evidence is unaffected by having already observed particular other pieces of evidence, an assumption that is satisfied, for example, by "random sampling with replacement" of balls out of a bag containing a finite number of balls. But, on the contrary, scientific experiments are often deliberately designed in the light of experimental evidence already obtained. Rather than conducting a large number of independent tests of the same hypothesis, scientific hypotheses are often themselves modified in the light of evidence, and new tests devised for the new hypothesis, with the results of previous evidence assumed. Especially where the hypothesis is universal rather than statistical in form, and may therefore need modification as a result of *falsifying* evidence, a limited number of experiments structured in the light of previous evidence, and directed to progressively modified hypotheses, is a much more efficient procedure than blindly conducting a long series of independent tests of the same hypothesis, such as might be appropriate for statistical hypotheses concerning large populations.

Comparison of "convergence of opinion" with the conditions of scientific inference therefore yields the following results: where the convergence theorem applies there is indeed convergence of opinion with regard to hypotheses which have nonzero probability, and this convergence is independent of the actual distribution of initial probabilities and of Bayesian transformations. But the conditions of the theorem, especially the randomness and independence assumptions, are not valid for typical examples of scientific inference in limited domains and from controlled experiments.

We have now investigated three types of attempt to minimize constraints upon confirmation functions, namely, decision theory, non-Bayesian methods, and convergence of opinion. All of them have turned out in some respects inadequate. It is time to make a bolder approach and to reformulate the confirmation problem as one of finding some theory which is *sufficient* (though not necessary) to explicate intuitive criteria of confirmation in science. In this reformulation we shall not be afraid to adopt both the probability axioms and the Bayesian transformation and to draw further constraints from intuitive criteria of adequacy for inductive inference. The justification of such a confirmation theory will be in its consequences rather than its premises.

V. Criteria of Adequacy

1. Confirmation as Positive Relevance

What should be expected from a confirmation theory which is regarded as one among the possible explications of inductive inference rather than as an a priori and necessary framework for such inference? There are three types of conditions it should satisfy.

1. It should explicate and systematize inductive methods actually used in science, without grossly distorting any of them. This will generally imply clarifications of confused inductive intuitions and of the relations between them, and insofar as such clarification is provided, confirmation theory will escape the charge of being an ad hoc rehearsal of intuitions which gives out no more than was put in.

2. A confirmation theory should suggest fruitful new problems of inductive inference and new insights into the structure of science, which may even issue in revision of the intuitively accepted methods themselves. For example, problems about criteria for confirming generalizations have led to fruitful discussions, and it will be suggested below that the probabilistic approach to these problems leads to a revised view of the function of generalizations and theories in scientific inference which is of general philosophical importance.

3. A confirmation theory may indicate that biological learning and inference are at least isomorphic with the model of inference provided by the theory, since the existence of the model will at least show that there is a self-consistent way of satisfying certain inductive and epistemological requirements. This suggestion should at present be taken only in a heuristic sense, however. It should be interpreted neither as giving license to confirmation theory to dictate the results of psychological and physiological research nor as issuing in a "discovery machine" for future scientific progress. In its present stage of development, confirmation theory can have only the more modest aim of providing one logically possible model of scientific inference. Even this will leave much to contingencies, such as the basic language presupposed and the presuppositions of learning, which, as has already been argued, cannot be expected to be wholly normative. There are in fact likely to be at least as many contingencies in a confirmation theory as there are in respect of actual human learning; hence no confirmation theory can be expected,

at least at the present stage of research, to describe a more efficient discovery machine than the human brain itself.

The strongest possible arguments for adoption of a Bayesian confirmation theory will not be a priori approaches of the kinds already considered, but the demonstration that such a confirmation theory does satisfy these three conditions. In the past there have been many objections to Bayesian theory on the grounds that it cannot in principle explicate this or that piece of inductive practice: confirmation of universal generalizations, preference for theories of great power or simplicity, and so on. But it follows from the decision theoretic formulation of Bayesian theory already considered that all such objections are beside the point. There will always be *some* distribution of initial probabilities which will provide the required explication in particular cases; the significant question is whether such a distribution appears at all perspicuous and illuminating for general aspects of theory structure or whether it is entirely implausible and ad hoc. It is only in the latter case that cogent objections can be brought against Bayesian theory, and not enough effort has yet been put into development of the theory to permit summary rejection of it on these grounds.

Objections to Bayesian theory have also been made from the point of view of empirical study of learning, as we saw in connection with restricted Bayesian methods proposed by Suppes. But initially a Bayesian theory may have much more modest aims than are contemplated by these objections. If a Bayesian model of even some aspects of inference can be shown to be self-consistent, this may be sufficient to elucidate important epistemological problems, and whether this proves to be so or not can only be discovered by pushing the theory as far as it will go. Moreover, at present there are no workable alternatives. Study of learning machines and rats has as yet yielded no explicit body of theory such as could be used for a confirmation logic, and in any case the learning problems involved are elementary compared to those of scientific inference. Confirmation theories of nonprobabilistic character have indeed been suggested, but it remains the case that only probabilistic theory has a sufficiently developed logical structure to be easily applicable to even elementary inductive intuitions.

Bayesian theory can lay claim to some immediate successes which have not come to light in other approaches. In this section I shall list some of these, and go on in the next section to discuss the requirements of confirmation of scientific laws and theories which are more problematic and also more illuminating.

What relation has to hold between data and hypothesis if the data are to count as *evidence for, confirmation of,* or *support for* the hypothesis? We have to try to pick out what we regard as the essential features of that relation which may be expressed as, "Assuming the evidence expressed in true propositions, under what conditions do we regard the evidence as increasing the reliability of a hypothesis, or as making the hypothesis more acceptable as a guide to truth?" We have first to decide how to interpret "confirmation" in terms of probability functions. There are two obvious possibilities here. We may regard a hypothesis h as confirmed by evidence e if and only if the probability of h on e attains at least some fixed value k between 1/2 and 1. Thus we obtain the *k-criterion*

(5) e confirms h *iff* $p(h/e) \geqq k > 1/2$.

Roughly speaking, we then believe that h has at least a better than even chance of being true. Such a belief might well dispose us to *act upon* or *accept* h, particularly if k is near 1, or if a choice between h and $\sim h$ is forced upon us. But it may be questioned whether it is a suitable measure of the confirmation of h *relative to* e, for (5) may be satisfied even if e has *decreased* the confirmation of h below its initial value, in which case e has relatively *dis*confirmed h. For the relative measure of confirmation it seems better to adopt the condition, which I call with Carnap the *positive relevance criterion,*[18] which requires the posterior probability of h on e to be greater than its initial probability:

(6) e confirms h *iff* $p(h/e) > p(h)$.

I shall also say that e is negatively relevant to h, or disconfirms h, *iff* $p(h/e) < p(h)$, and that e is *irrelevant* to h *iff* $p(h/e) = p(h)$.

There are several useful corollaries of these criteria.

(R.1) e confirms/is irrelevant to/disconfirms h according as h confirms/is irrelevant to/disconfirms e.

For $p(h/e) = p(e/h)\,p(h)/p(e)$, hence $p(h/e) \gtreqless p(h)$ according as $p(e/h) \gtreqless p(e)$.

[18] R. Carnap, *Logical Foundations of Probability* (Chicago: University of Chicago Press, 1950), ch. 6.

(R.2) If e confirms h, then e disconfirms $\sim h$.

For $p(\sim h/e) = 1 - p(h/e) < 1 - p(h) = p(\sim h)$.

(R.3) If e confirms h, then $\sim$e disconfirms h.

For then h confirms e, from (R.1); h disconfirms $\sim$e, from (R.2); and hence $\sim$e disconfirms h from (R.1).

All these properties of positive relevance are intuitively desirable for a relation of confirmation.

Some elementary intuitive conditions of adequacy for inductive inference can also be seen at once to be satisfied by positive relevance, independently of any assignment of initial probabilities. Not all of these conditions are uncontroversial, but in line with the self-correcting explicatory approach adopted here let us first state them and then explore their consequences.

(i) *Equivalence.* A condition that seems fairly unproblematic is that logically equivalent expressions should have identical effects in confirming logically equivalent expressions. That is to say, we have the *equivalence condition*

If $g \equiv g'$, and $h \equiv h'$, then if g confirms h, g' confirms h'.

This is satisfied by the positive relevance criterion of confirmation immediately in virtue of probability axiom (P.2).

The main criticisms of this condition have been in connection with the so-called "raven paradoxes." There are, as we shall see, methods of dealing with these paradoxes other than abandoning the equivalence condition, which would, of course, entail also abandoning a probabilistic theory of confirmation, since the condition follows directly from the probabilistic axioms.

(ii) *Entailment.* Another condition that seems inescapable for the relation of confirmation is that any entailment of a proposition h must be confirmed by h. Thus we have the *entailment condition*

If $h \rightarrow g$, then h confirms g.

This is satisfied immediately for empirical propositions by probability axiom (P.3):

If $h \rightarrow g$, then $p(g/h) = 1$,

hence for any g whose initial probability is less than maximal, $p(g/h) >$

$p(g)$, and h confirms g. It is also satisfied by the *k-criterion*, for $p(g/h) = 1 > k$.

(iii) *Converse entailment.* This is a more controversial condition. In the deductive model of theories it is generally assumed that if, given some initial conditions, a hypothesis entails an experimental law or singular prediction, and the entailment is accepted as true on the basis of observation, the hypothesis is thereby confirmed. This is indeed the only way in the deductive model in which the reliability of hypotheses can be increased. It is also assumed that a generalization is confirmed by a substitution instance which it entails. Thus we have the *converse entailment condition*

If $h \rightarrow g$, then g confirms h.

We have immediately from Bayes's theorem, if $h \rightarrow g$, where neither $p(h)$ nor $p(g)$ are 0 or 1, then

$$p(h/g) = p(h)/p(g) > p(h).$$

Hence under the stated conditions the converse entailment condition is satisfied in Bayesian theory.

One objection to this condition is that, together with the equivalence condition (i) above, it entails the so-called "raven paradox" of confirmation first stated by Nicod and discussed by Hempel. The paradox will be examined in a later section, and a resolution will be suggested that involves abandoning neither the equivalence nor converse entailment conditions.

It has been further objected to the converse entailment condition that sometimes a single positive substitution instance of a general law is enough to confirm it and that scientists do not usually go on accumulating positive instances of the same kind as those already observed, since such instances would not be regarded as adding significantly to the evidence. What does add to the confirmation, it is held, is observation of positive instances in a variety of circumstances, the elimination of rival hypotheses, or the placing of the law in a logical relation with other laws and theories, which are themselves well confirmed. I shall consider elimination and variety immediately, and later we shall see how far a Bayesian theory can account for the greater confirmation expected from incorporation of a law within a general theory. With these supplementations I shall maintain the condition that any observed entailment provides some confirmation for a hypothesis whose initial probability is nonzero, however little that confirmation may be under some circumstances.

A useful consequence of converse entailment may be noted here. That is, that it ensures transitivity of confirmation in a deductive system in the following sense. If evidence f confirms a generalization or low-level theory g in virtue of being entailed by g, then f should also confirm a higher-level theory h of which g is an entailment. Since then h entails f, this confirmation follows at once from converse entailment by the transitivity of entailment.

(iv) *Eliminative induction and variety.* Bayesian theory provides an immediate explication and generalization of eliminative induction.

Suppose $S \equiv \{h_1, h_2, \ldots, h_N\}$ is a finite, exhaustive, and exclusive hypothesis set. Then since $p(h_1/e) + p(h_2/e) + \ldots + p(h_N/e) = 1$, refutation, or more generally disconfirmation, of any proper subset of S by evidence e increases the sum of the probabilities of the remaining subset. Without further information one cannot conclude that the probability of any particular remaining h_r is increased, since the evidence may in fact reduce it. The only conclusive result that can be obtained is in the case of elimination of *all* competing hypotheses, in which case $p(h_r/e)$ becomes 1, and h_r maximally certain. This result is, of course, ideally the aim of crucial experiments, in which the alternatives are supposed to be reduced to two, one of which is refuted by the experiment, although it is rare to find circumstances in which no other alternative hypotheses are possible.

The requirement of variety of instances can also be seen as a consequence of eliminative induction. Consider the generalization "All vegetarians are teetotallers," which would generally be said to be better confirmed by a sample of vegetarians taken from all sections of the population than by a sample taken, for example, exclusively from Methodists. General explication of this intuitive inductive rule turns out to be more difficult than might be expected, no doubt because what seems reasonable in examples of this kind depends heavily upon extra information, such as that many Methodists are teetotallers, and the more complex the background evidence assumed the more unmanageable becomes the explication.[19] We can, however, regard the requirement as a corollary of eliminative induction by supposing that the more varied the popula-

[19] Cf. P. Achinstein, "Variety and Analogy in Confirmation Theory," *Philosophy of Science*, 30 (1963), 207, and Carnap's reply, *ibid.*, p. 223.

tion from which samples are taken, the more rival generalizations about teetotallers are refuted if they are false, and hence the greater is the confirmation obtained for those generalizations that remain unrefuted.

(v) *Initially unexpected evidence.* It is generally assumed that if evidence e_1 is entailed by a hypothesis *h*, and e_1 would have been initially unexpected without consideration of *h*, then actual observation of e_1 does more for the confirmation of *h* than would initially expected evidence e_2 which is also entailed by *h*. The rule of searching for variety of instances can also be seen as an example of this consideration, for it is initially more unexpected that a general law, such as "All crows are black," should be true for a great number of different kinds of instances of crows (of all ages, for example, and observed in very various geographic regions), than that it should be true only for old crows in the south of England. Again, Einstein's relativity theory entailed that light rays should be detectably bent in the neighborhood of large gravitating masses. This result was highly unexpected in the context of theories and evidence available before relativity theory, and when the bending was actually observed in the neighborhood of the sun in the eclipse of 1919, it provided more support for relativity theory than would have been provided by evidence that was also accounted for, and therefore already expected, on older theories.[20]

The requirement of unexpected evidence can be rather easily explicated by probabilistic confirmation as follows. Interpret "initially unexpected" to mean "initially improbable," that is, improbable on the evidence available prior to e_1 or e_2. We can then compare the effect on the posterior probability of *h* of observing e_1 or e_2, where e_1 is initially much less probable than e_2, and $h \rightarrow e_1 \,\&\, e_2$. We have

$$p(h/e_1) = p(h)/p(e_1)$$
$$p(h/e_2) = p(h)/p(e_2).$$

But $p(e_1)$ is much less than $p(e_2)$, hence $p(h/e_1)$ is much greater than $p(h/e_2)$. Thus, the confirming effect of e_1 on *h* is much greater than that

[20] A requirement similar to this seems to be what Popper has in mind in speaking of the "severity" of tests. In *Logic of Scientific Discovery* he asserts (in italics) that "*the probability of a statement (given some test statements) simply does not express an appraisal of the severity of the tests a theory has passed, or of the manner in which it has passed these tests*" (p. 394, cf. p. 267). But in *Conjectures and Refutations*, p. 390, he suggests a measure of severity proportional to the initial improbability of the test, as is also suggested here.

of e_2, as required, and in a situation of limited experimental resources it would be wise to do the experiment to test e_1 rather than that to test e_2.

2. *Transitivity of Confirmation*

Some of the easy successes of Bayesian theory have now been outlined. However, it is well known that difficulties soon arise, even in relation to the comparatively elementary requirements of enumerative induction. These difficulties concern, first, the possibility of assigning finite initial probabilities to all the universal generalizations that may be confirmed by evidence, for if there are more than a strictly finite number of these, assignment of finite probabilities, however small, will violate the probability axioms. Hence arises the problem of choosing which finite set out of all possible hypotheses to permit to be confirmed by suitable evidence. Secondly, problems arise about what is to count as confirming evidence for a generalization in order to evade Hempel's paradoxes of confirmation. As we shall see, difficulties about lawlike generalizations are multiplied when we come to consider confirmation of scientific theories and their resulting predictions.

I shall not consider in detail the various attempts that have been made to solve these problems, but I shall rather use the problems, in conjunction with Bayesian confirmation theory, to suggest an interpretation of the structure of scientific theory which conflicts to some extent with the generally received deductivist model of science. In this way I hope to indicate the potentially *revisionary* function of Bayesian theory, that is to say, the way a systematic attempt to explicate inductive intuitions in science may interact with and modify these intuitions themselves. The intuitions concerned here are those relating to the *universality* of scientific laws and theories. I have argued elsewhere[21] that it is possible to interpret all of a scientist's behavior with regard to lawlike generalizations without assuming that he has any nonzero degree of belief in their universal truth, and indeed that it is more plausible to suppose that he does not have any such belief. One motivation for this argument is the difficulty in a Carnap-type confirmation theory of assigning finite initial probabilities to enough universal generalizations. I now want to provide another motivation for the argument by considering some more general

[21] M. Hesse, "Confirmation of Laws," in S. Morgenbesser, P. Suppes, and M. White, eds., *Philosophy, Science, and Method* (New York: St. Martin's, 1969), p. 74.

features of Bayesian confirmation theory, which seem to indicate that the function of general theories in science is not what it is assumed to be in the deductivist model and that theoretical inference, like Mill's eductive inference or instance confirmation in Carnap's theory, can better be regarded as going from case to case than by way of general theories and their entailments.

It is convenient to begin the examination of confirmation relations in scientific theories by a discussion of conditions of adequacy given by Hempel which have been shown to give rise to paradoxes. We have already seen that the equivalence, entailment, and converse entailment conditions are intuitive criteria of adequacy for scientific inference and that they are all satisfied by Bayesian theory under certain conditions of nonzero probabilities. These are three of the conditions discussed by Hempel in his pioneering "Studies in the Logic of Confirmation,"[22] in which they are initially presented, along with other suggested conditions, independently of whether confirmation should be explicated by a probabilistic or any other type of axiomatic system. Hempel's paper gave rise to the well-known "paradoxes of the raven," which will be discussed in the next section, but some of the other conditions of adequacy give rise to another paradox, which I shall discriminate as the "transitivity paradox." My terminology and notation will here be a little different from Hempel's.

Consider first the general question of how far the confirmation relation can be expected to be *transitive*, that is, should it satisfy the condition "If f confirms h, and h confirms g, then f confirms g"? In this general form transitivity is obviously not satisfied by the usual notions of confirmation or support. For example, the foundations of a house support the walls which support the roof, and it is true that the foundations support the roof, but although the pivot supports the seesaw, and the seesaw the child at one end, it is not in general true that the pivot supports the child. Again, if a man has disappeared, to find him shot dead confirms the hypothesis that he has been murdered, and that he has been murdered in itself confirms that there is a strangler at large, but that he has been shot does not confirm that there is a strangler.

[22] C. G. Hempel, "Studies in the Logic of Confirmation," *Mind*, 54 (1945), 1 and 97, reprinted with a new postscript in *Aspects of Scientific Explanation* (New York: Free Press, 1965), p. 3. See also B. Skyrms, "Nomological Necessity and the Paradoxes of Confirmation," *Philosophy of Science*, 33 (1966), 230.

We have, however, already noted that in the case of the entailment relation $g \rightarrow h$, and $h \rightarrow f$, there is transitivity of confirmation as the converse of transitivity of entailment. There is also another special case in which transitivity seems both plausible and indispensable, in which just one of the confirmation relations of general transitivity is replaced by an entailment. Consider the condition Hempel calls *special consequence*:

If f confirms h, and $h \rightarrow g$, then f confirms g.

This condition seems to be required for the following sorts of cases which are typical of scientific inference:

(a) Positive instances f of a general law h together with given antecedents i are taken to confirm a prediction of the next instance g as well as confirming the law itself. In this case $h \& i \rightarrow f \& g$. Hence by converse entailment f confirms $h \& i$, and by special consequence f confirms g.

(b) In a more complex theoretical deductive system the condition might be stated in the form "If a hypothesis is well confirmed on given evidence, some further as yet unobserved consequences of the hypothesis ought to be predictable with high confirmation." Consider Newton's theory of gravitation, in which we have the theory confirmed by some of its entailments, for example the laws of planetary orbits and the law of falling bodies. The theory predicts other as yet unobserved phenomena, such as the orbits of the comets or of space satellites. We are not *given* the truth of the theory, only of its original data from planets and falling bodies, and yet we want to say that this evidence confirms predictions about comets and satellites. Special consequence would allow us to transmit confirmation in this way in virtue of the entailment of predictions by the theory. There is also an even stronger intuitive type of theoretical inference. Not only do we require theories to yield strong confirmation to their predictions before the predictions have been tested, but sometimes we use a theory actually to *correct* previously held evidential propositions, if the theory is regarded as sufficiently well confirmed by its other consequences. For example, Galileo's "observation" of constant acceleration of falling bodies and Kepler's third law of planetary motion (about the relation of the periodic times of the planets to the parameters of their orbits) were both corrected by Newton's theory in virtue of its supposed confirmation by other evidence. If the confirmation here is regarded as transmitted by converse entailment and special consequence to the corrected laws, it must have higher degree than was originally ascribed to the pretheoretical laws on the basis of "observation," since the corrected laws are preferred to the laws previously confirmed by observation.

It seems clear that any adequate account of confirmation must be able to explicate our intuitions and practice in cases such as these. Unfortunately, however, it can easily be shown that the conditions of converse entailment and special consequence just described will not do so, because these conditions together yield further unacceptable consequences. To see this in terms of a specially constructed counterexample, put h equivalent to $f \& g$. Then f confirms h by converse entailment. Also h entails g, hence f confirms g by special consequence. But f and g may be any propositions whatever, and the example has shown that the two conditions together entail that f confirms g. Any relation of confirmation which allows any proposition to confirm any proposition is obviously trivial and unacceptable, for it contradicts the tacit condition that confirmation must be a selective relation among propositions. A paradox has arisen by taking together a set of adequacy conditions, all of which seem to be demanded by intuition; hence it is appropriate to call it the *transitivity paradox*.

That converse entailment and special consequence are not always satisfiable together can be made more perspicuous than in the artificial counterexample just given. If h entails f, and h consists of a conjunction of premises representable by $h_1 \& h_2$, of which only h_1 is necessary to entail f, it is clear that f may confirm only h_1. If in addition h_2 is that conjunct of h which is necessary to entail g (for example, g may be h_2 itself) there is no reason why f should confirm either h_2 or g. What seems to be presupposed in the examples mentioned as cases of the special consequence condition is a more intimate relation between the parts of the hypothesis which entail f and h respectively than mere conjunction. We have to try to determine what this more intimate relation consists of.

Before doing so it is important to notice how general and powerful the transitivity paradox is. Nothing has been assumed about the relation of "confirmation" or "support" except conditions that seem to arise directly out of commonly accepted types of theoretical argument. As soon as the relation of theory to evidence is construed in deductive fashion the conditions seem natural presuppositions of any relation of confirmation within such a system. In particular, no assumption has yet been made about the relation of *probability* to confirmation. Probability is the best explored measure of the confirmation relation, and many difficulties have been found in it. But that it cannot incorporate the intuitive conditions exemplified here in simple theoretical inferences cannot be held against probabilistic confirmation any more than against its rivals. For the conditions as they stand are inconsistent with any adequate confirmation theory, and this inconsistency may indeed drive us into deeper examination of the structure of theories than the disputes about probabilistic or nonprobabilistic theories of confirmation. We shall, however, first consider what might be done about the paradox in a Bayesian theory.

Expressed in Bayesian language the paradox is represented as follows. We are interested in the value of $p(e_2/e_1)$, where e_1 is an observed or otherwise given consequence of a theory h, and e_2 is an as yet unobserved or otherwise problematic further consequence of h, that is, an untested prediction made by h. Now $p(e_2/e_1)$ is a single-valued probability function of its arguments alone, and its value cannot depend on whether or not we are interested in some particular h from which e_2 and e_1 are deducible. Consequences e_1 and e_2 are in general deducible from a number of different sets of premises in the language, whether we have thought of them or not. Some of them are perhaps potential theories. But whether or not they turn out to be viable as theories, $p(e_2/e_1)$ has its unique value independent of them.

If this result still appears counterintuitive, we can make it a little more transparent in probabilistic terms. We want to show that $p(e_2/e_1) > p(e_2)$, which is the condition that e_1 and e_2 are not probabilistically independent of each other, but are related by positive relevance. We may be disposed to argue that if $h \rightarrow e_1 \,\&\, e_2$, then we should be able to show both that (1) $p(e_2/e_1) \geqq p(h/e_1)$ since $h \rightarrow e_2$, and also that (2) $p(h/e_1) > p(e_2)$, since h can be imagined to be increasingly confirmed by a greater and greater amount of evidence e_1 until its posterior confirmation is greater than the initial confirmation of e_2. Then it would follow immediately that $p(e_2/e_1) > p(e_2)$ as desired. But this intuitive argument is fallacious in general. (1) entails $p(e_1 \,\&\, e_2) \geqq p(h \,\&\, e_1) = p(h)$ since $h \rightarrow e_1$. (2) entails $p(h) > p(e_2)p(e_1)$. Hence both conditions cannot be satisfied together unless initially $p(e_1 \,\&\, e_2) > p(e_1)p(e_2)$, which is equivalent to the result which was to be proved, namely, $p(e_2/e_1) > p(e_2)$.

The transitivity paradox was first noticed by Hempel and discussed by Carnap, and has been approached in various ways in the literature.[23] Carnap, and later Hempel, proposed to adopt the positive relevance criterion of confirmation, which entails converse entailment and hence entails abandoning special consequence, but neither Carnap nor Hempel considered the prima facie importance of special consequence in theoretical contexts, and therefore they did not discuss the implications of its abandonment. Other suggestions for resolving the paradox have included restricting the standard rule of entailment presupposed in the application of special consequence; replacing the positive relevance criterion of confirmation by the k-criterion; and supplementing confirmation by concepts of "acceptance" or "explanation," both of which stand in need of explication at least as badly as does confirmation itself. All these suggestions appear undesirably formal, for the paradox does seem to be a sufficiently fundamental and material consequence of our inductive intuitions to require a type of resolution suggested by concrete cases rather than by formal considerations.

In attempting to resolve the transitivity paradox in an illuminating way, I shall examine three types of inductive inference which can be regarded as special cases of the intuitions leading to the proposed special consequence condition. These are as follows: inferences concerned with *enumerative induction*, namely, general laws, instance confirmation, and De Finetti's concept of exchangeability; theoretical inference from *analogies* and *models*; and the requirement of *simplicity*. I shall argue that these can be interpreted in terms of a unified theory of Bayesian inference which evades the transitivity and raven paradoxes and results in reinterpretation of the function of general laws and theories.

VI. Generalizations, Analogy, and Simplicity

1. *Positive Relevance as Clustering, and the Raven Paradox*

Cases in which the special consequence condition seem to be required

[23] Carnap, *Logical Foundations*, p. 479; Hempel, *Aspects*, p. 33; B. A. Brody, "Confirmation and Explanation," *Journal of Philosophy*, 65 (1968), 282; H. Smokler, "Conflicting Conceptions of Confirmation," *Journal of Philosophy*, 65 (1968), 300; W. Todd, "Probability and the Theorem of Confirmation," *Mind*, 76 (1967), 260. A related problem of nontransitivity in connection with the concept of "indirect inductive support" is discussed by W. C. Salmon, "Verifiability and Logic," in P. K. Feyerabend and G. Maxwell, eds., *Mind, Matter, and Method* (Minneapolis: University of Minnesota Press, 1966), pp. 361ff.

are cases where it is desired to show that two consequences of a hypothesis are positively relevant to each other, and the transitivity paradox consists in the apparent irrelevance of the hypothesis to this dependence. This immediately suggests that Carnap's notion of *instance confirmation* of general laws[24] or Mill's *eduction* from particulars to particulars are paradigm cases of the sort of inference concerned here. Let us therefore first examine the problems surrounding confirmation of general laws.

Take a universal generalization h in the form "All P's are Q" and consider first only the class of individuals $a_1, a_2, \ldots, a_n$ which are P's, thus evading for the moment the "raven paradox" to be discussed later, in which individuals which are neither P nor Q are considered as evidence for h. Call those individuals in the P-class which are also Q's *positive instances* of h, and individuals which are P but not -Q *negative instances*, and assume for simplicity Q and not -Q are initially equally probable. A single negative instance falsifies h and sends its posterior probability to zero. It is usually assumed that a positive instance should confirm h, and an increasing number of positive instances should increasingly confirm h. A clause must of course be added to the effect that there is no other relevant evidence, to avoid confirmation of generalizations like "I wake up every morning at 7 o'clock" by evidence consisting of sixty year's worth of mornings on which I have woken up at 7 o'clock.

Assume the initial conditions $P(a_1), P(a_2), \ldots, P(a_r), \ldots$ to be given, and put their conjunction equivalent to i. Also put $e_r \equiv Q(a_r)$, and $E_r \equiv e_1 \,\&\, e_2 \ldots e_r$ $(r = 1, 2, \ldots, n)$. Then, provided neither $p(h)$ nor $p(E_r)$ is zero, since $h \,\&\, i \rightarrow E_r$, and i is given, by converse entailment h is confirmed by E_r. It also follows that if $p(e_{r+1}/E_r) < 1$, all r, then h is increasingly confirmed as r increases. This increasing confirmation does not of course have limit 1 unless h is *true*, for if h is false a negative instance will eventually be reached and will send the posterior probability of h to zero.

So far the intuitive requirements of enumerative induction are satisfied, but at the cost of assuming $p(h) > 0$. It should be noticed at once, however, that if we are interested in predicting that the next instance of a P will also be a Q, increasing confirmation of h does not necessarily help us. For the argument that since $h \,\&\, i$ is confirmed by E_r, and $h \,\&\, i \rightarrow e_{r+1}$, therefore e_{r+1} is confirmed by E_r is a fallacious argument of

[24] Carnap, *Logical Foundations*, p. 571.

just the kind that has been exposed in connection with the special consequence condition. However, since $h \,\&\, i \rightarrow e_{r+1}$ but not conversely, and therefore $p(e_{r+1}/E_r) > p(h \,\&\, i/E_r)$, $p(e_{r+1}/E_r)$ may attain a high value even when $p(h) = 0$.

We have e_{r+1} confirmed by E_r given i if and only if

$$(10) \qquad p(e_{r+1}/E_r \,\&\, i) > p(\sim e_{r+1}/E_r \,\&\, i)$$

or alternatively,

$$p(e_{r+1} \,\&\, E_r \,\&\, i) > p(\sim e_{r+1} \,\&\, E_r \,\&\, i).$$

Whether this is the case or not will depend only on the initial probabilities of the e_r's and their conjunctions given i. We have already assumed $p(e_r) = p(\sim e_r)$, all r. Then two further conditions are jointly sufficient for (10). First it is generally assumed in the case of instances of universal generalizations that the tests for each instance being positive or negative are *symmetrical* or *exchangeable*,[25] that is, that the probability of obtaining r instances of a specific kind depends only on r and is independent of the order of the individual tests. In particular, the initial probability of a given outcome of a single test of a *P*-object for *Q*-ness or for not -*Q*-ness will respectively be the same for each individual *P*-object, that is, $p(e_1) = p(e_2) = \ldots = p(e_r) = p(\sim e_1)$, and so on. The tests are in this sense random over the individuals.

Secondly, however, to satisfy (10) we still need the initial probability to be such that a conjunction of $r + 1$ positive instances is more probable than a conjunction of r of them with one negative instance. This does not follow merely from exchangeability, which tells us only that it was equally likely that the negative instance might have turned up anywhere in the sequence, not that its coming at the end is less probable than a positive instance coming at the end. Exchangeability is concerned only with the order of tests, not with the specific outcomes of those tests. Thus the assumption that the world is more likely to be homogeneous than heterogeneous, in the sense that $r + 1$ positive instances are initially more probable than r positive and one negative, must be explicitly built into the initial distribution and cannot be derived from the probability axioms and exchangeability alone. It follows, for example,

[25] Cf. Carnap, *Logical Foundations*, p. 484; B. De Finetti, "Foresight: Its Logical Laws, Its Subjective Sources," trans. in H. E. Kyburg and H. E. Smokler, eds., *Studies in Subjective Probability* (New York: John Wiley, 1964), p. 118.

that the counterinductive instruction to predict negative instances after some finite fixed number of positive instances can be represented by Bayesian methods just as well as enumerative inductive methods can. To give finite initial probability to universal generalizations in an infinite universe (which means giving vanishing probability to an infinity of possible statistical generalizations) is one way of giving a preference to homogeneity in the initial distribution. But it is not the only way, for a weaker requirement on the distribution is the strictly finite condition that, of r instances of P's, it is more probable that none or all will be positive instances of "All P's are Q's" than that there will be any other proportion.[26] This assumption, it will be noticed, is directly contrary to the result that would be obtained if the P's were supposed to consist of collection of *independent* Q's and not Q's equally weighted, for then the most probable composition of r P's would be $r/2$ Q's and $r/2$ not Q's. In these matters of induction it must be assumed that God is not a coin-tosser.

I shall call the assumption of homogeneity the *clustering postulate.*[27] It has more general application than merely to enumerative induction, as we shall see presently. But first it will be illuminating to consider the *raven paradox* of confirmation from the point of view just developed. This paradox arises from the equivalence and converse entailment conditions for confirmation. The equivalence condition gives

If $h \equiv h'$, then if e confirms h, e confirms h'.

Consider $h \equiv$ "All ravens are black." This is confirmed by its entailed positive instances, black ravens. The generalization h is also equivalent to $h' \equiv$ "All non-black things are non-ravens," which is confirmed by its positive instances: white shoes, red herrings, blue books, and many other

[26] This is essentially the assumption underlying Carnap's c* distribution. *Logical Foundations*, appendix. See also his discussion of the continuum of inductive methods in terms of their degrees of preference for more or less homogeneous universes. *Continuum*, sec. 22.

[27] This postulate is similar to J. M. Keynes's "principle of limited independent variety." *A Treatise on Probability* (London: Macmillan, 1921), ch. 22. The principle may be paraphrased, "If groups of properties are present together in a number of instances, in the absence of other evidence it is more probable than not that they will be present together or absent together in further instances." Attempts to incorporate a clustering postulate of this type into Carnap's theory are made in R. Carnap and W. Stegmuller, *Induktive Logik und Wahrscheinlichkeit* (Vienna: Springer-Verlag, 1959), appendix B, and M. Hesse, "Analogy and Confirmation Theory," *Philosophy of Science*, 31 (1964), 319.

objects. But by the equivalence condition anything which confirms h' confirms h, hence this motley collection of objects confirms "All ravens are black" just as black ravens do. This appears paradoxical, for we should not generally regard observations of red herrings, etc., as support of any kind for "All ravens are black."

Let us, however, along the lines previously suggested, forget the confirmation of universal generalizations, which may in any case be unobtainable, since $p(h)$ may be identically zero. The only finite concern we can have with h in a finite time is to confirm finite sets of predictions of the kind "*a* is a raven, hence *a* is black." Suppose we have observed, as we may well have, many black ravens and no nonblack ravens, and, as we certainly have, many nonblack nonravens. How much of this evidence supports the prediction "Raven *a* is black"? First, it is unlikely that we regard the test of raven *a* for blackness as *exchangeable* with color tests of herrings, polar bears, mailboxes, and the like, since there is no reason to suppose that the probability of obtaining a given proportion of black to nonblack objects in such a sequence of tests depends only on their number and not on their individual character as ravens, bears, etc. Indeed this is just a way of formulating part of the intuition which led to the paradox in the first place, namely that nonravens are *not* exchangeable for ravens in testing "All ravens are black." But now our assignment of probabilities explicitly depends on such intuitions and is not dictated by probability-equivalences of universal generalizations. Secondly, and more cogently, there is no reason whatever to assume that a motley collection which contains only one raven (the next instance) which is black is initially more probable than a collection of the same size and composition, except that it contains that raven and it is not black. No considerations of homogeneity of the universe dictate this, for homogeneity is more naturally represented by considering the properties of specified similar objects which are obvious candidates for exchangeability and for natural "clusters."

Suppose, however, we are concerned with a finite universe in which it is easy to conceive "All ravens are black" having a finite initial probability. It may then be thought that the paradox reappears, since positive instances now certainly confirm their respective generalizations. But the discussion of the abortive special consequence condition has shown that nothing follows from this about confirmation of *next* instances. Confirmation that the next raven is black cannot be obtained from the motley collection unless the relevant initial probabilities satisfy exchangeability and clustering conditions as before, and there is nothing in the confirmation conditions or the probability axioms to force such conditions upon us in this case. So long as we restrict consideration to next instances, therefore, the paradox can be evaded by suitable choice of initial probabilities of particulars, whether the universal generalization has finite initial probability or not.[28]

Whatever may be thought of these considerations as a resolution of the raven paradox, they do at least bring out the contrast between the orthodox view of scientific generalizations and the view advocated here. We are concerned here not with the relation of singular instances to generalizations, or of entailments to hypotheses, but with the relations subsisting among the instances or among the entailments themselves. So far the relations that have emerged as relevant are those of exchangeability and similarity or homogeneity. We shall now see how far these will carry us in considering inferences within theoretical systems.

2. *Positive Relevance as Analogy*

The examples that have been given of cases where the special consequence condition seems to be required can be seen to be cases where the clustering of instances of a generalization is replaced by the more general notion of an *analogical* relationship between two or more entities, systems, or empirical domains. The theory in these cases can be considered as confirmed by evidence in a given domain and as yielding predictions in another domain antecedently believed to be analogous to the first. Such predictions are, for example, derivation of the orbits of moon satellites from a Newtonian theory tested only in the neighborhood of the earth; prediction of the properties of a new chemical compound from knowledge of its elements and of other compounds; prediction of the properties of a full-sized object from experiments on a replica in, say, a wind tunnel; and prediction of the properties of micro-objects and events from those of macroscopic models.[29]

[28] Methods of this kind will not suffice to resolve Goodman's "grue" paradox, which I believe to be a deeper paradox than those considered here, in the sense that it depends on choice of language basis rather than choice of initial probabilities for a given language. For my proposed resolution of "grue" see "Ramifications of 'Grue,'" *British Journal for the Philosophy of Science*, 20 (1969), 13.

[29] I have discussed some of these examples in more detail in "Consilience of Inductions," in I. Lakatos, ed., *The Problem of Inductive Logic* (Amsterdam: North

Predictions of next instances of universal generalizations are elementary special cases of this kind in which the notion of "analogy" between a model whose behavior is known in domain i_1, and the predictions in domain i_2, reduces to the notions of exchangeability and clustering of instances in the different initial conditions i_1, i_2. This suggests that the clustering postulate, which was invoked to provide an adequate initial probability distribution in the case of next instance confirmation, may be generalized to deal with analogical inference in theories also. Indeed De Finetti himself suggests such a generalization:

Let us consider, to begin with, a class of events (as, for example, the various tosses of a coin). We will say sometimes that they constitute the *trials* of a given phenomenon; this will serve to remind us that we are almost always interested in applying the reasoning that follows to the case where the events considered are events *of the same type*, or which have *analogous* characteristics, without attaching an intrinsic significance or a precise value to these exterior characteristics whose definition is largely arbitrary. Our reasoning will only bring in the events, that is to say, the trials, each taken individually, the analogy of the events does not enter into the chain of reasoning in its own right but only to the degree and in the sense that it can influence in some way the judgment of an individual on the probabilities in question.[30]

In other words, according to De Finetti, the judgment that the events of a given class are exchangeable is a subjective judgment made prior to and independently of the probability calculations and is capable only of influencing our initial beliefs. One point in De Finetti's remarks needs clarification here: although a relation of analogy between events may suggest that they are exchangeable, exchangeability does not imply that they are analogous, and to obtain next instance predictions in cases of ravens or tossed coins, we need the clustering as well as the exchangeability postulate, as we saw in connection with counterinduction and the raven paradox.

If we consider the kind of theoretical inference of which examples have been given as a species of analogical argument, we must conclude that the judgment of analogy is itself a judgment that the theory in question applies to diverse domains, that the differences between the do-

Holland, 1968), pp. 232, 254, and in "An Inductive Logic of Theories," in M. Radner and S. Winokur, eds., *Minnesota Studies in the Philosophy of Science*, vol. 4 (Minneapolis: University of Minnesota Press, 1970), p. 164.

[30] "Foresight: Its Logical Laws, Its Subjective Sources," p. 120.

mains are probabilistically independent of their similarities, and that their similarities are sufficient to justify regarding experiments in any domain of the theory as typical of all its domains. Samples of planets and falling bodies are exchangeable with and analogous to the tests of any other consequences of the gravitational law; macroscopic experiments confirming, say, Maxwell's theory, are analogous to tests of that theory in microphysics; and in general analogical arguments are justified when sufficient analogy is recognized between events.

This approach to the confirmation of theoretical inference has its attractions even though it is certainly incomplete. It will appeal to those who regard the adoption of a theory as a "way of looking at" phenomena rather than as an inference based upon phenomena. It evades the practically insurmountable problem of specifying total evidence which afflicts all global confirmation theories of Carnap's type, for the background evidence is here simply assimilated into the judgment that a set of events is analogous in certain respects and hence probabilistically independent of the rest of the evidence which thereafter need not be explicitly mentioned. From this point of view theories which specify and summarize analogy judgments may be regarded as simplifications of the mass of our total experience which are essential to allow us to operate with it. The significant content of the theory is then seen to be, not so much the assertion of a set of universal generalizations, but rather the assertion that these generalizations describe the essential character of a certain finite set of domains which are held to be analogous in respect of that theory. If such a theory were said to constitute an "explanation," this would not be in the deductive sense, but in the sense of specifying what are the significant *causal* relations in a given domain and distinguishing them from those that are irrelevant and accidental with respect to that theory. It would be to assimilate all the domains to which the theory is applied to a particular explanatory model which is already found to be pervasive in experience (as for example mechanism in seventeenth-century science), or which is regarded as significant in more limited domains (for example, natural selection or field theories). Recognition of similarities and differences in observable instances would be primitive, prior to theories, and ultimately unanalyzable in terms of theories.

It should be remarked in passing that this view of theories does not commit us to instrumentalism, although its emphasis on observable data and predictions may seem to do so. But the theory has an essential function in relation to data and predictions, for theory does have a material content, which consists in the assertion that certain observable events, objects, or systems judged analogous *are* really analogous in respects that permit successful prediction from one system to another. The universality which is usually held to be an essential constituent of theories is seen in this view as rather a convenient method of summarizing case-by-case analogy judgments. But there are no conclusive grounds for associating lack of universality with instrumentalism. Real relations of analogy can subsist between events or objects without the events or objects constituting infinite classes.

On the other hand, this analogical view of explanation seems to preclude the possibility of asking, within the context of confirmation, what *justifies* adoption of the theory, or, what amounts to the same thing, what justifies us in regarding all features of a domain, other than those specifically mentioned in the theory, as independent of the theory and hence discardable. Surely this judgment is itself based on evidence, and hence is in principle capable of inclusion in the process of confirmation? It is clearly not sufficient to regard it as, in De Finetti's words, "largely arbitrary." We must try to specify in further detail the presuppositions of the independence judgments involved, which will then involve appropriate restrictions upon the choice of initial probability distributions.

The clustering postulate directs us to regard as more probable that state of the world which is more homogeneous. As between $f \& g$ and $f \& \sim g$, this means that state of the world which exhibits greatest analogy between the systems described by f and by g on the one hand, or by f and by $\sim g$ on the other. This analogy is what would be expressed by a general theory claiming that in certain features systems f and g are identical, that is, they are instances of the same causal laws. It is not, however, the generality of the theory which is in question here, but rather the specific application of it to f and g as analogous instances. But since "analogy" is not yet a very clear notion, it still remains to answer the question: What precisely constitute analogous instances in cases more complex than the recognition that certain observable particulars are sufficiently similar to count as substitution instances of the same generalization? I shall now attempt to answer this question in terms of a more general and universally recognized set of criteria for good theories, namely various kinds of *simplicity* criteria. These will be shown to yield criteria of positive relevance both in cases of analogical argument from familiar and observable models and in more general examples where there is no analogical relationship with an observable model.

3. *Positive Relevance as Simplicity*

There are two major contending concepts for the task of explicating the simplicity of hypotheses, which may be described respectively as the concepts of *content* and *economy*.[31] First, the theory is usually required to have high *power* or *content*; to be at once *general* and *specific*; and to make precise and detailed claims about the state of the world; that is, in Popper's terminology, to be *highly falsifiable*. This, as Popper maintains against all probabilistic theories of induction, has the consequence that good theories should be in general *im*probable, since the more claims a theory makes on the world, other things being equal, the less likely it is to be true. On the other hand, as would be insisted by inductivists, a good theory is one that *is* more likely than its rivals to be true, and in particular it is frequently assumed that simple theories are preferable because they require fewer premises and fewer concepts, and hence would appear to make *fewer* claims than more complex rivals about the state of the world, and hence to be more probable. Since Popper has annexed the notion of simplicity to the first of these sets of conflicting requirements,[32] it has been difficult to do proper justice to the second.

There are, however, two irreducibly different concepts operating here which ought to be considered separately. That this is so can be made obvious in the following example. Suppose the explanandum j is a conjunction of empirical laws of the form

$$j \equiv (x)\ (PQx \supset Sx. \&. QRx \supset Sx)$$

and take k, k' to be theories both entailing j, where

$$k \equiv (x)\ (Qx \supset Sx),$$
$$k' \equiv (x)\ (Px \supset Sx. \&. Rx \supset Sx).$$

Assuming that expressions of the form $(x)(Qx \supset Sx)$ all have the same initial probability and are logically independent, the probability ordering

[31] I am indebted to Elliott Sober for useful discussions on the notion of simplicity, although he will not agree with my proposed explication.

[32] *Logic of Scientific Discovery*, ch. 7. See, however, the countersuggestion in N. Goodman, "Safety, Strength, Simplicity," *Philosophy of Science*, 28 (1961), 150.

is $p(j) > p(k) > p(k')$, and this is the order of the number of states of the world in which the respective expressions can be true, and hence the inverse order of content or power. Thus if content is the measure of simplicity, simpler theories are less probable, as Popper asserts. But the ordering of intuitive simplicity s here is neither this nor its converse, but rather $s(k) > s(k') > s(j)$. And k would surely be preferred to k' as an explanatory theory for j on grounds of economy of predicates and simplicity of logical form. The greater content of k' does not in this case count in its favor, since the explanation it gives of j depends on two logically unrelated laws, and it gives no insight into what may be the common cause of the two laws conjoined in j.

This example is enough to show that the orderings of content, preferability, and economy are not directly correlated and that content does not exhaust the desirable characteristics of a theory. However, in the situation outlined at the end of the last section, we are not concerned directly with the content of theories, but with comparing two states of the universe $f \& g$ and $f \& \sim g$, where it has been assumed that these two state-descriptions have the same content, that is, they imply conflicting results of the test for g or $\sim g$, each result being assumed to have the same initial probability. We are therefore free in this situation to consider the comparison of $f \& g$ with $f \& \sim g$ only for economy.

Take first the clustering postulate which has already been invoked as the presupposition of analogy arguments. This is also in a sense an economy postulate. Consider evidence consisting of individuals a_1, a_2, . . . , a_n, all of which have properties P, Q, and R. Now consider an individual b with properties $\sim P \& Q$. Does b have R or not? If nothing else is known, the clustering postulate will direct us to predict Rb, since, ceteris paribus, the universe is to be postulated to be as homogeneous as possible consistently with the data, and a world in which $PQRa$, where a is one of $a_1, a_2, \ldots, a_n$, and $\sim PQRb$, is more homogeneous than one in which $PQRa$ and $\sim PQ\sim Rb$. But this is also the prediction that would be made by taking the most economical general law which is both confirmed by the data and of sufficient content to make a prediction about b. For $h \equiv (x)\ (Qx \supset Rx)$ is certainly more economical than the conflicting

$$h' \equiv (x)(PQx \supset Rx) \ \& \ (x)(\sim PQx \supset \sim Rx),$$

or even than a weaker particular hypothesis

$$h'' \equiv (x)(PQx \supset Rx) \;\&\; {\sim}PQ{\sim}Rb.$$

Notice that h and h' have the same content, namely all Q's, whereas h'' has lower content as well as being less economical than h.

Here the clustering postulate gives the same result as adoption of the most economical general law, but the probabilistic justification for choosing the prediction Rb must be given in terms of clustering of individuals rather than economy of the law, since the probability of the law in itself is untransferable to its consequences in virtue of nontransitivity.

Many examples of economy do, however, seem to involve hypotheses directly. A straight line hypothesis is preferred to a higher-order curve when both are consistent with the evidence, and, in general, hypotheses introducing fewer extralogical predicates and fewer arbitrary numerical parameters are preferred to those that introduce more. Popper and H. Jeffreys[33] have both given accounts of the simplicity of curves that depend essentially on counting the number of arbitrary parameters and assigning simplicity inversely as this number. Let us see how this general proposal can be applied to the predictive inference from data points to another specific point satisfying the hypothesis of a straight line rather than a higher-order curve.

Let f be the assertion that two data points (x_1, y_1), (x_2, y_2) are obtained from an experiment. These points can be written equivalently $(x_1, a + bx_1)$, $(x_2, a + bx_2)$, where a, b are calculated from x_1, y_1, x_2, y_2. The two points are consistent with the hypothesis $y = a + bx$, and also of course with an indefinite number of other hypotheses of the form $y = a_0 + a_1x + a_2x^2 \ldots$ where the values of $a_0, a_1, \ldots$ are not determined by x_1, y_1, x_2, y_2. What is the most economical prediction of the y-value of a further point g, where the x-value of g is x_3? Clearly it is that prediction which uses only the information already contained in f, that is, the calculable values of a, b, rather than a prediction which assigns arbitrary values to parameters $a_0, a_1, \ldots$ of a higher-order hypothesis. Hence the most economical prediction is $g = (x_3, a + bx_3)$, which is also the prediction given by the "simplest" hypothesis on almost all accounts of the simplicity of curves. Translated into probabilistic language, that is to say that to conform to intuitions about economy we should assign higher initial probability to the assertion that points

[33] H. Jeffreys, *Scientific Inference*, 2nd ed. (Cambridge: At the University Press, 1957), pp. 37, 60; *Theory of Probability* (Oxford: Clarendon, 1961), ch. 1.

$(x_1, a + bx_1)$, $(x_2, a + bx_2)$, $(x_3, a + bx_3)$ are satisfied by the experiment, than to that in which the third point is inexpressible in terms of a and *b* alone. In this formulation economy is a function of finite descriptive lists of points rather than general hypotheses, and the relevant initial probability is that of a universe containing these points rather than that of a universe in which the corresponding general law is true (which probability may in any case be zero).

A similar interpretation can be given to theories which are held to be simple or economical in virtue of exhibiting *invariance* between different systems. For example, Einstein was originally convinced of the preferability of relativity theory over the extended classical theory of Lorentz on grounds of this kind of economy. In the period immediately following Einstein's 1905 paper Lorentz's theory was consistent with all existing experimental results, including the Michelson-Morley experiment, and even had greater content than Einstein's theory, in that it contained a micro-theory of electrons which Einstein did not attempt.[34] With regard to Lorentz invariance and the consequent radical revision of space-time concepts in Einstein's theory, however, Lorentz's theory involved postulation of an entity, the ether, which Einstein abandoned, and a parameter, the velocity of a system within the ether, which became meaningless in Einstein's theory. Invariance to Lorentzian transformations in space-time was considered by Einstein (and in 1909 by Lorentz himself) as an argument for adoption of Einstein's theory. Such invariance can be interpreted as economy in the description of systems in relatively moving space-time frameworks, since where the two theories conflict, prediction of the result by Lorentz's theory involved the velocity

[34] For discussion of "simplicity" comparisons of Einstein's and Lorentz's theories see A. Einstein, "Autobiographical Note," in P. A. Schilpp, ed., *Albert Einstein: Philosopher-Scientist* (Evanston, Ill.: Library of Living Philosophers, 1949), p. 23; G. Holton, "On the Origins of the Special Theory of Relativity," *American Journal of Physics*, 28 (1960), 627, "Mach, Einstein, and the Search for Reality," *Daedalus*, 97 (1968), 636, "Einstein, Michelson, and the 'Crucial' Experiment," *Isis*, 60 (1969), 133; A. Grünbaum, *Philosophical Problems of Space and Time* (New York: Knopf, 1963), ch. 12; K. F. Schaffner, "The Lorentz Electron Theory and Relativity," *American Journal of Physics*, 37 (1969), 498; "Outline of a Logic of Comparative Theory Evaluation with Special Attention to Pre- and Post-Relativistic Electrodynamics," in R. Steuwer, ed., *Minnesota Studies in the Philosophy of Science*, vol. 5 (Minneapolis: University of Minnesota Press, 1970), p. 311. For a useful analysis of simplicity and related concepts of "invariance" and "optimal coding," see H. R. Post, "Simplicity in Scientific Theories," *British Journal for the Philosophy of Science*, 11 (1960), 32, "A Criticism of Popper's Theory of Simplicity," *ibid.*, 12 (1961), 328.

of systems relative to the ether, whereas prediction by Einstein's theory did not.

In probabilistic terms, finite conjunctions of invariant systems should therefore be assigned higher initial probabilities than those not exhibiting invariances. This is the justification in terms of a probabilistic confirmation theory of the preference for general theories entailing invariance as more economical, even though the theories themselves may have to be assigned zero probability if they are expressed in strictly universal form.

4. Conclusion

The claims made here for Bayesian methods in confirmation theory are at once more specific than is usual in the literature and also more modest. They are more modest in that I do not claim these methods are adequate for a global inductive theory for scientists or any other learning organisms. The theory developed here is an explication of some elementary inductive and theoretical inferences applicable when the language basis remains sufficiently stable. In spite of much recent discussion of "meaning variance," I presuppose that language, as well as its meanings, remains locally invariant throughout such pieces of inference, for if it does not it is impossible to understand how communication, let alone rational argument, is possible either in science or in everyday life. I have not attempted to identify the elements of the language or to assign probability weightings to them, since I assume that once a model of inference is given which explicates adequately enough our inductive intuitions, its application to actual cases will suffice to indicate sufficiently well what the language basis in those cases is. To the charge of possible circularity here, I would reply that this is the virtuous circularity of the process of explication itself, in which different parts of the exercise should be mutually adjustable. Confirmation theory should not in any case be so rigorous and inflexible that it becomes open to the opposite charge of dictating the methods of science a priori.

I have not suggested stronger constraints on initial distributions of probability than are required for explication of certain familiar types of inference. Consequently nothing like numerical assignments of probability are provided by these constraints, but only comparisons of probability in cases where comparisons of reliability of hypotheses and predictions are normally assumed to be possible. To the objection that such weak constraints on initial probabilities only repeat these intuitive assumptions, it must be replied that the analysis, if correct, has had far-reaching consequences for the deductive model of science. Discussion of generalizations and theories has led to the proposal that confirmation in science should be seen primarily as an application of inference from case to case rather than as confirmation of universal laws and theories. I suggest, however, that far from being an inadequacy in the concept of confirmation, this consequence is highly illuminating for the general problem of theoretical inference in science. In particular it provides resolutions of some so-called "paradoxes of confirmation," and it also enables a unified inductive account to be given of enumerative induction, analogical inference, and the criterion of simplicity of theories. This inductive account differs in important respects from the usual deductive model of science, but its very systematic character provides an argument in its favor, for the deductive model has no such systematic account of the process of confirmation and related desiderata for theories such as analogy and simplicity. A modest Bayesian confirmation theory therefore has revisionary as well as descriptive implications for our understanding of the structure of scientific theories.

GROVER MAXWELL

Induction and Empiricism: A Bayesian-Frequentist Alternative

The theory of confirmation sketched herein is subjectivist in a manner that will be explained. According to it, however, the *degree of confirmation* of a hypothesis is an objectively existing *relative frequency* (or a *propensity*, if one prefers). The resolution of this apparent paradox is simple, but its implications are, I believe, profound. I am convinced that they provide the means for resolving all of the current "paradoxes," "riddles," and "puzzles" of confirmation and induction. This is an overstatement only if one assumes, *contra hypothesis*, that all is pretty much all right with contemporary theories of confirmation and that resolution of difficulties will consist only of a little patching up here and there. On the contrary, I argue, these "paradoxes," "riddles," etc., are legitimate reductios ad absurdum of most current confirmation theories, and their resolution requires the rather radical approach indicated by the first two sentences of this essay.

Since the *solution* that I advocate *is* radical, I begin with a rather detailed and elementary discussion of some of the crucial aspects of the problem, including a few of the more relevant historical ones. This will explain, I hope, the extremely refractory nature of the problem as well as the surprisingly small amount of progress that has been made toward its solution since it was posed by Hume[1] and, finally, the necessity for the drastic measures that I have been led to propose. This discussion will be followed by a brief outline of the proposed contingent (though not "empirical") theory of confirmation that allows us to escape the skepticism of Hume. This includes a discussion of the nature of the *prior* (or

NOTE: Support of research by the National Science Foundation, the Carnegie Corporation, and the Single-Quarter Leave program of the University of Minnesota is gratefully acknowledged.

[1] As is emphasized at greater length below, this is in no way to disparage the crucially important though, in my opinion, mostly negative results of a host of tireless workers in the area, several of whom will be discussed presently.

antecedent, or *initial*, etc.) probabilities that must be used in estimating degrees of confirmation. I shall mention some of the philosophical and scientific implications of the failure of other theories especially those of empiricist persuasion and some implications of the theory advocated here. Finally, in an appendix, several current approaches to the problems of confirmation that were not previously discussed are considered.

It would doubtless be foolhardy to try to persuade most contemporary philosophers that this notorious problem, the problem of induction, or the problem of nondeductive inference, or the problem of confirmation or corroboration — term it how ever we may — the problem posed so poignantly by Hume, has occupied far too little of their attention. Most would probably reply that, on the contrary, it is one of those typical traditional philosophical problems which, in spite of interminable, ad nauseam discussion, remains as apparently refractory as when first considered. Therefore, they would continue, perhaps we should begin to suspect that it is not a genuine problem; at any rate, they would say, the very most that should be accorded it is *benign neglect*. Those who reject such a comfortable rejoinder and continue to take seriously Hume's critique *do* constitute a substantial minority. They include most if not all of the contributors to this volume as well as such luminaries as Carnap, Feigl, Feyerabend, Grünbaum, Hempel, Lakatos, Popper, Reichenbach, and Russell, to mention (alphabetically) only a few. An attempt to give a complete list would be tiresome and inevitably invidious anyway, and, certainly, some whose names are not included here have done work as extensive and important as that of those whose are. However, the names of Strawson, Edwards, and others who support the "ordinary-language" solution (or *dissolution*) of the problem are advisedly omitted because they arrive, quite honestly, of course, and apparently to many quite persuasively, at one variety of the "comfortable rejoinder" position mentioned above.

The view for which I shall argue is that few if any of us, excepting, possibly, Hume himself, have hitherto realized the enormity of the import of the problem, not even those of us who have devoted a large portion of our philosophical lives to it. What is at stake is not just a theory of evidence or confirmation, although this certainly *is* at stake and is a vitally important matter. But also in question is the basis, indeed the very nature, of epistemology and of knowledge itself. Involved are radical and pervasive implications concerning scientific methodology, the very nature of science *and* of philosophy as well as results that illuminate specific philosophical and scientific issues such as realism vs. phenomenalism (or other varieties of idealism, or instrumentalism, or operationism, etc.), the alleged line(s) of demarcation that separate(s) science from nonscience (or from metaphysics, or from pseudoscience, etc.), "empiricist meaning criteria," and the mind-body problem. In this paper I shall try to support, clarify, and explain some of the details of the prima facie extravagant and dramatic claims just made. (I have discussed some of these matters in other places.[2])

Hume's problem is, I contend, insoluble. To qualify this by saying that it is insoluble on empiricist grounds borders on redundancy, for it is a serious problem *only for* empiricists. Nonempiricists can circumvent it or dispose of it in a number of ways. Some might declare a set of principles of induction or of nondemonstrative inference to be valid by synthetic a priori warrant, while some would hold that statements of the basic regularities of the universe can be known directly and are themselves synthetic a priori; or one may take the position for which I shall eventually argue and accept a more relaxed, more realistic view about the nature of knowledge and the standards one should employ for its certification.

Such a "relaxed" position is not one that I have taken easily. Honest empiricism is certainly more than just a philosophical fad or fashion. Prima facie, it is both plausible and attractive, and the epistemological motives for holding it are not only understandable but, at least some of them, eminently admirable as well. Moreover, most philosophers, including myself, accept it as a truism that such developments as the applications of non-Euclidean geometry and indeterministic theories in physics, etc., have forever refuted Kant's reply to Hume, as well as all of the more drastic varieties of rationalism.

What indeed could be more reasonable and desirable than to build the entire edifice of knowledge using only bricks and mortar consisting

[2] "Theories, Perception, and Structural Realism," in R. Colodny, ed., *The Nature and Function of Scientific Theories, Pittsburgh Studies in the Philosophy of Science*, vol. 4 (Pittsburgh: University of Pittsburgh Press, 1970); "Russell on Perception: A Study in Philosophical Method," in David Pears, ed., *Bertrand Russell: A Collection of Critical Essays* (New York: Doubleday Anchor Paperbacks, 1971); "The Later Russell: Philosophical Revolutionary," read at the Bertrand Russell Centenary Conference at Indiana University, March 1972, and published in George Nakhnikian, ed., *Bertrand Russell's Philosophy* (Bloomington: Indiana University Press, 1974); and "Corroboration without Demarcation," in P. A. Schilpp, ed., *The Philosophy of Karl Popper* (LaSalle, Ill.: Open Court, 1974).

of the two and only two kinds of knowledge claims that are pretty much secure and unproblematic: (1) singular propositions known by *direct observation* (whatever we take direct observation to be) and (2) propositions certifiable solely on the basis of the principles of logic (or of "pure" mathematics). And since, it would seem, the items in both of these categories can be known *objectively*, or, at least, intersubjectively, this highly prized property will be exemplified by any and all portions of the edifice. Our knowledge, fashioned then from these two impeccable materials, might not, at any one time, be perfect or certain, but it could be objectively certified as the best possible, given the bricks (evidence) available at the time.

To switch metaphors, such an epistemology and the methodology to which it gives rise are ideal for warding off or for chastising subjectivists, metaphysicians, takers on faith, and any *unscientific* charlatans that threaten to clutter up the philosophical, the scientific, and, even, in principle at least, the practical scenes. As one raised (philosophically) in the empiricist tradition, I speak *after* many years of experience and *from* the heart. To have become convinced that empiricism is gravely defective, both in letter and in spirit, has been a painful thing indeed. On the other hand, the anguish has been considerably mitigated both by the intellectual excitement of beginning to see definite answers, if somewhat negative ones, to the notorious problem of Hume and by the intellectual freedom made possible by shedding the epistemological straitjacket that empiricism turns out to be. Such freedom is, of course, one of the very things that, as empiricists, we feared from the beginning. Does it not open the floodgates to almost any sort of crackpotism imaginable? I shall say more on this later; but be that as it may, if empiricism is untenable, it is untenable. It is not a matter for legislation, no matter how nice it might be if only it *were* viable. To those, including a *part* of myself, who are still inclined to mourn its moribundness, I can only paraphrase (once more) Bertrand Russell and say that I am sorry, but it is not my fault.

Such rhetoric, I realize, risks offense or, at least, stylistic criticism, and certainly to some it will seem that I put my thesis in an extreme and unnecessarily provocative fashion when it becomes, in effect, the assertion of the poverty of empiricism.[3] But I believe that continued

[3] Forgive me, Sir Karl! Cf. the title of K. R. Popper's book, *The Poverty of Historicism* (Boston: Beacon Press, 1957).

investigation and reflection will vindicate my so stating it, if, on nothing else, on didactic and pedagogical grounds. The results that contemporary studies in confirmation theory thrust upon us *are* extreme. Indeed, they are much more drastic, I think, than even most of the involved scholars have realized — so drastic that what at first seems to be an extreme assertion may well turn out to be an understatement.

To turn now from rhetoric and to the business at hand: What *is* empiricism, and what is its stake in the problem of induction? When I speak of empiricism here, I am concerned only with what some have called "judgment empiricism,"[4] which is a doctrine on how knowledge claims are "validated," confirmed, supported by evidence, or otherwise certified as being real knowledge, or probable knowledge, etc. In other words *judgment empiricism* is part of a theory of *confirmation*. Using the same terminology, *concept empiricism* is part of a theory of *meaning*, is logically independent of judgment empiricism, and will not be our concern here at all; hereafter "empiricism" will mean *judgment empiricism* unless otherwise stated. As a first approximation, we may begin by saying that empiricism is the doctrine that our (contingent) knowledge claims can be confirmed only by (1) *observational knowledge*, plus (2) *logic*. (I shall take the notion of observational knowledge to be unproblematic here, although of course it is not; however, I believe that what I have to say can be easily adapted to fit into *almost* any theory of observation.) The immediate problem is to decide what should be included under "logic." In addition to deduction,[5] Hume taught us that some forms of nondeductive *inference* (or nondeductive *logic*, many would say) must be classified as legitimate if our knowledge is to go beyond observations of the moment. Empiricists, however, understandably have wanted the word "logic" to retain as much rigor and "toughness" as possible here. More specifically, it is vital for empiricism that even the nondeductive or "inductive" logic be, *in some important sense*, free of *contingent* elements. Whether this is possible and, if so, in just what sense, is, of course, the nub of the problem; but why empiricism

[4] E.g., John Hospers, *An Introduction to Philosophical Analysis*, 2nd ed. (New York: Prentice-Hall, 1967), pp. 102ff.

[5] I am assuming here that the principles of *deductive* logic *are* unproblematic. This was, I believe, an assumption of Hume and certainly is one of most contemporary empiricists; and the criticism of empiricism in this essay is, in the main, internal criticism. Moreover, I am convinced that this contemporary empiricist philosophy of deductive logic is essentially correct.

must impose this condition has been obvious ever since Hume: logical principles are universal principles and for this (and other) reasons cannot be known directly by observation (or "experience") and, therefore, if any one of them were contingent it would be just as much in need of validation, confirmation, etc., as the most mundane "empirical" generalization. And, of course, any contingent generalization requires nondeductive inference for its support, and we are immediately caught in Hume's vicious circle of "using induction to justify induction," etc. Empiricists, of course, have recognized and even insisted on this point, but I think it is fair to say, also, that they have required that *logic*, even "inductive logic," retain whatever rigor and toughness that derives from the following: (1) Analogously to deductive arguments owing their validity to their purely formal properties, legitimate nondeductive arguments owe whatever legitimacy they have to *their* purely formal properties: for example, one of the allegedly legitimate arguments, *induction by simple* enumeration, always has (something like) the form x_1 is A and x_1 is *B*, x_2 is A and x_2 is *B*, x_3 . . . . , ... , therefore everything that is A is *B*. (2) The legitimate forms of inductive inference (and *only* these forms) are legitimate in every possible world; i.e., they provide the "rational" way to infer nondeductively, or the "best" way, or the best advised way, etc., for every possible world, although, of course (being nondeductive) they *will* fail in some possible worlds.

These two requirements are, obviously, pretty tough. However, it is difficult to see how they can be significantly relaxed without reopening the gates to Hume's devastating critique. Even something so apparently innocuous as the "principle of total evidence," to which most inductive logicians (and others) subscribe, introduces immediate trouble. On the one hand its use seems necessary for avoidance of certain intolerable "inductive paradoxes"; but, on the other, it seems highly questionable whether it is a *formal* principle and, certainly, there are possible worlds in which knowing subjects make much better inferences when they ignore a certain portion of the evidence that they happen to have at hand.[6]

Let me pause here to beg indulgence for laboring elementary points in such an elementary fashion — points many of which were made by Hume in a more elegant, more persuasive manner and points, moreover, that

[6] Whether they would be well advised to do so and, if so, under what (specifiable?) conditions is, of course, arguable. But this is just my point; it is arguable and, thus, questionable in both directions.

are made by most contemporary philosophers and methodologists who write on these problems. It does seem to me, however, that many such philosophers and methodologists, after explaining these elementary difficulties with persuasive clarity, then proceed to forget, ignore, or repress some of them at crucial junctures in the developments of their own "solutions" to Hume's problem. When we recall again, however, that it is the very life of empiricism that is at stake, such repeated instances of apparent perversity in intellectual history become much more understandable and easier to condone. Indeed it seems very nearly certain to me that the numerous and persistent attempts to "justify induction" or otherwise dispose of the problem will, *in the long run*, turn out to produce invaluable benefits in several areas of philosophy and, especially, in scientific methodology. The ingenuity, originality, and technical competence exemplified in many of these studies not only are intrinsically admirable but have produced results of lasting importance for other areas of thought as well as for the one of explicit concern, no matter what its future courses of development may be. In the meantime, however, given the current chaotic situation, I do not see any way to avoid continuing to reiterate some of Hume's basic insights.

One of the most ingenious and best known attempts to rescue empiricism from Hume is due to Herbert Feigl and Hans Reichenbach. Although they were acquainted and were most congenial, both personally and philosophically, with each other, their initial work on this was, as far as I can determine, begun independently; and it used similar but distinctly different approaches. In an article first published in 1934, Feigl contended that the principle of induction is not a *statement* but, rather, a *rule*.[7] Thus, it is neither true nor false and, a fortiori, it is not contingent (or necessary). He felt that this enables inductive logic to avoid contingent principles. Nevertheless, the rule must sanction *ampliative* (nondeductive) inferences and therefore, as he recognized, it stands in obvious need of justification, since an unlimited number of alternative rules giving mutually incompatible results can be formulated. He also recognized, at least implicitly, that such justification could not be accomplished solely by means of (other) rules alone but must contain propositional elements, elements that were either contingent or neces-

[7] "The Logical Character of the Principle of Induction," *Philosophy of Science*, 1 (1934), 20–29. Reprinted in Herbert Feigl and Wilfrid Sellars, eds., *Readings in Philosophical Analysis* (New York: Appleton-Century-Crofts, 1949).

sary. This would *seem* to run the empiricist afoul of Hume again, for if the justificatory propositions are contingent we are immediately caught in Hume's circle, and if they are necessary, then, given the empiricist philosophy of logic (and of necessity), which is not in question here, they are logically true (or at least *analytic*) and, thus, factually empty. This would seem to render them impotent insofar as selecting among different rules that give different sets of contingent results, each set incompatible with every other one. But Feigl, at the time, seems to have disagreed. He thought that a tautologous justification could select our "normal" inductive procedures as being the best advised kinds. He admitted that it was, in a sense, a trivial justification, but it was, he thought, nevertheless illuminating. And its very triviality is, in a way, its great virtue, for it enables us to avoid the fatal introduction of contingent elements. As far as I know, this was the first of a number of interesting attempts to give deductive or *analytic* justifications of induction.

In a later, justly renowned article, Feigl made his famous distinction between *validation* and *vindication*, the two kinds of justification.[8] Surely this is a very familiar matter, so I shall only say enough about it here to help set the stage for what is to follow. In a *vindication*, what is justified is an action, or a procedure, or a rule that sanctions a procedure or a course of action. A successful vindication is one that shows that the action or rule, etc., is one that is well advised relative to the goal or aim of the action, procedure, or rule. Preferably, the vindication will show that the procedure, etc., in question is a *good* method for attaining the goal desired and, *most* preferably, that it is the *best* method. Feigl, and soon others, hoped and evidently believed that induction (our "normal" inductive procedures) could be given such a vindication relative to our goals of obtaining knowledge of laws of nature, prediction of the future, etc.

Note that in talking about a *good* method and the *best* method an evaluative component is introduced *in addition to* the necessarily evaluative one connected with selecting final goals such as predicting the future, obtaining general knowledge. This might seem to open the way for troublesome complications. For example, someone might have such

[8] "De Principiis non Disputandum . . . ? On the Meaning and the Limits of Justification," in Max Black, ed., *Philosophical Analysis* (Ithaca, N.Y.: Cornell University Press, 1950; republished, Englewood Cliffs, N.J.: Prentice-Hall, 1963); reprinted in part in Y. H. Krikorian and A. Edel, eds., *Contemporary Philosophic Problems* (New York: Macmillan, 1959).

an overwhelming preference for very easy methods that for him, armchair guessing would be the "best method" even if he could be persuaded that its chance of predicting the future, etc., was only a small fraction of what it would be if he applied, say, "Mill's Methods." A vindication, then, seems to be relative not only to a goal but also to preferences, or criteria, etc., concerning intrinsically the course of action itself. Once these are stated, or once such criteria are agreed upon, then what is a good method or the best method can be treated as a purely cognitive question; and the premises used to support such a vindicatory argument will all be propositional, true or false, and, thus, either necessary or contingent. Thus, while it is undoubtedly important and certainly true to point out that inductive procedures require vindication rather than validation, it is not at all clear how the shift from the latter to the former alters the situation one whit as far as Hume's problem is concerned. Nevertheless, Feigl and others certainly appeared, again, to hope and believe that the *truth* of the cognitive proposition that induction ("our" kind of induction) satisfies the criteria for a "good" method or, even, the "best" method for predicting the future, securing general knowledge, etc., could be (deductively) *demonstrated*. To those unkind enough to remark that this is an obvious attempt to transform induction into deduction, there is a ready reply: No, if we succeeded in such a vindication we would not have shown that induction is always truth preserving as, of course, deduction is; we would have shown only that our inductive procedures provide us with a good (or the best) method for discovering those nondeductive inferences that lead from true premises to true conclusions although they will sometimes fail us in this. But after such a modest and apparently impeccable retort, Fate herself becomes unkind. Much more difficult obstacles arise when one attempts to give acceptable criteria for "good method" or "best method," or, even, for "a method." "A method" can certainly not be taken to mean a method that *will* attain the goal (of predicting, etc.) or even one that is likely to succeed. Such criteria cannot be *known* to hold for *any* method; the shade of Hume will not be denied. Coming up with criteria for "good" or "best" methods is beset with similar, even greater difficulties.

On the matter of induction, philosophers have been ready to settle for considerably less than half a loaf. There has often been much rejoicing and, perhaps, excessive gratitude over whatever crumbs may seem to be falling from the table. For example, when it was realized that we could not vindicate (normal) induction by showing that it was the best method (using nonperverse criteria of "best") since there could easily exist other (abnormal) procedures which might turn out to give better predictions, etc., Reichenbach opted for a weaker but ingenious kind of vindication. (He used the term "pragmatic justification" rather than "vindication.") According to him, we *can* show that if any method will succeed, induction will succeed (or, contrapositively, if induction fails, then all other methods will fail). Now this still *sounds* pretty strong. The weakening, the nature and the handling of which constitute a great deal of the ingenuity, is twofold. First, the goal of induction is grossly weakened in a manner to be discussed; here I shall just note that it is no longer stated in such general terms as we have been using and that successful prediction, in effect I shall contend, is no longer a part of it. Second, it turns out that Reichenbach really means that if any method will succeed, then induction will succeed *eventually* (or if induction fails *in the long run,* then all other methods will fail *in the long run*); and it also turns out that *eventually* and *in the long run* may, insofar as the vindication goes, both turn out to be, literally, *forever.* This would be of little consequence in the epistemology of God or Omniscient Jones but for the rest of us it could mean a total epistemic debacle.

As we all know, the more modest goal that Reichenbach sets for induction is the determination of values of limits of relative frequencies in certain ordered classes.[9] Reichenbach provides us with a method (actually a family of methods) which, *if applied long enough,* is *deductively* guaranteed to produce the correct value of the limit *provided a limit exists.* Again, this sounds impressive; if true, it would seem to provide us with the coveted deductive vindication of induction. Its defects, however, are well known. A sympathetic but thorough and incisive discussion of them is, again, provided by Wesley Salmon, a former student of Reichenbach and one of the outstanding figures in confirmation

[9] For anyone not familiar with this approach — or for anyone at all, for that matter — an excellent source of information is the work of Wesley Salmon, some of which will be discussed here. See, for example, his *Foundations of Scientific Inference,* Pittsburgh Paperbacks (Pittsburgh: University of Pittsburgh Press, 1966); reprinted from Robert G. Colodny, ed., *Mind and Cosmos, Pittsburgh Studies in the Philosophy of Science,* vol. 3 (Pittsburgh: University of Pittsburgh Press, 1966).

theory today.[10] Nevertheless, a brief discussion of some of them here is necessary for reasons similar to ones given above.

First of all, as already hinted, as far as the deductive vindication goes, there is absolutely no reason to suppose that the "long enough" qualification (also signaled above by "eventually" and "in the long run") will not stretch into *forever*. Thus, as far as the vindication takes us, there is absolutely no reason to suppose that Reichenbach's method will ever provide any knowing subject with a value that is anywhere near the correct value, unless he assiduously applies it until, as Carnap once said, he reaches the ripe old age of infinity. Moreover, the vindication provides no reason for doubting that there are other methods, incompatible with Reichenbach's, that will provide much better estimates of the relative frequency unless, again, they are applied to portions of the sequence which are, for us, impossibly large; and this holds not only for values of the limit but, and this is the practical payoff, for values of the relative frequency for any unobserved portions of the sequence. This brings to mind another fatal practical defect of the vindication; even if the limit is known, there is no deductive relation between it (in either direction) and the value of the relative frequency in any finite portion of the sequence, so that, for example, the true value of the limit could do nothing, as far as the vindication is concerned, toward predicting frequencies in finite classes (the notorious "problem of the short run"). As just mentioned a "perverse" method might well be free of this defect, again, as far as Reichenbach's vindication is concerned. Thus for practical purposes such as prediction, and even for a *practical* method of getting a good estimate of the limit, the vindication is no vindication at all. For something labeled "the pragmatic justification of induction," this seems amiss.

There are other important difficulties with Reichenbach's program, but those given above will suffice here. The reader is again referred to Salmon, whose attempts to give a vindication of induction (specifically, a vindication of Reichenbach's "straight rule") that avoids the errors but retains the spirit of Reichenbach's work have been valiant, indeed.[11] I have learned a large portion of whatever I know about confirmation

[10] See *ibid.* and the references therein.

[11] The *straight rule* enjoins us to use the value of the relative frequency determined for observed portions of the sequence as the estimate of the value of the limit: e.g., if one one-thousandth of observed opossums have been albinos, infer that one one-thousandth of all opossums are albinos.

theory from his admirable work, and the value of its by-products, especially the critiques of other approaches to confirmation theory, is beyond estimate. The attempts themselves, however, I believe to be doomed to failure for general reasons that I shall give presently, so I shall not discuss them specifically now.

Before leaving the subject, it *will* be instructive to look at a general moral to be drawn from Reichenbach's attempt. His *deductive vindication* of the *straight rule* amounts to the triviality that, if values of the relative frequency in a sequence do converge to a limit and if the value of the frequency in the sequence up to the point of convergence is determined (by observation of members of the sequence up to this point), then this determined value will be the value of the limit (plus or minus ϵ, for any ϵ) or, in simpler language: if the frequency converges to a limit and we observe the value at the point of convergence, then we will have observed the value of the limit (although, of course, even when, if ever, we do succeed in observing the value of the limit, we never *know* that it *is* the limit). It is difficult to believe that anything very significant — anything that is, even psychologically or therapeutically speaking, news — could follow from such a triviality, let alone something that is supposed to rescue empiricism from Hume's formidable skepticism.

In propounding their philosophy of logic (and their theory of necessity which they develop from it), empiricists have emphasized, rightly in my opinion, that from trivialities only trivialities follow. How, then, can one account for the curious lapse whereby many of them have hoped to dispose of the justly celebrated problem of induction by employing a few logical and mathematical trivialities? Well, for one thing, we have noted that the defining postulates of empiricism seem to preclude any other kind of solution. But apart from obvious human frailties such as wishful thinking and hope springing eternally, I believe that there is a more charitable explanation deriving from a more subtle factor, one that, again, was clearly anticipated by Hume. We are constitutionally so irresistibly disposed to employ certain "normal" modes of nondeductive reasoning that we often do so tacitly, implicitly, perhaps unconsciously — call it what you will. This built-in predisposition and prejudice may operate in at least two significant ways: (1) it may shed a spurious plausibility on attempts to justify such modes of inference and (2) it may subtly introduce an unnoticed circularity of the type exposed by Hume, whereby induction is "justified" inductively.

Reichenbach's straight rule, for example, is so congenial with our inductive animal nature that, when he assures us that, in spite of Hume, if we only apply it *long enough* it is bound to produce the desired results if such results are attainable at all, we breathe a sigh of relief. We felt all along that it must be so; how nice that Reichenbach has been able to prove it! And when the full chilling force of the qualification, "long enough," finally strikes us, we may feel that, anyway, it was a nice, plausible attempt and that perhaps the matter can be remedied by adding another triviality or two to the one used by Reichenbach. (Note that I have not claimed that either our inductive predispositions or the straight rule is "wrong," but only that neither has been deductively vindicated.)

As for the circularity, consider the following example. It is often claimed that, no matter what nondeductive procedure we might employ, we would and/or should test the procedure itself by "normal" induction, by, as a matter of fact, something like the straight rule. Suppose we employ a certain rule *R*, let us say, to make predictions. After a number of applications, we would and/or should note the number of successes and the number of failures of *R*. If it has been always or often successful, we would and/or should infer that it will continue to be always or often successful — that it is a "good" rule and if mostly unsuccessful that it is a "bad" one. This, the claim continues, shows the primacy of the straight rule and, it is sometimes added, again, shows that if any rule succeeds so will our normal inductive procedures such as that given by the straight rule. Again, this may sound very plausible initially, but, after the forewarning above, the circularity must now be obvious. It is, no doubt, a psychological fact that we *would* test *R* in such a manner; but this shows absolutely nothing of significance, except, perhaps, that Hume was right in claiming that we do have virtually irresistible animal (or other kinds of) impulses to use "normal" induction. And I do not see what warrant there is for claiming that we *should* so test *R* unless we assume what is at question: that "normal" induction or the straight rule does have primacy — is legitimate, etc. — for making nondeductive inferences, whether about successes of rules or about other kinds of entities. It would seem that the only things vindicated by this prima facie seductive line of thought turn out again to be the remarkable ruminations of Hume.

Another feature of the preceding example leads into what I consider to be the clincher of my argument. Such a test of *R* by using induction by simple enumeration or the straight rule is demonstrably worthless unless it is bolstered by very strong additional assumptions. For there will always be an infinite number of rules each of which gives results identical with those of *R* for all applications to date but which will give results incompatible with those of *R* (and with those of each other) for future applications. Thus, if such testing favors *R*, it will favor equally well an infinite number of other rules incompatible with *R* (and with each other). This can be easily proven *deductively*, as can be the following, related theorem:[12] *For each simple inductive argument with true premises and true conclusions, there exists an indefinitely large number of inductive arguments each with the same logical form as the first and each with all premises true but with conclusions incompatible with that of the argument in question (and with those of each other)*. The phrase 'simple inductive argument' is used here in a technical sense to include induction by simple enumeration and its generalization into the "straight rule," most other commonly encountered statistical inferences, "eliminative induction," "Mill's Methods," and other similar "normal" inductive procedures. Thus, statistically speaking, as Russell has put it, *induction leads infinitely more often to false, unacceptable results than it does to true or acceptable ones.*

Now it is tempting to claim immediately that this demonstrates the worthlessness of all such forms of "normal" induction, for, even if Reichenbach or someone else "vindicated," say, the straight rule, he would have merely vindicated, for any application, an indefinitely large family of possibilities only one of which could do us any good.[13] But while the second main clause of the preceding is true, it does not follow that the first is also; and it should immediately be acknowledged that interpretation of this rather spectacular theorem requires a large measure of restraint by "anti-justificationists" such as myself, a measure perhaps

[12] See, for example, Bertrand Russell, *Human Knowledge: Its Scope and Limits* (New York: Simon and Schuster, 1948), pp. 400–12, and my "Theories, Perception, and Structural Realism," in Colodny, *The Nature and Function of Scientific Theories*, pp. 3, 7–9. Proofs will also be given later in this paper when a more general theorem concerned with hypothetico-deductive inference is discussed.

[13] The existence of such a *kind* of family is acknowledged by Reichenbach and has received extended treatment by Salmon. It is, however, a different family from the one under consideration now. For example, it is not generated by the straight rule alone but by a family of *rules*, only one of which is "straight."

already exceeded, even, by the remark just attributed to Russell. I shall not claim, for example, that it has thus been shown that there is on a given occasion a zero or only a very small probability of hitting upon a good inductive argument out of the indefinitely large number of mutually incompatible possibilities. To do so would be to assume, tacitly, some kind of principle of indifference; and apart from the general difficulties with such a principle,[14] I believe, as will appear presently, that such an assumption, in this case, would be false. The correct position seems clearly to be that any assumption about the distribution of probabilities of success among the members of the family of mutually incompatible alternatives is a *contingent* assumption. It may be a true assumption, for example, that those applications of the straight rule that human beings employ have a much greater probability[15] of success than other possible applications. As a matter of fact, *I believe that it is true* and of some importance, although I do not think that the straight rule has the essential role attributed to it by Reichenbach. The crucial point at hand now is that, without some such assumption, there *is* no satisfactory reason to suppose that any simple inductive inference will ever give us knowledge, where "satisfactory reason" means *reason from a deductive vindication.* The Reichenbachian may reply that this is not disturbing and that it would have been true even if Reichenbach's vindication had accomplished all that he hoped. But the heart of Reichenbach's attempt was to give us a unique, specified procedure that would succeed if any method would; and this program has collapsed, at least temporarily. An alternative application of the straight rule might well have succeeded where the one we selected failed. Now, I do not know how to *prove* that there is no deductively vindicable procedure that will provide a unique kind of application of simple inductive procedures — for example, a kind that would succeed if any other would. I am strongly convinced that no such vindicable procedure exists, and I am inclined to believe that its nonexistence *can* be proved, remembering earlier considerations about nothing but trivialities following from trivialities, etc. But I am not too concerned that no such proof has been discovered, nor am I especially motivated to look for one. Proofs about such matters that are apparently rigorous can often be evaded by tactics such as somewhat modifying the conclusion — for example, a slightly different kind of vindication or a

[14] See, e.g., Salmon, *The Foundations of Scientific Inference.*

[15] I am using 'probability' in the sense of relative frequency (or propensity).

somewhat different kind of "inductive" procedure might be proposed. Even such a celebrated proof as that of the impossibility of trisecting an angle using only compass and straight edge has to be "protected" by very precise and restrictive specifications on what is to count as a *legitimate use* of compass and straight edge. One of Mary Lou's (my wife's) psychotherapy patients (John C. Gustafson, a sane and intelligent person if I ever knew one) showed us a method whereby one may, by rapid and easily applied successive mutual adjustments of compass and straight edge, trisect an angle, approximating the accuracy of any "legitimate" use of compass and straight edge as closely as one pleases. Could one only do as well for vindicating induction, the problem would be solved. So, even if impressive impossibility proofs about vindication were discovered, I doubt very much that they would suffice to convince a confirmed vindicationist that his task is even as difficult as trisecting the angle or squaring the circle.[16]

Before leaving simple inductive inferences for other nondeductive procedures, consideration of a matter connected with Goodman's "new riddle of induction" may help to round out the discussion. Goodman's "pathological" predicates provide merely one among a number of ways of proving the theorem about the multiplicity of mutually incompatible inductive inferences discussed above. (More accurately, such predicates generate *one* of a number of infinitely large families of mutually incompatible inferences.) I have given such a proof,[17] using, instead of Goodman's 'grue', 'bleen', etc., perfectly general predicates free of any explicit or tacit time dependence or other incorporation of individual constants. Wesley Salmon has attempted to provide a selection rule that eliminates all but one member of the family of mutually incompatible inferences that results from such predicates.[18] He imposes a very strong condition that restricts such inferences to the use of "ostensive" (direct observation) predicates. Thus predicates such as 'red,' 'green,' 'warm' could be used in arguments by simple enumeration, but ones such as

[16] I am indebted to Keith Gunderson for the analogy between justifying induction and squaring the circle. This, in turn, brought to mind our friend's trisection of the angle. For a fascinating account of other geometrical proofs, proof evasions, new "strengthened" proofs, new proof evasions, etc., etc., see Imre Lakatos, "Proofs and Refutations" (in four parts), *British Journal for the Philosophy of Science*, 14 (1963), 1–25, 120–39, 221–45, 296–342.

[17] "Theories, Perception, and Structural Realism."

[18] "On Vindicating Induction," in H. E. Kyburg and E. Nagel, eds., *Induction: Some Current Issues* (Middletown, Conn.: Wesleyan University Press, 1963).

'grue', 'bleen', and, presumably, 'diamagnetic' and 'ultraviolet' could not. Apart from the difficulty this would seem to create for many widely employed scientific inferences, Salmon's proposal has been criticized as being arbitrary and ad hoc. I am suspicious of criticisms that depend mainly on the use of the epithet "ad hoc,"[19] but it seems undeniable to me that such a strong requirement, especially when it has such serious consequences, is as much in need of vindication as the original inductive inference forms themselves or more so. It is true that Salmon's rule would remove the outrage to our inductive intuitions that is perpetrated by Goodmanesque, "pathological" predicates. But, again, surely Hume teaches us that, if inductive intuitions sufficed, our normal inductive procedures would have stood vindicated from the onset, and all the toil and sweat of Reichenbach, Feigl, a host of others, and especially Salmon himself would have been attempts to crash through open doors. It seems to me that, on his own grounds, Salmon's rule must at least fulfill Reichenbach's requirement: it must be shown that the kind of inference it selects will, in general, succeed if the kind exemplified by any other member of the family in question will. This may be true, but I do not believe that it can be demonstrated.[20] In fact, whether or not it is true seems, certainly, to be a *contingent* matter. Surely there are possible worlds in which the color of emeralds changes suddenly at times and in which plummage pigmentation is a function of the period of rotation of the planet of habitation.

Is it, then, arrant anthropomorphism to suppose that, in the *actual* world, there is something very special about properties and features that happen to attract our attention and hold our interest or, even, those that our particular perceptual facilities enable us to observe directly, or something special about the expectations that such properties arouse in us and the kinds of inferences in which we are disposed to use these properties, something so special that our expectations are fulfilled and our inferences are successful a surprisingly large portion of the times?

[19] See my "Corroboration without Demarcation," in P. A. Schilpp, ed., *The Philosophy of Karl Popper, The Library of Living Philosophers* (LaSalle, Ill.: Open Court, 1974), and Adolf Grünbaum, "The Special Theory of Relativity as a Case Study of the Importance of the Philosophy of Science for the History of Science," in B. Baumrin, ed., *Philosophy of Science,* vol. 2: *The Delaware Seminar* (New York: John Wiley, 1963).

[20] Goodman's restriction of inductive inferences to the use of "entrenched" predicates is, of course, just as arbitrary and just as much in need of justification as Salmon's rule, but I am sure that this does not disturb Goodman.

Yes, outrageously anthropomorphic as far as logic is concerned, including, I maintain, any kind of "inductive logic" that can be vindicated deductively. Nevertheless, to peek ahead at the happy ending, I am going to maintain that, as a matter of *contingent* fact, the world does have these happy characteristics. Salmon's restriction to directly observable properties is much *too* restrictive, but otherwise, his (*contingent!*) intuitions were in the right direction: there *is* (*contingently*) something special about those properties of the world about which we concern ourselves and about the regularities in which our predictions or *projections* (more properly, *our theories*) locate them. This is why knowledge is possible. True, this can be a happy ending only for those who abandon the empiricist goal of justifying all knowledge claims by means of observation and logic. For I certainly cannot deductively vindicate this very strong contingent assertion. Whether we are in any way justified in "believing" it is a matter to be discussed later in the paper. Even now, however, it is important to remark that these contingent assumptions do not provide a "method" or a specified mode of nondeductive inference. Those seeking escape from Feyerabend's "methodological anarchy" will find scant comfort here.[21]

It is now time to remind ourselves that the predominant mode of reasoning both in everyday life and certainly in science is *hypothetico-deductive*, although the *simple* inductive modes discussed above are used at times and are of some importance. This is generally acknowledged today even by most empiricist philosophers of science although the knowledge has not yet trickled down to those who set the most popular fashions in methodology for psychology and the social sciences. The mistaken belief that what I have been calling "simple inductive procedures" provide the *only*, or at least the *predominant*, or at the very least the *fundamental*, means of confirming significant knowledge claims has been responsible for a large number of the many dead ends reached in work on induction and confirmation as well as for much of the grief suffered in general scientific methodology and other areas of philosophy. For this error fosters the view that the accretion of knowledge, both in common-sense contexts and in "scientific" ones, proceeds according to a set pattern, a pattern consisting of the collection of (homogeneous,

[21] Paul K. Feyerabend, "Against Method: Outline of an Anarchistic Theory of Knowledge," in M. Radner and S. Winokur, eds., *Minnesota Studies in the Philosophy of Science*, vol. 4 (Minneapolis: University of Minnesota Press, 1970).

observational) data or evidence which, when subjected to the appropriate simple inductive rule, automatically and mechanically leads us to the correct (and eo *ipso*, thusly confirmed) knowledge claim; for example, the evidence, consisting of the observation of ninety-six black crows (and no nonblack ones) is properly handled by induction by simple enumeration (or the "straight rule") and, thus (inductively) confirms and, moreover, leads us univocally to the one proper conclusion that all crows are black (or, if we are interested in singular conclusions, say, for predictions, to conclusions that the next crow will be black, the next ten crows will be black, etc.). We have already seen that no such univocality exists, but that is not the point of interest now. The important thing is that, in the more usual case, our data are not homogeneous nor do we proceed according to the simple inductive rules. Our data might consist of a number of different aspects of the behavior of gases, or a number of different properties of a newly prepared chemical compound, or a number of different clues connected with a certain crime. In each case, even if we wished to apply a mechanical (inductive) inference rule, none is available. What we do, in each case, is to try to dream up a theory or hypothesis which if true would *explain* the data, where "explain" means to infer from a group of propositions some of which are general in character.[22] The appropriate theory for the first example we believe to be the kinetic theory of gases, in the second case our theory would be the (tentative) assignment of a certain molecular structure to the new compound, and in the third case we would propose a singular hypothesis about who committed the crime, how it was accomplished, etc. (Let us note, even at this point, what I shall emphasize later, that, in each case, our proposed theory or hypothesis will be one among an indefinitely large number each of which will "explain" (imply) the data equally well, *as far as logic is concerned.*) In each case, we would probably want to claim that the data constitute evidence that confirms the theory or hypothesis in question *by virtue of the fact of the data being explained by*

[22] For an account of this theory of explanation see, for example, K. R. Popper, *The Logic of Scientific Discovery* (London: Hutchinson, 1959), and Carl Hempel, *Aspects of Scientific Explanation and Other Essays* (New York: Free Press, 1964). For criticisms, see, for example, Michael Scriven, "Explanations, Predictions, and Laws," in H. Feigl and G. Maxwell, eds., *Minnesota Studies in the Philosophy of Science*, vol. 3 (Minneapolis: University of Minnesota Press, 1962), pp. 170–230. For anyone who does not subscribe to this analysis of explanation, I can give an account of hypothetico-deductive (better: *hypothetico-inferential*) reasoning without using the concept of explanation.

(inferable from) the theory or hypothesis. It should be remarked that the (explanatory) inference *from* theory (or hypothesis), even when the appropriate auxiliary theories, initial conditions, etc., are conjoined, *to* data is sometimes statistical or otherwise nondeductive. Therefore, 'hypothetico-inferential' is a better, less misleading term than 'hypothetico-deductive'.

Although lip service is quite generally paid to most of the important features of this mode of reasoning, many of them are quite often forgotten or ignored. For this reason, I have to apologize again for explaining the obvious both above and in some of what is to follow.

First of all, the necessary prevalence of the hypothetico-inferential mode of reasoning accounts for the obvious "fact" that significant progress in the pursuit of knowledge is difficult and involves creativity. If "scientific method" consisted of using simple inductive inference rules, then all one would need to do is make observations to provide the data simply by looking or perhaps by performing experiments and then mechanically apply the appropriate inductive rule in order to reap, automatically, the correct (or most "rational," etc.) conclusion. In the usual case, that is, the case involving hypothetico-inferential reasoning, there *is* no rule to take us from the data (more generally, to take us from the problem at hand) to the theory or hypothesis. The proposal of the theory — often a humble singular hypothesis, even — is, as Einstein once said, a free, creative leap of the mind. And, again, although this is generally acknowledged (not by *everybody*, however) and is perhaps obvious, we should note that, from it, it follows that there is no such thing as scientific method or inductive logic if either of these is supposed to produce out of the data at hand the "correct conclusion" (or the "most rational," "best advised," etc., conclusion). There are many "Yes, but —" replies possible to this point, some of which will be considered later, but, for now, I shall be content if granted the point itself.

As far as the *justification* or *vindication* of hypothetico-inferential reasoning as a mode of confirmation or as a means of nondeductive inference is concerned, it is in pretty much the same boat as simple inductive procedures. The crucial point is that, given any amount of data or evidence whatever, there will always be an infinitely large number of mutually incompatible theories or hypotheses from which (together with appropriate conjuncts of auxiliary theories, etc.) the data may be inferred and which, therefore, as far as the *logic* of the situation is concerned, are all equally well confirmed by the data. Hypothetico-inferential confirmation, then, is no better off, vindication-wise, than simple inductive inference but, on the other hand, it does not seem to be any worse off either, since both appear to be infinitely *bad off* as far as deductive vindication is concerned. In fact some argue that hypothetico-*deductive* confirmation[23] is just a special case of induction by simple enumeration, for, they point out, to assert that a theory or hypothesis is true is to assert that all its consequences are true, which, in turn, induction by simple enumeration would confirm by noting that *this* consequence is true, *that* consequence is true, etc. Although I believe that there are crucially important differences between hypothetico-inferential reasoning and simple inductive procedures, I shall not contest this claim for their equivalence *insofar as their* (deductive) *vindication as modes of confirmation is concerned.* In this regard they are both hopeless. For example, every set of consequences of one theory (or hypothesis) is also a set of consequences of an infinite number of other, mutually incompatible theories. As far as logic is concerned, both simple induction and hypothetico-inferential reasoning confirm too much, so much that they are completely worthless unless bolstered by nonlogical, unvindicable (I claim) assumptions. Since both modes are equally bad off, it would seem that we are just as well advised to embrace hypothetico-inferential reasoning in all its fullness as to limit ourselves to simple inductive procedures. I can see no good reason, logical or otherwise, to hold that simple inductive procedures are more fundamental or more basic or that they have any kind of priority at all, except, perhaps, a psychological one for those of us raised in the empiricist tradition.[24] I shall assume then what is borne out by a careful examination of our common-sense knowledge and by even the most cursory and superficial attention to the history of science — that the most important, most fundamental, and most prevalent mode of nondeductive reasoning is hypothetico-inferential.[25]

[23] And, presumably, all other hypothetico-*inferential* confirmation, as well.

[24] There is a legitimate prima facie objection against hypothetico-inferential reasoning that involves unobservables, but, following suggestions from the work of Russell and Frank Ramsey, I have shown how this objection loses its force (in "Structural Realism and the Meaning of Theoretical Terms," in Radner and Winokur, eds., *Minnesota Studies in the Philosophy of Science*, vol. 4, and "Theories, Perception, and Structural Realism").

[25] Turnabout is fair play. If hypothetico-deductive reasoning can, in some sense, be reduced to induction by simple enumeration, the converse is also true. For example, the datum that *this is black* can be inferred from the "theory" that *all crows*

This expository digression has been necessary for reasons that will, I hope, emerge presently. One of them is that it has prepared us for a brief examination of the approach of Sir Karl Popper, which, according to him, "preserves empiricism" (but *not* induction) from the Humean onslaught.[26] And it happens to turn out that the discussion of Popper's proposals leads quite naturally to the heart of the matter with which this essay is concerned. Popper's views are well known and I have discussed them elsewhere at some length[27] so I shall be very brief here. There are, I believe, two independently sufficient reasons for the failure of Popper's attempt: (1) Theories (the interesting and important ones) are not, in general, falsifiable, nor *should* they be so *made* by "methodological rules." (2) No matter how many theories we falsify (in a given case) there will always remain an infinite number of unfalsified theories that imply the data, *only one of which can be true.*

As to (1), Popper explicitly acknowledges that because of the necessity for using auxiliary theories ("background knowledge"), etc., theories are not, in general, deductively falsifiable by "basic" (observation) statements *unless* certain "methodological rules" are adopted. I do not believe that it is a distortion or an unfair simplification to say that his methodological rules amount to the injunction to assume that the auxiliary theories in use are true, in order that the theory in question be, thus, rendered falsifiable. I have argued, as have others that the history of science shows that not only have such methodological rules been honored more in the breach than in the observance, but also that this is as it should be; to have done otherwise would have been detrimental to scientific progress.[28] Moreover, it is very misleading to say that the assumption of the truth of a (nonobservation) statement "makes" a theory deductively falsifiable by observation statements. Surely the *assumptive* component is transmitted to the (ostensibly) "falsified"

are *black* and the initial condition that *this is a crow.* (This, together with the "facts" that hypothetico-deductive reasoning is reducible — in the sense discussed earlier — to induction by simple enumeration and that any set of data is inferable from an infinite number of mutually incompatible theories, provides another proof that induction by simple enumeration leads to error infinitely more often than to truth.)

[26] *Conjectures and Refutations* (New York: Basic Books, 1962), pp. 54–55.

[27] "Corroboration without Demarcation."

[28] *Ibid.* See also, for example, I. Lakatos, "Falsification and the Methodology of Scientific Research Programs," in I. Lakatos and A. Musgrave, eds., *Criticism and the Growth of Knowledge* (New York: Cambridge University Press, 1970).

theory. It would be no more misleading to say that in view of the (observed) falsity of the appropriate basic statement we *assume* that the theory is false in the light of our assumption that the auxiliary theory is true. But be that as it may, an *empiricist* must certainly recognize that the assumption enjoined by the "methodological" rule is contingent and, thus, must either be confirmed (which would mean running afoul of Hume's charge of circularity) or be somehow "vindicated" (presumably, as before, deductively).[29] Although he does not use this terminology, Popper seems to offer an attempt at a deductive vindication when he says that his methodological rules provide (at least a part of) a *definition* of science. But we must immediately ask why we should accept such a definition especially since, as noted above, it does not seem to correspond either to the *actual* practice of science or to science as it *ought* to be practiced.

Even more seriously, and this brings us to (2), we must ask what reason there is to suppose that "science" as defined by Popper ever produces *knowledge* or, even, that it is a "good" (vindicable, etc.) method for pursuing knowledge. I do not know of any such viable reason whatever. Suppose, contrary to what is the case, that theories were, generally, falsifiable, or suppose that we "make" them falsifiable by adopting Popper's methodological rules. We are reminded by (2) that no matter how many theories we falsify (relative to a given problem or set of data) or how many we eliminate with methodological rules, there remains an infinite number of mutually incompatible theories each of which is as falsifi*able* (not falsifi*ed*, of course) by the data as any other. Getting rid of a small (or even a large) number of false theories does not reduce this bewildering multiplicity one whit. And what reason is there to sup-

[29] I do not wish to claim, of course, that we are never justified in rejecting a theory on the basis of the evidence provided by the experiment or observation in question *plus* our belief that we also have good evidence that the relevant auxiliary theories, etc., are true. Surely this *is* how we reject many of the theories that we believe to be false. See, for example, A. Grünbaum, *Falsifiability and Rationality* (Pittsburgh: University of Pittsburgh Press, forthcoming), as well as examples in his recently revised and greatly expanded *Philosophical Problems of Space and Time*, 2nd ed. (Boston: Reidel, 1973) and "Can We Ascertain the Falsity of a Scientific Hypothesis?" in M. Mandelbaum, ed., *Observation and Theory* (Baltimore: Johns Hopkins University Press, 1971) (reprinted in *Philosophical Problems of Space and Time*, 2nd ed.). Such an approach is not, of course, available to Popper, since it assumes the possibility of giving theories positive confirmation on the basis of evidence, a possibility that he emphatically rejects, and, as we shall see presently, his proposed purely negative reasons for tentatively accepting theories that have so far withstood falsification attempts do not suffice.

pose that the as-yet-unfalsified theory (or theories) that our feeble efforts have been able to devise is the one true one (or is any "closer to the truth" than any of the others)? It is fair to say, I believe, that Popper, in spite of his "anti-justificationism," tries to answer this question. And well he might. Otherwise, why go to all the trouble of proposing theories and trying to falsify them at all? In giving his answer Popper says that *we aim at truth* (*Conjectures and Refutations*, pp. 33–65). We are justified [*sic*!], he says, in accepting tentatively, even for practical purposes, a theory which has resisted our earnest attempts at falsification and which *for all we know may be true*. Well, returning temporarily to (1), *for all we know* the theories we have "falsified" *may be true* as well, since they have not really been falsified but only rejected by means of the contingent assumptions enjoined by the "methodological rules."[30]

But again, be that as it may, let us once more for the sake of argument suppose that falsification is possible and that the problem is the remaining infinite number of unfalsified theories only one of which can be true. Again, let us not make the mistake of (tacitly) applying a principle of indifference and assigning equal probabilities to each of these bewildering possibilities and, thus, concluding that we have zero probability of hitting upon the theory that is true.[31] Popper not only could point out such a mistake but would also be completely correct in saying that for all we know, we *may* hit upon the true theory (or one "close to the truth") after a few falsifications (or even after none) in a sizable portion of our attempts to expand our knowledge. Put in this way, this unexceptionable statement might seem austere enough to avoid offending even the

[30] It will do no good to require that *these* assumptions themselves be testable (falsifiable) in other contexts. This would require use of other assumptions and we would be off on an infinite regress or into a circle. Popper maintains that such regress is not vicious, but this seems wrong to me unless there is something *intrinsically* exoneratory about individual attempts at falsification. But this, indeed, *is* the point at issue, and thus a vicious circle becomes explicit.

Why then (it might be asked) does the *falsifiability-falsification* account of scientific methodology have such great intuitive plausibility? Well, if it is (contingently!) true, as I contend, that the kinds of theories and hypotheses that we happen to propose (including auxiliary theories) have a much greater probability than the others that would also explain the data, then it *will* be important and productive to eliminate the false (or *probably* false) ones from this *relatively small class*, so that the probability of the remaining one(s) will be rendered even greater thereby (by virtue of their being assigned to a different, "better" *reference* class). *If and only if something like this is the case can* I see any point whatever to falsification attempts, "crucial experiments," etc.

[31] Zero *probability* in a frequency theory does not, of course, mean *impossibility*.

most rabid anti-justificationist. But reflection soon forces upon us *the central, most crucial*, and no doubt for many, the *most awful fact of the entire epistemological enterprise: if we are to have an appreciable amount of nontrivial knowledge (or, even, of true beliefs), we* MUST *hit upon fairly often the true theory or hypothesis* (or one reasonably close to the truth) *after a relative tiny number of trials.* As was the case with our earlier, exactly parallel considerations about the bewilderingly diverse results yielded by the "straight rule," etc., *time is too short for us to sift through* (using falsifications to eliminate the chaff) *more than an insignificantly small number of the infinite number of possibilities always sanctioned by the data we happen to have.* It *should* be noted that this central assertion *does* presuppose that our attempts to increase our store of nontrivial, general knowledge and our attempts to predict the future, etc., proceed *hypothetico-inferentially*. (Recall also that hypothetico-inferential procedure includes simple inductive methods as special — rather degenerate — cases.) It is, of course (logically), possible that one might arrive at general knowledge (or, at least, at true beliefs) and make successful predictions using "pure" guesses or divine revelation quite apart from any kind of hypothetico-inferential strictures. And, emphatically, I must admit and insist that, although I subscribe to it, I do not believe that hypothetico-inferential procedure can be *vindicated* (in the strong sense required by empiricism) relative to the goals of obtaining general knowledge (or true belief) and making successful predictions. That is, I do not believe that it can be *shown* to be the "best" procedure or to be a procedure that will succeed if *any* procedure will.

This statement of this "awful" fact (surely it is, rather, a wonderful one) is equivalent to saying that, if significant knowledge is possible, the *relative frequency* of successes among human attempts at knowledge accretion must be very high indeed, given the possibilities of failure. This is, in turn, to say that the *probability* of success must be surprisingly high.

Let me interrupt this train of thought briefly to emphasize that this illustrates the utility and, indeed, the primacy of a relative frequency interpretation of probability for dealing with at least some of the important areas concerned with the truth of statements, including universal statements, and, perhaps, it may open the door to a general, comprehensive relative frequency treatment of probability of statements — including their *prior* (though *not* a priori) probabilities. It seems inescapable that the "payoff," from not only the practical but also the purely epistemic viewpoint, of the probability of a statement must eventually be in terms of frequencies. Suppose, for example, that we *begin* with interpreting "probability" as "degree of belief" (or "degree of rational belief"). It would still seem that the only reasonable warrant for having, say, a degree of belief of one-half in a certain statement would be the belief that statements of *this kind* turn out to be true about half of the time. The fact that we only tacitly fix the *reference class* by use of the phrase "this kind" does not seem to me a crucial objection to the claim that a frequency interpretation is really what is operative here. How else could the "(fair) betting odds" explication of "rational degree of belief" make any sense? Surely we would agree that the fair betting odds on the truth of proposition *p* is 50–50 if and only if we believed that propositions like *p* in certain relevant, important respects turn out to be true about half of the time.

Purely "logical" interpretations of probability, such as one based on some kind of inverse of the logical content of statements, may have their uses; but they have nothing to do with the likelihood of statements being true — except in certain degenerate and usually unimportant cases, cases which, moreover, by virtue of the calculus of probability, hold in frequency (as well as all other) interpretations. Popper is quite correct, for example, when he says that science aims at low probabilities (*other things being equal*, he should always add) when he is talking about probability in the sense of the *inverse of the content of statements*. But it would be grotesque to conclude from this that, as far as science is concerned, the lower the (*objective*) *likelihood of a statement's being true* the better, i.e., to conclude that science aims at low probabilities in the sense of *relative frequency* of truth in appropriately selected reference classes. It may also be true, as Popper claims, that the probability of a (contingent) universal statement is always zero *in the sense of logical probability*, although I am dubious about this. But if so, it *must* be such that this has nothing to do with the likelihood (in the sense of relative frequency, as above) of a given universal statement's being true. Clearly it is a contingent matter about which logic is silent concerning how often statements that resemble in certain important, relevant respects "All swans are white" turn out to be true. Popper explicitly recognizes that science aims at truth (or closeness to truth), as does all epistemic activity. So if he seriously contemplates the possibility of the existence of knowledge, I do not see how he can avoid assent to the considerations about probability of statements, in the sense of relative frequency of truth (or relative frequency of successful selection of true hypotheses), that have just been outlined. (It is true that Popper has argued against the viability of using truth-frequency interpretations for the probability of statements. However, I do not believe that his arguments touch the kind of interpretation that I, following Salmon, advocate herein. [See Salmon, *The Foundations of Scientific Inference*, and my "Corroboration without Demarcation."]).

Finally, I should note that my claim of the primacy of relative frequency does not entail a simple *limiting* frequency interpretation. Indeed, I believe that, in order for a frequency interpretation to be applicable with any utility, quite strong (contingent) conditions of distributions of frequencies *in finite segments or subclasses* of the reference class must hold. I have proposed (in "Corroboration without Demarcation") a (frequency) "definition" of 'probability' that imposes one such set of conditions, although I am not positive that a better one may not be given.

The relatively high probability of success of our attempts to attain knowledge (or true beliefs) cannot, of course, be guaranteed, nor can the assumption of its holding be "justified," in the sense of *deductive vindication*. Or, at any rate, I do not believe that it can be. About this I am completely with the anti-justificationists: neither the general assumption that we have (and shall continue to have) knowledge (or even true belief) nor specific knowledge claims about the world can be justified in the strong sense that has preoccupied empiricists in general and vindicationists in particular. Any justification must be in terms of other contingent assumptions, and this of course is of no comfort to the empiricist. What many philosophers and most empiricists have held to be the central concern of epistemology must, therefore, be discarded. *We must discard the question* How are our knowledge claims to be justified? *They cannot be justified at all in the sense required by those who usually ask the question.*

On the other hand, the necessary subscription of the anti-justificationist falsificationists to hypothetico-inferential procedure leaves most of them no better off than the justificationists, for they are forced to face the dilemma: Either we do not have an appreciable amount of significant general knowledge (or true beliefs) and henceforth shall not attain an appreciable degree of success in our predictions or the probability (relative frequency) of success in these matters is relatively very high. It is highly irresponsible, I believe, to take a totally agnostic attitude toward this dilemma. For if we want seriously to contemplate the possibility of knowledge (or true belief) and successful predictions, *we must be willing to examine the consequences of this possibility for other central concerns of philosophy, science, and everyday affairs. One of many of these crucial consequences, I contend, is the untenability of empiricism.*

It is time now to summarize the central contentions of this paper. The fundamental question of theory of knowledge is *not* How are our knowledge claims to be justified (or how should they be)? It is rather, *What are the conditions that if fulfilled can or will result in the existence of significant knowledge* (or true belief)? I have put the matter vaguely and perhaps ungrammatically because I want to leave it open, for the present, whether the conditions are necessary ones, sufficient ones, or both. The answer to this central question that has emerged, assuming, as always, that the search for knowledge proceeds hypothetico-inferentially, is this: The theories and hypotheses that humans propose (i.e., the guesses or knowledge claims that they make) to meet problem situations and account for data or "evidence" at hand must be true (or reasonably "close to the truth") a relatively great portion of the time, given the number of false though *otherwise adequate* possibilities that always exist. (To say that they are "otherwise adequate" means not only that they are *logically* adequate but that they fill *all* of the reasonable requirements of hypothetico-inferential procedure.) Since, in any given case, the proposed theory (or hypothesis) and all of the other adequate theories are *equally well supported by the evidence* as far as hypothetico-inferential requirements (and, a fortiori, as far as simple inductive requirements) are concerned, it seems quite natural — indeed we seem to be inexorably led — to *equate the frequency of success* (*i.e., of truth or reasonable closeness to truth*) *of the kind of theory we have proposed in a given situation to the prior* (NOT *a priori*) *probability of the theory* (*or hypothesis*). Since this is probability in the sense of *relative frequency* (or if one chooses *propensity*), problems about the nature of the "best" reference class, of course, arise. These are difficult but not, I believe, insuperable. I shall discuss them later but, for the time being, shall just assume that they have been solved in some reasonably satisfactory manner. Notice that in order to attain knowledge (or, at any rate, true beliefs, or true proposed hypotheses) it is not necessary for the knowing subject to know the values of these probabilities. Strictly speaking, it is not logically necessary, even, for him to estimate or guess at their values. All that is necessary is that a significant portion of the theories and hypotheses that he proposes, believes, uses for predictions, etc., actually *have* fairly high prior probabilities. The actual values of the probabilities, *being relative* frequencies (or propensities) are, of course, "objective"; they are what they are completely independently of whether anyone ever knows what they are, ever "measures" them, or ever even makes guesses about or otherwise contemplates them.

But estimation of prior probabilities by the knowing subject, although not logically necessary for true belief, success in prediction, etc., may, nevertheless, turn out to be desirable. And although such estimation is virtually never done *explicitly* in everyday and in scientific contexts, it surely occurs *tacitly* fairly often. When we suddenly feel quite sure that the theory or hypothesis that has just occurred to us must be the correct explanation of some hitherto puzzling set of circumstances, whether we be detectives faced with strange, apparently unrelated clues or scientists trying to account for whatever data there are at hand, we are tacitly estimating a fairly high value for the prior probability of our proposed hypothesis or theory.[32] We may (or may not) signal this by telling ourselves or others that *this* hypothesis "just sounds right" or that we "just feel it in our bones that it must be the right one." Or, again, when we select among several competing hypotheses or theories proposed for a given set of data, we are ranking them on the basis of our tacit estimation of their prior probabilities.[33] In such a case we might say something like "*This* hypothesis seems so much more plausible than any of the others (although as far as the evidence is concerned, i.e., as far as hypothetico-inferential [and/or simple inductive] requirements are concerned, all of the hypotheses fare equally well)."[34] Or we might come even closer to

[32] We are also tacitly estimating a fairly low or, at any rate, not-too-high value for the prior probability of the data or the clues — the evidence; but more of this later.

[33] Here it is not necessary to consider the prior probability of the evidence since it (the evidence) is the same for each hypothesis.

[34] It might be pointed out that both in this case and in the previous one, our strong partiality toward the selected hypothesis might be based not only on the current evidence and our guesses about prior probabilities but on past experience, coherence with other, accepted theories, old but relevant evidence, etc. — let us call all such factors "background knowledge." This is true, but it merely pushes the need

being explicit by saying something like "Hypotheses *like this one* (adding perhaps 'in cases like *this*') surely turn out to be true (or quite close to the truth) *more often* than hypotheses like the others that are before us." Here the reference to frequency *and thus to probability* breaks out into the open as does one of the central contentions of this paper: *Prior probabilities are* (or, at any rate, *ought* to be) *objectively existing relative frequencies* of the kind discussed both above and below. Our efforts to estimate them may often fall way off the mark, but there is no other viable way of computing or estimating the extent to which our knowledge claims are confirmed by the evidence at hand. The fact that the *reference class* signaled by the phrase "like this one" is not explicitly delineated is not a conclusive objection; it may be, indeed I claim that it is, true that we have the ability to group theories and hypotheses on the basis of intuited similarities and differences even when we are unable to state explicitly what the similarities and differences are.

The claims just made are, of course, contingent, as are virtually all claims that I have been making about our estimating prior probabilities. I am emphatically *not* turning my claims into trivialities by stipulatively (and "persuasively") defining "making an estimate of prior probabilities" as *meaning* the same as "proposing and strongly believing certain hypotheses or selecting certain hypotheses as opposed to others proposed." This would be a false report of an *actual* meaning and a bad proposal for a new one. Although it *is necessarily* true that if we are to have significant knowledge, the probability of the hypotheses we select must be relatively high (assuming hypothetico-inferential procedure), this does not get us very far toward a theory of confirmation — i.e., toward a theory providing for estimates of probabilities of theories and hypotheses (prior *and* posterior). As far as this necessary condition is concerned, it is *logically possible*, for example, that God implants in our minds a good hypothesis or guides us to make a good choice in a relatively high proportion of instances, quite apart from even a tacit concern on our part with prior probabilities. My contentions above imply, of course, the (contingent) falsity

for an estimate of prior probability back a step. For, again, there will be another indefinitely large family of mutually incompatible theories or hypotheses all of which fare equally well as far as the evidence *and* the background knowledge are concerned. So no matter how much background knowledge is operative, we eventually have to make proposals or make selections from among alternatives toward which it is neutral, and, according to the line of thought we have been pursuing, this involves making tacit or explicit estimates of prior probabilities.

of such a possibility. Moreover, I contend further that *we have the innate ability*, or, perhaps better, the *innate capacity to develop*, given appropriate environmental history, *the ability to make estimates of prior probabilities that*, in a significant proportion of trials, *are not hopelessly at variance with the true values. This too is*, of course, *contingent*, especially if one holds, as I do, that abilities and capacities are always due to intrinsic, structural characteristics of the individual.

We now have (most of) the ingredients for a contingent theory of *confirmation*. The expression 'P(X,Y)' is to be read 'the probability that something is an X, given that it is a Y'[35] and the value of the probability is, by definition, numerically equal to the relative frequency of X's among Y's (i.e., the number of things that are both X and Y divided by the number of things that are Y), *provided* certain distributions of relative frequencies hold in certain appropriate subclasses of Y.[36] Now consider the following form of Bayes's theorem (the "division theorem" of the calculus of probability):

$$P(B,C \cap A) = \frac{P(B,A)\ P(C,A \cap B)}{P(C,A)}$$

where '$\cap$', the "cap," indicates class intersection ($A \cap C$ is the class of all things that are [in] both A and C). Now, following Salmon (*The Foundations of Scientific Inference*) let A be the class of all theories (or hypotheses) like *H* (the theory [or hypothesis] in question) in (I add)

[35] I used to use the Reichenbach-Salmon notation with the arguments reversed, i.e., with 'P(Y,X)' to give the reading indicated, but I have yielded to numerous protests that this is confusing.

[36] More details are given in "Corroboration without Demarcation," and space does not permit their repetition here. Briefly it is required that the relative frequency in most appropriately selected (e.g., randomly selected) subclasses remains *fairly* uniform. (The vagueness due to the italicized words is intentional and, I argue, felicitous.) It *should* be noted here, however, that whether the distribution restrictions hold is, in every case, contingent. I argue that any satisfactory and workable theory of probability or of *statistical inference* requires *some*, fairly strong, contingent assumptions about distributions. It may be of some interest that, if the reference class is treated as a sequence (my definition does *not require* this) and if the restrictions I impose hold for all the sequence in question, then the probability as I define it will coincide with the *limit* of the relative frequency if the sequence is "long enough" (strictly speaking, infinite). However, all that is *necessary* in order for this concept of probability to have considerable applicability — especially for "practical" purposes — is for the distribution restrictions to hold for just those subclasses with which we happen to be concerned on a given occasion; as we have noted earlier, in order to have a viable interpretation of probability, *it is neither necessary nor sufficient that probability and the* LIMIT *of the relative frequency coincide* (although, admittedly, it is nice when they happen to do so).

certain relevant, important respects (whether these respects be intuited, stated explicitly, or a mixture of both). Let B be the class of true theories (and hypotheses).[37] My treatment of C is quite different from Salmon's (*ibid.*). Let 'O' designate all the existing and immediately projected evidence (the conjunction of all relevant data statements) that, on a given occasion, is being considered for our theory or hypothesis of interest, H.[38] Let us refer to such an occasion as a "test situation (for H)" and designate it by 'E'. (We assume that O meets all of the hypothetico-inferential requirements for H.) Now if the evidence really *is* at hand, that is, if O actually obtains, we say that H has *passed* the test (situation) E. This is not trivial, for O may consist in part or in whole of projected evidence, i.e., predictions and, thus, may turn out not to obtain (the predictions may be false), in which case H fails E. We now let C be the class of theories (or hypotheses) that pass (or *would* pass) tests of the same degree of severity as E.[39] The denominator $P(C,A)$ will be, thus, the relative frequency of hypotheses that pass tests of the degree of severity of E among hypotheses like H (i.e., the fraction of hypotheses like H that pass tests as severe as E). We now define the severity of E as the reciprocal of the probability that H will pass E. Since H's passing E is by definition O's turning out to be true, it follows that the probability of H passing E is the probability of O's being true given that the test conditions of E obtain. If we let F be the class of occasions when events *like* those that O states will occur *do* occur and G the class of occasions when test results *like* those of E obtain, then this probability, i.e., the probability that O will be true given that the test conditions of E obtain,

[37] More realistically B may be taken to be the class of theories (and hypotheses) whose "closeness to the truth" is above a certain reasonable degree. This introduces complications but not serious ones. The matter is discussed in my "Corroboration without Demarcation." Popper has rightly emphasized the importance of the notion of *closeness to the truth* (see his *Conjectures and Refutations*). However, I am uncertain whether his technical concept, "verisimilitude," is a viable explication of it. For the present, I am content with using homely examples such as "Thanksgiving (holiday) (in the U.S.A.) is always on Wednesday" (though false) is closer to the truth than "Thanksgiving always comes on Monday," or "New Orleans has a tropical climate" (though false) is closer to the truth than "New Orleans has a frigid climate."

[38] Immediately projected evidence consists of positive outcomes of impending experiments or other observations — outcomes that are (hypothetico-inferentially) *predicted* by H.

[39] This amounts to letting C be the class whose members pass tests at least as severe as E, for if a theory would pass a test more severe than E, it would, in general, also pass E.

which is surely the *prime candidate* for being the *prior probability of the evidence* O, will be the probability that something is a member of G given it is a member of *F*, i.e., P(G,*F*), and thus equal to the relative frequency of G's among *F*'s. Now, obviously, given the definition of 'severity of *E*', the probability that a theory or hypothesis like *H* will pass a test as severe as *E*, i.e., P(C,A), must be equal to P(G,*F*), which is the *prior probability of the evidence*.

The left-hand term of Bayes's theorem now is the relative frequency of true hypotheses among hypotheses like *H* that have passed tests as severe as *E*. Surely this is an appropriate value for the posterior (relative to *E*) probability of *H*. Let us now consider first those cases in which either *H* (together, perhaps, with unproblematic initial conditions) entails O or in which we take it to be unproblematic that *H* at least *implies* O (because, for example, we take the operative auxiliary theories to be unproblematic). In these cases, the second factor of the numerator of the right-hand side, $P(C,A \cap B)$, will be equal to one. For it is the relative frequency with which hypotheses like *H* that are true pass tests as severe as *E* and, according to our definitions, a *true* hypothesis like *H* will always pass a test like *E*. For such cases, then, the expression becomes

$$P(B,A \cap C) = \frac{P(B,A)}{P(C,A)}$$

The numerator, (P(*B*,A), is the relative frequency with which hypotheses like *H* are true (irrespective of all other factors such as evidence situation) and, therefore, deserves to be designated the *prior probability of H*. Thus, for these cases, we have arrived at the familiar result that the *posterior* probability of the theory or hypothesis of interest, *H*, is equal to the *prior* probability of *H* divided by the prior probability of the evidence.[40] The distinguishing and not-so-familiar feature of this result is that we use a relative frequency interpretation of *probability* in

[40] Cases where the second factor of the numerator is not unity or near thereto are, of course, more complicated but, often, not hopelessly so. For example, when *H* is a statistical hypothesis (theory) about the long-run relative frequency (probability) of some attribute of interest in some reference class and O is a statement of (an interval) of the relative frequency in a subclass, the value of $P(C,A \cap B)$ may be calculated, using the binomial theorem. (The distribution requirements that my definition of "probability" impose are *almost* equivalent to requiring randomness, so that, for all practical purposes in virtually all cases, the binomial theorem will apply.) Still more complicated cases require more guesswork, but, I shall argue, this is not a fatal objection.

our application of Bayes's theorem instead of a directly propositional interpretation such as some so-called "logical" interpretation or *degree-of-belief* (or degree-of-*rational*-belief) interpretation.

Partisans of directly propositional interpretations[41] may justly claim that, for *practical* purposes, there is little difference between the two kinds of interpretations — *justly*, that is, if they grant my earlier contention that both the theoretical *and* practical significance of *any* probability ("degree-of-belief," "logical," or whatever) must be grounded in *likelihood*[42] *of truth* (or a likelihood of closeness to truth) which, in turn, must be in terms of (*objectively existing* though perhaps never precisely known) *relative frequencies* (of the truth of propositions of such and such kinds in such and such circumstances, etc.). To put this qualification in other words: Even if we use a degree-of-belief approach, the following crucial point must be recognized — *only if the positive correlations of our degrees of belief with such (objectively existing) truth frequencies is not hopelessly low can our attempts to accumulate significant knowledge (or true beliefs), predict the future, etc., meet with appreciable success.* (I have claimed, moreover, that it is a psychological fact that we often think in terms of such relative frequencies, sometimes implicitly, sometimes explicitly.) Thus, for epistemological purposes, for the purposes of the foundations of a theory of nondemonstrative inference, the foundations of a theory of scientific and common-sense knowledge,[43] in general, and the foundations of such disciplines as statistics, in particular, the relative frequency approach must be taken as basic.

According to this theory of confirmation, then, the procedure followed when we wish to estimate the probability (or the "degree of confirma-

[41] A *directly propositional* interpretation is one in which the argument variables, say, 'X' and 'Y' of the probability functor 'P(X,Y)', take (directly) as their values individual propositions (theories, hypotheses, etc.); for example, an expression such as 'P(q,r)' would be read 'the probability [in this case *not* relative frequency, of course] that q is true given that r is true'. In a *relative frequency* (or in a propensity) interpretation, the values of these variables are, of course, always *classes*. Moreover, I contend, there is a very wide latitude in the kind of members such classes may have; in the main cases we have been considering, the members of the classes are propositions (or statements or sentences for those who so prefer); this yields what *could* be called an "indirectly propositional interpretation."

[42] Common-sense *likelihood*, not just *likelihood* in the statistician's sense.

[43] I have used the qualifying phrase "a theory of" advisedly, for I certainly do not want to contend that *knowledge* (or nondemonstrative inference or science) has a foundation in the sense of a *firm*, unproblematic base. Even the observational basis of knowledge may be, and of course has been, challenged, although it is not a purpose of this essay to do so.

tion")[44] of a theory (or hypothesis) of interest, *H*, relative to the evidence at hand, O, is something like the following. (For simplicity, let us assume that $P(C,A \cap B)$ is close enough to unity to be neglected.) We make, explicitly or implicitly, estimates of the values of $P(B,A)$ and $P(C,A)$ and divide the former by the latter to yield the value of $P(B,A \cap C)$. That is to say, we divide our estimate of the prior probability of the theory (or hypothesis) by our estimate of the prior probability of the evidence to yield the posterior probability ("degree of confirmation," etc.) of the theory (or hypothesis).

Regarding this procedure, I wish to defend a *claim* and a *recommendation*. I claim that it is a pretty accurate "reconstruction" of the reasoning that we often do perform — sometimes explicitly, more often, perhaps, implicitly[45] — both in science and in everyday life. Much more importantly, I *recommend* that this theory be adopted explicitly in those cases where a theory of confirmation is needed; that is, I say that our (nondeductive) reasoning *ought* to conform to this model.

To defend the *claim*, I would cite some of the familiar arguments of the "personalists" and other so-called "Bayesians." Surely, for example, it is a psychological fact that we have a high degree of confidence in a hypothesis or theory (a high estimate of the frequency with which such theories are true) when the theory accounts for "facts" (evidence) that otherwise would be very *surprising or mysterious*, i.e., when we estimate the prior probability of the evidence to be quite low (*provided* that our estimate of the prior probability of the theory is *high* enough to prevent us from discounting it in spite of the evidence). And I have already given arguments to the effect that we often *do* account for our degrees of confidence in terms of estimated (truth) frequencies and that, if they are to have practical or theoretical utility, we are obliged to do so.

Part of the defense of the *recommendation* has been given already. I have argued for the inadequacy, indeed the total impotence, of purely "logical" approaches[46] to confirmation theory (or [tentative] hypothesis

[44] In a previous exposition, I called this the "confirmation index" ("Corroboration without Demarcation").

[45] It would be, perhaps, too simple-minded or even wrong to equate *implicit* reasoning with "unconscious" *reasoning*. Here, let me be rather noncommittal and say that *implicit reasoning* is reasoning that appropriate Socratic questions and challenges can often make explicit.

[46] Previously, I lumped all such approaches together under the label "strict confirmationism." Popper would, of course, object vehemently to being called a *confirmationist* of any kind, but I merely used the term to include approaches that try

acceptance) such as *vindicationism* and *falsificationism*. I take it to be unproblematic that common sense and scientific inquiry proceed hypothetico-inferentially (remembering that simple inductive procedures are special cases of hypothetico-inferential ones) and, thus, that it is now established that, if we are to have an appreciable amount of significant knowledge (or true belief), the relative frequency of our proposal and selection of true (or reasonably close to true) theories and hypotheses must be relatively high. Now I have made the *contingent* claim that, if this appreciable amount of significant knowledge (or true belief) does exist, it is explained by our constitutional abilities to propose, not *too* infrequently, pretty good theories and hypotheses, to make, fairly often, estimates of prior probabilities that are not hopelessly off the mark, etc. (This explanation is opposed to other contingent possibilities such as God's popping a good hypothesis or a good estimate into our heads not too infrequently.) From this claim arose the (*contingent*) theory of confirmation just proposed.

The contingent explanation just cited *is*, of course, problematic. In "Corroboration without Demarcation," I purported to give a *kind* of (very weak) "vindication" of it, but I wish now to withdraw any such pretension, especially in view of the high hopes that have become associated with "vindication" attempts. By way of defense now, I can only say that this claim along with the resulting theory of confirmation seems as adequate to its task as is possible and seems to avoid the fatal objections to all other such explanations and theories that have been proposed. Being contingent, it must, of course, admit of alternatives, and, when any of them is proposed, it should be carefully considered with open-minded interest. I certainly do not want to be inconsistent enough to contend that my theory or any alternative can ever have a conclusive justification or vindication. Needless to say, if someone offers an alternative and argues that it should be accepted because its implications are more congenial with his, or with the prevailing philosophical *tastes*, I shall be unmoved *unless* his theory also accounts for and provides for the other crucial aspects of the epistemological predicament as well as mine does. I can only end *this* defense on still another autobiographical note: So far, owing no doubt to lack of ability or lack of motivation or both, I have not been able to actually construct any *adequate*, *plausible* alterna-

to make do with observation statements plus logic (including the broad but still "tough" sense of "logic" explained earlier) alone.

tive. Presently, I shall try to answer some specific objections to the theory.

Now it must be admitted, even insisted, that the necessary reliance on personal estimates of prior probabilities introduces an irreducibly *subjective* component into every value of the probability or degree of confirmation that we are ever able to calculate for a theory or hypothesis. (This does not mean that the actual value of the probability [which, recall, is an "objectively existing" relative frequency] does not exist, or is itself subjective, or is something about which it is pointless to concern ourselves — *any more than* the fact that in order to forecast tomorrow's weather we must make subjective estimates, guesses, hunches, etc., means that tomorrow's weather [when or before it happens] does not exist or is subjective, or is something about which it is pointless for meteorologists to concern themselves.)[47] I have no desire to make a virtue out of this subjective component. But *there it is,* and unavoidably so. As noted earlier, we may be able to postpone the "bare" estimation of prior probabilities one or two steps backward by relying on old but relevant evidence, or rather, "background knowledge." But there will always be (ad nauseam) the troublesome multiplicity of hypotheses that fare equally well with any amount of old evidence and background knowledge. Eventually, we must make an (evidentially) unaided estimate or, if you choose, a pure raw guess about the relevant prior probabilities.

As noted earlier, part of my proposed interpretation of Bayes's theorem is due to the pioneering work of Wesley Salmon. He has proposed (in *The Foundations of Scientific Inference*) that the needed values for the prior probabilities be obtained (at least in part) by using induction by simple enumeration. Just how this is to be accomplished is not completely clear to me, but it does not matter: we *must* note that any application of induction by simple enumeration (like all other kinds of hypothetico-inferential procedures) is totally useless without some (at least tacit) use of prior probabilities. (E.g., for any theory or hypothesis supported by such application of induction by simple enumeration there will be our multiplicity of other theories equally well supported by equally "good" [logically speaking] applications of induction by simple enumeration.) So *simple inductive procedures can in no way provide means for avoiding "bare" estimates of prior probabilities.*

[47] See Salmon, *The Foundations of Scientific Inference,* and my "Corroboration without Demarcation."

Involved in this use of Bayes's theorem there is another irreducibly personal subjective factor. This concerns the selection of the reference classes needed for specification of the relevant relative frequencies (probabilities),[48] especially *A*, the class of hypotheses (or theories) *like H*, and *F*, the class of occasions when events *like* those that *O* states will occur *do* occur. To say that the members of *A* are *like H* in certain *important, relevant* respects does not, of course, go very far. We might like to be able to know and to state explicitly, in specific cases, just what these respects are taken to be. The difficulties are partly, though by no means altogether, ameliorated by one of the properties of (nicely behaved, as ours are required to be) relative frequencies: as long as our location of members in their respective classes is consistent, then, in the long run, our inferences about (as yet unobserved) relative frequencies (and, thus, probabilities) will, for the most part, be fairly accurate.[49] (Fortunately, if our distribution requirements hold, there is no danger of *the long run* becoming infinite. It is contingently possible that they do not hold, of course, in which cases "probability" would not be defined.) It must be admitted, however, that, in practice, *some* reference classes serve our ends considerably better than *others* and, on the other hand, that we would often find it difficult or impossible to say explicitly just what are the characteristics of our theory of interest, *H*, that, we believe, delineate a class of theories, a fairly high portion of which are true or close to the truth. We may be able to do this at times, but certainly, on many occasions, we must rely (1) on our personal conviction that there *is* some such set of *in-principle manageable*[50] character-

[48] The importance of this problem was also noted by Salmon, *The Foundations of Scientific Inference.*

[49] For example, no matter what reference class an insurance company uses, if its actuarial data are representative, it can adjust premiums and policy values to yield reliably any predetermined margin of profit. In practice, of course, companies select reference classes such that the margin of profit is high even though premiums are *relatively* low. In this way, they sell many more policies and also avoid the outrages against a "free society" that would result if they fostered conditions whereby the fortunate (the "good risks") bear part of the burden of the unfortunate (the "uninsurable"). (Actually this is more a description of principle than of practice. Insurance companies have almost succeeded in consistently collecting huge premiums with very little risk of paying out benefits, especially where insurance is virtually mandatory as with homestead and automobile insurance. The very real danger of policy cancellation or of even much *higher* premiums often prevents policyholders from reporting quite significant, legitimate, and innocent losses.)

[50] Without this or similar qualification, this requirement would always be (trivially) satisfied. For, owing to reasons similar to those given in connection with other trouble-

istics and (2) on our constitutional ability to classify theories according to such characteristics even though we do not have explicit knowledge of what they are.

Of those who may object to "all of this subjectivity" I can only ask, "What is the alternative?" It seems undeniable to me that we *do* proceed in something very much like the manner I have described in actual cases of significant knowledge accretion (assuming that such cases occur). There *are* logically possible alternatives, but the ones that have occurred to *me*, such as divine intervention in all or in most *individual* successful trials, do not seem very plausible, whether they be claims about what actually occurs or recommendations on what confirmation procedures, given the difficulties we have considered, we *ought* to employ. Again, however, I am open to suggestion.

The necessity for these two "subjective" activities, the estimation of prior probabilities and the grouping of theories and hypotheses into fairly appropriate reference classes, seems to me to constitute the considerable kernel of truth in the "new subjectivism" attributed to figures such as Paul Feyerabend, Thomas Kuhn, and Michael Polanyi. (I hope that I am not too far off the mark if, on the basis of his essay in this volume, I place Walter Weimer in the same camp.) But, although, as emphasized earlier, arguments from wishful thinking are of no avail, I do not believe that the consequences of this required amount of subjective activity are nearly as disastrous as empiricists fear nor, even, quite as extreme as, say, Feyerabend and Polanyi sometimes represent them.

Certainly skepticism is not a consequence of my views except for those whose standard of certification or "definition" of knowledge is unrealistically and misguidedly rigorous or who will not be satisfied with quite adequate though, of course, not completely conclusive support for holding that given beliefs are true or close to true. I certainly believe very strongly that we attain true beliefs fairly often (though not as often as we commonsensically believe that we do). Of course, I do not *know* this *for certain* or even believe it beyond all *shadow* of doubt. I also believe that we have a certain amount of significant *knowledge* (again, probably not as much as we think we do), and I am quite willing, even, to say that I *know* that we do, though I do not know it *for certain*. I am willing, even, to propose a definition of 'knowledge' with the under-

some multiplicities, there will always be many classes of theories with high truth frequencies each delineated by *some* subset of characteristics of *H*.

standing that it is only approximate, probably needs further qualification, will not cover all cases in all contexts, etc.[51] It is as follows:

X knows that *p* if and only if:

(1) *p* is true or fairly close to the truth

(2) X believes *p* (or, perhaps, X estimates that the probability of *p* is quite high, or X sincerely proposes *p* as an hypothesis that ought to be accepted tentatively, etc.)

(3) *p*, when conjoined with appropriate theories, auxiliary theories, etc., and/or "initial conditions," yields as a hypothetico-inferential consequence statements of the evidence that X has (or X knows that *p* "directly" [by observation, etc.])

(4) The (actual) prior probability of *p* and that of the evidence *as well as* X's estimates of these prior probabilities are such that the actual value and X's resulting calculated value of the posterior probability of *p* are both quite high (although, in general, they will not coincide).[52]

If someone were to say that this is just the old definition in new clothes, I would not object but would have to contend that *clothes make the definition.* I have no objection to saying that statements of conditions (3) and (4) amount to stating that X has *good* or *adequate* evidence for *p* or that X is justified in believing *p* or in proposing *p* for tentative acceptance. But it is imperative to note that these standards or "criteria" for *good* or *adequate evidence*, for *justified* belief or *justified* proposal, and, thus, for *knowledge*, are a far cry from those that most philosophers — certainly most empiricists — want to require. It may be of special concern that the subjective component is duly incorporated into the definition.

By now it is, no doubt, anticlimactic and redundant to criticize empiricism on the basis of the (contingent) confirmation theory proposed here. It was, indeed, at least partly because of the fatal *logical* defects of empiricism (discussed earlier here, although they were adequately and elegantly demonstrated by Hume) that the theory was proposed. Perhaps

[51] The wisdom of Bertrand Russell concerning the "nature" of knowledge is worth quoting over and over, but rather than repeat it here, I have reserved it for the final footnote of this paper (the final footnote of the appendix).

[52] I should have added that X's selection of the reference classes or X's (implicit) classification of *p* and the evidence is adequately appropriate and, no doubt, still more besides, but the definition is too long already.

it is not amiss, however, to try to gather up a few loose ends on this score. We are not foolish enough to deny that we learn at least a great portion of what we know by experience, nor do we wish to minimize the crucial epistemological role of (observational) evidence and, even, of *justification* in our attenuated sense. But we do have to maintain that testing by confrontation with evidence is crucially important *because* of certain contingent but *untestable* propositions that must be true if knowledge exists. We have seen that unless the relative frequency of success after a fairly small number of knowledge trials is considerably greater than infinitesimal we cannot obtain an appreciable amount of significant knowledge. Unless this condition holds, no amount of testing, no amount of evidence, will, in general, be of any avail. On the other hand, if the condition does hold, i.e., *if we are going to have to consider only a relatively small number of hypotheses,* then it is valuable, perhaps imperative, to have evidence that may eliminate some that are false *if* (it is often a big "if") we have relatively unproblematic background knowledge (background theories with high prior and/or posterior probability) and knowledge of initial conditions, etc. (the kernel of truth in falsificationism), and it is perhaps equally important to have evidence which, when properly evaluated for its prior probability and when combined with the prior probabilities of the remaining theories or hypotheses of interest will give us an estimate of the posterior probabilities of these hypotheses or theories of interest. It thus turns out that we cannot learn from experience unless empiricism is false. *"Empirical" testing is worthless unless empiricism is false and unless, moreover, certain unempirical though contingent propositions are true. If* the conditions in question *do* hold (including) our ability to estimate prior probabilities and to group propositions into useful reference classes, then it follows obviously and necessarily that evidence can produce the desirable results just mentioned.

It is sometimes claimed that Bayes's theorem has very limited applicability or that in some contexts, including attempted calculation of the probability of theories, its use is illegitimate. However, with any relative frequency interpretation, all the theorems of the calculus of probability are (trivial) truths of arithmetic, or rather, of set theory; there can be no question of legitimacy or of the validity of the results when relative frequencies are substituted into Bayes's theorem, and, thus, no objection to the use to which we have put Bayes's theorem.

There is, however, a legitimate and prima facie troublesome question regarding the efficacy of evidence. It might be asked: Since the calculation of the posterior probability of *H* is made solely from prior probabilities (of the evidence and of *H*), what reason is there to *prefer* the value of the posterior probability over the prior probability of *H*, i.e., why prefer the reference class $A \cap C$ to the reference class A? Both values are values of relative frequencies and, as we saw earlier, predictions (of frequencies) on the basis of one (long-run) frequency will, in the long run, turn out as well as predictions made on the basis of any other (long-run) frequency. To state the puzzle thus adequately is to solve it. We are interested not only in being right about (truth) frequencies but in being right *as often as possible* about the truth of individual hypotheses. If we can locate our hypotheses of interest in "good" reference classes, the classes corresponding to $A \cap C$ that have, we may hope, higher frequencies of truth than those used in estimating prior probabilities, those corresponding to A, then when we act on assumptions like the assumption that *H* is true, we will be right more often than when the frequency has the lower value corresponding to A. Our knowledge of the evidence provides us with the knowledge that *H* is, in addition to being a member of A, a class with truth frequency, $P(B,A)$, it is also a member of $A \cap C$, which has the higher truth frequency, $P(B,A \cap C)$. It is also obvious, *at least in the cases where H implies* O, why we want to collect as much evidence as possible, ceteris paribus.[53] New evidence,

[53] It is also easy to explain why new evidence *of the same kind* as we *already have* usually counts for very little (and thus why new *kinds* of evidence count for much more) *and* why the probability of a universal proposition (theory) is not necessarily zero, contra Popper *and* Carnap. (In this connection, of course, Carnap and Popper use [different kinds of] "logical" interpretations of probability. I argued earlier that such interpretations are worse than useless for the epistemic purposes with which we *and they* are concerned. The fact that such interpretations yield such a strange and undesirable result [zero probability for *any* universal statement] seems to me to provide another strong argument against their legitimacy.)

Let O be some evidence statement and let O′ be a statement that is true if and only if a repetition of the "same" experiment or the same relevant conditions of observation produce the same results as those stated to obtain by O (i.e., same experiment, same result). Now even if the prior probability of O is quite low, the prior probability of the conjunction of O and O′ will, in general, be only negligibly less than that of O alone, or, using a somewhat different approach, the probability of O′ given that O obtains (the *prior probability relative to the new evidence*) is relatively high. *It is merely a shibboleth that the (relevant) probabilities of* O *and* O′ *must be independent of and equal to each other*. Using our frequency approach, this is easy to see. We use for the prior probability of O, of course, the frequency of occasions that events like those "predicted" by O occur among occasions like those of the experiment or occasion of observation. Now obviously, although such frequency may be low, it may be not much greater than that of occasions when *two* such events like those predicted by O occur when conditions like those of the experiment hold twice. Or, to proceed somewhat differently, the frequency with which we observe events like the one stated to occur by O among the relevant observational occasions may be quite low, but once we observe such an event to occur, we should put other events including, of course, the one corresponding to O′ in a *different reference class*; specifically, the frequency with which occasions in which events like those corresponding to O′ occur among occasions in which the relevant observational conditions obtain *and which occur after a very similar event* (the one predicted by O) has occurred — this frequency will, in general, be relatively high. *That this is true is of course contingent*, and this is as it should be; it seems to be a contingency that is well supported by the evidence. (The latter contention may seem very problematic; it is discussed in fn. 56.) These much higher prior probabilities (prior *relative to a given evidential situation*) of repeated instances mean, of course, that they will raise the posterior probability of the theory by a very small, perhaps a negligible amount; and since the relevant probabilities of the instances of a universal statement, as they occur, or as they will occur, or as they would occur, are neither equal to nor independent of each other, we cannot conclude that the *probability* of the universal statement is zero; i.e., this probability *cannot* be calculated by taking the product of the probabilities of the instances and even if it could it would not be zero because of the rapid increase (approaching unity) of the probabilities of repeated instances.

These considerations, along with previous ones, also lead us to reject or, at best, severely qualify the widely held contention that values for prior probabilities are not so important after all because widely differing estimated values will produce converging results for posterior probabilities as the amount of evidence collected increases, or, as it is sometimes put, the prior probabilities "wash out" as more evidence rolls in. *Some* qualifications are generally recognized in connection with this claim, such as requiring that the estimates for the prior probability of the theory of interest be fairly low but not *too* near zero and that the estimates for that of the evidence not be *too* near to unity or to that of the theory (in which, latter case, the convergence will proceed much *faster* than is reasonable). Even this weakens considerably the original claim. For example, surely there *are* many cases in which the prior probability should be estimated as being near unity. Moreover, the two following qualifications must be added: (1) the prior probability of the theory of interest can never "wash out" relative to other competing theories, and (2) in order for there to be *any* appreciable "washing out," the new evidence that "rolls in" must always be *different in kind* from the old, otherwise its prior probability certainly should be estimated as being close to unity.

On the other hand, it should be admitted and insisted that our estimates of prior probabilities will virtually never yield anything like precise values to be used in Bayes's theorem. This is no great handicap. In many cases, especially when we have a considerable amount of varied evidence, we will be in good shape if our estimates are within a few orders of magnitude of the true values; there *is some* "washing out" effect and it *is* of some importance. In fact, in most cases we will simply say something like "I estimate the prior probability of the theory of interest to be not too low, say between 0.05 and 0.5, and that of the evidence to be not *too* high, say between 0.06 and 0.6." (Obviously, when *H* implies O, the probability of O must be at least as great as that of *H*.) Also, when the problem is to select among competing hypotheses (relative to the same evidence), all that is necessary is to order the hypotheses according to their prior probabilities; *no* estimate of numerical values are required.

if favorable to H (if H passes the test, in our sense), will tell us that H is a member of a class with an even higher truth frequency; if it is unfavorable, we will reject H as false, and the more false hypotheses we eliminate from our (contingently) limited number of viable possibilities, the more likely it is that the next one for which we collect a store of favorable evidence will be true. All of this may be painfully obvious by now, but it clears up a number of puzzles, I believe. It *explains* why and how it is that we "learn from experience" or from confronting our theories with evidence rather than having to *postulate* that this is how knowledge grows.[54] And for the simple cases (where H implies O) our theory justifies the "principle of total evidence" as well as the "conditionalization" of probabilities (conditionalized upon evidence or conditionalized again upon new(er) evidence) — a problem for the personalists and perhaps other directly propositional interpreters of probability[55] — and certainly our approach does not need any further justification of the use of the calculus of probability for dealing with evidence, etc., such as the possibility of "making Dutch book" if one doesn't.[56]

Space does not permit discussion of the more complicated cases where H does not imply O (except "probabilistically") or the less rigorous, more complicated but better advised policy of taking B to be the class

[54] A fault shared, it seems to me, by the systems of both Carnap and Popper (among others) even though they may widely diverge otherwise.

[55] See, for example, Paul Teller, "Conditionalization, Observation and Change of Preference," *Synthese*, forthcoming, and I. Hacking, "Slightly More Realistic Personal Probability," *Philosophy of Science*, 34 (1967), 311–25.

[56] I do not want to take the space, in an essay already becoming too long, to list and explain all the cases in which, I believe, this theory clears up rather notorious difficulties in confirmation theory. In almost every case the difficulty disappears owing either to the abandonment of the mistaken idea that the confirmation relationship between evidence and theory (in *that* direction) is *purely logical* or to the use of the (truth) frequency interpretation of probability (or to both). For example, I have explained in "Corroboration without Demarcation" how this theory (or almost any Bayesian confirmation theory) "solves" Goodman's "new riddle of induction" and Hempel's notorious "paradoxes of confirmation." (The first case should be obvious by now. "All emeralds are grue" is just one of the infinite number of "theories" that compete with "All emeralds are green," relative to the present evidence (lots of observed green [*or grue*, etc.] emeralds); we reject it and the other competitors on the basis of the much higher prior probability accorded by our estimates to "All emeralds are green." In the latter case, the prior probability of the next crow's being black is low *relative* to the prior probability of the next nonblack thing's being a noncrow so that the latter piece of "evidence" confirms "All crows are black" (and, of course, its logical equivalent, "All nonblack things are noncrows") much, *much* less than does the next crow's being black.)

of hypotheses whose closeness to the truth is equal to or above a certain desirable value instead of the class of true hypotheses. The results are similar to those for the simple cases, though not as neat, and they rely, at times, even more heavily, on something like the distribution requirements imposed by our definition of 'probability'.

Nature is, of course, by no means necessarily bound to be so obliging as to fulfill such requirements. But one prime result to which we have returned again and again is that, unless nature is kind in *some* vital respects, knowledge is impossible. The vital respect that I have stressed and in which I have assumed such kindness to obtain has been the existence in us (a *part* of nature) of constitutional abilities to cope with the epistemic predicament. In principle, this *could* be sufficient. No matter how complicated nature may be and even if our distribution requirements do not obtain, given any set of data (including any set of distribution data), there will always be some function (and, thus, some theory), indeed an infinite number of them, that will account for the data *and* accurately predict future occurrences in the domain in question; this holds, a fortiori, for distributions of frequencies. If our distribution requirements do not hold we could, in principle, drop them and *possibly* discover, in each case, a distribution function that *would* do the job. As a contingent matter of fact, however, it seems probable that the existence of knowledge, if it does exist, is due both to our constitutional abilities *and* to a certain (additional) amount of simplicity[57] in nature — enough, for example, so that the distribution requirements hold: I must admit that, at present, I see no satisfactory way of constructing my (contingent!) theory of confirmation without them. In fact, more generally, it must be admitted that, in addition to our contingent assumptions *about* our constitutional capacities, we also, in effect, are assuming that the comparatively limited amount of evidence that we are able to accumulate is not hopelessly unrepresentative. Some have remarked that it would be perverse or self-stultifying not to so assume. This is in all likelihood true, but it should be recognized that such an assumption, though *contingent*, is *not* an *empirical* one.

These (contingent!) assumptions, just discussed, are, in effect, the postulates of my proposed theory of confirmation. Additional (contin-

[57] This by no means implies that nature exhibits *maximum* simplicity in all or even in any cases, or that we should always or usually choose the simplest hypothesis or theory.

gent) "postulates," of a kind, enter, of course, into each individual case of confirmation (or disconfirmation); they are our estimates of the appropriate prior probabilities. These postulates may be compared and contrasted with Bertrand Russell's "postulates of scientific inference" (*Human Knowledge: Its Scope and Limits*). As important as is Russell's pioneering criticism of empiricist approaches to confirmation and as much as I am indebted to him, we part ways in our constructive approaches to confirmation. His "postulates" may very well reflect an important part of our scientific and common-sense knowledge, but they are, I believe, neither sufficient nor necessary for a viable theory of confirmation. But, if my arguments ad nauseam have been of any avail, it is clear that *some* postulation *is* absolutely necessary if confirmation is to proceed and be of any use whatever. I must, therefore, defend Russell against my dear friend Wes Salmon, who twits him for succumbing to the "temptation" to use postulates and, thus, to his (Russell's) own bête noire of choosing the "advantages of theft over honest toil."[58] For, just as we know now that an infinitude of "honest toil" could never derive Euclid's parallel postulate from his others and that accepting it as a real, operative postulate (for *Euclidean* geometry) is *not* theft but bare necessity, we must now recognize that *abandonment* of the search for confirmation without postulation, far from smacking of larceny, is the abandonment of futile wishful thinking for the honest confrontation of epistemic reality.

Does the untenability of empiricism and the unavoidable reliance on the importance of subjective elements in confirmation open the floodgates to all sorts of crackpotism, dogmatism, pseudoscience, etc., or at best, to either scientific chaos, on the one hand, or intellectual "elitism," on the other? Well, as remarked earlier, there is no way of preventing the chips from falling where they may. I do not believe, however, that things are as bad as some may fear; and, again, this is a contingent matter but, this time, one regarding which it is proper to cite evidence at hand, and, if desired, to collect more. *The evidence seems to support the view* that, in general, the best success is obtained, *so far* at least, and also as far as such things as success in prediction and having theories that fare best as more evidence is collected are concerned, by those who

[58] W. Salmon, "Russell on Scientific Inference or Will the Real Deductivist Please Stand Up?" in G. Nakhnikian, ed., *Russell's Philosophy* (London: Duckworth, 1974).

"keep an open mind," who propose theories that are amenable to confirmation and disconfirmation by evidence, who collect as much evidence as is feasible and invite others to do likewise, as opposed to those who scorn evidence altogether or who accord extremely high prior probabilities to their "pet" theories, extremely low prior probabilities to evidence that supports them and extremely high ones to evidence that disconfirms them, etc. (There are, of course, important and brilliant exceptions in the history of science. The existence there of a little, though not a great deal, of "elitism" seems to be indicated by the evidence.) This claim is also supported by our theory of confirmation, for, as we have seen, it gives a rationale for collecting as much and as varied evidence as is feasible, attempting either to increase the degree of confirmation or to eliminate false (or not close enough to true) theories by falsification or disconfirmation. It is time not only to admit but to insist that it is a *contingent* fact (if it is a fact), apparently well supported by the evidence (and other good contingent and logical considerations), that "good scientific practice" produces, in general, better results than such alternatives as dogmatism and reliance on authority or (purported) "revelation." This claim assumes of course that our estimate of the prior probability of the "theory" (that "good" scientific practice is better than dogmatism, etc.) is not *extremely* low and that our estimation of prior probabilities for the various bits of evidence is not extremely high.

We must now face an important and inevitable question: What, if anything, can be done to resolve disagreements due to interpersonal differences in (subjective) estimates of prior probabilities? Well, one can recommend that one's opponent reexamine the "background knowledge," that he take into account alternative auxiliary theories or "old" evidence of which he may not have been aware, etc. If this is of no avail, we might try collecting more evidence *of a different kind*, hoping that his estimate of its prior probability will be quite low. Finally, we might try reconciling our selection of appropriate reference classes. If none of this is of any avail, I do not see anything to do but to recognize the disagreement. We may *feel* supremely confident that our opponent is wrong on this matter or, even, that he is stupid — that he is such a poor estimator of prior probabilities and such a poor selector of reference classes that he will never attain enlightenment — but I do not know of any additional means whereby such a "feeling" can be established. I hope that it will, before too much longer, be considered a mark of epistemological maturity to have relinquished the quest for "logical" or "methodological" whips with which to chastise those so misguided as to disagree with one's own cherished conclusions.

I should like now to rest both the case against empiricism and the case for the alternative theory of confirmation that I have proposed. I hope that it is unnecessary to add a personal note warning against misinterpretation of my criticism, some of it perhaps too severe, of the work of others, especially that of Feigl, Popper, and Salmon. If any of my proposals have any value at all, it is due mainly to the foundations laid by their vital pioneering efforts in the field (plus those of Bertrand Russell). And I take it that it is a truism that, if some or even many of their basic contentions require modification or abandonment, this is what one should expect and, indeed, welcome as a virtually necessary condition for progress in both science and philosophy (I argue that the two differ, in the main, only in degree).

The paper will now be brought to a close with a discussion of some of its implications for common sense, science, and philosophy. I have written on this subject at some length elsewhere,[59] and shall, therefore, limit myself here to a brief discussion of a few main points. In an appendix to the paper, a few of the other currently popular approaches to confirmation, probability, and induction will be considered briefly.

I shall not dwell at all on the wearisome and depressing matters having to do with restrictive and truncating methodologies such as *operationism, behaviorism,*[60] and related radically empiricist positions. Although they totally lack any viable rationale and are disastrously impeding progress, especially in psychology and the other "behavioral"[61]

[59] In my "Theories, Perception, and Structural Realism," "Russell on Perception: A Study in Philosophical Method," "The Later Russell: Philosophical Revolutionary," and "Corroboration without Demarcation."

[60] Not to be confused with what is sometimes called "behavioralism" — the view seeking to exclude value judgments from science or, at any rate, to distinguish them sharply from the main body of the science. I do not wish to take sides on this issue here.

[61] The almost universal acceptance and usage of this question-begging, pseudoscientific label (along with facts such as B. F. Skinner's being, according to a recent survey by the popular magazine *Psychology Today* the most admired psychologist by those in the profession today) indicate how pervasive is the occurrence of these methodological diseases. Fortunately, with a sizable portion of scientists, the infection is relatively superficial. They give lip service to operationism or behaviorism but it does not seem to affect their behavior [!] in their scientific (or everyday) practice. They go on acting, and talking, and theorizing as if they believed that people (and animals) *do* have ("inner") feelings, thoughts, pains, pleasures, etc. Apparently they

sciences, their refutation does not require the material developed in this paper. Sufficient to destroy completely any prima facie justification that such views might seem to enjoy are the realizations that (1) hypothetico-inferential procedures are as legitimate as, and indeed, more prevalent and fundamental than, simple inductive ones, a fact now acknowledged by most empiricist philosophers, and (2) we can *refer* to unobservable things, events, and properties without *naming* them — such reference being accomplished by means of definite *and* indefinite descriptions a la Russell and Ramsey.[62] (In most parts of psychology and the social sciences, we do not even need the Russell-Ramsey devices; we can refer *directly* to the thoughts and feelings of others because we know what our own thoughts and feelings are and, according to theories best accounting for the evidence that we have, these are of the same kinds as those of others.)

Obviously, restrictive bludgeons such as "meaning criteria" based on verifiability or, even, confirmability of any kind acceptable to empiricism must fall by the wayside along with other empiricist nostrums. This also holds for Popper's "criterion of demarcation," for we have seen that, in general, interesting and important theories and hypotheses neither *are* nor *ought to be* falsifiable. In fact, as far as methodology is concerned, virtually all the important implications are negative ones. I am so strongly tempted that I shall yield and say that *the less methodology the better* and support something very similar to Feyerabend's "methodological anarchism."[63] Whenever and wherever methodological needs *do* arise, they are almost certain to be due to *contingent* facts about the area of

are misled into the lip service because they are misled to infer from the *truism* that (most of) our *evidence* in these sciences consists of behavior the false conclusion that the (main or only) *subject matter* must also be behavior rather than emotions, thoughts, pains, pleasures, etc. (Compare the erroneous inference: (most of) the *evidence* in contemporary physics consists of pointer readings of measuring instruments; therefore the (main or only) *subject matter* of contemporary physics must be pointer readings rather than electrons, neutrinos, mesons, energy, heat, etc.; or: (most of) our *evidence* for occurrences in the Vietnam war consisted of words in newsprint and sounds from electronic devices such as radios, T.V. sets, etc.; therefore what happened in Vietnam consisted of words in newsprint and sounds from electronic devices rather than bombed-out villages, dead children, habitual military and administrative lying, etc.)

[62] See, for example, my "Structural Realism and the Meaning of Theoretical Terms," and "The Later Russell: Philosophical Revolutionary."

[63] Paul Feyerabend, "Against Method," in M. Radner and S. Winokur, eds., *Minnesota Studies in the Philosophy of Science*, vol. 4 (Minneapolis: University of Minnesota Press, 1970).

investigation, and developments in response to them should proceed along with other developments in the science in question.

Does the abandonment of empiricism land us in the arms of its arch-rival, *rationalism*? No, certainly not in the classical sense. We *do* accept, at least until something better comes along, the postulates required for our theory of confirmation *and* the estimates of required prior probabilities even though they are *not* justified by experience (bcause they are not justified in *any* classical sense at all). But though they are *prior* to experience in *this* sense, they are not a priori in the classical sense at all. We have seen that they are *nonnecessary* (*contingent*) even though they are nonempirical (not justifiable by experience) and may, as far as a priori knowledge (or any kind of *certain* knowledge) is concerned, be false.

I have written at length elsewhere[64] about the implications of confirmation theory for questions about the nature of philosophy (and of science and common sense). The issue is so important, however, that a few remarks about it will be included here. My contention is that a great many, though not all, problems that are usually considered to be philosophical, differ, if at all, only in degree and not in kind from problems that are considered to be purely scientific ones. Among these are the mind-body problem, many of the "philosophical" problems about perception and reality, etc., some crucial components of the controversies between realism and instrumentalism, realism and phenomenalism, etc., important aspects of theories of meaning and other parts of "philosophy" of language, and, as we have seen, several vital issues concerning confirmation theory itself. As long as one holds the belief that observations plus logic (including "inductive" logic) are sufficient to confirm or disconfirm our contingent knowledge claims (i.e., as long as one remains a judgment empiricist) it does seem quite natural to consign all these contingent claims to the realm of the "empirical," which, in turn, is held to include only scientific and common-sense knowledge. Such a move leaves for philosophy only those truths (or falsehoods) which are noncontingent and, given the empiricist view of *contingency* and *necessity*, only those statements that are *logically* true (or false) or analytic. Philosophical activity could consist then, only of "logical analysis," or "con-

[64] Theories, Perception, and Structural Realism," "Russell on Perception: A Study in Philosophical Method," "The Later Russell: Philosophical Revolutionary," and, to a lesser extent, "Corroboration without Demarcation."

ceptual analysis," or the "analysis of language," etc. — certainly a very natural view given prevailing philosophical fashions but one that collapses immediately once the false assumption of judgment empiricism is abandoned.

Even when I held such a view, I was very uncomfortable at having to defend the thesis that a question such as whether there existed an external, mind-independent world or whether all of reality was "in our minds" or consisted of certain arrangements of sense impressions, etc., was either a "pseudoquestion" resulting from a "misuse" of language or a question that, properly interpreted, was about the language we use, or, perhaps *ought to use*, or "find more convenient" to use, etc. — and even more uncomfortable about similarly "explaining away" the question whether the mind and the body are separate and distinct or whether they are in, some sense, one. I am sure that I have been as guilty as anyone else, but the machinations of philosophers supporting these contorted interpretations are indeed sights to behold; especially remarkable are the intuitively outrageous, *extraordinary* interpretations that the "*ordinary* language" philosophers find it necessary to put upon *ordinary* language itself. I am not concerned to deny that logical, conceptual, and linguistic "analyses" are sufficient to "solve" *some* "philosophical" problems or that they play a role, sometimes a vital one, in the problems just discussed. Indeed they do, but they also play equally important roles in many problems generally acknowledged to be *scientific* ones — special and general relativity are spectacular examples, but there are many more less dramatic ones.

Some philosophers, like Quine and Goodman, found this view of the function of philosophy and the resulting certainty which a correct "philosophical analysis" should be able to bestow upon those few significant conclusions left in its province difficult to accept. They also recognized some of the implications of some of the difficulties in confirmation theory. Their (notorious) solution was to blur the distinctions between contingent and necessary, between analytic and synthetic, etc. This move also deemphasized the importance of analyses of meanings and creates a continuum containing philosophical, scientific, and common-sense problems (or statements). I believe that their move, however, is a *bad* way of reaching *good* conclusions. Here I stand with the empiricists against it. However, I do not want to debate the matter here[65] but, rather, to point out that the same "good" conclusions follow from the theory of confirmation advocated in this paper, leaving the analytic-synthetic distinction untouched.

Once we recognize that *just* the evidence (plus logic) not only does not produce general knowledge but does not even select from among competing knowledge *claims*, our views about the classification of theories and hypotheses change drastically. We realize that empiricist criteria of meaning based on confirmation classify even all significant scientific problems as pseudoproblems (since confirmation *in the empiricist sense* is impossible) and that Popper's *criterion of demarcation* would make them all "metaphysical" (since scientific theories are not, in general, falsifiable). I believe that the most important principle, if one must have a principle for distinguishing science from philosophy, is that questions that seem to be more easily decidable on the basis of the evidence are ones that we are inclined to call "scientific" and those which seem more tenuously related to the evidence are ones that have been, traditionally, classified as *philosophical*. One or more, or all, of the following usually hold for propositions more commonly classified as philosophical than as scientific: (1) They are conjoined with a greater than usual number of auxiliary theories, hypotheses about unobservable initial conditions, etc. (2) General competing propositions (theories) are usually under consideration none of which has an obvious or widely agreed upon advantage vis-à-vis its prior probability. (3) It is relatively easy to adjust, replace, etc., the auxiliary theories and assumptions about initial conditions to protect, in turn, each of the competing theories against prima facie difficulties with the evidence. (4) They are usually propositions about matters that seem quite basic and fundamental to us for one reason or another. They may be concerned, either pro or con, with our most cherished or most tenaciously held beliefs, sometimes tacit beliefs of which we become aware only after we fall into the company of philosophers, such as our belief in the existence of the external world and of other minds. Note that (1) and (3) have to do with our abilities

[65] I *have* done so at length, with a fervor that is now incomprehensible to me, in my "Meaning Postulates in Scientific Theories," in H. Feigl and G. Maxwell, eds., *Current Issues in Philosophy of Science* (New York; Holt, Rinehart, and Winston, 1961), and "The Necessary and the Contingent," in Feigl and Maxwell, eds., *Minnesota Studies in the Philosophy of Science*, vol. 3. I am now much more sympathetic with the view of Hilary Putnam ("The Analytic and the Synthetic," in *ibid.*,) that not as much hangs on the issue, one way or the other, as most philosophers seem to think.

in these special cases to think of auxiliary theories and possible (unobservable) initial conditions that *seem* not too unreasonable to us (to which we assign not *too* low prior probabilities), while (2) is totally and explicitly concerned with *our* estimates of prior probabilities of the competing theories of interest. The anthropocentrism involved in (4) requires no comment. It begins to appear that even the difference in degree between philosophy and science or common sense depends to a large extent on our (subjective) estimates, interests, and beliefs. Again, I am not claiming that this is the whole story. Perhaps, for example, relative to the main body of our knowledge (whatever this may be) there are *some* logical differences, at least in degree, between philosophical problems and scientific ones, but they are not the ones that have been proposed by contemporary philosophy. As far as logic (including "inductive logic," if only it existed) relative to actual *and* possible evidence is concerned, scientific and philosophical problems are on a par. By skillful manipulation of auxiliary theories and assumptions about unobservable initial conditions, not only philosophical theories but also the most mundane scientific theory not only can be "saved" no matter what the evidence, but can also be made to account for any possible (relevant) evidence. In practice, of course, we soon balk and refuse to consider what we judge to be *silly, contorted, unnatural,* etc., auxiliary theories and assumptions. However, silliness, contortion, unnaturalness, and even *extremely low prior probability* are *not logical* notions. It is mainly because of the constitutional, perceptual, and mental capabilities we happen to have (plus, perhaps, some historical accidents) that we consider the competition between, say, realism and phenomenalism to be philosophical and that between relativity and Newtonian mechanics to be scientific. In each case each competitor accounts for all the evidence and can be made to do so come what evidence there may be. Some of us may feel that realism has a much higher prior probability than does phenomenalism and that it accounts for the disappearance of the burning logs in the fireplace in an observer-empty room in a "more natural," less convoluted manner than does phenomenalism — that the auxiliary theories and assumptions used by realism in this case are more "reasonable" or have a higher prior probability than those that a phenomenalist would have to employ. But exactly the same sorts of considerations could be applied to the comparison of relativity's explanation of the advance of the perihelion of Mercury's orbit with the one that could be (and, indeed, has been)[66] advanced by the Newtonian. In these, as in all cases, we cannot make decisions without tacitly or explicitly making estimates of the prior probabilities that are involved.

To bring the main body of this paper to a close, I repeat a perhaps somewhat oversimplified but elegant and succinct explanation by Bertrand Russell of the fact, if it is a fact, that we have any general knowledge and are sometimes able to make successful predictions: "*Owing to the world being such as it is*, certain occurrences are sometimes, *in fact*, evidence for certain others; and owing to animals being adapted to their environment, certain occurrences which are, *in fact*, evidence of others tend to arouse expectation of those others"[67] (italics not in the original).

Appendix

A few more of the more popular approaches to confirmation theory will be briefly considered in this appendix, although no attempt is made to provide anything like a complete survey.

Many philosophers seem to find appeals to simplicity very attractive. "Simply select the simplest theory," they say; "what simpler way could one desire to eliminate all but one of whatever competing theories there may happen to be around?" Perhaps. But such a procedure does not seem to move one whit toward answering the problem posed by Hume nor does it fare any better with what we have argued is the prime question of epistemology: What are the conditions to be fulfilled if knowledge (or true belief) is to exist? Unless it is contingently postulated to be the case, what reason under the sun is there to suppose that the simplest theory is any more likely to be true than any of the other infinite number of competitors? A contingent postulate that asserts such a higher prior probability for simpler theories may be true — or may be well advised, etc. But it *is* contingent, and if we fall back upon it to support our inferential procedures, we have certainly abandoned empiricism. On the other hand, if we disallow such a postulate, but claim that

[66] See, e.g., Feyerabend, "Against Method."

[67] *Human Knowledge: Its Scope and Limits*. This quotation appeared as part of the epigraph of my "Theories, Perception, and Structural Realism." I emphasized the adaptive and (what is not necessarily the same thing) the evolutionary aspects of epistemology in "Corroboration without Demarcation" but, for reasons of brevity (believe it or not), have omitted consideration of them from this paper.

the likelihood of truth of simpler hypotheses is *supported* to some extent *by* decades of *evidence* from the history of science, this is at least reasonably arguable;[68] however, then it cannot be used, on pain of vicious circularity as a basis for confirmation theory. Simplicity may be beautiful, and it certainly makes things easier. Moreover, a certain *minimum* amount of it in nature is no doubt a necessary condition for the existence of knowledge. But none of this provides a justification for making it the basis of a decision procedure for selecting among competing theories. For example, the existence of the bare *minimum* of simplicity necessary for the existence of knowledge is perfectly consistent with the *simplest* theory's always being false. If a congress of scientists or philosophers (or even elected politicians) legislated that such a decision procedure was henceforth to be employed, I would practice civil disobedience and politely but firmly demur, preferring to use, rather, my feeble hunches in each case about which competing theory had the highest prior probability, and I would also want to learn about the estimates of prior probabilities by other investigators in the area instead of seeing them muzzled by simplicity requirements. To resort to banalities, I would much rather trust the results of the free competition of diverse ideas than to trust any uniformity that might result from slavishly following a "rule of simplicity." By this time it should be abundantly clear that the search for "*the* scientific method" whereby a given body of evidence univocally selects *the* best theory or hypothesis is a search for a chimera at *least* as elusive as the philosopher's stone or the fountain of youth.

Be all this as it may, however, I do not believe that, in most cases, there is any satisfactory, univocal way of deciding which theory should be called *the simplest*. It may be easy to decide, when we are dealing with points on a graph, that a straight line is simpler than a very "curvaceous" Fourier curve. However, neither theories and hypotheses nor the evidence for them comes neatly packaged, very often, as curves or series of points on a graph.

If someone proposed that, when we have proposed before us two or more viable competing hypotheses or theories and *our estimates of their prior probabilities are all equal*, we should *then* select the simplest one (provided we can decide which *is* simplest), I would probably agree for

[68] Though as so stated, false, I believe. Simple theories usually require subsequent qualifications and, thus, complications. Moreover, they are usually applicable only to "ideal" and, strictly speaking, nonexistent systems.

reasons of convenience or of laziness, but, as far as deciding which to bet on as being more likely to be true or, even more likely to stand up best under further testing, I would just as soon toss a coin.

Regarding the approach of the *personalists*, I have already discussed some of the similarities and some of the differences between their approach and the one defended in this paper; and I should like to emphasize and gratefully acknowledge the importance and precedence of their work.[69] As explained earlier, taking a relative frequency interpretation of the probability of propositions to be the fundamental one, I give a (contingent!) rationale for supposing (or hoping) that our estimates of prior probabilities bear some, not-too-hopeless relationships to reality. To put it in another way, I try to remove the ontological, though not the epistemic, subjectivity from confirmation theory. We may think of knowing subjects (human beings and at least *some* other animals) as being very imperfect, though not completely hopeless, *measuring instruments* for determining prior probabilities. Inaccurate though they may be, the relative frequencies (or, perhaps better, propensities to realize relative frequencies) that they attempt to measure are just as *objective* and mind-independent as is tomorrow's weather, whatever it may turn out to be.[70]

As strange as it may seem, I do not feel that a detailed consideration of Carnap's celebrated attempts to systematize "inductive logic" has a place in this paper. The reason is that I believe that his approach is a *nonempiricist* one. The axioms for his system are admittedly and avowedly not logically valid in the sense of deductive logic and are, thus, in this sense *synthetic*. They are, moreover, according to him, a priori. This paper has just assumed without discussion that classical rationalism is untenable, so I shall not bother with the obvious implications of these two points. Carnap's move is to claim that these basic principles of inductive logic are validated and their use made possible by our *inductive intuitions* just as, he claims, we must have *deductive intuitions* in order to employ successfully (and to see the validity of) deductive principles. My position is, of course, that at least *some* of the axioms (or tacit assumptions) of his system are not only synthetic but also *contingent*.

[69] For example, that of L. J. Savage, *Foundations of Statistics* (New York: Knopf, 1963). Abner Shimony, "Scientific Inference," in Colodny, ed., *Studies in the Philosophy of Science*, vol. 4, and Paul Teller (see, e.g., his article in this volume).

[70] For more detailed discussion of some of the weaknesses of the personalists' approach and of the desirability of a frequency interpretation, see Salmon, *The Foundations of Scientific Inference*.

For example, I believe that this is clearly true of the principle that "structure-descriptions" have equal "measures." "Measure," here, is an obvious euphemism for *prior probability*, and my earlier arguments concerning these and any attempted application of anything like a *principle of indifference* hold, I believe, with full force here.[71] Carnap's system, then, seems to emerge as a *contingent* theory of confirmation and to be, moreover, in competition with the one developed in this article. Of my many objections to it, I shall discuss only one here — one that seems to me decisive. In order to apply Carnap's system, given any instance of evidence to be used, we must already know what are the relevant individuals and properties in the universe (as well as the cardinality of the relevant classes of individuals). In other words, we must already have the universe "carved at the joints" and spread out before us (otherwise we could not perform the necessary counting of state-descriptions, structure-descriptions, etc.). But just how to "carve at the joints," just what are the important relevant properties, and just which segments of the world are to be designated as individuals are all contingent questions — or, at least, they have crucial contingent components — and, moreover, are often the most important questions that arise in scientific inquiry. Before any application of Carnap's system can begin, we must assume that a great deal — perhaps *most* — of that for which a confirmation theory is needed has already been accomplished. We may, as might have been expected, put the objection in a form that, by now, is familiar ad nauseam: Given any body of evidence, there will be an infinite number of mutually incompatible ways of "carving the world" (setting up state-descriptions, structure-descriptions, etc.) each of which will give different results for predictions, "instance confirmations," etc. I do not believe that any set of Carnapian axioms can select a unique method without tacitly relying on the same kinds of estimates of prior probabilities that are explicitly recognized in my theory. Carnap's system cannot circumvent our result that, if the (hypothetico-inferential [or simple inductive]) search for knowledge is to meet with any significant success, then we must be able *by extralogical means* to select theories that have a relatively high prior probability.

The preceding statement applies with full force to the "ordinary language solution" or "dissolution" of Hume's problem. Even if one ac-

[71] For more detailed and forceful objections to Carnap's approach, see, again, Salmon, *The Foundations of Scientific Inference*.

cepted completely its "analysis" of *rationality* — which I, for one, do not — the results and implications of the paper would stand unaffected.[72] If "being rational" just *means* using those kinds of reasoning that we, in fact, usually do employ, then, *if I am correct* in my contention that, both in science and in everyday life, we *do* do something very much like applying the frequency version of Bayes's theorem, estimating appropriate prior probabilities, etc., it follows that my theory provides the standard of (nondeductive) rationality and that to apply it is, *by definition*, to be rational.[73] But I would take scant comfort from this, even if I accepted this definition of 'rational'. A rationale, or a justification, or a solution or "dissolution" provided by a *definition* is bound to be pretty trivial. How can the practice of labeling our commonly practiced inferential procedures with the (soporific?) word 'rational' do anything toward answering the searching questions of Hume? Linguistic philosophers are inclined to retort that Hume merely showed that induction is not deduction and that we should have known this all along. Well and good, provided they go ahead and recognize the consequences of this. If they recognize that these consequences mean giving up strict inductivism *and* strict confirmationism, in short, *giving up empiricism* and acknowledging the irreducibly subjective epistemic components that must be present in whatever remains of the "context of justification"[74] and if they recognize that our "ordinary" modes of reasoning involve, at least tacitly, something very much like the (frequentist) application of Bayes's theorem, then the only dispute that will remain will be the relatively minor one about what the word "rational" means in ordinary speech.

This *does* suggest, however, a question that is rather fervently debated

[72] Once again, Wesley Salmon has provided a devastating critique of this popular "solution." See his "Should We Attempt to Justify Induction," *Philosophical Studies*, 8 (1957), 33–44.

[73] Strawson, on the other hand, seems to hold that our ordinary methods of (nondeductive) reasoning are what I have called "simple inductive" ones (see his *Introduction to Logical Theory* (New York: John Wiley, 1952), ch. 9). (He does not even mention hypothetico-deductive reasoning, but this does not significantly affect the main point at issue here.) In view of what we have seen earlier, this would mean that the "rational" way to reason is totally impotent or, rather, that if it is not to be so, it must be bolstered by the powerful contingent assumptions that, in effect, convert it into the set of confirmation procedures advocated in this paper. In other words, if Strawson is right, then what he would call "rational behavior" would be, for me, *irrational* in the extreme unless my assumptions — which he does not consider, of course — are added.

[74] In all fairness, one way or the other — or both — I should note that "context of justification" is a term of Reichenbach's rather than one of the British Analysts.

in some circles today: Are there (objective and permanent) "standards of rationality" or is "rational" more like a label for those modes of reasoning that happen to be fashionable in a given era (within, perhaps, a given group such as "the scientific community," etc.)? As Imre Lakatos has put it: Should we espouse epistemological authoritarianism (or dogmatism, etc.) or epistemological anarchy?[75] If pressed for an immediate or unqualifying answer, I would, of course, have to opt for anarchy. But this would be unfair, for the question is, obviously, loaded. Surely there are positions intermediate between total anarchy and total regimentation. Few of us, I suppose, would be so anarchistic as to advocate disregard of the principles of deductive logic, and most of us, I should hope, would require that proposed theories and hypotheses satisfy or give promise of satisfying the requirements of hypothetico-inferential explanation. How many or how few more requirements or "standards of rationality" *I* would endorse is not a matter to be discussed here, nor is it a particularly crucial one for the present issue. What must be emphasized at this point is the other side of the ledger: no acceptable set of 'standards of rationality' can be strong enough to eliminate the existence, for any given amount of evidence, of an infinite number of viable, mutually incompatible theories: *the proposal and choice of theories from this number must be made on an extralogical, nonrational basis.* The *proposal* of theories is, of course, the (subjective) "creative leap" of the intellect of the knower; this much is fairly uncontroversial even among (enlightened) empiricists. We have seen, however, that if knowledge is possible, the frequency of success of such leaps must be surprisingly high given the number of possibilities. Furthermore, to select from among *proposed* theories or to make even a wild guess to what degree a given theory is confirmed by the evidence at hand, we have seen that we must rely on (subjective) extralogical, nonrational estimates of the appropriate prior probabilities. Thus, the importance, indeed, the indispensability of the

[75] Private communication. Lakatos's widely discussed approach, admittedly and avowedly a development beginning with Popper's views, is based on what he calls "progressive and degenerating research programs." (See his "Falsification and the Methodology of Scientific Research Programs.") It is an improvement, at least in some respects, on Popper's system, because it recognizes the impossibility of viable falsifiability in the important and interesting areas of science. Like Popper's, however, it remains within what I have called the "strict confirmationist" and, therefore, within the empiricist fold. It, thus, fails to come to grips with Hume and leaves unanswered the fundamental epistemological question: What are the conditions to be fulfilled in order for knowledge to be possible?

subjective, the extralogical, the nonrational (but not irrational) for the growth of knowledge (both in the "context of discovery" *and* in the "context of justification") cannot be overly stressed. All attempts to eliminate this element and to establish "standards of rationality" that provide a univocal procedure for selecting theories on the basis of evidence at hand are doomed just as surely as was the specific instance of such an attempt, based on simplicity, that we considered earlier. The search for *rationality*, then, in this *strong* sense, must be abandoned. But just as we saw that there are senses, less stringent than those desired by empiricists and others, but still plausible and useful senses, of 'knowledge'[76] and 'justified', we could also easily adduce a similar sense for the word 'rational'. However, the concern of this paper has been the fundamental question of epistemology: What are the conditions to be fulfilled in order for significant knowledge (or true belief) to exist? And the answers to this question, which are the fundamental principles of epistemology, are mainly descriptive rather than normative. Therefore, the redefinition of the essentially honorific term 'rational' can be left for another occasion.

[76] As a final footnote *and* as a close to this paper I should like to quote Russell's wonderful remarks on the nature of knowledge contained in the introduction to his *Human Knowledge: Its Scope and Limits*: ". . . 'knowledge' in my opinion, is a much less precise concept than is generally thought, and has its roots more deeply embedded in unverbalized animal behavior than most philosophers have been willing to admit. The logically basic assumptions to which our analysis leads us are psychologically the end of a long series of refinements which starts from habits of expectations in animals, such as that what has a certain smell will be good to eat. To ask, therefore, whether we 'know' the *postulates of scientific inference* is not so definite a question as it seems. The answer must be: in one sense, yes, in another sense, no; *but in the sense in which 'no' is the right answer we know nothing whatever, and 'knowledge' in this sense is a delusive vision. The perplexities of philosophers are due, in a large measure, to their unwillingness to awaken from this blissful dream.*" (Italics added.)

PAUL TELLER

Shimony's A Priori Arguments for Tempered Personalism

1. Introduction

In his essay, "Scientific Inference," Abner Shimony presents an analysis of scientific confirmation which ties together two twentieth-century proposals for treating the justification of our scientific methodology. To characterize the first of these proposals, to be referred to as the method of pragmatic justification, scientific inference is analyzed as an activity which we undertake in order to achieve certain more or less specifically described ends. It is admitted that we cannot be sure of achieving those ends. But, on this view, if we desire the ends and if we have no reason to suppose them unattainable, we are justified in assuming that the ends can be achieved. This assumption is then used as a premise for an argument concerning the nature or structure of scientific investigation. Peirce originated this pattern of argument, and Reichenbach, discovering it independently, presented it in its best known form as his familiar "vindication" or "pragmatic justification" of induction.

One might refer to the second proposal as the "seafarer's doctrine" since it is so suggestively summarized by Neurath's simile which compares our efforts to study and correct methods of scientific inference with the efforts of sailors rebuilding their ship at sea. On this view, the only means we have for judging which methods of investigation will result in reliable beliefs are our currently accepted methods of argument and our scientific theories (and perhaps other beliefs) about the nature of the universe. These methods of argument might turn out to be unreliable, and the theories might turn out to be wrong; but at any given time they are the only things we have as a basis for judging, modifying, and extending our view of the world and our methods of evaluating beliefs. Quine suggests some view of this kind at the end of "Two Dogmas of Empiricism" (1963, pp. 42–46), and the doctrine bears more than a

AUTHOR'S NOTE: Work on this article was supported by National Science Foundation graduate and postdoctoral fellowships.

superficial similarity to Goodman's view in *Fact, Fiction and Forecast* (1965, pp. 62–65) that "inductive rules" are both developed and justified through a process of mutual adjustment with inference cases. Closely related points are made by Suppes (1966, pp. 62, 64), and the whole position has many points of similarity to Dewey's views in his *Logic* (1949).

Shimony uses his versions of the pragmatic justification and seafarer's doctrine in tandem. He describes a set of conditions which one might hold to be reasonable constraints on the structure and interpretation of scientific investigations. He then hopes to argue that

> scientific inference, thus formulated, is a method of reasoning which is sensitive to the truth, whatever the actual constitution of the universe may be. In this [first] stage very little will be assumed about the place of human beings in nature, but the result of such austerity is that scientific method is only minimally justified, as a kind of counsel of desperation [because it is only shown to hold out hope of reaching true beliefs if human beings are capable of reaching true beliefs]. In the second stage . . . the justification is enlarged and made more optimistic by noting the biological and psychological characteristics of human beings which permit them to use induction as a discriminating and efficient instrument. Thus the second stage . . . evidently presupposes a considerable body of scientific knowledge about men and their natural environment . . . " (1970, p. 81).

Shimony's paper presents a valuable contribution toward the goal of developing an adequate version of the seafarer's doctrine by describing many specific ways in which scientific conclusions and methodological principles may support and supplement one another. I believe that a program of this kind may provide the best way out of many of our present-day problems in confirmation theory, and I hope that the view will be further developed. On the other hand, I believe that Shimony's attempts to incorporate versions of Peirce's and Reichenbach's arguments are misguided and that his view would be improved if freed from all attempts to argue in favor of his outline of scientific investigation on strictly a priori grounds, that is, without appeal to contingent premises. The aim of the present paper is to sustain this belief by showing that those of Shimony's arguments which can be said to be a priori, and their plausible modifications, fail to support his conclusions. In preparation for this criticism, the next two sections will be devoted to describing tempered personalism and Shimony's proposed analysis of scientific inference, and to explaining Shimony's strategy for justifying tempered personalism a priori.

2. Tempered Personalism

2.1. *The Bayesian analysis of Carnap and the personalists.* Tempered personalism is a Bayesian analysis of scientific inference. It will be well to summarize the features common to all forms of Bayesianism and to state certain distinctive characteristics of other Bayesian theories; this will help to make the important features of Shimony's analysis clear, by giving both points of agreement and points of contrast.

A Bayesian description characterizes reasonable belief or credence as a variable quantity, the magnitude of which can be represented by a function from propositions (or sentences) onto numbers between zero and one, inclusive. The function, $P(h/e)$, has two arguments, h and e. The value of the function is usually interpreted as a conditional betting rate; if P describes the beliefs of an agent, $P(h/e)$ determines the odds at which the agent would be indifferent between accepting either side of a bet on the truth of h, where the bet will be called off if e should turn out to be false. If e is a tautology, such as $p \vee \sim p$, the bet is unconditional, $P(h/e)$ is interpreted as the betting rate of an unconditional bet on h, and "$P(h/e)$" is abbreviated to "$P(h)$." $P(h/e)$ is read as "the degree of rational belief (or credence) in h given (the truth of) e."

In a Bayesian description P satisfies the axioms of conditional probability, of which Bayes's theorem is a trivial consequence:

$$(1) \qquad P(h/e) = \frac{P(h)}{P(e)} P(e/h).$$

$P(h)$ is often referred to as the *prior degree of belief in* h, or simply the *prior probability* of h. $P(e/h)$ is often referred to as the *likelihood of* e *on* h. If h is taken to be a hypothesis and e fully describes newly acquired evidence, then Bayes's theorem gives a plausible account of how one's degree of belief should change from P to a new function P'; Bayesians hold that $P'(h) = P(h/e)$ if e has been found to be true. In other words, the new degree of belief in h is taken to be the old degree of belief in h given the truth of e. This process is called conditionalization. However, for a Bayesian theory fully to characterize the degrees of belief which arise by conditionalization, the theory must specify the belief function from which to start. This initial function is called the *prior probability function.*

A major difficulty faced by any Bayesian analysis is the problem of giving an acceptable specification of prior probability functions; and variants of Bayesianism can be classified by the answers they propose to this problem. One form of Bayesianism, known as personalism, holds that the prior probability function is to be determined by the agent's subjective judgments about what odds he would take in bets, subject to the sole constraint that the prior probability function must satisfy the probability axioms. The subjective judgment of different people may give rise to very different prior probability functions, but the personalists argue that as long as the functions satisfy the probability axioms there is no judging one to be more or less rational than any other (cf. Savage, 1967b, p. 597, and *passim*). Another view, advanced by Carnap, holds that it should be possible to put additional constraints of rationality on the prior probability function. Ideally, according to Carnap, we should be able to find sufficiently strong constraints on all rational belief to enable us to define the prior probability function completely. The resulting function would give a "logical" concept of the degree of belief it is rational to have on any given evidence, in the sense that the rational degree of belief on any given evidence could always be calculated from the explicit definition.[1]

If it would be possible to find justifiable constraints on a prior probability function sufficient to define it completely, all changes in rational belief could henceforth be described by conditionalization on new observations. In other words, the initial constraints would once and for all determine confirmation relations between any possible hypothesis and any possible observations.[2] Many personalists differ from Carnap in this respect also. They recommend their doctrine only for use in restricted, relatively well-defined situations in which an agent is confronted with a practical decision-making problem in face of uncertainty (cf. Savage, 1954, pp. 16, 82–91; 1967a, pp. 306–7). Examples are the wildcatter's problem of where to drill an oil well and the manufacturer's problem of how much to produce. The personalists advise the wildcatter and the manufacturer to estimate their initial degrees of belief subjectively for

[1] Cf. Carnap, 1963, p. 971ff. Also 1962, p. 307. It should be added that Carnap seems to show some vacillation on these points in his *Continuum of Inductive Methods* (1952, pp. 53–55).

[2] Cf. references of the last footnote.

any one such problem, but if a similar problem arises years later under considerably different circumstances they are advised to make new evaluations rather than to try to conditionalize their old estimates on the vast body of intervening observations of uncertain relevance to the problem. Furthermore, personalists recommend that in preparation for a Bayesian analysis, a problem should be structured in terms of possible states of the world, possible available acts, and possible consequences of these acts. In order to make the problem manageable this structuring must be highly idealized, and personalists recommend that one simply ignore states, actions, and consequences, the possibility of which have not occurred to the agent or which seem, prima facie, to be irrelevant to the problem. Savage has attempted to describe how such a "localization" and specialization of a problem-description should proceed (Savage, 1954, pp. 82–91; 1967a, pp. 306–7), but to the best of my knowledge no personalist has tried to describe conditions which determine when an agent should continue conditionalization on his old probability function and when he should start from scratch, restructuring his problem and making new subjective evaluations of a prior probability function.

2.2. *The first distinctive feature of Shimony's Bayesian analysis*. Shimony's first departure from the views of both Carnap and the personalists results from his desire to give an account of scientific inference, viewed as an enterprise undertaken to achieve the "theoretical [end] of learning the truth about various aspects of the world . . . " but not the practical ends of special concern in problems of making decisions in face of uncertainty (1970, p. 81). For the purpose of giving such an account Shimony rejects the interpretation of $P(h/e)$ as a betting quotient for h given e: "The source of the difficulties [with this interpretation] is the indispensability of general theories in natural science as it is now studied and, indeed, as it conceivably could be studied if the ideal of general insight into the nature of things is not to be abandoned" (1970, p. 93). The personalists could interpret $P(h/e)$ as a betting quotient insofar as they restricted application of their theory to practical decision-making problems in face of uncertainty, in which the outcome of acts are observables such as a drilling operation resulting in a gusher or a dry hole. But if the subject of concern is the truth of theories, the betting quotient interpretation is no longer directly applicable, as plainly there is no betting on the truth of theories such as quantum mechanics or general relativity. Moreover, even if the function $P(h/e)$ is freed from its interpretation as a fair betting quotient, in many cases it cannot plausibly be interpreted as the rational degree of belief in the *truth* of *h*, given evidence e (1970, pp. 93–94). For if *h* is a general and precise theory, such as quantum mechanics, it does not seem rational to have a high degree of belief in its literal truth: the history of science has had too many upsets for that. Yet there is a sense in which such theories can be said to be highly or poorly "confirmed," and Shimony would like his interpretation of $P(h/e)$ to be as close to this sense as possible.

The history of science suggests an alternative explicandum to Shimony. He proposes that $P(h/e)$ be called a "rational degree of commitment to h given e" and that this should be interpreted as the rational degree of belief in the claim that the proposition *h* is "near to the truth" when the evidence is *e*. A proposition whose truth can be established by observation is "near to the truth" if and only if it is true. Any other proposition, *h*, is "near to the truth" if it "is related to the truth in the following way: (i) within the domain of current experimentation h yields almost the same observational predictions as the true theory; (ii) the concepts of the true theory are generalizations or more complete realizations of those of h; (iii) among the currently formulated theories competing with h there is none which better satisfies conditions (i) and (ii)" (1970, pp. 94–95). This characterization is far from complete and Shimony presents an interesting discussion of the difficulties to be expected in filling out the description (1970, pp. 95–97). But since the points I have to make will quite clearly be independent of any plausible attempt to clarify further the interpretation of "near to the truth," we may rely on the intuitive understanding of the phrase suggested by conditions (i), (ii), and (iii).

By way of abbreviation, and in conformity with Shimony's usage, I will henceforth refer to $P(h/e)$ as the probability of *h* on e, when it is clear that "probability" is to be understood as "rational degree of commitment."

2.3. *The second distinctive feature of Shimony's Bayesian analysis.* Shimony gives his Bayesianism its second distinctive feature by stipulating that probability evaluations are to be made only for the propositions of a limited and well-defined investigation. An investigation is determined by:

(1) A set of hypotheses $h_1, \ldots, h_n$ (of which the last may be a "catch-all" hypothesis equivalent to $\sim[h_1 \vee \ldots \vee h_{n-1}]$) which have been "suggested as worth investigating." [Henceforth, hypotheses which have been "suggested as worth investigating" will be referred to by Shimony's frequently used expression, "seriously proposed hypotheses."][3]

(2) A set of possible outcomes $e_1, \ldots, e_m$ of envisaged observations; and

(3) The information, *i*, initially available. This information in actual circumstances is very heterogeneous, consisting partly of vague experience concerning the matter of interest, partly of experience which may be sharp but is of dubious relevance, partly of sharp evidence which does not seem to bear directly on the question at hand but is relevant to other questions in the same field, and partly of propositions which are regarded as established even though they go beyond the actual evidence and, therefore have been accepted because of previous investigations (1970, p. 99).

In Shimony's account, probabilities are specified for all the propositions of an investigation, namely the proposition in *S*, the truth functional closure of $\{h_i, \ldots, h_n, e_1, \ldots, e_m\}$ and for nothing else. An investigation begins with the specification of its prior probability function *P*. For every member, *d*, of *S*, $P(d)$ must represent, at least approximately, the investigator's intuitively judged initial degree of commitment to *d*, subject to the restriction that *P* satisfy the probability axioms and to one further restriction to be discussed at length below. The planned observations are then made, and the investigation terminates by determining the posterior probability of the investigation. This is done by conditionalizing the prior probability function, *P*, on the observed outcome of the observations, e. A hypothesis, h_i, is said to be preferred to a second hypothesis, h_j, at the end of an investigation with observed outcome e if $P(h_i/e)/P(h_j/e)$ is large.[4]

[3] Although Shimony here says that h_n *may* be a catch-all hypothesis, in the rest of his discussion he seems to assume that it always is.

[4] Since the publication of "Scientific Inference" Shimony has decided that h_n, the catch-all hypothesis, should be treated differently from h_i, $1 \leq i < n$. This shift is necessitated by difficulties which arise in explicating the notion of "near to the truth"; since the catch-all hypothesis is only specified as the negation of the disjunction of the other hypotheses of the investigation, it is highly unclear in what

Shimony does not exclude the possibility that the concept of rational degree of commitment might be extended to broader applications, and he makes some general remarks about such a possibility (1970, pp. 118–21). But tempered personalism itself deals only with the specification of rational degrees of commitment to the propositions of specific investigations, in the form of the prior and posterior probabilities described above.

Shimony attributes to Jeffreys the suggestion that scientific investigations be localized for the evaluation of probabilities. As I remarked earlier, the personalists have similarly recommended application of a Bayesian theory only to isolated and well-defined practical decision-making problems; and they have attempted to explain how such localization of a problem might best be described. It was also noted that personalists have said little about deciding when old probability functions should be abandoned in favor of a new subjective evaluation, while guidelines for such decisions are at least implicit in the formulation of tempered personalism: an investigation ends when the investigator has made the observations that he envisaged in setting up the problem. If further examination of any hypotheses is desired, new observations should be planned and a new subjectively evaluated prior probability function determined.

Shimony's main motivation for stipulating that probability calculations be limited to narrowly specified investigations is that it "permits probability to be a manageable instrument in scientific inference" (1970, p. 99; see also p. 100). If not restricted in some such way the problem of assigning and updating probability evaluations would be hopelessly difficult; moreover there is no reason to suppose that an individual investigation would gain from a global assignment of confirmation relations. Quite to the contrary, it is surely overoptimistic to suppose that men could, following Carnap's prescription, once and for all settle on a perfectly general prior probability function which henceforth need only change by conditionalization.[5] From time to time we seem to be in need of revising our opinion about the confirmation relation between evidence and hypotheses. This is made possible by Shimony's localized investiga-

sense the catch-all hypothesis can be said to be "near to the truth," or whether this makes sense at all.

[5] In fact Putnam has proved that there cannot be such a function. See Putnam (1963).

tions with their freshly assigned prior probabilities, which give scientific investigations "greater openness to the contingencies of inquiry than possessed by Carnap's c-functions" (1970, p. 97).

2.4. *The third distinctive feature of Shimony's Bayesian analysis.* As so far described, Shimony's position differs from that of the personalists only in that he rejects the betting quotient interpretation of the probability function. The theory is to apply only to more or less narrowly specified problem situations, and the prior probability function is to be determined by the investigator's subjective judgment. Both Shimony and the personalists agree that the prior probability evaluation must be subject to the constraint of the probability axioms. But orthodox personalism holds that the probability axioms are the *only* rational constraint that can be placed on *P*, while Shimony goes so far in agreeing with Carnap as to say that there is at least one other rational constraint:

The only kind of irrationality which [the personalists] explicitly discuss is incoherence [failure of the probability axioms], but there are surely other properties of belief systems which are characterizable as irrational, and which would inhibit the progress of natural science if they were universal. In particular, the assignment of extremely small prior probabilities to unfamiliar hypotheses is compatible with coherence, but may be irrational in the sense of being the result of prejudice or of obeisance to authority or of narrowness in envisaging possibilities. But unless a hypothesis is assigned a non-negligible prior probability, Bayes' theorem does not permit it to have a large posterior probability even upon the grounds of a very large body of favorable observational data, and thus the theorem does not function as a good instrument for learning from experience (1970, p. 92).

In keeping with these remarks Shimony imposes the following condition on the prior probability evaluation: for every seriously proposed hypothesis of an investigation, *h*, $P(h)$ "must be sufficiently high to allow the possibility that it will be preferred to all rival seriously proposed hypotheses as the result of the envisaged observations . . . " (1970, p. 101). This condition is called the tempering condition, and accordingly Shimony calls his description of scientific inference "tempered personalism."

In effect, the tempering condition requires the prior probabilities and likelihoods of an investigation to satisfy the condition that for each seriously proposed hypothesis h_i of the investigation, there is some possible outcome, e_j, of the envisaged observations, so that conditionalization of the hypotheses on e_j will result in h_i being preferred to all rival seriously proposed hypotheses. This is to require that

TC For each h_i of an investigation, there is a possible outcome, e_j, of the envisaged observations such that $P(h_i/e_j) \geqq 1 - \epsilon$

where ϵ is the number such that a hypothesis, h, is said to be preferred to its rivals at the end of an investigation just in case its posterior probability is as large as $1 - \epsilon$.

Although the tempering condition is stated as a requirement on the prior probabilities of the seriously proposed hypotheses, it should be noted that it is really a collective requirement on the prior probabilities *and* the likelihoods. The tempering condition places the strict requirement on the prior probabilities that $P(h_i) \neq 0$, for all h_i; for if $P(h_i) = 0$ then $P(h_i/e_j) = 0$, whatever the likelihoods. But even if $P(h_i)$ is quite large, it does not follow that there is some e_j for which $P(h_i/e_j)$ is sufficiently large to guarantee that h_i will be preferred to all rival hypotheses. As a glance at (1) will show, if $P(h_i) \neq 0$ calculation of $P(h_i/e_j)$ palpably depends on the value of $P(e_j/h_i)$.[6] It is easy to give examples in which a prior probability distribution fails to satisfy the tempering condition and in which the distribution can be modified to satisfy the requirement by changing either the prior values for the seriously proposed hypotheses or the values of the likelihoods.

The tempering condition is imposed to ensure that every hypothesis of an investigation has a possibility of receiving a high evaluation at the end of the investigation; but it is worth remarking that the condition also implies a parallel "exclusion condition" on the possibility of a hypothesis receiving a low evaluation:

EC For each hypothesis, h_i, of an investigation there is an e_j among the possible outcomes of the investigation's envisaged observations such that $P(h_i/e_j) \leqq \epsilon$.

The exclusion condition follows from the tempering condition and the

[6] $P(h_i/e_j)$ appears to depend on $P(e_j)$. But since

$$\sum_{i=1}^{n} P(h_i/e) = 1$$

for any conditional probability measure P, the values of $P(h_i/e_j)$ can be calculated without using $P(e_j)$ by normalizing $P(h_i) \times P(e_j/h_i)$ taken as a function of h_i. Henceforth I shall speak of $P(h_i/e_j)$ being fully determined by $P(h_i)$ and $P(e_j/h_i)$ by Bayes's formula, although strictly speaking normalization is also required. The possibility of normalization or the method of normalizing may be affected by the changes mentioned in fn. 4.

assumption that for every hypothesis, h_i, of an investigation there is a rival hypothesis, h_j, i.e., a hypothesis, h_j, such that $h_i \& h_j$ is logically false. In the present formulation of tempered personalism this condition is always satisfied since the catch-all hypothesis (the negation of the disjunction of the other hypotheses) rivals all the other hypotheses. Although this might no longer be the case for the revised treatment which Shimony plans to give the catch-all hypothesis, the assumption will generally still be satisfied, since as a rule an investigation considers rival hypotheses.

To show that the exclusion condition follows from the tempering condition, the probability axioms, and the assumption given above, suppose that the exclusion condition, EC, fails. Then there is a hypothesis, h_i, of the investigation such that for all envisaged observations, e_j,

$$(2) \qquad P(h_i/e_j) > \epsilon.$$

Let h_k be a hypothesis which rivals h_i. Then for all e_j

$$P(h_i/e_j) + P(h_k/e_j) \leq 1,$$

by the axiom of total probability. Then by (2)

$$P(h_k/e_j) < 1 - \epsilon, \text{ for all } e_j,$$

which contradicts TC, the tempering condition.

3. An Outline of Shimony's Proposed A Priori Justification of Tempered Personalism

3.1. *Antecedents in the "pragmatic" form of argument.* The impasse created by Hume's skeptical arguments led Reichenbach and Feigl to develop ideas originally suggested by Peirce into a form of justification which seemed to them appropriate for use in discussion of nondeductive means of reaching conclusions about the unobserved. They saw induction as a specific policy or program of action and interpreted the "problem of induction" as the question of how one could justify following the policy. Moral problems aside, one's reason for following a course of action is to achieve some desired end, which in the case of induction is taken to be the end of obtaining true beliefs. Reichenbach and Feigl conceded to Hume that it is impossible to give a noncircular argument in support of the claim that following the policy will lead to true beliefs or frequently true beliefs. But a weaker claim seemed to them to justify a person in undertaking the proposed course of action. Let us call the end of reaching true or at least frequently true beliefs the "inductive aim." Let us suppose, as is usually the case, that we desire to achieve the inductive aim. Suppose we knew with certainty that if the inductive aim can be achieved by any course of action at all, then it can be achieved by following the proposed inductive course of action.[7] Suppose, further, that we have no reason to believe the aim cannot be achieved. Then, according to Reichenbach and Feigl, we are justified in choosing the inductive course of action over any other course of action (intended to realize the same aim) of which we do not have similar knowledge.

3.2. *Shimony's form of argument.* Shimony indicates the nature of the justification he intends by writing that "my treatment of scientific inference is instrumentalistic, in the sense that the justification for certain prescriptions is their conduciveness to achieving certain ends, but the ends admitted are only the theoretical ones of learning the truth about various aspects of the world . . . (1970, p. 81). Thus, Shimony's proposed pattern of justification is like the last in that he hopes to justify tempered personalism as a course of action, by appealing to certain ends which one might hope to achieve by following the tempered personalist's prescriptions for the conduct and interpretation of scientific investigations. The ends concerned are, as he says, "the theoretical ones of learning the truth about various aspects of the world." But unlike Reichenbach and Feigl, he does not argue that this end can be achieved by his proposed course of action if by any; rather he argues that tempered personalism is justified by virtue of certain of its properties which make it "conducive to achieving" the stated ends. I take Shimony to be making the following claim: of two possible courses of action, if the first has one of the properties in question while the second does not, then, other things being equal, there is more reason to suppose that the first course of action will lead to the stated end than will the second. Shimony will argue that tempered personalism has such properties. Consequently, other things being equal, there is more reason to suppose that the tempered personalist course of action will achieve the stated end than will some other course of action which lacks these properties. Finally, Shimony holds that these claims can all be established without appeal to contingent premises, and in that sense his justification is a priori.

[7] Reichenbach's method of proving such a conditional statement will be described in sec. 5.

The properties in question are described by the following three statements:

(1) Investigations conforming to the prescriptions of tempered personalism are open minded. That is to say, when true hypotheses (or hypotheses close to the truth) are seriously proposed, they have a chance of being accepted as true (or close to the truth): "the tempering condition is only a way of prescribing open mindedness within a probabilistic scheme of inference; and open mindedness is a necessary condition for the acceptance of true hypotheses in case they should ever be proposed" (1970, p. 158; see also pp. 130–33).

(2) Investigations conforming to the prescriptions of tempered personalism maintain a "critical attitude" (1970, p. 130) toward proposed hypotheses. That is, no hypothesis receives a high posterior evaluation or is accepted without careful comparison with competing hypotheses in their agreement with observations. Tempered personalism's "receptivity toward seriously proposed hypotheses is adequately balanced by a capacity to evaluate them critically" (1970, p. 133; see also pp. 133–36, *passim*).

(3) Investigations conforming to the prescriptions of tempered personalism can incorporate "any methodological device which analysis or experience indicates to be valuable" (1970, p. 122; cf. also pp. 130–33). In other words tempered personalism is claimed to be highly flexible.

Shimony holds that "one may fairly sum up these claims by ascribing 'sensitivity to the truth' to the tempered personalist formulation of scientific inference and also to the informal processes of confirmation used by scientists from which tempered personalism is extracted" (1970, p. 122), and that "a sensitivity to the truth can be claimed for scientific inference on a *priori* grounds, though the exact content of this sensitivity was not determined [by his examination of *a priori* considerations]" (1970, p. 158).

3.3. *Antecedents of conditions (1) and (2).* Conditions (1) and (2) might be seen as attempts on Shimony's part to incorporate and make more precise several attractive ideas appearing in the work of Peirce and Reichenbach. Condition (1) can be seen as motivated by Peirce's observation that it hardly makes sense to undertake the task of discovering general truths about the world unless we suppose ourselves to have the power to propose hypotheses which are true or at least sufficiently near to the truth to suit our purposes. This is so, according to Peirce, "for the same reason that a general who has to capture a position or see his country ruined, must go on the hypothesis that there is some way in which he can and shall capture it" (*Collected Works*, 7.219). But, speaking intuitively, the power of suggesting true or nearly true hypotheses is of no value unless when proposed they are treated with an open mind. The tempering condition and statement (1) can be seen as an attempt to make this intuition more precise.

Secondly, Shimony reads in Peirce the recommendation that scientific method should systematically and self-critically correct current beliefs in the light of experience (1970, pp. 126–27). This idea is refined (at the expense of narrowing its application) by Reichenbach and others who argue that the straight rule of induction, the rule of always positing observed frequencies to be long-run frequencies, has the advantage of correcting our beliefs about long-run frequencies by adjusting them in the light of new observations, in a way guaranteed to be successful in the long run. For reasons detailed by Black (1954, pp. 168–73) past efforts to make out these arguments have all failed. But statement (2) can be seen as another attempt to make good use of this prima facie appealing idea.

3.4. *The program for examining Shimony's main a priori arguments.* I do not dispute that investigations conforming to the prescriptions of tempered personalism have properties like those indicated by (1), (2), and (3) and that having such properties does in fact constitute grounds for preferring a tempered personalist formulation of scientific investigations to others. But I maintain that Shimony has not succeeded in supporting these claims by arguments which appeal to no contingent premises, and that it is doubtful whether this task could be accomplished. Statements (1), (2), and (3) are far from perfectly clear, so that it is not yet evident whether or not a method having the three intended properties is conducive to achieving the stated end, in the sense of there being more reason to suppose the method to be able to achieve the stated end than could an alternative method which does not have the properties. In particular I shall argue that when (1), (2), and (3) are clarified, in the way Shimony clearly intends, to leave no doubt that tempered personalism is conducive to achieving the end by virtue of having these properties, then it is not possible to argue a priori that tempered personalism has the properties in question. On the other hand, when (1), (2), and (3) are interpreted in a manner that makes such a priori arguments possible, there is no longer any reason to suppose that the properties make tempered personalist investigations conducive to the desired end. To facilitate the exposition of these arguments I will henceforth use the phrase, "Q is a desirable property for a method of scientific investigation" exchangeably with "Q, as a property of scientific investigations, is conducive to the end of reaching true or nearly true beliefs of the sort usually considered the concern of science." "Q is conducive to end E" is in turn to be interpreted, as explained above, as "there is more reason to suppose that end E can be attained by using scientific investigations having property Q than by using ones not having property Q, other things being equal."

4. Consideration of the Main A Priori Arguments

4.1. *Examination of the first condition.* Tempered personalism is claimed to have the first property, that of being open-minded, because tempered personalist investigations must conform to the tempering condition, that a seriously proposed hypothesis be given a sufficiently high prior probability to allow it a chance to be preferred at an investigation's termination to all other rival hypotheses of the investigation. Shimony argues for the desirability of this condition by the observation that it rules out "the skepticism toward human abductive powers [in Peirce's sense] implicit in any formal scheme which treats on the same footing seriously proposed hypotheses, frivolously proposed hypotheses, and unsuggested hypotheses. By giving preferential treatment to seriously proposed hypotheses but insisting on open mindedness within this preferred class, the tempering condition provides a safeguard against one of the major types of error that could be committed by a method of confirmation: the error of rejecting, because of *a priori* commitments, a true hypothesis which someone has been fortunate enough to put forth" (1970, pp. 132–33).

To evaluate the tempering condition and the suggested argument for its desirability we must examine the application and effect of the condition in further detail. From the infinite collection of hypotheses, stated and unstated, the various existing processes of proposing hypotheses select a subset, comprising hypotheses such as those proposed by scientists, crackpots, eccentrics, small children, and so on. From this set a smaller subset is designated as the set of seriously proposed hypotheses; tempered personalism prescribes that all its members are to be considered by scientific investigations. A full statement of this condition would require an explicit characterization of the properties determining these two nested subsets, however flexible the properties might be.[8] For example, the property determining the subset of seriously proposed hypotheses might be as follows: a hypothesis is seriously proposed if (a) someone with a Ph.D. regards it worthy of research, or (b) it is categorized as worthy of research by some method of hypothesis proposing which has been highly confirmed to be useful by previous investigations. In order for the reference to something called "seriously proposed hypotheses" to be a nonvacuous part of the characterization of tempered personalism, not only must the condition be explicitly described, but as described it must be genuinely restrictive, in the sense that from all hypotheses suggested or recognized by any human agency whatsoever the condition must pick out a proper subclass.

Clearly, the tempering condition is desirable only if the restriction made by the reference to "seriously proposed hypotheses" is a desirable one for scientific investigations. The only plausible consideration which would make the restriction desirable would be that "seriously proposed hypotheses" are, in some sense, more likely to include hypotheses which are true or closer to the truth than hypotheses proposed by other means. But how can we establish that this is the case, that "seriously proposed hypotheses" are more likely to include close approximations to the truth than frivolously or randomly proposed hypotheses? Grounds for this conclusion are to be found, no doubt, in the long history of man's search for knowledge, but grounds of that kind are clearly a posteriori.[9] If we are to forgo reliance on facts about which methods of proposing hypotheses have proven particularly successful in the past or on other facts about the world which lead us to have greater trust in one rather than some other method of generating hypotheses, it is to be doubted whether we are any longer in a position to justify the claim that one set of proposed

[8] In sec. III, D (1970, pp. 110–14) and sec. V, D (1970, pp. 150–58) Shimony discusses at length the considerations which may be relevant to defining the class of seriously proposed hypotheses. But he does not formulate a specific definition.

[9] Shimony discusses these considerations in sec. V, B, 3 (1970, pp. 143–48).

hypotheses is more likely to contain close approximations to the truth than some other set. That such a choice cannot be justified a priori follows as a special case of Hume's conclusion that propositions about the unobserved are logically independent of propositions about what is known by observation. Let K be the class of hypotheses which do not logically conflict with what is known and which have been suggested or recognized by any human agency whatsoever. Let C be the subset of K which constitutes seriously proposed hypotheses, and let $C' = K - C$. The foregoing remarks make it reasonable to assume that both C and C' are nonempty. By the very fact that the members of C' are logically consistent with what is known, it is logically possible that more truths (or better approximations to the truth) will be found in C' than in C. Whether or not this is the case is a matter of nondemonstrative inference from our present body of knowledge. In particular, whether the intuitive screening of hypotheses performed by Ph.D.'s is advantageous in admitting a greater number of true (or nearly true) hypotheses than some other method of screening depends on the contingency of trained scientists having good intuition for good hypotheses. A similar comment holds for methods of hypothesis-proposing confirmed to be useful by previous scientific investigations.

The foregoing argument can be recast in the form of a dilemma which is easily stated by considering the passage from Shimony (1970, pp. 132–33) quoted above: "By giving preferential treatment to seriously proposed hypotheses but insisting on open mindedness within this preferred class, the tempering condition provides a safeguard against one of the major types of error that could be committed by a method of confirmation: the error of rejecting, because of *a priori* commitments, a true hypothesis which someone has been fortunate enough to put forth."

We must ask, to begin with, whether the proposed condition really gives preferential treatment to some class of hypotheses. If not, the condition is not at all what it is set out to be, and we may fairly ask for a restatement before considering the question of justification. If there is a class of hypotheses to which the tempering condition gives preferential treatment, we must further ask on what grounds is this preferential treatment given. If the grounds are a priori it seems that, after all, hypotheses which might be true are rejected (or at least excluded from present investigations) as a result of a priori considerations. Thus any attempt to justify the condition a priori undermines the very advantage that is claimed for it. This does not show the condition to be undesirable, for the alternative answer to the second question may be, and I believe is, perfectly acceptable — a given preferential treatment can be justified on a posteriori grounds.[10] At one stage of inquiry some hypotheses are excluded from consideration, but on principles which are subject to revision precisely because they are not justified on a priori grounds. Since the principles of selection may be revised, errors made in selecting hypotheses for consideration at the present stage of inquiry may be corrected at a later stage.

Although the distinction between proposed and seriously proposed hypotheses is of no help in making out an a priori justification of scientific methodology, one might retreat to the first horn of the dilemma and restate the tempering condition in a way which makes no pretense of giving preferential treatment to any class of proposed hypotheses.[11] I suppose the suggestion to be spelled out along these lines: from the field of all statable hypotheses very few are actually thought of. Of those that are thought of or explicitly formulated, many are not really proposed as scientific hypotheses. For example, they may have only been put down as didactic or philosophical examples, or they may have been discarded in the process of formulation. Even taking a very liberal attitude in deciding which hypotheses are to classify as having been actually proposed, our innate capabilities, limitations, and intuitions severely restrict the range of hypotheses that we in fact put forward. Let us use the term "actually proposed hypotheses" to refer to hypotheses which are fully formulated and put forward for consideration.

We may now inquire whether anything is to be gained by revising the tempering condition in the following way: replace the requirement that all seriously proposed hypotheses are to be assigned nonnegligible prior probability with the requirement that all *actually proposed* hypotheses are to be assigned nonnegligible prior probability. I will call the revised condition the total tempering condition, and investigations which satisfy it totally open-minded investigations. It might be supposed that this form of the tempering condition constitutes true open-mindedness for a Bayesian formulation of scientific confirmation and that such true

[10] Shimony discusses this possibility in *ibid.*, V, D (1970, pp. 150–58).

[11] Professor Shimony has suggested this as a possible way out of the present difficulty.

open-mindedness is prima facie desirable, in the sense of being conducive to reaching scientific truths. Shimony appears to have something of this sort in mind when he writes, "The tempering condition is only a way of prescribing open mindedness within a probabilistic scheme of inference, and open mindedness is a necessary condition for the acceptance of true hypotheses in case they should ever be proposed [N.B. *not* 'seriously proposed']" (1970, p. 158).

I believe that such a revision would be a step in the wrong direction. Indeed, if one were to agree to test all hypotheses which were actually proposed, most investigations could quickly be brought to a halt by proposing spurious correlations which we intuitively judge not to be worth subjecting to direct examination. To make the point, I hereby propose (not seriously of course) that the observed frequency of all pulsars would be double their presently observed values if the U.S. Senate had two hundred members instead of one hundred. If we were required to give all truly proposed hypotheses a nonnegligible prior probability, this hypothesis would stand in competition to our present, well-confirmed belief in the near constancy of pulsars' frequency, and the progress of astrophysics would have to await constitutional reforms. Such examples support the claim that only seriously proposed hypotheses merit serious consideration, even though it may be far from clear how the term "seriously proposed hypothesis" is to be defined. In part V of "Scientific Inference" Shimony further supports this claim with an interesting discussion of relevant contingent facts about man and his environment. For example: "to first approximation we have a good sense of the relevance and irrelevance of various factors to phenomena of interest. This crude sense of relevance is often wrong, and some of its errors, such as over estimating the influence of 'wonders' in the heavens upon terrestrial events, have hampered the development of knowledge. Nevertheless, when this sense of relevance is controlled by critical intelligence, it makes crucial observations possible by enabling men to disregard the innumerable details that are potentially distracting in experimentation" (1970, p. 143). Were we required to be rigidly open-minded about all actually proposed hypotheses, such considerations would be beside the point. But in point of fact they are extremely important in understanding, explaining, and justifying our confidence in our methods of scientific investigation. Consequently, the total tempering condition and the "open-mindedness" to which it could be said to give rise would not be a desirable feature of tempered personalism.

I have argued from contingent premises that the total tempering condition is, as a matter of fact, not a desirable feature of scientific investigations; and if in fact the condition is not desirable, it is not an a priori truth that it is desirable. A different line of analysis clarifies these considerations, and provides an independent argument in support of the claim that the total tempering condition can not be shown to be desirable on a *priori* grounds. In the total process of an investigation we start with the set of all actually proposed hypotheses and conclude with some selection among them. In other words, the actually proposed hypotheses must go through a filtering process which excludes some hypotheses by giving them a very low valuation, and retains others by giving them a relatively high valuation. In a Bayesian formulation this filtering can take place in two ways: by assignment of prior probabilities to the hypotheses and by the processes of conditionalization (which depends in part on the assessment of likelihoods and in part on the results of the observations). However one chooses to measure selectivity, if the degree of final selectivity is fixed, then a decrease (increase) in the first filtering process must be accompanied by an increase (decrease) in the selectivity of the second filtering process. If we are equally ignorant about the efficacy of both filtering processes as means to the end of true or nearly true beliefs, then there is no grounds for opting to eliminate one filter in favor of the other. And as long as we forswear appeal to contingent premises, we are completely ignorant about the efficacy of both filters. Above, we considered the possibility of determining the first filter by reference to seriously proposed hypotheses, and we saw it to be logically possible that it would select false hypotheses or hypotheses far from the truth. Consequently, contingent premises would be required to conclude that such a filter is desirable. The same argument is easily seen to work for other criteria which might plausibly be substituted for "seriously proposed." Below I will argue that we have also been given no reason to suppose, on strictly a priori grounds, that the second filter is desirable.

The total tempering condition seemed to be a desirable feature of scientific investigations because it ensures that if a true or nearly true hypothesis is proposed it has a chance of surviving. But it is now clear that imposing this condition has a price; for if it is imposed, and if an investigation is to result in a selection of hypotheses, the selection must be fully carried out by the second filter. Thus the total tempering condition is desirable only if it is a desirable feature of scientific investigations that the process of selection among proposed hypotheses is fully carried out by conditionalization, the second filtering process. I turn now to consider whether we have been given any a priori reasons to suppose that conditionalization is desirable, that is, whether the process of selecting hypotheses by conditionalization is conducive to the end of reaching true or nearly true beliefs. It will follow from the complete absence of any such reasons that there are no a priori reasons for the desirability of selecting hypotheses by conditionalization alone.

4.2. *Examination of the second condition.* As explained in section 2, the hypotheses which have been admitted to an investigation are evaluated by conditionalization of the prior probability on the outcome of the observations. We will first ask whether Shimony succeeds in giving any a priori support for the claimed desirability of using conditionalization of a probability measure, rather than some other method, as the means of critically evaluating hypotheses. Then, granting that evaluation is to be carried out by conditionalization of a probability measure, we will ask what conditions must hold to warrant description of the evaluation as "critical" in a clearly desirable sense, and we will inquire whether it is possible to argue without appeal to contingent premises that these conditions obtain for the probability measures of tempered personalist investigations.

As for the first question, Shimony maintains that an investigation should have a Bayesian evaluation structure by arguing that the function specifying the degree to which it is rational to be committed to the truth of propositions must satisfy the axioms of conditional probability. Bayes's theorem follows from the probability axioms, thus making conditionalization the natural means to describe change of belief within an investigation.[12] Shimony holds that "the axioms of probability are necessary conditions for orderly thinking about propositions with unknown truth values" (1970, p. 158). He supports this claim with an ingenious combination of the Cox-Good argument and the Dutch book argument which avoids many of the shortcomings of each and makes it

[12] It is open to question whether conditionalization can be justified on the grounds that the belief function satisfies the probability axioms (cf. Hacking, 1967). Although Shimony says nothing on this point, I will not press the issue since I believe there are ways the claim can be substantiated.

possible to derive the probability axioms from exceedingly weak assumptions (1970, pp. 104–9). Still several assumptions are required and it is at least unclear how one could argue, strictly a priori, that these assumptions must be true of anything we would call "orderly thinking about propositions with unknown truth values." This is most clearly the case for the fifth and sixth assumptions from which Shimony argues, and which are stated as follows (with Shimony's numbering): as before, $h_1, \ldots, h_n$ are the hypotheses of an investigation and $e_1, \ldots, e_m$ the possible outcomes of the investigation's envisaged observations. Then for any e from among $e_1, \ldots, e_m$ and any c and d from the truth functional closure of $\{h_1, \ldots, h_n, e_1, \ldots, e_m\}$:

(5) There is a function F_e such that $P(c \,\&\, d/e) = F_e\,(P(c/d \,\&\, e), P(d/e))$.

(6) There is a function G_e which is continuous and monotonically increasing in both its variables such that if e logically implies $(c \,\&\, d)$ then $P(c \vee d/e) = G_e(P(c/e),\ P(d/e))$ (1970, p. 105).

As Shimony observes (1970, pp. 106–7), these conditions are intuitively very compelling for any plausible interpretation of the notion "rational degree of commitment"; but it is hard to see how we might turn this strong intuitive appeal into an a priori argument in support of the claim that any method for which (5) and (6) fail will be less conducive to the aim of learning about the world than a method of which (5) and (6) are true.

Let us suppose, however, that the proposed hypotheses of an investigation are to be evaluated in the Bayesian manner, by conditionalization. Let us then ask what conditions must hold to warrant description of such an evaluation as critical in a clearly desirable sense. And let us ask whether it is possible to argue a priori that these conditions do hold. In a trivial sense posterior evaluations of the hypotheses of an investigation are *determined* by the outcome of the envisaged observations, since the posterior probability of a hypothesis, h, is determined by conditionalization on the outcome, e; more exactly by multiplying the prior probability $P(h)$ by the likelihood $P(e/h)$. But the investigation can insofar hardly be said to be *critical* of the proposed hypothesis; nor can such a determination of $P(h/e)$ be claimed without further comment to be a desirable feature of the investigation. For all that has been said so far, the likelihoods might increase the probability of many false hypotheses and decrease the probability of many true hypotheses. To make plausible any such claims about the desirability of the hypotheses' evaluation some further requirement must be imposed on the likelihoods.

Consider the following condition on h and e:

(a) If h is true (or near to the truth) then e will always be, or will most frequently[13] be, the outcome of the envisaged observations. If h is false (or is far from the truth) then e will always not be, or will most frequently not be, the outcome of the envisaged observations.

and the condition on $P(e/h)$:

(b) If (a) is true then $P(e/h)$ is high. If (a) is false, $P(e/h)$ is low.[14]

If the likelihoods of an investigation satisfy some such condition we will say they are (objectively) reliable. If likelihoods are reliable in this sense, hypotheses which are true (or near to the truth) will always have, or will most frequently have, a higher posterior probability than prior probability; and other hypotheses will have a lower posterior probability than prior probability. If this condition holds for a series of investigations, the investigations can be said to be critical of their proposed hypotheses in the sense that false hypotheses (or hypotheses far from the truth) will generally receive low posterior probabilities and true (or nearly true) hypotheses will generally receive high posterior probabilities. And this state of affairs is by definition a desirable one for scientific investigations, as we have taken "desirable" to mean "conducive to achieving the end of reaching true or nearly true beliefs." It is clear from Shimony's discussion in section IV, D (1970, pp. 133–36) that this interpretation of tempered personalism's critical properties is the one he has in mind. But how are we to know that the likelihoods of an investigation are reliable? Their reliability must be established without appeal to contingent premises if there is to be any substance to the claim that tempered personalism can be seen a priori to be critical in the desirable sense we have just described.

[13] The reference class for the frequency count is the sequence of repetitions of the observations which it would be possible for the investigator to make, under conditions which the investigator subjectively judges to be similar to the original investigation.

[14] An obvious generalization of this is required for statistical hypotheses.

In section IV, D, Shimony attempts to show that tempered personalism's "receptivity toward seriously proposed hypotheses is adequately balanced by a capacity to evaluate them critically" (1970, p. 133). His arguments do support the claim that the likelihoods of tempered personalist investigations may be expected to be reliable on the whole, but an examination of these arguments reveals that none of them are a priori. Shimony begins by observing that the evaluation of likelihoods is, prima facie, a subjective process, since the likelihoods are to be determined along with the prior probabilities of hypotheses by the investigator's intuitive judgment at the beginning of an investigation. The only stated restrictions are that the overall prior probability assignment satisfy the probability axioms and that the tempering condition hold. However, he argues that this prima facie subjectivity can be limited by application of the tempering condition. In practice, an investigator's evaluation of the likelihoods will largely be determined by the antecedent beliefs he brings to an investigation. If there is any reason to suppose that these antecedent beliefs are inaccurate in a way which would lead to an unreliable evaluation of the likelihoods, the suspected inaccuracy can be seriously proposed and tested in an auxiliary investigation. Thus, to cite Shimony's example (1970, p. 135), if an investigator dismisses the method of collection of a sample as irrelevant to the properties of interest in the sample, someone might seriously propose that in fact the method of collection biases the data. The investigator would then be obliged to devise an independent investigation to establish whether or not such a bias exists. In this way most reasonable doubts about the reliability of an investigator's background beliefs can be effectually eliminated.

But the fact that it is always possible to examine critically an investigator's evaluation of likelihoods in no way shows that the reliability of his evaluation can be or ever is supported a priori. There are two reasons why likelihood evaluations always might be and sometimes are unreliable in spite of the constant possibility of checking. First of all, even when possible sources of systematic error are investigated, their examination depends on the subjective evaluation of the likelihoods of the auxiliary investigation. A chain of auxiliary investigations can be pushed back quite far, of course; but it must be cut off at some point if we are ever to come to any definite conclusions in any investigations. There is no reason to think that the subjective evaluation of likelihoods in initial investigations can be supposed, on strictly a priori grounds, to be reliable.

Consequently, we are given no reason to think that any investigations' evaluation of likelihoods can be supposed, on strictly a priori grounds, to be reliable.

The second reason why likelihood evaluations always might be and sometimes are unreliable in spite of the constant possibility of checking is that for any investigation there will be possible sources of systematic error which go unchecked. Shimony makes two remarks about this contingency. In the first he points out that if we were to seriously regard every contingency as a possible source of systematic error "no evidence could be utilized as reliable grounds for critically judging the seriously proposed hypotheses. A certain amount of tentative trust is a prerequisite for critically probing and testing, whereas sweeping skepticism is methodologically sterile" (1970, pp. 135–36). This is undoubtedly true, but it in no way circumvents the difficulty with the claim that the reliability of an investigator's likelihood estimates can be justified a priori. The problem was that, for all we can say a priori, there may very well be unsuspected biasing correlations. The fact that if we are to go on doing science we cannot endlessly worry ourselves about the possibility of unsuspected systematic errors in no way shows or supports the claim that there are no unsuspected systematic errors.

In his second remark about the constant possibility of unsuspected sources of systematic error, Shimony comments,

> Even after all serious proposals concerning possible bias have been checked and appropriate corrections in the sampling procedure (or in the analysis of sampling data) have been made, there always remain special circumstances concerning a sample which are unnoticed by investigators or which are habitually dismissed as trivial, and one might skeptically suspect that a systematic error is associated with them. The proper answer to this kind of skepticism is that a possible source of systematic error which no one has proposed for serious consideration is merely one of the infinite set of possible but unsuggested alternatives to the seriously proposed hypotheses in the investigation, and hence, it should be treated as a component of the catch-all hypothesis (1970, p. 135).

Do these observations help with the problem of giving a priori support for the claim that the likelihoods of tempered personalist investigations are, on the whole, reliable? It is true that if a false hypothesis is highly confirmed because of some unsuspected systematic error, then the true hypothesis either will be one of the other proposed hypotheses or will be a component of the catch-all hypothesis; thus the true conclusion which has been highly disconfirmed as the result of the unsuspected systematic error is in some way included in the alternative hypotheses of the investigation. But this fact does not bear on the question at hand since it does not provide any a priori reason to suppose that the likelihoods of the investigation are reliable or unreliable; nor does it provide any a priori reason to suppose that a systematic error has led to confirmation of a false hypothesis, should that be the case.

However, the passage seems to claim that hypotheses specifying each possible source of systematic error are themselves always covered by the catch-all hypothesis whenever they have not been explicitly suggested and included among the specific hypotheses of an investigation, and that consequently the danger of unsuspected bias is always directly taken into account. But this claim is mistaken; for the possible existence of an unsuspected bias is, with one exception discussed below, never in any sense "included" in the catch-all hypothesis. If h_1 and h_2 are the proposed hypotheses, the catch-all hypothesis, h_3, is $\sim(h_1 \vee h_2)$. If h_1 and h_2 do not assert the existence of biasing conditions, a bias may exist whether h_3 is true or false. For example, suppose h_1 is the hypothesis that pulsars are rotating neutron stars and h_2 the hypothesis that pulsars are oscillating gaseous masses. Then h_3 might be either true or false in face of the presence or absence of the spurious correlation with the U.S. Senate's size, suggested above on p. 184. In no sense is the Senate-correlation hypothesis "included" in h_3, and confirmation or disconfirmation of h_3 ordinarily throws no light on it. On the other hand, suppose that h_1 and h_2 were suggested for an auxiliary investigation intended to examine the possibility of misleading correlations. Thus h_1 and h_2 might respectively assert a correlation of a pulsar's frequency with its distance from the earth and with the amount of matter between it and the earth. In the same way as in the previous example, the Senate-correlation hypothesis will under most circumstances be independent of h_3, both logically and with respect to confirmation relations. The problem is that the logical contrary to the existence of a specified correlation is its nonexistence, not the nonexistence of other correlations. The only way out of the impasse is for one of the hypotheses of the auxiliary investigation, the catch-all or some other, to assert the absence of *all* unsuspected sources of systematic error. But how is such a hypothesis to be tested? And even if it could be tested, the difficulty with the a prioristic claim for the reliability of a set of likelihoods would stand; for the rejection of all possibility of unsuspected systematic error would not be a priori but would rely on a scientific investigation.

It should be observed in passing that although the constant possibility of subjecting an investigator's antecedent beliefs to an auxiliary investigation is no help in establishing a priori that tempered personalism is critical in the indicated sense, this possibility may play a very important part in other acceptable forms of argument which support the tempered personalist method of conducting investigations. For example, critical examination of an investigator's antecedent beliefs might be pushed back far enough so that it need assume no more than the rough reliability of a person's prescientific view of the world, in a sense which would have to be explained. If, along the lines of the seafarer's doctrine mentioned in the introduction, it is possible to develop a view of rationality which would allow us to appeal to such considerations even though they cannot be established a priori, it would then be possible to give persuasive arguments in favor of one or another method of evaluating beliefs. In the hope that such a program will be successful I think it worthwhile to pursue Shimony's discussion (1970, section V) of considerations depending upon facts about the world which support a tempered personalist formulation of scientific inference.

4.3. *Examination of the third condition.* As stated in section 3 of the present paper, the third aspect of tempered personalism's sensitivity to the truth was that

> (3) Investigations conforming to the prescriptions of tempered personalism can incorporate "any methodological device which analysis or experience indicates to be valuable."

As I understand it, the desirability of this property of tempered personalism derives from the first two properties. New methodology is seen in new ways of seriously proposing hypotheses or new ways of testing the ones proposed. The desirability of any such methodological tool can itself be proposed as a hypothesis, tested within the existing framework, and evidence permitting, accepted. However, this is a desirable event only if the methodological change is in fact for the better, in the sense that the change results in a method of inference which is more conducive to achieving the end of true beliefs than was the original method of inference; any argument for the claim that tempered personalism can be seen on a priori grounds to be desirable because (3) is true of it must show that the methodological changes adopted as a result of tempered personalist investigations are, on the whole, for the good, in the sense just described. But we have seen an all-round failure of the attempts to argue without appeal to contingent premises that (1) and (2) are at the same time true of tempered personalism and clearly desirable for scientific investigations; consequently we are given no a priori considerations in favor of tempered personalism on the grounds of (3).

The same point can be made in a different way. Beyond question, (3), the statement that tempered personalism can incorporate any methodological device which analysis or experience indicate to be valuable, seems to be a desirable property for a method of scientific inference. But (3) is deceptively incomplete in statement in a way which conceals the grounds on which it may be called into question. To be accepted, a new methodological device must be *indicated to be valuable by experience or analysis*. This indication of value must arise in some system of confirming hypotheses, which is all along understood to be tempered personalism. But if the investigations conforming to the prescriptions of tempered personalism should happen to be pernicious, suggesting and confirming bad methodology, justification of tempered personalism on the grounds of (3) would be a mockery. The claim that (3) stands to tempered personalism's credit rather than discredit must rest on independent justification of tempered personalism. I believe that such a justification may be made out along the lines which Shimony presents in section V, but none of these supporting arguments are a priori.

5. Attempts to Justify Tempered Personalism by Further Modification of the Pragmatic Form of Argument

5.1. *Motivating considerations.* Occasional remarks in "Scientific Inference" suggest several alternate approaches to the problem of giving an a priori justification for tempered personalism. Common to these suggestions is the idea that by seeking to show less with a pragmatic argument, the argument may succeed where it failed as an attempt to support a stronger claim. In spite of their great prima facie plausibility, pragmatic arguments do not seem to bear the weight of a full justification of scientific inference; but perhaps they may be used to provide a partial justification of some kind.

5.2. *An attempted modification of Shimony's arguments.* The first of these suggestions is to give an a priori argument in support of the claim that tempered personalism is a better means to the end of reaching true scientific conclusions than is any other *Bayesian* method of scientific inference. If we cannot show, a priori, that tempered personalism is desirable when compared with any other method of scientific inference, we might nonetheless give it a partial a priori justification by showing it to be desirable when compared with any other method which changes degree of belief by the application of likelihoods to prior degrees of belief. That Shimony at times has such a form of argument in mind is suggested by his acknowledged omission of comment on non-Bayesian treatment of scientific inference (1970, pp. 86–87) and by remarks which seem intended to justify tempered personalism as preferable to other *Bayesian* analyses of scientific inference. For example he writes that "the tempering condition is only a way of prescribing open mindedness *within a probabilistic scheme of inference,* and open mindedness is a necessary condition for the acceptance of true hypotheses in case they should ever be proposed" (italics mine, 1970, p. 158).

However, Shimony's a priori justificatory arguments will not support the weaker claim; in fact they will not support, a priori, a justification of tempered personalism as preferable to all the members of *any* restricted class of methods of scientific inference, even if the class has only one member. This conclusion can be sustained by reexamining Shimony's form of a priori pragmatic argument, which he described by saying, "My treatment of scientific inference is instrumentalistic, in the sense that the justification for certain prescriptions is their conduciveness to achieving certain ends, but the ends admitted are only the theoretical ones of learning the truth about various aspects of the world" (1970, p. 81). Study of Shimony's arguments has revealed that, for each prescription which defines tempered personalism, its conduciveness to achieving the end in question can be supported only by appeal to contingent considerations. If, restricting ourselves to a priori arguments, we have not succeeded in supporting the claim that tempered personalism is conducive to achieving the intended end, then the claimed conduciveness cannot be cited in an a priori argument as grounds for preferring tempered personalism to *any* method of scientific inference, whether the method is taken from the class of all methods, the class of Bayesian methods, or the one-membered class of a single competing method.

One might hope that the apparent failure of Shimony's arguments to show the preferability of tempered personalism to any of a narrow class of competing methods is merely an accident of the quoted statement of his overall strategy and that something can be salvaged by examining the details of his analysis. But a review of the reasons we gave for rejecting his a prioristic claims shows that there are little grounds for such a hope. Since Shimony seems to have had the class of Bayesian methods in mind, let us examine the possibility of justifying tempered personalism in preference to any other Bayesian method of scientific inference, appealing to the considerations which Shimony has suggested. (Similar comments can be made if one considers some other competing class of methods.) First, tempered personalism is held to be preferable bcause it is open-minded. But, however one characterizes open-mindedness, for all we can say a priori, a close-minded policy might, by luck or some incredible quirk of nature, stumble upon more truths than an open-minded policy. In particular, if open-mindedness is characterized by the prescription that seriously proposed hypotheses be given nonnegligible prior probability, then for all we know a priori a Bayesian method which gave nonnegligible prior probability to all and only facetiously proposed hypotheses might confirm more truths than tempered personalism, if the latter class of hypotheses happened to contain more truths than the former class. If open-mindedness is characterized by the prescription that all actually proposed hypotheses be given nonnegligible prior probability, then tempered personalism will be a more successful method than its competitor only if its likelihoods are reliable, for which we likewise have no a priori reasons.

Secondly, tempered personalism is held to be preferable to competing methods because it has the "capacity to evaluate hypotheses critically." We first argued that Shimony does not succeed in giving an a priori justification for the stipulation that critical evaluation be carried out by conditionalization of a prior probability function. Then we agreed to assume that critical evaluation was to be carried out in this way, and we inquired at length whether Shimony has given any a priori reason to believe that the likelihoods of tempered personalist investigations are reliable. We saw that the reliability of tempered personalist likelihoods depended on the reliability of the investigator's subjective judgment and that the checks which one might place on this subjective judgment likewise depended on the subjective judgment used in other investigations. Since justification of the reliability of tempered personalist likelihoods depends on the reliability of subjective judgments used at some stage of investigation and since we have no a priori grounds for trusting these subjective judgments, we have no a priori grounds for believing that the likelihoods of tempered personalist investigations will be at all reliable, or more or less reliable than the likelihoods of some alternative Bayesian method, say that of Carnap.

Finally, the third ground for preferring tempered personalism, that the method is flexible, was seen to depend entirely on the first two grounds. As neither of these gives an a priori reason for preferring tempered personalism to any other Bayesian method of scientific inference, neither does the third.

5.3. *Summary of Reichenbach's justificatory argument.* The second and third suggestions for modifying a pragmatic justification return to the form of argument developed by Reichenbach and are summarized as follows (cf. Reichenbach, 1949, p. 474): Reichenbach tried to show, for a stated end E and a program of action M, and without appealing to contingent premises, that the program of action M is certain to achieve the end E if there is any program of action by which E could be achieved. He then observed that if we desire to achieve E and if nothing else relevant is known, then this conclusion *in itself* constitutes a practical reason for adopting method M as a means of achieving E. This observation is an application of what decision theorists call the dominance principle: if, as far as one can tell, a course of action, M, will under all circumstances yield at least as desirable results as would any alternative, then it is reasonable to choose M as one's course of action.

Let us write "$F(X)$" for "end E will certainly be achieved by following program of action X." Then Reichenbach's conclusion can be represented by the expression

(1) $(\exists X)\, F(X) \supset F(M)$.

Reichenbach demonstrated (1) by stating a condition C and then demonstrating (2) and (3) to be logical truths:

(2) $(\exists X)\, F(X) \supset C$ (C is a necessary condition for the existence of a method by which E can be achieved).

(3) $C \supset F(M)$ (C is a sufficient condition for the attainability of E by method M).

(1) follows logically from (2) and (3). (3) can be weakened to

(4) $(\exists X)\ F(X) \supset (C \supset F(M))$ (C is a sufficient condition for the attainability of E by M, if E can be attained by any means at all; or, for short, C is conditionally sufficient for the attainability of E by method M).

It is to be noted that in giving an argument of this form we do not need to know or to have any reason to believe that C is true, as long as we do not know that it is false and as long as we do not know that F(M) is false. But for the argument to be a priori it is essential that (2) and (3) (or (4)) be demonstrated to be logical truths so that no contingent premises are needed in demonstrating (1).[15]

5.4. *The second attempted modification of a pragmatic argument.* Applying Reichenbach's argument to the problem of justifying tempered personalism, we take E to be the end of reaching true scientific conclusions and M to be the tempered personalist method of performing scientific investigations. For a first proposal for adapting Reichenbach's argument one might suggest the same kind of modification examined above: limit the range of the quantifier to some subset, B, of possible methods of scientific inference. In other words the proposal is to show

(5) $(\exists X)\ (X \epsilon B \,\&\, F(X)) \supset F(M)$

to be a logical truth. Since (5) has a stronger antecedent than (1), one might hope to show (5) to be a logical truth, even though (1) is not.

[15] Reichenbach's justification failed, not because of any fault in his logic, but because he succeeded in demonstrating (2) and (3) to be logical truths only by giving an unacceptable interpretation to the statement that E has been attained. He held that we succeed in attaining the end of "finding" the limiting value of relative frequencies in a sequence of observations when we come blindly to posit the correct limit value, without having any idea whether what we posit is the correct value or not. And he held that all the practical and theoretical aims of science can be adequately interpreted in terms of "finding" limit values in this sense. Many of the more familiar difficulties with his justification can be seen to depend, at least in part, on this fundamental flaw.

I am not able to give a general argument in support of the claim that (5) is not a logical truth for any acceptable interpretation of "reaching true scientific conclusions" and for any collection B of methods of scientific inference (except, of course, $B = \{M\}$), but examination of plausible examples leaves little hope that such a B can be found. For instance, taking B to be the set of all Bayesian methods of scientific inference, it is easy to argue that (5) could, as a logical possibility, be false. For it is logically possible that men should seriously propose only false hypotheses but facetiously propose true ones. If so, $F(M)$ would be false, but it is logically possible that at the same time there exist a Bayesian method, X, which assigns nonnegligible prior probabilities to facetiously proposed hypotheses, such that "$F(X)$" is true. Or, if tempered personalism is taken to be stated in terms of the alternate tempering condition, giving nonnegligible prior probability to all actually proposed hypotheses, then again $F(M)$ may be, in fact probably will be, false. As I argued in section 4.1, facetious suggestion of implausible hypotheses which are hard to test would quickly bring use of such a method to a halt. No scientific conclusions would be reached. At the same time some other Bayesian method may work perfectly well, and this possibility again shows (5) not to be a logical truth. Similar remarks could be made by appealing to the logical possibility that tempered personalism fails because the likelihoods resulting from human subjective judgment are systematically unreliable; while in the same circumstances another Bayesian method, say Carnap's, might succeed.

Presumably, someone wishing to give a priori reasons in support of (5) would try to use Reichenbach's method and find a condition, C, satisfying (2) and (3), (or (4)), with the range of the variable X restricted to B. But again, in the case of plausible candidates for B, there seems to be little reason to expect that such a condition, C, can be found. Let us again examine the case in which B is taken to be the collection of all Bayesian methods. What conclusions which might be used in Reichenbach's argument schema can we draw from the assumption that there exists a successful Bayesian method of scientific inference? It follows from this assumption that

(i) there exists a method that men can use for assigning prior probabilities which gives some true hypotheses nonnegligible prior probabilities, and

(ii) there exists a method which men might use for evaluating likelihoods such that the likelihoods resulting from this method are, on the whole, reliable.

Let the conjunction of (i) and (ii) be the condition *C*. *F*(*M*), the statement that tempered personalism is a successful method of scientific investigation, can hardly be said to follow from *C*. To conclude that *F*(*M*) we require not only the condition that there exist some humanly applicable methods of assigning satisfactory prior probabilities and likelihoods, but that these methods be the ones described by tempered personalism, appealing as it does to the condition of serious proposal and to subjective evaluation. There are, after all, other methods of assigning prior probabilities and likelihoods. Carnap has tried to develop one such method which many philosophers take seriously, and the range of all possible Bayesian methods must be taken to include methods which we know, on contingent grounds, to be foolish, such as any random number generator taken together with a mechanical procedure for transforming an arbitrary sequence of numbers into a distribution satisfying the probability axioms. The assumption that there exists some method of assigning satisfactory prior probabilities and likelihoods is logically compatible with the assumption that the method happens to be one of these alternatives to tempered personalism.

5.5. *The third attempted modification of a pragmatic argument.* A second proposal for weakening Reichenbach's argument is suggested by a remark at the end of Shimony's discussion of his a priori considerations:

[I]t is tempting to take Reichenbach's pragmatic justification as a paradigm and to make a conditional apodictic claim — i.e. *if conditions 1, 2 . . . are satisfied, then* [*the tempered personalist formulation of scientific inference is a method which, if persisted in indefinitely, can discriminate true from false hypotheses*] — and to do so in such a way that a methodological argument can be given for acting as if conditions 1, 2, etc., are true, whether or not their truth can ever be checked. Indeed, a step in that direction was already taken in section IV. C, in reasoning that "unless we act as *if* good approximations to the truth will occur among the hypotheses which will be seriously proposed within a reasonable interval, we are in effect despairing of attaining the object of inquiry." This reasoning supplies a natural condition 1 . . . (1970, p. 137).

Shimony clearly states that he does not see how the list is to be completed and that it seems futile to him to try to make the apodictic claim without qualification (1970, p. 137). Yet someone might suppose that in partially specifying the list of conditions 1, 2 . . . one succeeds, in some sense, in giving a partial pragmatic justification of tempered personalism.

If there is a suggestion for modifying Reichenbach's argument implicit in the foregoing passage, it is the following: Let conditions C_1, C_2, . . . , C_n be conditions which are individually necessary for the existence of some program of action by which *E* can be attained and which are jointly conditionally sufficient for the attainability of *E* by *M*. We suppose it is not known that *E* cannot be achieved by following program of action *M*. Clearly, the conjunction C_1 & C_2 & . . . & C_n will serve as a condition, C, for Reichenbach's form of argument. Unfortunately, we have only found a condition C_1 which we do not know to fail and which is a necessary condition for the existence of some program of action by which *E* can be attained; but at the present time we see no way to complete the list of conditions, nor do we have any concrete reason to suppose that the list can be completed. Nonetheless, in providing a first member of a list which might be completed, we have carried out a first step in Reichenbach's argument, and in providing *part* of the argument we have provided some reason to favor program of action *M* over alternatives.

Any such argument should be rejected out of hand. The claim was to have given part of a valid argument and thereby to have given some reason for the conclusion. But to provide only *part* of an argument which ordinarily supports a conclusion is not to support the conclusion at all. Part of a good or acceptable argument simply need not be a good or acceptable argument. Part of an argument provides no reason for its conclusion, unless (which is not the case in the present situation) it is obvious from context or by common knowledge how the argument is to be completed. Therefore, as things stand, there is no reason to suppose that arguments of the form we have been considering provide reason to think that a program of action, *M*, is preferable to alternatives as a means to an end, *E*.

6. Concluding Remarks: The Alternative to an A Priori Justification

I have shown that none of Shimony's arguments support tempered personalism a priori, and by an extensive consideration of alternative arguments I have cast strong doubt on the claim that any argument can accomplish this aim. What repercussions should these conclusions have for the evaluation of tempered personalism?

I take the answer to be: no repercussions. No proposed method of scientific inference, not independently known to be unacceptable, has ever been supported by a priori arguments. There are strong — some would say conclusive — reasons to think that no such method can be justified a priori. So methods of scientific inference must be compared and evaluated, if at all, on the basis of a posteriori considerations; and the most powerful method we have for pursuing this objective is to make a proposed method of scientific inference an object of study and evaluation of our currently accepted scientific framework, in the spirit of the seafarer's doctrine described in the introduction. Shimony begins such an investigation of tempered personalism in the last section of "Scientific Inference," to which I have made passing references incidental to the main issues discussed here.

Whatever analysis is given of circular argument, the process of subjecting a method of scientific inference to scientific examination will surely be counted as circular, and on that ground many philosophers would dismiss as "vicious" any supportive argument resulting from such a process. With a few notable exceptions, such as Goodman (1965, pp. 62–65), and Quine (e.g., 1966, pp. 239–41), contemporary philosophers have been too quick, indeed totally unreflective, in dismissing all circular arguments as incapable of giving support to their conclusions. It may well be that an analysis of different types of circular arguments will help to unpack the metaphors of the seafarer's doctrine and disclose conditions under which a circular pattern of justification and adjustment is, in Goodman's words, "virtuous" rather than "vicious."

In his closing paragraph Shimony gives his best defense of a scientific examination of methods of scientific inference. He considers the objection to such a justificatory pattern that it provides no safeguard against the possibility that our beliefs and our methods of changing belief might fit together in a way which makes it impossible to discover that certain beliefs are false:

Conceivably . . . such a meshing [of scientific conclusions and methodology] could occur in spite of deep-lying errors in the scientific picture, which failed to be exposed in spite of strenuous experimental probing. There would, so to speak, be a probabilistic version of the Cartesian deceiving demon. Despite this conceivable eventuality, a full commitment to a methodology and a scientific picture which mesh in this way would be rational, for no stronger justification is compatible with the status of human beings in the universe. After all, in spite of the advances which have been made upon Hume's treatment of inductive inference, his basic lesson remains valid: that the conditions of human existence which force us to resort to inductive procedures also impose limitations upon our ability to justify them (1970, p. 161).

One might take this comment as a last attempt to justify some method of scientific inference without appealing to any contingent premises. But this would be a mistake; for the comment does not *in itself* justify commitment to any one method, as opposed to some other. Rather the comment defends the rationality of commitment to whatever method it may happen to be that results from a process of criticism and adjustment in the spirit of the seafarer's doctrine. The defense is far from conclusive; one might suggest in opposition that commitment to a method of scientific inference is rational only if the principles to which the method appeals are true. Russell claimed something of this sort in *Human Knowledge* (1948 cf. pt. VI, especially pp. 495–99). But the defense has considerable initial plausibility, giving us one more reason for clarifying existing statements of the seafarer's doctrine and for attempting a more exact analysis of the related circular forms of justification, so that we may more competently sustain or reject these attractive suggestions.

REFERENCES

Black, M. (1954). "Pragmatic Justifications of Induction," *Problems of Analysis.* Ithaca, N.Y.: Cornell University Press. Pp. 157–90.

Carnap, R. (1952). *The Continuum of Inductive Methods.* Chicago: University of Chicago Press.

Carnap, R. (1962). "The Aim of Inductive Logic," in E. Nagel, P. Suppes, and A. Tarski, eds., *Logic, Methodology, and the Philosophy of Science.* Stanford, Calif.: Stanford University Press. Pp. 303–18.

Carnap, R. (1963). "Replies and Systematic Expositions," in P. A. Schilpp, ed., *The Philosophy of Rudolph Carnap*. La Salle, Ill.: Open Court. Pp. 966–79.

Dewey, J. (1949). *Logic, the Theory of Inquiry*. New York: Henry Holt.

Goodman, N. (1965). *Fact, Fiction, and Forecast*. 2nd ed. New York: Bobbs-Merrill.

Hacking, I. (1967). "Slightly More Realistic Personal Probability," *Philosophy of Science*, vol. 34, pp. 311–21.

Peirce, C. S. (1931). *Collected Papers*. C. Hartshorne and P. Weiss, eds. Cambridge, Mass.: Harvard University Press.

Putnam, H. (1963). " 'Degree of Confirmation' and Inductive Logic," in P. A. Schilpp, ed., *The Philosophy of Rudolph Carnap*. La Salle, Ill.: Open Court. Pp. 761–83.

Quine, W. (1963). "Two Dogmas of Empiricism," in *From a Logical Point of View*. New York: Harper & Row. Pp. 20–46.

Quine, W. (1966). *The Ways of Paradox and Other Essays*. New York: Random House.

Reichenbach, H. (1949). *The Theory of Probability*. Berkeley: University of California Press.

Russell, B. (1948). *Human Knowledge, Its Scope and Limits*. New York: Simon and Schuster.

Savage, L. (1954). *The Foundations of Statistics*. New York: John Wiley.

Savage, L. (1967a). "Difficulties in the Theory of Personal Probability," *Philosophy of Science*, vol. 34, pp. 305–10.

Savage, L. (1967b). "Implications of Personal Probability for Induction," *Journal of Philosophy*, vol. 64, pp. 593–607.

Shimony, A. (1970). "Scientific Inference," in Robert G. Colodny, ed., *Pittsburgh Studies in the Philosophy of Science*, vol. 4. Pittsburgh: University of Pittsburgh Press. Pp. 79–172.

Suppes, P. (1966). "Probabilistic Inference and the Concept of Total Evidence," in J. Hintikka and P. Suppes, eds., *Aspects of Inductive Logic*. Amsterdam: North Holland.

ABNER SHIMONY

Vindication: A Reply to Paul Teller

1. One of the few things about which I am optimistic is that we are close to a clear understanding of scientific inference. I think that, as a result of the work of many contributors, all the elements for a complete theory of scientific inference are at hand — although it is risky to say so, especially in view of the dramatic history of deductive logic after it was commonly supposed to have been a complete science. However, even if the requisite elements have all been discovered, there is still hard work to do in properly combining them, for what is needed is not eclecticism but a piece of intellectual architecture.

Professor Paul Teller has criticized[1] the architectural design that I proposed in my essay "Scientific Inference."[2] I tried there to justify scientific inference in two stages. The first stage was a priori, utilizing a kind of vindicatory argument. The second stage was a posteriori and relied upon certain factual propositions about the world, which themselves have been incorporated into our scientific world view as a result of scientific inference. Teller claims that nothing was accomplished in the first stage, and he suggests that the only justification which is possible or needed for scientific inference will be a posteriori. Teller's criticisms of my a priori arguments are forceful and point to weaknesses in my design. It probably was a mistake to make a sharp separation between the a priori and the a posteriori stages in the justification. I do not concede, however, that we must dispense with vindicatory arguments. Indeed, I feel sure that when he works out the details of his a posteriori justification of scientific inference, he will inevitably find that he also must resort to them. The following examination of his criticisms leads

[1] Paul Teller, "Shimony's A Priori Arguments for Tempered Personalism," this volume.

[2] Abner Shimony, "Scientific Inference," in R. G. Colodny, ed., *Pittsburgh Studies in the Philosophy of Science*, vol. 4 (Pittsburgh: University of Pittsburgh Press, 1970).

to the conclusion that a complete and adequate theory of scientific inference must present a subtler meshing of a priori and a posteriori considerations than one finds in my essay.

2. Some remarks on the development of the scientific method will be useful before replying to Teller's specific criticisms. As Peirce pointed out, the essence of the scientific method is the deliberate attempt to arrive at beliefs which are shaped by objective reality.[3] The scientific method is the systematic outgrowth of our natural intellectual propensities toward seeking the truth, which are operative even though we are also endowed with contrary tendencies toward fantasy and delusion, and even though cultural factors almost always curtail objective inquiry. The achievement of systematic procedures for seeking the truth required not only the conquest of these psychological and cultural impediments, but also the discovery of effective procedures. Such procedures as searching for correlations and invariants, proposing and checking hypotheses, introducing theoretical concepts, taking samples, utilizing control groups, taking into account systematic and random errors, and employing probabilistic reasoning are all embryonically present in prescientific native intelligence. Yet their explicit formulation, and their separation from other procedures which are also embryonically present but turned out to be misleading, must be reckoned as discoveries. Consequently, there has been a long and intricate history of scientific method, intertwined with the history of science itself, for the successful procedures were formulated in the course of struggling with substantive scientific problems or in reflection upon accomplishments. The historical development of the scientific method is as good an example of the dialectical uncovering and refinement of knowledge preexistent in the soul as any which Plato himself offers in his dialogues, though the dialectic process required not only discussion and analysis but also experimentation.

My essay, which concerned only the inferential part of the scientific method, proposed a central principle called "the tempering condition": that a seriously proposed hypothesis should be assigned a sufficiently high prior probability to allow it a chance to be accepted at the conclusion of an investigation. Although this principle was first articulated only about thirty years ago, by Harold Jeffreys,[4] the openmindedness which it

[3] Charles S. Peirce, *Collected Papers*, Charles Hartshorne and Paul Weiss, eds., vol. 5 (Cambridge, Mass.: Harvard University Press, 1934), para. 384.

[4] Harold Jeffreys, *Theory of Probability*, 1st ed. (Oxford: Clarendon Press, 1939), pp. 107–8.

prescribes surely evokes in many people the sense of recognition that occurs when a deep-rooted intellectual propensity is stated explicitly. For this reason, the vindicatory argument proposed in my essay, which turns crucially upon the tempering condition, has applicability beyond my particular (tempered personalist) formulation of scientific inference.

3. Teller accepts the tempering condition but offers two objections to my treatment of it. First, he points out that the phrase "seriously proposed" is not fully explicated in my essay, and therefore one cannot tell how restrictive the condition is. Second, he says that no matter how the phrase is explicated, so long as it is restrictive in some way, it is a factual matter whether the class of seriously proposed hypotheses is more likely to include close approximations to the truth than does its complement; and factual matters can never be settled a priori.

On page 110 of "Scientific Inference" I anticipated the first objection by stating a dilemma: either one gives the phrase "seriously proposed" a precise explication, which would undoubtedly be arbitrary, or one gives no precise explication and thereby leaves too much to subjective judgment. The methodological guidelines which I then suggested in order to escape from this dilemma appear to me to be correct, as far as they go, but they should be strengthened by a posteriori considerations. Something has been learned about controlled activities which foster the native human propensity for seeking the truth. For example, there are rules governing experimental design, which ensure that competing hypotheses are exposed in an evenhanded manner to the facts of nature. Although no strict rules have been discovered governing retrodiction (the proposal of hypotheses), at least some heuristic principles have been learned from millennia of experience: e.g., become immersed in the subject matter, learn the art of using the work of great predecessors as paradigms, employ analogies, try to achieve perspective and to see through irrelevant complications. Social institutions have evolved for the primary purpose of cultivating the techniques which systematize the truth-seeking propensity; and even after proper allowance is made for other motivations, which tend to corrupt universities and professional societies, the existence and scientific success of these institutions are facts to marvel at. Reasonable criteria for classifying a proposal as seriously proposed should take into account the accumulated knowledge concerning the process of inquiry, and therefore these criteria are subject to change with the development (and, one hopes, the improvement) of the procedures of scientific method. There is, I think, a vague but powerful criterion of this sort operative in the subjective judgments of most research workers — that a serious proposal is one made in circumstances which have in the past been favorable to good retrodictions. This criterion could be narrowly construed, to admit only hypotheses proposed by persons with credentials attesting to approbation by those institutions which have been successful in advancing knowledge. It could also be widely construed, so as to admit any hypothesis made in circumstances satisfying the standards of which those institutions are the custodians. For the methodological reasons given in "Scientific Inference" I favor the broader construction, but also I do not believe that imprecision of definition on points of this sort is crucial. (See pp. 110–111 of "Scientific Inference" and also the epigraph.)

In order to answer Teller's second objection, the vindicatory argument which I gave should be modified to take into account a posteriori considerations. I argued that my formulation of scientific inference "is in no way 'a logic of discovery' but rather supports whatever powers human beings may have for making intelligent guesses" (*ibid.*, p. 132). This argument can now be sharpened by recognizing that the accumulated experience with inquiry provides rough criteria for delimiting the class of serious proposals. If we do not give preferential treatment to seriously proposed hypotheses over frivolously proposed and unsuggested ones, where the class of seriously proposed hypotheses is defined by reference to the general character of past scientific achievements, then effectively we are skeptical of human retrodictive powers and of the possibility of systematic increase of knowledge. If our goal is systematic knowledge about the world, then nothing will be gained, and possibly the supreme object will be lost, by such skepticism. Here, then, is a sketch of a vindicatory argument which makes use of a posteriori considerations, instead of trying (as in "Scientific Inference") to avoid an appeal to them.

Teller wishes to dispense with vindicatory arguments such as the one just sketched and to defend the tempering condition on a posteriori grounds alone. The grounds for the conclusion that the class of seriously proposed hypotheses is more likely to include close approximations to the truth than its complement, he claims, "are to be found, no doubt, in the long history of man's search for knowledge" ("Shimony's A Priori

Arguments for Tempered Personalism," p. 181). I am not sure what detailed strategy Teller has in mind for an a posteriori justification. He may, like Black,[5] take the inductive arguments of the past to constitute a sample from the entire population of inductive arguments and estimate the percentage of successes in this sample; then, using statistical inductive inference, he could extrapolate from sample to population. Or he may continue the line suggested in section V of "Scientific Inference" of investigating the contingencies of the actual world which favor our inferences. Whatever his strategy, however, I cannot conceive of his arriving at the conclusion he desires about the class of seriously proposed hypotheses unless he makes enormous extrapolations from all his evidence. This is particularly clear if one considers that the truths sought in the natural sciences are mostly universal generalizations, each of which transcends any finite body of data; hence, statements about the distribution of approximations to these truths among various classes of hypotheses are very remote from experimental data. Some kind of vindicatory argument, to the effect that nothing that we desire will be lost and much may possibly be gained by permitting tentative extrapolations, seems to be required in order to complete Teller's reasoning. Otherwise, by maintaining Hume's standards for justification, he will not be able to avoid Hume's skepticism.

4. Teller also objects to the reasons given for the claim that in my formulation of scientific inference "receptivity toward seriously proposed hypotheses is adequately balanced by a capacity to evaluate them critically." He seems to agree that the claim is correct, but on a posteriori grounds; whereas my defense fails because I am unable to give a priori reasons for expecting the tempered personalist evaluations of likelihoods to be on the whole reliable. His criticisms seem to me to be penetrating, even though I cannot accept his conclusion.

The rationale for the reliability, on the whole, of these evaluations is more complex than I indicated in "Scientific Inference" and also more complex than he seems to recognize. First of all, it is essential to recognize the character of the tempered personalist likelihood $P_{X,b}(e/h \,\&\, a)$, where X is the person making the evaluation, b is his background information, a is his total body of auxiliary information and assumptions,

[5] Max Black, "The Raison d'Être of Inductive Argument," *The British Journal for Philosophy of Science*, 17 (1966), 286.

e is a body of explicit experimental data, and h is the hypothesis in question. Although the tempering condition constrains subjective judgment somewhat in evaluating likelihoods (see pp. 134–35 of "Scientific Inference"), it nevertheless allows a great difference between $P_{X,b}(e/h \,\&\, a)$ and $P_{X',b'}(e/h \,\&\, a)$ (where a is the same in both, but different persons and different background information are involved). In case the difference is great, at least one of these two likelihood evaluations will be unreliable as a guide to actual occurrences. The difference in subjective judgment may be compatible with intellectual integrity on the part of both X and X'. For example, suppose that e is a proposition about a statistic in a sample. Then X may suspect, because of his background b, and X' may not suspect because of b', that under hypothesis h and body of assumptions a there is a bias operative in the selection of the sample. Now if X and X' are seriously interested in the truth and wish their opinions to be shaped by objective reality, then the clash of their subjective judgments can often be resolved by changing the experimental situation. For instance, an auxiliary investigation may be performed to check the possible source of bias that X suspects, and if a' is the total body of auxiliary assumptions and information at the conclusion of this checking operation, then $P_{X,b}(e/h \,\&\, a')$ and $P_{X',b'}(e/h \,\&\, a')$ may be very close to each other. In fact, once two investigators tentatively agree that sampling is random, their likelihood evaluations usually agree quite closely. (If either X or X' is dominated by motives other than desire for the truth in the matter at hand, the situation is yet more complex, and case studies of flat-worlders etc. show that questions of scientific method and of psychopathology become intermingled. A complete epistemological theory must come to grips with these complexities, or else it will fall into smugness about the wisdom of the scientific establishment. But in this brief reply it does not make sense to discuss them.) Now, of course, bias may remain in the sampling procedure even at this stage, despite the agreement between X and X' that it does not occur. At this point two different considerations are relevant. The first is a posteriori and was not mentioned in "Scientific Inference": that once serious investigators have attempted to detect sources of bias and are satisfied that none occur in the case at hand, then there is a propensity to reliability in their likelihood evaluations. This propensity is indicated by the large proportion of successful judgments in past cases of this kind. Whether there is more to propensities than frequencies is yet another complex question. I think the answer is positive, but I do not think the answer is crucial in the epistemological analysis of scientific inference. The second consideration is vindicatory and was discussed on pages 135–36 of my essay: that if we allow suspicion of possible but unspecified biases to undermine our belief in the validity of every process of checking a hypothesis, then in effect we have let skepticism block the road to inquiry.

The foregoing is only a sketch of an answer to Teller concerning the reliability of tempered personalist likelihood evaluations, but it suffices, I believe, to show that a posteriori and methodological considerations must be conjoined in an intricate manner.

5. Teller's final criticism is directed against my statement that tempered personalism is sensitive to the truth because it can incorporate any methodological device which analysis or experience indicates to be valuable. He says that I have made an unjustified a priori claim and that one can only judge a posteriori whether methodological changes adopted in accordance with tempered personalist procedures are conducive to achieving true beliefs.

However, my statement about sensitivity to the truth was definitely not intended to be a claim to any kind of synthetic a priori knowledge. Rather, it was a claim that the tempered personalist formulation of scientific inference has neither more nor less structure than can be specified on methodological grounds alone. This formulation is sufficiently open to allow the incorporation of techniques of scientific inference which experience has shown to be successful, where success is judged by accuracy of predictions and by the coherence of the resulting world view. In particular, this openness permits the criteria of "serious proposal" to change, since new evidence may accumulate regarding the circumstances for good retrodictions. On the other hand, the tempered personalist formulation is sufficiently structured to provide the minimal machinery (probability theory and guidelines for assigning prior probabilities) for evaluating the achievements of novel techniques.

We do not know with certainty, however, that successive refinements of procedures of scientific inference in the light of a posteriori judgments about success will lead to a highly reliable instrument and thereby to a true picture of the world. Even if the refinements continue until there is exact meshing of methodology and the scientific picture of the world, it will still be necessary to fall back upon a vindication.

> Conceivably . . . such a meshing could occur in spite of deep-lying errors in the scientific picture, which failed to be exposed in spite of strenuous experimental probing. There would, so to speak, be a probabilistic version of the Cartesian deceiving demon. Despite this conceivable eventuality, a full commitment to a methodology and a scientific picture which mesh in this way would be rational, for no stronger justification is compatible with the status of human beings in the universe ("Scientific Inference," p. 161).

When Teller attempts to give posterior justifications of principles of scientific inference by referring to the history of science, I think that he will arrive at the same juncture and will therefore be forced to rely upon a similar vindicatory argument.

RONALD N. GIERE

The Epistemological Roots of Scientific Knowledge

1. Introduction

The objective of this paper is to complete one part of a broad program for resolving all important issues involving probability, inductive logic, and scientific inference generally. I shall begin by sketching the general program and locating the specific problem of interest in the general scheme. After the specific problem has been resolved I will conclude with some further discussion of the general program.

Though metaphilosophy belongs last in the order of discovery, there are good reasons for putting it first in the order of exposition. Let us begin, therefore, with the meta-question: What is the ultimate goal of inquiry into probability and inductive inference? Some have argued that scientific, nondeductive reasoning is necessarily an intuitive piecemeal affair, more art than theory. The most we can accomplish, on this view, is some conceptual clarification and codification of a variety of inductive methods.[1] The goal of the majority, on the other hand, has been to construct a unified, comprehensive account of scientific reasoning that could win the kind of widespread acceptance enjoyed by deductive logic.[2] This is the goal adopted here.

AUTHOR'S NOTE: This work was completed during my tenure as associate research scientist at the Courant Institute of Mathematical Sciences, New York University, and has been supported by the National Science Foundation under grants GS 2525 and GS 28628. A special acknowledgment is due Professor Allan Birnbaum of the Courant Institute both for much helpful criticism and for making it possible to pursue this work in such exciting surroundings.

[1] Among philosophers this viewpoint is generally associated with the "ordinary language" school exemplified by Strawson (1952) and Black (1971). Popperians (1959) also generally reject the idea of a systematic theory of inductive inference, though for very different reasons. Among statisticians, Tukey (1961) is well known for advocating "data analysis" over standard methods of testing and estimation.

[2] This was certainly the aim of both Carnap (1950) and Reichenbach (1949). It is also the aim of later philosophical students of inductive reasoning, e.g.,

The core of any overall account of inductive inference would be a precise *characterization* of the form (or forms) of inductive inferences. The system thus characterized would be an inductive logic in the broad sense that it would be a canon for validating scientific reasoning. But that it would be anything like the inductive logics of Carnap (1950) or Hintikka (1966) is an open question. Indeed, the central question for any account of inductive reasoning seems to me to be whether inductive logic is to be characterized in terms of a *relation*, e.g., degree of belief, between evidence and a hypothesis, or in terms of a *method* for using data to reach a conclusion or to choose a course of action. My understanding of 'inductive logic' is intended to be broad enough to cover both possibilities.

While a precisely characterized inductive logic is the heart of any comprehensive account of scientific reasoning, there are several other crucial components. One is an *interpretation* of the mathematical probability calculus. The problem of interpreting the probability calculus and characterizing an inductive logic are quite distinct, though often closely related. For Reichenbach (1949) or Salmon (1966) the distinction is clear: probabilities are interpreted as limiting relative frequencies while induction is characterized by a rule of simple enumeration. Carnap's system (1950) employs both inductive (logical) probabilities and statistical probabilities, the latter being interpreted as limiting relative frequencies. Even subjective Bayesians like de Finetti (1937) and Savage (1954) face both an interpretation problem and a characterization problem, though here the two are very closely entwined.

Another vital component of a unified view of induction is an account of how inductive reasoning may be used in *practical decision-making*. Some philosophers, e.g., Popper (1962) and his followers, would deny any connection between theoretical scientific reasoning and "engineering." But this seems to me only to show the untenability of the strict Popperian view of science. There has got to be an intelligible connection between knowledge and action; any overall view of scientific inference that does not give an account of this connection is defective. Accounting for the connection, however, is not an easy task. Indeed, the difficulties in constructing a suitable decision theory on other characterizations of induc-

Kyburg (1961), Jeffrey (1965), Salmon (1966), Hintikka (1966), and Levi (1967).

tive inference constitute one of the major attractions of a subjective Bayesian characterization.[3]

Finally there is the philosophical problem of the *justification of induction*. Historically this problem was most forcefully exposed by Hume. He assumed an inductive logic (basically simple enumeration) in which singular or general statements are inferred from singular statements alone. Noting that the connection between premises and conclusions is nondeductive, Hume asked how one can justify drawing the conclusion without making some other nondeductive step for which a similar justification is needed. He concluded that no rational justification is possible. For those wishing to avoid such skepticism, the problem is to characterize a nondeductive form of inference and then to justify using it.

Though it may be couched in logical or psychological terms, Hume's problem is essentially epistemological. The problem is to justify claiming that one knows, or has good reason to believe, the conclusion, given that one knows the evidence. The same sort of problem arises for relational characterizations as well. Thus one who claims that scientific inference has anything to do with acquiring scientific knowledge must provide some kind of justification for using his particular inductive logic to achieve this end. Yet many kinds of justification are possible, even for a single inductive logic.

Granting that one must justify using any particular inductive logic to validate claims to empirical knowledge, there is an even more general meta-problem to be faced. Suppose one has several inductive logics, each with a different kind of epistemological rationale. How is one to choose among these different comprehensive systems? For example, how would one decide between Carnap's system and Reichenbach's system? I call this the problem of determining the *adequacy* of a comprehensive account of inductive reasoning. In what follows it will be essential to distinguish this most general meta-problem from the narrower problem of justifying a particular characterization of inductive reasoning. I take it as a defect of most discussions of "the justification of induction" that these

[3] It is quite clear that for Savage (1954) one of the primary motivations for introducing subjective probabilities is that they automatically reduce the problem of decision-making under uncertainty to that of decision-making under risk. For a good background discussion of these issues see ch. 13 of Luce and Raiffa (1957). A number of philosophers, e.g., Jeffrey (1956), Carnap (1963), and Bar-Hillel (1968), have used the problem of constructing a practical decision theory as the basis for rejecting any characterizations of inductive logic in which hypotheses are 'accepted' or 'rejected'. For a contrary view see Kyburg (1968) and Levi (1967).

two problems are not clearly distinguished. Indeed, many authors tackle the justification problem with little or no attention to the assumed characterization of inductive inference. It is to be expected that such discussions would confuse the problems of justification and adequacy.

How then do we evaluate the adequacy of a unified account of scientific inference? I will discuss this problem further in section 8, but several conclusions are needed here. My view is that judgments of adequacy cannot be reached by *rational* argument without begging the question. Such judgments can, however, be reached through a *causal* process, the appropriate process being scientific research itself. It may be objected that one should no more give in to unformalized intuition, even collective intuition, at the meta-level than at the object level. But not everything can be formalized and rationally justified. It is sufficient for science that there be a well-developed inductive logic for validating scientific results. That the adequacy of the whole system must rest on judgments that are ultimately only caused and not reasoned is unavoidable. Those inclined to lament this state of affairs should consider whether the situation is any different regarding rival systems of *deductive* logic.

It follows from the above that the only judgments of adequacy we can make are evaluations of the *relative* adequacy of unified accounts of scientific inference which are sufficiently well developed to be applied in actual research contexts. This requirement effectively eliminates most philosophical accounts of scientific reasoning as candidates for our inductive logic since these are invariably so schematic and idealized as to be inapplicable even to the simplest cases of actual research. For sufficiently rich inductive logics one must look to the last half century's achievements in theoretical statistics. But here one faces the reverse problem. Statistical methods are widely applicable, and widely applied but there exists no really unified characterization of the logic, only partial characterizations which are not even mutually consistent.[4] Thus before raising any questions of adequacy, one must first reconstruct a suitably rich logic of statistical inference. The same is true of the statis-

[4] Formulating widely held principles of statistical inference in terms of evidential irrelevance, Birnbaum (1969) demonstrates several striking inconsistencies. Birnbaum is troubled by this result and seems pessimistic regarding the prospects for any unified inductive logic. Others, who, like Tukey (1961), refer to themselves as 'data analysts', seem content to pick and choose among apparently inconsistent methods. For an example of this growing eclectic approach, see the recent text by Kempthorne and Folks (1971).

tician's interpretation of probability. There is a lot said in standard treatises about relative frequencies, but few theoretical statisticians hold a strict limiting frequency interpretation a la von Mises (1957) or Reichenbach (1949). Again, some statisticians, following Wald (1950), would define statistics as the theory of practical decision-making using statistical data. But this description is hard to square with the practice of these same statisticians in scientific contexts.[5] Finally, theoretical statisticians, like scientists, have little stomach for questions about ultimate justification. The statistical literature contains little that could be regarded as even an implicit answer to Hume. On the other hand, at least some statisticians have an intuitive appreciation for the problem of judging the relative adequacy of competing systems, though the problem is not precisely stated and there is little meta-discussion of how it might be resolved.[6]

With the background presented above it is possible to give a brief statement of my general program for resolving the fundamental issues concerning scientific inference. It is to develop rival approaches to scientific inference into systematic, unified accounts of scientific reasoning so that they may be compared in typical scientific research contexts. Beyond this one can only encourage theoreticians and practitioners of scientific inference to make the same comparisons and hope that there is a de facto convergence of opinion on the relative adequacy of the rival systems.

Judgments of adequacy require precisely developed and comprehensive accounts of scientific inference which as yet do not really exist. So the immediate task is one of construction, or reconstruction, and development. Now there are several extant prototype inductive logics which can deal with interesting scientific cases. There is, first of all, the Fisherian tradition which includes maximum likelihood point estimation, hypothesis testing, and fiducial inference. Then there is the Neyman-Pearson account of testing and interval estimation. The latter is

[5] See, for example, ch. 1 of Lehman (1959). Neyman must be regarded as the founder of this school. Yet an examination of his scientific work in such fields as astronomy and meteorology reveals little explicit adherence to the decision theoretic point of view. For references to these scientific papers see the bibliography in Neyman (1967).

[6] Savage (1967) is an exception to the rule that statisticians do not take Hume seriously. Also, Birnbaum (1971) has explicitly advocated the use of scientific case studies to help resolve fundamental conflicts regarding the foundations of statistics.

closely related to the Neyman-Wald account of statistical decision theory. The most systematic inductive logic and decision theory is that of subjective Bayesians like Savage, but extensive scientific applications of the Bayesian approach are hard to find. Finally, there are those who, like Tukey, practice 'data analysis' without utilizing any systematic logic.[7]

Of these possibilities, the Neyman-Pearson-Wald tradition seems to me most promising. This is largely because, like Peirce and Reichenbach, I see induction in terms of a method for using data to reach conclusions and not merely in terms of a quantified relation between data and hypotheses. As indicated above, however, there is no single unified account within this tradition. Thus one must reconstruct a plausible inductive logic and decision theory, develop a congenial interpretation of probability, and supply a philosophical justification for using the reconstructed logic to certify scientific knowledge claims. Characterizing the logic of Neyman's approach is the most urgent and difficult of these problems, but I am not prepared to do justice to this task here. I have already done some work on the interpretation of probability. This leaves the justification problem. Of course one cannot discuss justification without presupposing a logic and an interpretation of probability, but enough can be given to provide a basis for raising interesting questions about justification. Moreover, for those not familiar with current statistical theory, and this unfortunately still includes most philosophers — even most philosophers of science — a discussion of the justification problem may provide the better motivation for taking seriously the kind of general program outlined above.

In the following section I will sketch all of scientific inference and argue for the epistemologically fundamental position of statistical inference. I will also give an interpretation of physical probability. The third section contains a brief characterization of a Neyman-like theory of statistical hypothesis testing. This is followed by a discussion of the

[7] Fisher's views are developed in several books (1925, 1935, 1957) and in many papers (1956). Neyman and Pearson's fundamental papers are now reprinted in Neyman and Pearson (1967) and in Neyman (1967). The bible of the subjective Bayesian school is Savage (1954). See also Good (1950, 1965). Less technical presentations, together with considerable bibliography, may be found in Savage (1962) and in Kyburg and Smokler (1964). For a recent exposition by the foremost British Bayesian, see Lindley (1971). The 'data analysts' include Tukey (1961), Birnbaum (1962), and Kempthorne (1971). Finally, one group not mentioned above consists of those, e.g., Hacking (1965) and Edwards (1971), who attempt to build a theory of statistical inference on likelihood ratios alone.

epistemological structure of hypothesis testing. After looking in some detail at a semirealistic example, I examine, in section 6, the prospects for employing a traditional 'foundationist' justification for the indicated logic of testing. Rejecting the foundationist approach, I go on to develop a Peircean 'methodological' justification. In section 8 I return to the relation between justification and adequacy. The conclusion is devoted to some discussion of the metaphors of ultimate justification, i.e., the traditional notion of the "foundations" of knowledge, Quine's "web of belief," and the idea that scientific knowledge may have "roots" even though it has no "foundations."

2. Probability and Physical Systems

The world is a complex physical system. Any aspect of a physical system may be described by a quantity called a *state* of the system. Thus the total state of a system may be represented by a vector in the appropriate *state-space*. A theory of a kind of system includes at least a description of the relevant state-space together with a designation of certain states as physically possible (kinematics) and a function giving the development of the system in time relative to any possible initial state (dynamics). The history of any system is represented by a trajectory in the state-space.[8] Roughly speaking, a system is *deterministic* if truly described by a theory which, for any arbitrarily given initial state, specifies a unique state for any time. A system is *probabilistic* if truly described by a theory which, for any arbitrarily given initial state, specifies a probability distribution over possible states at later times.[9]

In a deterministic system, some states are *physically necessary* relative to a given initial state. I am inclined to hold that this necessity is real, i.e., a feature of the physical system itself and not merely a creature of language. Similarly, I wish to say that a *probabilistic* system has *physical propensities* to evolve into certain states relative to a given initial state. Such propensities are like physical necessities, but somehow weaker. Moreover, on this view there is no direct logical connection between propensities and relative frequencies. The strongest connection is given

[8] This general view of the nature of theories follows van Fraassen (1970), though here more in spirit than in letter.

[9] For an appreciation of the kinds of difficulties that develop when trying to give a precise account of deterministic theories see Earman (1971), Glymour (1971), and van Fraassen (1972).

by an interpretation of the Bernoulli theorem. If a long series of relevantly similar independent systems begin with the same initial state, then the compound system has a very strong propensity to exhibit final states with relative frequencies very near the corresponding propensity. But this relative frequency is not necessary, no matter how long the sequence.[10]

The above is mainly ontology, and the subject here is epistemology. But since I find this ontology congenial and shall assume it for the purposes of exposition at least, it is best to make it explicit. On the other hand, there are probably no necessary connections between my ontology and epistemology, so that the following discussion may easily be reinterpreted by anyone with other, e.g., Humean, ontological commitments.

Turning, then, to epistemology, how does one test theories about physical systems? The traditional view is simply that from our theory of a kind of physical system we deduce characteristics of particular systems whose operations are open to observation. If the observed systems behave as the theory indicates, that is taken as evidence that the overall theory is correct. This characterization of theory testing is extremely general and incomplete. It does not even tell us whether positive results confer probability on the theory or whether they make the theory 'acceptable' without assigning probabilities. On the other hand, its very generality makes the traditional account agreeable to almost everyone, and, fortunately, no more precise account is necessary for present purposes.

It is a commonplace that an experiment never tests a theory alone. Rather, what is tested is the theory together with assumptions about the experimental setup. But putting this point in terms of "auxiliary assumptions" masks the nature of the situation. It is better, as Suppes (1962) seems to suggest, to think of a hierarchy of interacting physical systems described by a corresponding hierarchy of theories, including a theory of the experimental apparatus. Making these theories "mesh" appropriately raises some nice logical and methodological problems, but these are of no special concern here. The crucial point is that the lowest level system is always a *probabilistic* system, i.e., a data generator. The epis-

[10] I have discussed a "single case" propensity interpretation of probability in Giere (1973a), (1975). The most extensive development of any propensity theory to date is due to Mellor (1971). See also my review, Giere (1973c).

temologically relevant hypotheses concerning the experimental setup are therefore *statistical* hypotheses. This makes the testing of statistical hypotheses epistemologically prior to the testing of general theories. One may thus consider the justification of a method of statistical inference as basic to the justification of any complete inductive logic.

The claim that experimental setups are always probabilistic systems need not be understood as a claim about reality. Even if we have good general grounds for taking a particular data generating arrangement to be strictly deterministic, we in fact never know all the factors operating in any particular measurement. The combined action of these unknown factors produces a pattern of outcomes that is indistinguishable from that produced by a genuinely indeterministic system. Thus, from a practical point of view, we can simply treat the setup as being indeterministic. This form of "methodological indeterminism" at the experimental level is fully compatible with methodological determinism regarding theories, i.e., with the admonition always to seek a set of state variables which make the system deterministic. In fact, one can even espouse methodological determinism regarding experimental setups since we never do eliminate all unknown influences.

My view, then, is that theories are judged by the outcomes of statistical inferences regarding particular experimental systems. This apparently conflicts with much that has been written in recent years about the dependence of observation on theory.[11] Indeed, I seem to be advocating a version of the traditional three-level view of scientific knowledge and inquiry. Singular observations form the inductive basis for statistical hypotheses about experimental setups and these hypotheses in turn form the inductive basis for theories concerning broad classes of physical systems. In fact I will deny an important tacit assumption of this view, namely, that any statistical hypothesis can be inferred from singular statements *alone*. But this is not enough to avoid conflict with the more extreme advocates of 'theory-ladenness'. However, as I do not wish to debate this matter here, I will simply state my conviction that such views are wrong, and, moreover, that they are so because of a mistaken theory of meaning.

[11] Here, of course, I am referring to the vast literature which surrounds and to a large extent grew out of the work of Hanson (1958), Feyerabend (1962), and Kuhn (1962). I have dealt in a general way with some of this literature in Giere (1973b).

While rejecting views implying the impossibility of observationally testing scientific theories, I do agree with much that has been written about the conceptual and methodological importance of theory in scientific inquiry. In particular, the theories we hold may determine the *categories* of things we perceive without determining our *judgments* about them. In this sense theory is truly the midwife of observation. Moreover, it is generally the case that what a scientist investigates, and when and how, is a function of his theoretical commitments, vague though they may be. I give no hostages to that Popperian bogeyman, Baconian inductivism. Nevertheless, when it comes to validating those commitments before the scientific community, the logic to be employed follows the general pattern outlined above. That, at least, is the assumption that guides the rest of this study.

3. The Logic of Statistical Hypothesis Testing

As indicated above, there is no single, unified characterization of statistical inference that would be acknowledged in either the theory or practice of contemporary statisticians and scientists. In fact, the trend in recent years has been toward increasing eclecticism — in spite of fundamental incompatibilities among all the methods employed. In order to follow my general program, therefore, it will be necessary to depart from current statistical practice by isolating and reconstructing a single approach to statistical inference. The logic adopted here is a version of the theory of statistical hypothesis testing developed by Neyman and Pearson in a fundamental series of papers published between 1928 and 1936. This choice is based partly on my personal conviction that the Neyman-Pearson (N-P) viewpoint is basically correct. But it is also justified by the tremendous influence this work has had for nearly a half century. The N-P approach must count as one of the three or four major candidates for a comprehensive inductive logic. To produce a relatively complete and philosophically adequate reconstruction, however, would be a major undertaking. The following sketch is geared to the problem at hand, which is justification, not characterization, and is as simple as the demands of clarity and pedagogy will allow.[12]

[12] The basic original papers are now reprinted in Neyman and Pearson (1967). For a personal account of the formative years see Pearson (1966). Currently the definitive presentation of the theory is due to Lehman (1959). For an elementary exposition see Hodges and Lehman (1964). Hays (1963) has given a highly in-

The basic conception in my account is that of a *chance setup*.

CSU is a chance setup iff CSU is a physical process with one or more physically possible final states (or "outcomes").

This definition includes deterministic processes as a limiting case, but henceforth I shall be concerned only with systems having more than one physically possible outcome.[13] Let A_i be a possible outcome of CSU and U the set of all possible outcomes, i.e., $U = \{A_i\}$. Now let E_i be a subset of U and $F = \{E_i\}$ the family of all such subsets. It is these subsets of U, not the basic outcomes themselves, that mathematicians call 'events'. Following standard mathematical practice we shall assume there is a quantity $P(E_i)$, the probability of E_i, which obeys the standard axioms for probability spaces. For finite spaces:

(1) $P(E_i) \geq 0$

(2) $P(U) = 1$

(3) $P(E_1 \cup E_2) = P(E_1) + P(E_2)$ if $E_1 \cap E_2 = \Lambda$.

In keeping with my ontological preferences, I shall interpret $P(E_i)$ as the measure of the propensity of the physical system, CSU, to produce one of the outcomes constituting event E_i. And I assume that propensities operate on each individual trial of a CSU. Those who find such a conception excessively metaphysical or obscure may attempt to interpret $P(E_i)$ in terms of the relative frequency of event E_i in a (perhaps infinite) series of trials of the relevant CSU.

A precise characterization of statistical hypothesis testing requires a definition of a statistical hypothesis.

H is a statistical hypothesis iff H is a statement ascribing a distribution of probabilities to the outcomes of a CSU.

Taking a traditional example, conceiving a child might be considered a chance setup with the set of outcomes (male, female). The statement that $P(\text{male}) = .51$ is then a statistical hypothesis. Following standard mathematical practice we shall assume the existence of a real valued function (conveniently, but misleadingly, called a *random variable*) over the outcomes of any chance setup. A statistical hypothesis is then any statement describing the probability distribution over a random variable.

formed and sometimes critical presentation geared to applications in psychological research. My own exposition tends to follow Neyman (1950, now out of print).

[13] The term 'chance setup' has been borrowed from Hacking (1965). The usual term among statisticians is 'random experiment', but as the indicated processes may be neither random nor experiments, another terminology seems preferable.

For example, if X (male) $= 0$ *and* X (female) $= 1$, then $P(X = 0) = .51$ expresses the statistical hypothesis given above.

Many discussions of statistical inference take the concept of a *population* as fundamental. On such a view, a statistical hypothesis gives the relative frequency of a characteristic in some (perhaps infinite) population. The viewpoint above is more general and, I think, more fruitful. Populations may be dealt with in terms of a chance setup consisting of a sampling mechanism operating on the indicated population. How the distribution of population characteristics and the probabilities for different outcomes are related depends on the sampling mechanism, e.g., whether it is random or not.

On the Neyman-Pearson account it is crucial to distinguish simple from composite statistical hypotheses.

> H is a simple statistical hypothesis iff H specifies a unique probability distribution over a random variable.
>
> H is a composite statistical hypothesis iff H is a statistical hypothesis that is not simple.

The importance of this distinction will soon be apparent. We are now equipped to begin talking about *testing* statistical hypotheses. The intuitive notion of a statistical test is that one uses the results of a number of trials of the chance setup as evidence for or against the proposed statistical hypothesis. The N-P approach provides a precise rendering of this intuitive idea.

Imagine a series of n trials of a specified CSU, and let $P(X_j = x_j)$ be the distribution of probabilities over the jth trial. Representing each of the j random variables on a separate axis, the n trials determine an n dimensional *sample space* consisting of points representing each possible outcome of all n trials. That is, each point in the sample space is an n-tuple $x = (x_1, x_2, \ldots, x_n)$. Now let H be a simple statistical hypothesis ascribing a unique probability distribution to the outcomes of each trial. For example, H might give the *same* distribution to each trial. In this case there are well-known mathematical techniques for computing a probability distribution relative to H, $P_H(X = x)$, over the whole sample space. Continuing the standard example, the basic sample space for a series of n conceptions consists of the 2^n possible sequences of males and females. Given reasonable assumptions, however, it is sufficient for most purposes to consider only the sample space consisting of the $n + 1$ possible numbers of male offspring in n trials, i.e., $S =$

$\{0, 1, 2, \ldots, n\}$.[14] If $P(\text{male}) = p$ is the same on each trial, the probabilities of each of these possible sample points are given by the familiar binomial formula:

$$P(X = m) = \binom{n}{m} p^m (1 - p)^{n-m}.$$

Since at least the beginning of the eighteenth century, many authors have implicitly held that a statistical hypothesis is to be 'rejected' whenever the observed sample point would be highly improbable if that hypothesis were true.[15] But this intuition provides only a very incomplete characterization of a statistical test. In many cases *every* possible outcome is quite unlikely. For example, exactly 510 males in 1000 conceptions is highly unlikely even if $P(\text{male}) = .51$. Yet this result should provide greater support *for* the hypothesis than any other possible outcome.

By the second decade of this century, R. A. Fisher and others had focused on finding a *region* of the sample space (the 'critical' or 'rejection' region) for which the probability of occurrence is low if the test hypothesis is true.[16] If one resolves to reject the (simple) hypothesis under investigation if and only if the observed sample point falls in the rejection region, then there is only a small and controllable probability that we will by chance reject *H* when it is in fact true. This probability is called the *significance level* of the test of *H*, and it may be controlled by suitably adjusting the number of trials and the rejection region.

Though it is still a point of dispute, Fisher's characterization of hypothesis testing seems to me quite incomplete.[17] The most obvious gap is the lack of any systematic grounds for choosing among several possible rejection regions with the same significance level. Indeed, Fisher

[14] This reduction of the sample space is treated in discussions of 'sufficient statistics' found in any good text, e.g., Kendall and Stuart (1958–66).

[15] As early as 1710 Arbuthnot used data from the London registry to argue in effect that the hypothesis $P(\text{male}) = 1/2$ is to be rejected. He went on to argue, perhaps in jest, that this implies that sex determination is not a matter of chance and is therefore the result of divine intervention. For further details see Hacking (1965). Laplace also used an implicit theory of statistical hypothesis testing in his scientific work. Some examples are discussed in his popular *Philosophical Essay on Probabilities* (1814).

[16] The development of Fisher's views may be traced through his papers (1950). One should also read at least the early chapters of *The Design of Experiments* (1935). It is most illuminating to compare Fisher's discussion of "the lady tasting tea" (ch. 2) with Neyman's commentary (1950, ch. 5).

[17] Anscombe (1963), for example, argues that the basic Fisherian test of significance has important applications.

gives no systematic reason for not taking a small set of values near 510 males in 1000 trials as the rejection region for the hypothesis $P(\text{male}) = .51$. Moreover, though Fisher was careful to minimize the probability of accidentally rejecting a true hypothesis, he did not deal systematically with the problem of accepting a false one. This is in keeping with his statements that the only purpose of experimentation is to give the data a chance to reject a hypothesis.[18] But non-Popperians are not likely to be satisfied with only methods for rejecting hypotheses. An inductive logic that provides for rejection ought to provide for acceptance as well.

The developments initiated by Neyman and Pearson may be viewed as an attempt to fill in the gaps in Fisher's characterization of hypothesis testing. On their view, a test of H is to be characterized as a method of reaching a decision either to reject or to accept H. A good test is then characterized as one for which the probabilities of a wrong decision are not only low, but, more importantly, controllable. Now if both acceptance and rejection are possible, then there are not one but two kinds of possible 'mistakes'. There is the Fisherian case, or Type I error, of by chance obtaining a sample point in the rejection region even though H is true, and thus rejecting a true hypothesis. Conversely, one might obtain a sample point in the acceptance region even though H is false, and thus suffer a Type II error, accepting a false hypothesis.

If H is simple, the probability of Type I error is just the Fisherian significance level. But what about the probability of Type II error? Formally this is the probability of obtaining a sample point in the acceptance region if H is false. But since not-H is in general a *composite* hypothesis, this latter probability cannot be defined without a prior weighting of the component simple hypotheses. Let us postpone the question of how, or even whether, this is to be done, and simply assume that there are only two possible, logically exclusive, simple hypotheses, H and K. The probability of Type II error, i.e., accepting H if H is false, is then the probability, given K, of obtaining a sample point not in the rejection region of the sample space. The complementary probability, i.e., the probability of not making a Type II error is called the *power*

[18] "Every experiment may be said to exist only in order to give the facts a chance of disproving the null hypothesis" (Fisher, 1935, 4th ed., p. 16). I cannot help but speculate that Popper's sympathetic reception by British scientists may be partly the result of the widespread influence of Fisher's views. Gilles (1973) has recently attempted to develop a rule for rejecting statistical hypotheses along lines suggested by Popper (1959).

of the test. This is the probability that the test will detect the falsity of H if K is true.

We are now equipped to give a formal definition of a Neyman-Pearson statistical test.

> T is an N-P statistical test of hypothesis H relative to chance set-up CSU, sample space S, and alternative, K iff: T is an exclusive and exhaustive partition of S into two regions, one, R, designated as the rejection region.

A noteworthy philosophical consequence of this definition is that the N-P logic cannot be reconstructed as simply a logical relation between H and a description of the observed sample point. In addition to the CSU and sample space, there is essential reference to a set of *admissible hypotheses* which includes H and at least one simple, exclusive alternative hypothesis. The reference to alternative hypotheses, however, has more than general philosophical interest. It provides the means for giving a precise mathematical formulation of the problem of choosing an optimal rejection region. In some cases at least, this problem has a precise and intuitively appealing solution.

Continuing the case of testing a simple hypothesis against a simple alternative, it is natural to wish simultaneously to minimize the probability of both kinds of error, i.e., to minimize the significance level while maximizing power. Brief reflection, however, shows that for a fixed number of trials, these two probabilities are inversely related. Thus the statement of the problem given above is not well formed. Suppose, then, that we fix the significance level at some conveniently small value, say .05, and then seek to maximize the power of the test for a fixed number, n, of the trials. In this case one obtains the following elegant result: Let R consist of all those sample points, x_i, of S for which

$$P_H(X = x_i)/P_K(X = x_i) \leqq c,$$

where $c > 0$. The significance level of this test is then

$$P_H(X \epsilon R) = \alpha,$$

where α can be chosen arbitrarily by a suitable choice of c. It is then fairly easy to show that there is no other rejection region, R', for which $\alpha' \leqq \alpha$ and for which the *power* of the test is greater. In sum, the likelihood ratio of points in the sample space may be used to determine a rejection region which has the minimum probability for Type II error relative to a fixed maximum probability for Type I error. More briefly, the likelihood ratio determines a test which has maximum power for a fixed maximum significance level.[19] This test clearly accords well with the basic intuition that one can confidently reject a hypothesis in favor of an alternative if the observed result is much more probable on the alternative (even though both probabilities may be small).

Now suppose that the alternative hypothesis is composite, e.g., let $K = K_1$ or K_2, where K_1 and K_2 are both simple statistical hypotheses. As long as the test hypothesis H is simple, the significance level relative to any rejection region is well defined, i.e., $\alpha = P_H(X \epsilon R)$. Being a weighted sum of two probabilities, however, the overall *power* $P_K(X \epsilon R)$ is not determinable without further assumptions. That is:

$$P_K(X \epsilon R) = w_1 P_1(X \epsilon R) + w_2 P_2 (X \epsilon R).$$

The obvious interpretation of w_1 and w_2 is as the prior probabilities $P(K_1)$ and $P(K_2)$. This interpretation, however, is difficult to square with an objective, empirical interpretation of probability, except in those rare cases when a super CSU produces ordinary CSU's as outcomes. And even in these cases the probability distribution of the super CSU may not be known. For these reasons it is desirable to find a method which requires no prior probabilities for any hypotheses.

In recent years, of course, prior probabilities, in the form of coherent subjective degrees of belief, have become much more respectable than they once were. But if one adopts this view, the natural course is to characterize inductive inference as a subjective probability relation and to apply Bayes's theorem. There is no need for an independent account of hypothesis testing. Thus any appeal to subjective prior probabilities of hypotheses in reconstructing a Neyman-Pearson account of hypothesis testing would take us out of the Neyman-Pearson tradition altogether. Of course the Bayesian approach might in the end prove the more adequate, but that is another question.

The Neyman-Pearson response to composite alternatives is to avoid any talk about the overall power of a test. Instead, such a test is char-

[19] The above is an elementary statement of "The Fundamental Lemma of Neyman and Pearson," first proved in their 1933 paper "On the Problem of the Most Efficient Tests of Statistical Hypotheses." Reprinted in Neyman and Pearson (1967). Lehman (1959, ch. 3) gives a detailed proof and Hodges and Lehman (1964, pp. 351–52) an elementary version. My statement of the lemma is strictly correct only for continuous distributions, but this need not bother us here.

acterized by the power viewed as a function of a parameter which characterizes the set of simple alternatives. In certain cases this function has obviously desirable properties. For some sets of admissible hypotheses, for example, it is possible to find a rejection region which is simultaneously most powerful for each simple alternative in K, given a fixed significance level. Thus even though the power of the test varies for different simple components of the composite alternative, for no simple alternative is there a more powerful rejection region at the same significance level. Such regions are called Uniformly Most Powerful (UMP) regions. For example, the likelihood ratio criterion determines a UMP region for a test of $P(\text{male}) = .51$ *vs* $P(\text{male}) > .51$.

When a UMP region exists, it determines the best possible test for any number of trials. For many composite alternatives, however, there is no UMP region, i.e., $P(\text{male}) = .51$ *vs* $P(\text{male}) \neq .51$. In such cases Neyman and Pearson introduce a *new* condition on optimal rejection regions, namely, for all $K_i \in K$,

$$P_i(X \in R) \geqq P_H(X \in R).$$

This condition, called "unbiasedness," simply means that one is always more likely to reject the test hypothesis if it is false than if it is true. Often, as in the example above, there is a UMP region among the unbiased regions which is then taken as optimal. But while this condition is plausible, though hardly self-evident, there are sets of admissible hypotheses for which not even UMP unbiased tests exist.

Over the years other criteria for determining various classes of UMP tests have been proposed.[20] We need not consider them here. Two points, however, are important. First, the notion of a uniformly most powerful test largely preserves the idea of building a method of accepting and rejecting hypotheses which has known and controllable error probabilities. Second, the theory of testing simple hypotheses against composite alternatives is still open-ended. Depending on the set of admissible alternatives, one could introduce new criteria for optimal rejection regions in order to obtain a UMP test of some kind. This move is always open.

Turning finally to cases with both a composite test hypothesis and a composite alternative, it is clear that neither the significance level nor the power is uniquely determined without a distribution of prior prob-

[20] See, for example, Lehman's chapter on "invariant" tests (1959, ch. 6).

abilities over the component simple hypotheses. For such cases Neyman suggests an obvious extension of the methods used for simple hypotheses. Consider the ratio

$$\lambda(x) = P_i(X = x)_{max}/P_j(X = x)_{max}$$

where x is a particular sample point, H_i and K_j are component simple hypotheses of H and K respectively, and $P_i(X = x)_{max}$ is the least upper bound of the probability $P_i(X = x)$ for components H_i of H. The rejection region is then to consist of all sample points for which $\lambda \leq c$, where c is some small fraction.[21] The basic idea, of course, is that H is to be rejected if there is any simple alternative for which the observed outcome is much more likely than it is on any component of H. But here the rationale in terms of known error probabilities applies even less well than in the previous case. There at least the probability of Type I error was well defined. Here one is in effect reduced to comparing the *most favorable* member of *each* of the two sets of hypotheses. Lacking prior probabilities this is a reasonable compromise strategy, but it is all based on possible rather than actual error probabilities. For these reasons it is not surprising that one finds little literature on the development or application of such tests.

This completes my sketch of the logic of statistical hypothesis testing. Before turning explicitly to questions of epistemological justification, however, there are several points about the logic that require mention. The first is that, granting only some general mathematical conditions, any contingent statistical hypothesis can be tested by some N-P test. This follows because any statistical hypothesis must be either simple or composite, and one can always find some alternative, simple or composite, to form a test. One could even effectively avoid the λ-criterion by always choosing a simple alternative. Let us summarize this result by saying that the N-P logic is *weakly comprehensive*. The logic is also *strongly comprehensive* if the λ-criterion is regarded as fully legitimate. That is, for any H and any K, both either simple or composite, there exists an N-P test of H against K. But strong comprehensiveness does not hold if one is restricted to a fixed narrower class of tests, e.g., UMP

[21] Note that if H and K are both simple, the λ-method reduces to the likelihood ratio test described above. My statement of the λ-criterion differs from Neyman's in that his denominator refers to the whole set of admissible hypotheses, not just the alternative hypothesis. This difference is inconsequential here. See Neyman (1950, pp. 340–42).

unbiased tests. The general comprehensiveness of the N-P logic of testing will play an important role in the epistemological justification to be developed in section 7.

Finally, it must be acknowledged that the characterization of statistical hypothesis testing described above is itself incomplete in at least one important respect. For Neyman and Pearson, testing a hypothesis is a systematic way of reaching a decision to accept or reject that hypothesis. But what is it to decide to accept (or reject) a hypothesis? In his later writings, though not necessarily in his statistical practice, Neyman identifies accepting a hypothesis with choosing a course of action.[22] A statistical test thus becomes, in Wald's terminology, a statistical decision rule. For each possible outcome of a set of trials, the rule prescribes a particular course of action. Viewing hypothesis testing as a special case of decision theory not only tells us what "accepting a hypothesis" means, it also gives us a way of determining specific values for the significance level, power and number of trials for a test. The logic as characterized above only shows how these quantities are related; it does not determine an optimal set. In a decision theory context, these are in effect determined by the gains and losses associated with the possible results of the decision, i.e., a specific action in a certain actual situation.

Despite considerable initial plausibility, the Neyman-Wald approach has severe liabilities. It leads to the unresolved and perhaps unresolvable problem of finding a general criterion for decision-making under uncertainty. Minimax is not an adequate general strategy. Moreover the kinds of utilities typically found in practical decision-making contexts are nonexistent, or at least exceedingly hard to find, in scientific research contexts.

An alternative to a full-fledged decision theoretic approach is to take "accepting *H*" to mean something like adding *H*, tentatively, to the body of scientific knowledge. The choice of significance level and power would then turn not on anything like monetary profits and losses, but on the desire to increase the body of knowledge in content, precision, etc., as efficiently as possible. Hopefully one could use this goal to justify the decision rule implicit in at least some N-P tests without appealing to

[22] I take this to be a consequence of Neyman's concept of 'inductive behavior'. See his " 'Inductive Behavior' as a Basic Concept of Philosophy of Science" (1957). Pearson, by the way, is reputed to take a much less behavioristic view of acceptance.

any completely general strategy like minimax. Some work has been done along these lines, but much more is necessary before the logic of testing presented above could be suitable as part of an adequate overall account of inductive reasoning.[23]

4. The Epistemological Structure of Statistical Tests

Suppose that a statistical hypothesis, *H*, has been accepted (or rejected) as the result of an N-P test, *T*.[24] In this case the acceptance of *H* is *validated* by *T*. There are, however, several legitimate ways one could question whether the acceptance of *H* is "really" justified in some more ultimate sense. One could, for example, question whether the acceptance of any hypothesis should depend solely on the supposedly favorable error probabilities of an N-P test. But this question concerns the adequacy of the whole N-P approach and will not be considered here. Or, one might question whether the significance level and power were sufficiently stringent. But this question again concerns aspects of the N-P characterization which cannot be considered here. However, even granting that *T* represents an appropriate application of N-P methods, one might still question the acceptance of *H* on the grounds that the stated error probabilities may not in fact be correct. On what grounds, then, may it be maintained that the indicated error probabilities are indeed correct? This question gives rise to a regress typical of Humean arguments against the justifiability of induction.

Looking back at the characterization of N-P tests it is fairly easy to see that in general there are two, and only two, points at which *empirical* assumptions may enter into the formulation of an N-P test. One is due to the necessity for designating a set of admissible hypotheses since it is assumed that some member of this set, i.e., either *H* or *K*, is true. Thus only if the disjunction, *H* or *K*, is *logically* exhaustive will the need for a set of admissible hypotheses fail to yield an empirical assumption. A second possible source of empirical assumptions opens up whenever a test involves more than one trial of a CSU. Any claim of equality

[23] In a forthcoming paper, Giere (1975), I develop this approach both for tests of statistical hypotheses and for tests of theories. This later paper thus provides an answer to the arguments of Carnap and Jeffrey (cited in fn. 3 above) that it is impossible to incorporate 'acceptance rules' into an adequate inductive logic.

[24] In cases where no confusion should arise I will simply say "accepted" rather than saying "accepted or rejected."

(or difference) of probability distributions on two or more trials is an empirical claim. All but a very specialized class of N-P tests incorporate such claims.

I will refer to assumptions of the first type as *admissibility assumptions*. Assumptions of the second type will be called *multiple trial assumptions*. Finally, I will refer to any empirical assumption of an N-P test as a *presupposition* of that test. This term is natural and appropriate since the truth of the presuppositions is assumed whether the test hypothesis is accepted or rejected.[25]

We are now in a position to recognize a feature of the N-P characterization that has important consequences for the justification problem, namely, that all presuppositions of an N-P test are themselves statistical hypotheses. That admissibility assumptions are statistical hypotheses is obvious. Indeed, it is clear that the admissibility assumption of any N-P test is just the *composite* statistical hypothesis, H or K. That multiple trial assumptions likewise include only statistical hypotheses is less obvious. Given multiple trials, there are just two questions that must be answered if they are to be part of a single test. The first is whether the distributions of probabilities are the same or different on each trial. The second is whether the trials are independent. A statement of sameness or difference of distributions on a series of trials is just a conjunction of statistical hypotheses regarding the individual trials and thus itself a statistical hypothesis. To see that independence of trials also involves only statistical hypotheses, let X and Y be the random variables associated with two trials of a CSU. The case of more than two trials is more complex, but not in principle any different.[26] Now two trials are said to be independent iff

$$P(X = x \,\&\, Y = y) = P(X = x)\, P(Y = y)$$

for all x and y. Stated in terms of conditional probabilities this condition becomes

$$P(X = x \mid Y = y) = P(X = x).$$

[25] From now on it will be assumed that *all* empirical presuppositions of an N-P test are either admissibility or multiple trial assumptions. I am thus ignoring the fact that performing an N-P test requires such things as the observation and recording of data which are possible only under certain empirically given conditions. This seems perfectly legitimate in a study of the finer structure of scientific inference.

[26] For the definition of independence for more than two trials see any good textbook of probability or statistics.

It is clear that these conditions are just complex statistical hypotheses relative to the joint CSU consisting of two trials of the original CSU. Standard tests for such hypotheses are discussed in many texts and related tests will be treated in the following section. The intuitive idea, of course, is that the probability distribution for any one trial should have no influence on the distribution of any other.

Recall, now, our conclusion that the N-P characterization is comprehensive in the sense that any statistical hypothesis can be subjected to some N-P test. Given that any presupposition of an N-P test is a statistical hypothesis, it follows that any presupposition of an N-P test can itself be tested by an N-P test. This result may be summarized by saying that the N-P characterization of statistical hypothesis testing is *presuppositionally closed.*[27]

Let us now return to the situation in which the justifiability of accepting H as a result of performing test T is questioned on the grounds that some presupposition, P, of T may not be true. Given presuppositional closure, it is always possible that this question be answered without appealing to any other inductive method for testing statistical hypotheses. P may itself have been N-P tested and accepted. But here the possibility of a Humean regress sets in. In general the test T' of P will itself have presuppositions which may in turn be questioned, and so on. Thus, assuming we have only the N-P logic for testing hypotheses, at any particular time the ultimate basis for the acceptance of H has the sort of rootlike structure shown in the diagram on page 234, Figure 1. I will call P_1 and P_2 the *immediate* presuppositions of T. All others are *indirect* or *remote* presuppositions of T. (Note that P_3 and P_4 are immediate presuppositions of T'.) In the case of presuppositions like P_1 and P_4 which have themselves been tested, it is assumed that the tests were positive, i.e., P_1 and P_4 were accepted. This point will be discussed further in section 7. Here I will only note that this picture of the ultimate justifiability of accepting H does not presume the erroneous principle of the transitivity of conditional probabilities. We are not here arguing that P_4 makes H probable because P_4 makes P_2 probable and P_2 makes H probable. In the N-P logic there is no relation of conditional probability between the presuppositions of a test and the test

[27] I owe this apt term to Lewis Creary who commented on an ancestor of this paper at a session of the May 1970 meeting of the American Philosophical Association, Western Division, in St. Louis.

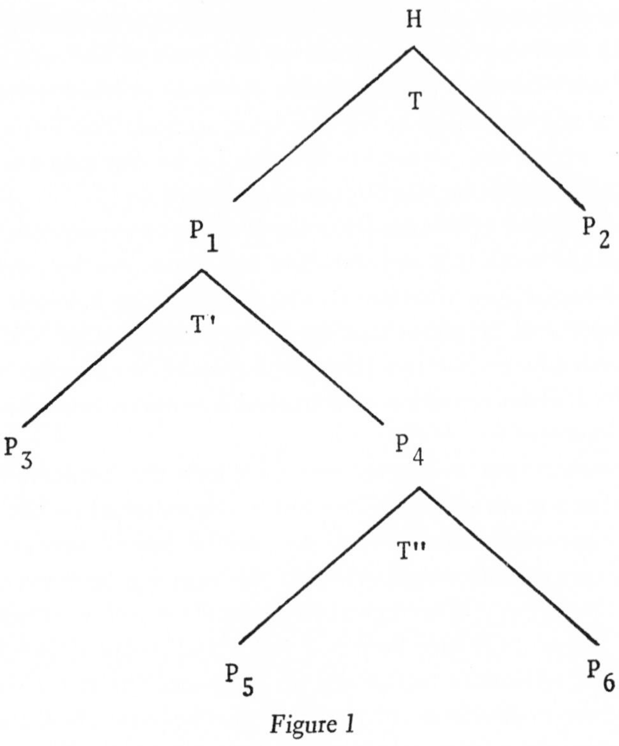

Figure 1

hypothesis. On the other hand, I could hardly claim that there are no further relations among hypotheses like P_1, P_4, and H that need further investigation.[28]

The root structure of statistical tests illustrated above suggests two different general programs for constructing an ultimate justification for accepting a statistical hypothesis. Both focus primarily on those presuppositions currently not connected to further presuppositions. Let us call these the *ultimate* presuppositions of T, notwithstanding the fact that the set of ultimate presuppositions of any N-P test may change from time to time. The first program, which follows what I call the 'foundationist' tradition, is to "seal off" the root structure by insisting that all ultimate presuppositions be subjected to N-P tests whose presuppositions are purely logical. The alternative program is to find an ultimate

[28] Indeed, all the questions about the applicability (practical or otherwise) of 'accepted' hypotheses may be raised here. There is also the possibility of a lottery paradox. For references see Kyburg (1970) and fns. 3 and 23 above.

justification allowing ultimate presuppositions which are so simply because they have not yet been tested and therefore have been neither accepted nor rejected. I will develop a "methodological" version of this approach in section 7.

5. A Semirealistic Example

An example may be helpful in conveying some intuitive feel both for the logic of testing and for related epistemological issues. One must be wary, however, of simple textbook examples which, by excessive abstraction from any realistic context, often distort more than they illuminate. The ideal example would be a miniature case study of a scientific inquiry. The following "semirealistic example" falls between the excessively abstract and the ideal.

Since ecology became a subject of public concern there has been increased sensitivity to questions about the effect of various substances on human and other organisms. A fairly crude example of such a question would be: Do certain low levels of DDT decrease the adult size of mammals? Of course a scientific study of this question would require precise criteria for size, DDT level, and other variables, but let us ignore such details.

I would regard the claim that DDT decreases size as a claim about a kind of probabilistic physical system, e.g., that exemplified by a particular strain of laboratory rat. At birth each animal has a propensity distribution over possible adult sizes. The general claim under question may be that the mean of this distribution is a decreasing function of DDT levels. This is not merely a statistical hypothesis but a theoretical function relating state variables of a probabilistic physical system to a probability distribution over other states. But the theoretical hypothesis has statistical consequences, for example, that the expected, i.e., most probable, adult size of rats exposed to DDT will be lower than that of rats otherwise similar but not so exposed. This kind of hypothesis is subject to direct statistical tests.

Before looking in more detail at the appropriate statistical tests, one should note the methodologically necessary but epistemologically minimal role played by high-level theory in cases like this. At present, biological theory is not so well developed as to yield the unequivocal prediction that DDT will tend to lower the adult size of rats. Still less can any theory yield a precise form for the functional relation between

DDT exposure and adult size. Yet existing theory is important because it tells us that DDT is the kind of substance that *could* lower the expected adult size of mammals. Without this minimal theoretical background it is very unlikely that the question of such a relationship would arise or be investigated in a serious scientific context. And with good reason. Scientific inquiry cannot efficiently be guided by consideration of what is merely logically possible. It is usually necessary to limit the field to what one regards as physically possible.

Turning to the problem of designing a statistical test, imagine two groups of rats of the appropriate type. The members of one, the experimental group, are subjected from birth to specified levels of DDT; the other group, the controls, are kept free of DDT. Now the hypothesis that the expected size of the experimental rats is lower than that of the controls is a composite hypothesis and thus not an ideal subject for statistical testing. On the other hand, the hypothesis that there is no difference in expected means is simple and thus better suited to our logic. To be more precise, let the expected sizes of the experimental and controls be μ_E and μ_C respectively. The test hypothesis, or 'null hypothesis', then, is H: $\mu_C - \mu_E = 0$. Realistically there are only two alternative hypotheses that might be considered, i.e., $\mu_C - \mu_E \neq 0$, and $\mu_C - \mu_E >$ 0. Which we use depends on whether or not our previous knowledge is sufficient to reject the possibility that DDT increases the expected adult size of rats. Assuming such knowledge we have K: $\mu_C - \mu_E > 0$.

In addition to a set of admissible hypotheses we require a suitable sample space and rejection region. Let us assume that at birth the probability distribution of each rat over possible adult sizes is normal with mean μ_C or μ_E and variance σ. In short, the only possible difference in distributions is that the experimentals may have their curves shifted downward. Finally, we assume that all distributions are unaffected by the distribution for any other member of the group. With these assumptions the standard sample space would be all possible values of the difference $d = m_C - m_E$ of the *observed* mean adult sizes of rats in the two groups. Now suppose there is some difference $\delta_o > 0$, which, for theoretical or practical reasons, is considered important. By requiring a fairly high power for this particular alternative, say .8, we determine the required sample size n and rejection region $d > d_o$. The resulting test, the well-known 't-test', is UMP for H against K.[29]

[29] The t-test is discussed in any text of applied statistics, e.g., Hays (1963).

The presuppositions of this test are apparent. The admissibility assumptions are (i) that $P(X)$ is normal with mean μ_C or μ_E and fixed variance σ, which need not be known, and (ii) that $\mu_C - \mu_E \geqq 0$. The multiple trial assumptions are (i) that $P_E(X)$ is the same for all experimentals and that $P_C(X)$ is the same for all controls, and (ii) that all distributions are independent of all outcomes.

Anticipating upcoming epistemological questions, let us consider briefly how one might justify some of these presuppositions. In practice one would obtain experimental animals from a laboratory that specializes in breeding genetically homogeneous strains of rats. That each rat has very nearly the same expected adult size and the same reactions to various chemical substances would then be justified both by genetic theory and by long experience in breeding that particular strain. Epistemologically, however, one might object to the extent to which this justification relies on a high-level theory which itself must have been validated through previous statistical tests.

There is a common procedure which both decreases reliance on high-level theory and also eliminates any direct ontological commitment to propensities of individual organisms. The procedure is to isolate a large population of similar rats. This may require some theoretical knowledge, but not nearly so much as breeding because it is not required that each rat have the same propensities. Rather, all that matters is that each rat have a definite adult size and that the distribution of these sizes in the population has a known form, e.g., normal. Whether the individual rats are deterministic systems or not is immaterial. The CSU necessary to determine a statistical test is introduced in the form of a mechanism which selects samples from the indicated population. It is convenient to think of the sampling mechanism as having propensities for selecting different members of the population. The simplest assumption is that the mechanism is perfectly random, i.e., each member of the population has the same propensity of being selected on each draw. A weaker but still sufficient assumption is that even if the sampling mechanism favors some rats over others, its biases are independent of the attained adult size of any rat. In either case, each trial of the sampling mechanism will have a propensity distribution over adult sizes that exactly matches the distribution of sizes in the population. Thus we need only select two samples of n rats, subjecting one to the specified DDT level, and the test proceeds as above. As usual, those who prefer not to think in terms of propensities may attempt another analysis of random sampling.

Suppose the above test results in a decision to reject the null hypothesis. Strictly speaking, all that has been decided is that the observed difference in average adult size indicates an important real difference in expected size and is not just an extreme statistical fluctuation. Assuming that the decision is in fact correct, it does not logically follow that the reduction in expected size was *caused* by DDT. This would follow only if *all* other relevant variables were the same for both groups. But in practice we do not even know what all the relevant variables are, let alone that they have all been held constant. It might be, for example, that the experimental animals were handled more in the laboratory than the controls, and that handling alone retards growth by the observed amount. Of course we try hard to keep every possible relevant factor constant, and here random sampling and random assignment help. But such procedures do not guarantee success.

Here we see, therefore, an example of the inferential gap that in practice always exists between statistical conclusions and theoretical or causal hypotheses. Unfortunately I must once again decline to discuss the detailed logic of the second-level inference. I only wish to emphasize that there are two levels and that failure at the second level does not necessarily indicate any defects in the characterization or the application of the lower level statistical logic.

6. The Rejection of Foundationist Justifications for Hypothesis Testing

Much of the twentieth century literature on the justification of induction begins with the tacit assumption that the goal is to find a justifiable method of ampliative inference which requires only singular observation statements as premises. Only such a method, it is held, could avoid the ultimate epistemological circularity of an inductive logic which required as premises statements that could themselves only be justified by an ampliative inference. Following Nicod (1930) and Reichenbach (1949), two authors in this tradition who were very much aware of their goals, I will refer to the desired form of inference as a *primary* inductive method. All others are *secondary*.

With a primary inductive method at hand one might hope to build all knowledge on a foundation of singular facts only. This is certainly what Reichenbach tried to do with his rule of simple enumeration. Let us now see whether this 'foundationist' program could be carried out within the framework of N-P hypothesis testing. It is clear that to complete such a program there must be some bona fide N-P tests which make no nontrivial admissibility or multiple trial assumptions. Let us therefore investigate the possibility of constructing such a test.[30]

The simplest distribution is the binomial. If there is no presupposition-free N-P test for a binomial parameter, then there are no such tests at all. Continuing the example of the previous section, pick a size S_0 for adult DDT-free rats. Then the adult size of each such rat has two possible values, $x \leqq S_0$ or $x > S_0$, and $P(X \leqq S_0) = 1 - P(X > S_0)$. Assuming n trials, consider the simple statistical hypothesis:

H: (i) $P(X_i \leqq S_0) = \frac{1}{2}$ for all trials.
(ii) All trials mutually independent.

For ease of exposition I have made the needed multiple trial assumptions part of H itself so that all presuppositions of the test may be treated as admissibility assumptions. If we are to avoid all contingent presuppositions, the alternative for our test of H must be the composite hypothesis:

K: not-H.

The union, H or K, is therefore logically exhaustive and the test has only logically true presuppositions. The question is whether with such a broad class of admissible hypotheses one can construct a bona fide N-P test.

The basic sample space underlying any test of H vs. K would be the set of 2^n n-tuples $x = (x_1, x_2, \ldots, x_n)$ where x_i has two possible values, say 0 or 1. Now let R be any possible rejection region, i.e., R is any nonempty and nonexhaustive subset of n-tuples in the basic sample space. Given the inclusiveness of K, it is easy to prove that there is some $K_j \in K$ for which $P_j(X \in R) = 0$. Since R does not exhaust the sample

[30] Both Reichenbach (1949, pp. 360–61) and Pap (1962, p. 238) have explicitly claimed that N-P testing is a secondary inductive method and therefore incapable of being a philosophically fundamental inductive logic. Neither author, however, gives a detailed argument why this should be so. Such an argument has recently been given by Rogers (1971). I am grateful to Professor Rogers for prodding me into thinking more deeply about this question.

space, there is some sample point $x' \notin R$. Suppose this point is $x' = (x_1 = 1, x_2 = 0, \ldots, x_n = 0)$. Now since K contains all possible hypotheses specifying unequal probabilities on different trials, there is a $K_j \epsilon K$ for which $P_j(X_1 = 1) = 1$, $P_j(X_2 = 0) = 1, \ldots, P_j(X_n = 0) = 1$. Thus, $P_j(X \epsilon R) = 0$.

It follows immediately that R does not provide a UMP test of H vs. K since there obviously exists another possible rejection region with the same or lower significance level, namely $R' = x'$, for which the power against some alternative, namely, K_j, is greater. It also follows that R does not provide a UMP unbiased test of H vs. K since for finite samples the significance level $P_H(X \epsilon R)$ must be greater than zero and unbiasedness requires that the power be greater than the significance level for all simple alternatives. (The definition of unbiased tests was given in section 3.) For most statisticians this would be sufficient proof not only that there is no "best" N-P test of H vs. K, but also that there is not even a "good" N-P test. In practice one might tolerate a test that is slightly biased for a narrow class of alternatives. But in principle it is hard to countenance a test with many alternatives for which one is more likely to reject the null hypothesis when it is true than when it is false. This is the case with the test above.

It is, of course, not logically impossible that someone should propose an intuitively satisfactory criterion for good though clearly biased N-P tests like the above. As noted in section 3, the theory of testing simple against composite hypotheses is still somewhat open. In the absence of a definite proposal, however, one need not take this possibility very seriously. Moreover, I am confident that most students of statistical theory would judge it almost inconceivable that there should be an appealing criterion that could be satisfied in the face of so broad an alternative as that in the test above.

Let us grant, therefore, that there is little hope of constructing a legitimate empirically presuppositionless N-P test of a simple hypothesis versus a composite alternative. It might be suggested, however, that one can do better testing a composite hypothesis against a composite alternative. This approach has the disadvantage that the λ-criterion is the least well motivated of all N-P test criteria. Still, the possibility of using such tests in a foundationist program is worth investigating.

Let the test hypothesis assert "random trials," i.e.,

H: (i) $P(X_1 \leq S_o) = P(X_2 \leq S_o) = \ldots = P(X_n \leq S_o) = p$.
(ii) All trials mutually independent.

The logically exhaustive alternative is K: not-H. H is composite because p may have any value between 0 and 1. In this case the rejection region is determined by the lambda principle, i.e.,

$$\lambda(x) = P_i(X = x)_{max}/P_j(X = x)_{max} \leq c.$$

Recalling the definition of K, it is easy to see that $P_j(X = x)_{max} = 1$ for any x. No matter what the pattern of 0's and 1's in a sequence of trials, there is a $K_j \epsilon K$ which has $P_j(X = 0) = 1$ for all trials with outcome 0 and $P_j(X = 1) = 1$ for all trials with outcome 1. Thus the value of $\lambda(x)$ depends solely on the value of $P_i(X = x)_{max}$. Here we run into difficulties. Assuming n even, the lowest value of $P_i(X = x)_{max}$ occurs whenever there are $n/2$ 0's, regardless of order, and H_i is

H(½): (i) $p_1 = p_2 = \ldots = p_n = ½$.
(ii) All trials mutually independent.

Thus the minimal nonempty rejection region contains all outcomes with $n/2$ 0's, and no others. As far as the formalities of the lambda rule are concerned, we have here a presupposition-free test. But in terms of the general guiding principles of the N-P approach, this test is completely unsatisfactory.

Suppose $H(½)$ is true and a series of n trials results in the most probable outcome, i.e., half 0's and half 1's. The test described above would reject the hypothesis. Even if the 0's and 1's were ordered in an intuitively random pattern, any pattern, the hypothesis would be rejected. On the other hand, suppose $P_j(X = 0) = 1$ for the first $n/4$ trials and $P_j(X = 0) = 0$ for the ¾ n trials. Given the outcome that this alternative hypothesis is certain to produce, the hypothesis of equal, independent distributions would *not* be rejected. In short, even though we cannot appeal explicitly to significance level and power in tests involving two composite hypotheses, it is clear that this particular test is not very good at distinguishing random trials from nonrandom trials. And this will remain true no matter how we enlarge the rejection region or reduce the sample space.

I conclude that there simply are no satisfactory empirically presuppositionless N-P tests. The foundationist program for justifying the logic of statistical hypothesis testing must, therefore, be rejected. The underlying reason for the nonexistence of legitimate presupposition-free tests should be evident. A logically exhaustive set of admissible hypotheses is just too broad to permit meaningful discrimination on the basis of empirical trials alone. A legitimate application of the N-P logic of hypothesis testing requires a more narrowly prescribed set of admissible hypotheses, and this requires empirical presuppositions.

7. Methodological Justification: The Epistemological Roots of Statistical Knowledge

Empiricist epistemology has been dominated by the foundationist viewpoint since at least the seventeenth century. For those believing in the possibility of rational inductive reasoning, foundationism leads immediately to the search for a justifiable primary inductive method. Thus, abandoning the attempt to ground the acceptance of statistical hypotheses on presupposition-free tests means abandoning the foundationist program altogether. There is, however, another possible approach, one which emerges explicitly in the writings of Peirce, Dewey and, more recently, Karl Popper.[31] According to this 'methodological' viewpoint, no primary method is needed. It is enough that our methods of validation permit the *process* of scientific investigation to proceed in an "appropriate" fashion. Thus the ultimate justification of scientific hypotheses is not determined by a relation to any "foundation," but by rules governing the processes by which the whole body of accepted scientific claims changes and grows.

The task before us, then, is to mold this general conception of a methodological justification into a detailed account of the ultimate grounds for accepting the results of standard N-P tests. As explained in section 4 above, underlying an N-P test is a root structure of presuppositions, some of which have been subjected to tests introducing their own further presuppositions. At any time the root structure for

[31] Peirce's methodological writings are found throughout his *Collected Papers* (1934–58), but especially in vols. 2 and 5. Dewey's most systematic exposition is in his much-neglected *Logic: The Theory of Inquiry* (1939). The essays in *Conjectures and Refutations* (1962) form the most accessible introduction to Popper's philosophy. Given Popper's dogmatic insistence that there is no such thing as inductive reasoning, he could hardly be expected to support my use of some of his ideas (which, however, are mostly to be found in Peirce anyway).

any particular tested hypothesis is finite, though in principle it could be extended indefinitely. Furthermore, since there are no completely presupposition-free tests, the roots of a given test will necessarily terminate at untested "ultimate" presuppositions. We shall thus be constructing an ultimate justification for accepting a statistical hypothesis in terms of methodological rules largely concerned with the ultimate presuppositions of an N-P test. It must be emphasized, however, that the rules to be developed give conditions which the presuppositions must satisfy in order that the acceptance (or rejection) of *H*, the tested hypothesis, may be fully justified. These rules in no way constitute a primary inductive method leading to the acceptance of the presuppositions themselves. The ultimate presuppositions, in particular, may remain untested and therefore be neither accepted nor rejected.

In practice, some and possibly all presuppositions of a particular test may be indirectly well supported by theories already confirmed through previous statistical inquiries. But if one holds, as I do, that the ultimate epistemological grounds for accepting a theory are eventually all statistical conclusions, the appeal to theories can play no essential role in an inquiry into the ultimate justifiability of scientific claims. The following discussion, therefore, ignores the role of theories in the context of ultimate justification.

Before proposing specific methodological rules, we must consider a prior requirement on the rules themselves, namely, that the applicability of the rules be ascertainable without assuming the truth of any statistical hypotheses. The basis for this requirement is the simple desire to avoid the regress involved in using methodological rules to provide ultimate justification for accepting statistical hypotheses while simultaneously assuming the truth of statistical hypotheses in applying the rules. This requirement, which derives from Hume's analysis of induction and is assumed in all foundationist programs, can be defended on more positive grounds as follows. We can never be as certain of the truth of statistical (or universal) hypotheses as we are of singular statements which describe individual events. The most we can do is to find methods of utilizing singular statements to make it "reasonable" or "rational" to accept statistical hypotheses. But at least we should be certain that we are being rational in accepting *H*, even though we cannot be certain that *H* is true. But this is not possible if in applying the criteria for rational acceptance we assume the truth of statistical hypotheses. For example, we cannot say that the acceptance of *H* is (ultimately) justified if the immediate presuppositions of the relevant test are true. Since we can be no more certain of the truth of these presupposed statistical hypotheses than we can be of the truth of *H*, we could be no more certain that our acceptance of *H* was justified than that *H* is true. This much of the foundationist tradition seems to me correct and I will adopt the Humean requirement stated above.[32] Note, by the way, that we are not making the disastrous demand that to know *H* one must know that one knows *H*. We demand only that to be justified in accepting *H* one must know that he is so justified.

Turning now to consideration of the desired methodological rules themselves, a very plausible rule for justifying the acceptance of *H* relative to test *T* is:

> (1) No presupposition, *P*, of *T* shall have been rejected by an appropriate statistical test.

This rule is directed only at the tested presuppositions of *T*, if any, since ultimate presuppositions are by definition untested and thus neither accepted nor rejected. It simply reflects the fact that the reasonableness of accepting *H* through *T* rests on the calculated error probabilities of *T* and that these depend on the truth of *T*'s presuppositions. To accept the falsity of *P*, therefore, is to admit that the error probabilities of *T* have been calculated incorrectly and thus to destroy the grounds for accepting *H*. This does not mean, of course, that the calculated error probabilities might not accidentally be correct after all — but justification cannot depend upon happy accidents.

Note that rule (1) only says that if *T′* is a test of presupposition *P* in the root structure underlying the acceptance of *H*, then the result of *T′* must have been positive. It does not say, on pain of regress, that the acceptance of *P* must itself by ultimately justified. It will turn out, however, that if the acceptance of *H* is ultimately justified, then so is the acceptance of any tested presuppositions of *T*.

Let us focus now on the ultimate (i.e., untested) presuppositions of a statistical test. The remaining conditions constitute an attempt to

[32] Even if the Humean requirement were to be rejected, it would still be worth knowing how far one can get without crossing this line because then we might discover precisely why and how the line must be crossed. In this and other matters I may be accused of being too much influenced by my staunchly foundationist former colleague Wesley Salmon. Perhaps so, but these views are mine nevertheless.

develop in a relatively precise context the Peircean maxims honestly to seek the truth and never to block the way of inquiry.[33] Suppose that a presupposition of test *T* turned out not to be subject to an N-P test. In this case the acceptability of *H* would depend, if only indirectly, on the truth of *P*, but one would have no means for empirically testing *P*. One avenue for further inquiry into the acceptance of *H* would be blocked. Indeed, one could claim that the degree to which the acceptance of *H* is an empirical question within the N-P framework is roughly inversely proportional to the relative number of ultimate presuppositions of *T* not testable by N-P methods. If it happened that no *immediate* presupposition of *T* were N-P testable, the acceptance of *H* would surely lack any kind of "ultimate" empirical justification.

These considerations suggest the following condition for the acceptance of *H* to be ultimately justified:

(2) No ultimate presupposition of *T* may be logically impossible to test by N-P methods.

Note that this rule covers only presuppositions that might be *logically* beyond N-P methods. The satisfaction of this rule can therefore be determined by logical considerations alone, thus meeting the Humean requirement.

That N-P methods satisfy rule (2) follows from the fact, proved in section 4, that the N-P logic is presuppositionally closed. The preceding discussion suggests that any "fundamental" inductive method should exhibit this characteristic. Moreover, presuppositional closure seems to be a prerequisite for any adequate methodological account of ultimate justification. Indeed, without presuppositional closure it is hard to see how the methodologist can honestly refuse the foundationist's demand for a primary inductive method.

The concept of presuppositional closure may be further clarified by noting whether or not it holds for several other common characterizations of inductive inference. Simple enumeration may be said to be trivially closed since applications of this method presumably require no presuppositions whatsoever. The subjective Bayesian approach would seem to be presuppositionally closed if one takes prior probabilities to

[33] These maxims are present throughout Peirce's methodological writings. For an explicit statement and discussion of the rule "Do not block the way of inquiry" see (1934–58, vol. 1, bk. I, ch. 3; especially 1.135).

be presuppositions. Any prior probability distribution could be a posterior distribution relative to some other prior and some body of evidence. Induction by elimination also seems to be closed since the presupposition that the true hypothesis is a member of a given set of possibilities may be reached by elimination from a larger set of presupposed alternatives. On the other hand, purely logical accounts of inductive reasoning, e.g., Carnap's, are not closed since every system of logical probability presupposes a measure function, but no system assigns probabilities to measure functions themselves. This result might not be undesirable if one could show that the measure function is *not* an *empirical* presupposition, but I doubt this can be shown. Finally, it seems to be true generally of methods incorporating global presuppositions, e.g., measure functions, the uniformity of nature, limited independent variety, etc., that they preclude presuppositional closure. If every inductive inference must presuppose the same global principle, then the method could be closed only by begging the question.

This brief survey of inductive methods illustrates a further point concerning presuppositional closure, namely, that it can hold only for methods whose presuppositions are 'local' rather than 'global'. (The converse is not true.) Thus a necessary precondition for a successful methodological account of justification is that one's characterization require only local presuppositions. I will not attempt a precise definition of a local or a global statement since it is clear that any statistical hypothesis is local in the intended sense. Roughly speaking, global presuppositions apply to the whole universe at all times and are presupposed in every application of the methods in question.[34]

Turning to the next condition, it is useful to recall earlier philosophical debates over whether empirical meaningfulness requires only the logical possibility of verification, or whether it requires physical and perhaps even technical possibility as well. The same sort of issue arises here, though the problem is one of justification rather than meaning. Suppose at some time, *t*, an ultimate presupposition, *P*, of *T* is logically N-P testable, i.e., *P* is a statistical hypothesis, but *P* cannot be tested at that time for one of the following reasons: (a) Any N-P test would

[34] Levi (1967) explicitly distinguishes "local" from "global" justification. This type of distinction is also central to Shimony's views as expressed in his rich and inspiring paper "Scientific Inference" (1970). Shimony, who is also strongly influenced by Peirce, builds his views on a type of subjective Bayesian logic which he calls "tempered personalism."

violate a physical law; (b) Any test is beyond currently available technology; (c) Any test would violate the morals or laws of the surrounding society.[35] Would one, at *t*, be fully justified in accepting *H*?

The argument against saying the acceptance of *H* is fully justified is an extension of that given for condition (2). If *P* cannot be tested at *t*, for whatever reason, then inquiry is blocked. One who wishes to question the acceptance (or rejection) of *H* by questioning the truth of *P* cannot at that time perform an empirical test on *P*. He can only be offered the question-begging assertion that *P* would be accepted if it could be tested or will be accepted when it is eventually tested.

The argument against requiring the physical, technical, or social possibility of testing *P* is that it relativizes ultimate justification to a body of physical laws, techniques, or social customs. The first is no doubt easier to swallow than the latter two. Is it right that the acceptance of *H* should be unjustified now and justified later solely because the invention of some gadget makes it technically possible to test *P*? Again, should accepting *H* be justified in one society and not in another when everything, including all evidence, is the same in both, except for the fact that the one society permits euthanasia, thus making possible certain experiments on humans not possible in the other society? I raise these questions here mainly to make clear what is involved in granting such relativization. The amount of relativization necessary or desirable in a methodological account of ultimate justification can be better judged when we have a full set of conditions before us.

The attempt to demand more than the mere logical possibility of N-P testing ultimate presuppositions leads immediately to a further epistemic relativization. The statement that testing *P* is possible physically, technically, and perhaps even socially could only be asserted on inductive grounds. Thus any rule demanding such possibilities violates the general Humean requirement imposed above. An obvious way to resolve this conflict is to demand not that the possibilities obtain, but only that the *impossibility* of testing not be *known*. More precisely:

(3) No ultimate presupposition of *T* may be accepted (at *t*) as being physically, technically, or socially impossible to test by N-P methods.

[35] Examples of cases (b) and (c) are easily imagined. An example of (a) might occur in astronomy if obtaining a proper sample were to require signals traveling faster than light.

Of course accepting something as physically impossible requires an inductive logic which goes beyond N-P testing, but rule (3) does not necessarily commit us even to the existence of such a method, let alone to providing a detailed characterization of its logic. And surely (3) can in principle be applied without using any inductive methods whatsoever.

Note that rule (3) introduces a whiff of the venerable but generally disreputable doctrine that knowledge may be derived from ignorance. For according to (3), the ultimate acceptability of *H* depends on our not knowing that *P* is now physically, technically, or socially impossible to test. However, while I would strenuously object to any inductive logic which, unlike the N-P theory, had a principle of indifference built into its internal characterization, it is not obvious that some appeal to ignorance is undesirable, or even avoidable, at the level of ultimate justification. Indeed it seems entirely natural that at some point in the quest for further justification one should simply reply that we have no reason for thinking otherwise. Rule (3) tells us precisely when and how such a response becomes appropriate.[36]

It is time we faced squarely the problem of stopping the Humean regress implicit in the epistemological structure of N-P tests. Since every test presupposes some nonvacuous statistical hypotheses and every statistical hypothesis might be tested, how do we justify selecting certain untested presuppositions as ultimate? According to the pragmatist tradition, the motive force of inquiry is active doubt. There is no inquiry without initial doubt, and inquiry ends when doubt is resolved. This suggests taking as ultimate presuppositions only statistical hypotheses whose truth is currently not seriously questioned.[37] Several points of clarification are necessary before this suggestion can be turned into an explicit rule.

[36] Kyburg (1961), for example, builds a principle of indifference into the characterization of his inductive logic by *defining* a random sample as one *not known* to be unrepresentative of the population. It seems to me that an adequate inductive logic will have to leave random sampling as a special empirical presupposition, always open to further investigation. Relativization to current knowledge should come later — at the level of *ultimate* justification.

[37] Recall that doubt plays a key role in Peirce's and Dewey's theory of inquiry. My notion of "seriously questioning" a hypothesis is very similar to Shimony's (1970) conception of a "seriously proposed" hypothesis, and much of his discussion is relevant here. In particular, both notions lead to Peirce's view that scientific inquiry is an essentially social phenomenon.

First, it must be emphasized that *P*'s not being actively doubted (or seriously questioned) is nothing more than a brute singular fact. The Humean regress is thus blocked by simple facts. Any attempt to make active doubt more than a mere fact opens the door to the familiar regress. For example, it might be argued that it is not enough for *P* not to be doubted — there should be *no reason* to doubt *P*. On one analysis, this formulation is unobjectionable. *P* is an empirical statistical hypothesis, and the only feature of *P* that concerns us here is its truth or falsity. Moreover, in the present discussion of ultimate justification, we may assume that the only reason for taking *P* to be true (or false) would be the outcome of an N-P test. But since ultimate presuppositions are by definition untested, it follows that we have "no reason" to doubt the truth of *P*. On the other hand, one might have intended the stronger claim that there should be reason not to doubt *P*. This interpretation, however, must be rejected because a reason not to doubt *P* would, in the present context, have to be a favorable N-P test. This way lies the regress.

A second clarification is that on my understanding of the pragmatist tradition, doubting or questioning is essentially a social phenomenon. This means that any particular ultimate justification must be relativized to some sort of scientific community, say the members of a subfield like solid state physics or learning theory. Such relativization raises a number of difficult questions. For example, what defines the relevant scientific community for test *T*? This is especially difficult since undoubtedly one of the criteria for admission will be a body of shared assumptions. If the methodological approach to ultimate justification is to be further developed, such questions will eventually have to be answered. For the moment, however, I will rest content with the following rule:

> (4) No ultimate presupposition of *T* may in fact be seriously questioned (at *t*) by members of the relevant scientific community.

The problem of providing a further rationale for this rule itself will come up again in the following section.

According to rule (4), no presupposition of *T is* currently under serious question. To keep faith with the methodological tradition of Peirce and Dewey, however, we should perhaps emphasize that any presupposition *can* be questioned. One may contend that this latter rule follows from rules (2) and (3) which guarantee that, to the best of our knowledge, any presupposition can be *tested*. But even if it does follow, it is worth stating separately:

(5) No ultimate presupposition of *T* may be accepted (at *t*) as being physically, technically, or socially impossible to question ("seriously").

This rule has been relativized to accepted knowledge to avoid a possible violation of the Humean requirement. Also it may turn out that the class of statistical hypotheses which are physically or technically incapable of being questioned is null. This depends on one's understanding of what it is for a member of a scientific community seriously to question a hypothesis. That a hypothesis may be socially immune to serious question seems fairly clear once it is granted that serious questioning is essentially a social activity. It would suffice, for example, that anyone attempting to voice doubts about a particular hypothesis be subject to immediate arrest and imprisonment.

One final condition seems necessary to eliminate the possibility that rule (4) is satisfied merely as a result of sheer indolence or lack of interest in finding the truth. Suppose, for example, that no one seriously questions the ultimate presuppositions of *T* because no one is really interested in the truth or falsity of *H*, or perhaps any statistical hypothesis, and therefore no one is concerned with the truth or falsity of *T*'s presuppositions. In such a situation the acceptance of *H* as a result of *T* could hardly be said to be justified in any philosophically satisfactory sense. Unless it is a general policy of the community to consider, question, and sometimes to test presuppositions, the unquestioned status of *T*'s ultimate presuppositions is irrelevant to any ultimate justification for the acceptance of *H*. This point is covered by the following suggested rule:

(6) The scientific community relevant to test *T* must be actively engaged in the pursuit of truth.

Here it is presumed that active pursuit of truth implies examining, questioning, and sometimes even testing presuppositions of statistical tests.

Rule (6) is again reminiscent of Peirce who sometimes writes as if a sincere desire to discover the truth were both necessary and sufficient for success, if only the object of the desire is pursued long enough.[38] Rule (6) also resembles Popper's claim that corroboration of a theory requires a "sincere attempt" to refute it. Popper's position, however, is complicated by the fact that while the notion of a sincere attempt at refutation is part of his characterization of corroboration, he insists that it cannot be formalized. Both opponents and supporters of Popper's position have found his views unsatisfactory and have attempted to demystify the notion of sincerely attempting to refute a theory.[39] Now while I would agree with the opinion that such a notion has no place in the *characterization* of any inductive logic, I disagree that it should be banished altogether. It seems entirely appropriate that such a notion should appear in a methodological account of ultimate *justification*. But since Popper will have no part of either induction or justification, I could hardly expect his endorsement of my program.

There is, finally, an important qualification that should be made explicit. Suppose *H* has been subjected to test *T* with ultimate presupposition *P* which is then at some later time seriously questioned. The difficulty here is that *P* refers to a series of trials that are now over, and the data that were originally sufficient for a good test of *H* assuming *P* would not in general have been sufficient for a good test of *P* as well. Any test of *P*, therefore, will require a new set of trials of an appropriate CSU. Moreover, to apply the results of this new test to the earlier experimental situation requires the assumption that the two contexts are similar in all relevant respects. But at any later time it is at least physically impossible to subject this statistical hypothesis to any N-P test. The hypothesis may, of course, be supported by a well-confirmed causal theory, but the same problem arises regarding the statistical tests that support the theory. Thus, if we are to use a methodological approach to ultimate justification in eliminating Humean skepticism, we must narrow our understanding of this approach.

There are two possibilities. One is to restrict the sense of the expres-

[38] In some passages (e.g., 1934–58, 2.709, 2.729) Peirce anticipates Reichenbach's pragmatic vindication of induction. The difficulty is that while Peirce wanted to apply these ideas to all forms of inductive reasoning, the argument holds only for the inference to a limiting relative frequency. This gap in Peirce's thought is discussed by Laudan (1973).

[39] For Popper's claim see (1959, p. 418). Kyburg's comment (1964, p. 269) that "all this psychologism is distasteful to the logician" expresses a widespread viewpoint. For "demystifications" see Salmon (1966) (anti-Popper) and Watkins (1964) (pro-Popper).

sion "can be tested" so that it means only "could have been tested at the time of the original experiment." But this restriction effectively eliminates the idea that a methodological justification is concerned with the *process* of inquiry. The second possibility is to suppose that the whole presuppositional structure of a given test is retested, with appropriate changes, whenever the acceptability of a questioned presupposition cannot be decided on the original data. In practice, of course, this is not done because we rely on causal theories to relate present and past data. But the fact that such retesting is always possible suffices to give substance to the concern with the process of inquiry that underlies the methodological approach to ultimate justification.

Ideally, fulfillment of the preceding conditions would be necessary and sufficient for the acceptance or rejection of H by test T to be justified to the extent that acceptance (or rejection) of any statistical hypothesis can be justified. Yet even a believer in the methodological program may doubt that this particular set of rules is ideal. The necessity for eliminating social blocks to the testing of presuppositions is especially controversial. There are also additional rules one might require. For example, one might wish explicitly to stipulate that no hypothesis, including H, appear more than once in any branch of a root structure. (There is obviously no reason to prevent the same presupposition from appearing in different branches.) However, repeating branches, e.g., $H, T, P, T', H, T'', P, \ldots$, may be ruled out on the understanding that to test a hypothesis is automatically to question it, and ultimate presuppositions must, by rule 4, be unquestioned. I will not, however, pursue such matters further. Enough has been said to make clear the kind of commitments involved in a methodological approach to ultimate justification. A more pressing issue is to discover how one could show that anything like the preceding rules provide a *sufficient* account of ultimate justification in the context of statistical hypothesis testing.

8. Justification and Adequacy

How might one argue that satisfaction of the suggested methodological rules constitutes a sufficient ultimate justification for the acceptance (or rejection) of a statistical hypotheses? There seem to be only two general possibilities. One is to argue that satisfaction of the rules contributes to the goals of statistical inference as given by the N-P logic, i.e., the acceptance of true hypotheses and the rejection of false ones. The second is to claim that the rules explicate at least part of the meaning of ultimate justification or rational inference. Let us consider the first possibility first.

It is immediately obvious that there is no *logical* connection between satisfaction of the rules and either the truth of an accepted hypothesis or the correctness of the calculated error probabilities. The diligent seeker after truth may never accept a true hypothesis while the despot invariably decrees the truth on advice from his astrologer. In short, no methodological rules are logically necessary or sufficient for success now or in the indefinite future. Thus, if there is a connection between the rules for ultimate justifiability and the actual acceptance of true hypotheses, it could only be established inductively. But if N-P tests constitute our most basic inductive method, then any attempt to justify the conditions inductively introduces circularity at the meta-level. Consider an N-P test of the claim that an accepted test hypothesis is more likely to be true when the suggested conditions for ultimate justification are satisfied than when they are not. As evidence one would need cases in which the conditions were satisfied and the test hypothesis is true. But determining the truth of the test hypothesis would require an N-P test. Thus we cannot even obtain data necessary for the meta-test without assuming the ultimate justifiability of N-P testing.

In spite of this very Humean argument, it is difficult not to believe that an accepted hypothesis would more often be true when the methodological rules are satisfied than when they are not. In particular, an untested hypothesis that is nevertheless doubted by the relevant scientific community would more often be false than one that is not doubted. We know, inductively of course, that a trained intuition will pick up clues that would not count as evidence in any inductive logic. But the Humean argument does not show that such connections do not exist. It only shows that they may not be used to support the kind of justification program developed in the previous section.[40] This leaves us with the second alternative.

Philosophers of the so-called "ordinary language" school, as well as

[40] Shimony (1970) follows Peirce in appealing to natural selection in his justification of certain inductive procedures. I agree that some intuitions reinforced through natural selection play a role in scientific inquiry, though their applicability to microphysics or molecular biology seems farfetched. But the attempt to use such considerations in an account of ultimate justifications still seems to me objectionably circular.

Carnap in his latest writings,[41] have maintained that certain inductive methods are ultimately justified because these methods determine what we mean by justified inductive inference. To ask whether certain inferences are justified, therefore, is to ask a meaningless question. Applying this general approach to the present inquiry, it seems farfetched to claim that either the N-P logic of testing or the suggested methodological rules constitute an explication of what we mean by justified inductive reasoning. On the other hand, suppose we had a comprehensive inductive logic, including a precise characterization of theory testing, an interpretation of probability statements, and methodological rules like those given above. And suppose this overall account of scientific reasoning were judged superior to all other proposed systems of similar scope and precision. In this case we might well say that this system expresses our present conception of scientific reasoning, and the included schema of justification what we mean by justification in any relevant sense. On this view the problem of justification ultimately becomes part of the more general problem of judging one overall system of induction reasoning more *adequate* than another. How can we make these more general judgments?

At this point one might begin a search for still more general "criteria of adequacy" for complete systems of scientific inference. But this is a fruitless quest. The desired criteria cannot be purely logical principles, nor can they be supported inductively without circularity. Of course *after* one system has been judged superior to other proposed systems, then we might be able to formulate principles reflecting our judgments of relative adequacy, and these might help us better to understand our decisions. But such principles could not be taken as providing rational grounds for our decisions. Such decisions would have no rational grounds at all. They would merely be a causal result of our collective experience in working with the various systems in a variety of actual research contexts. Once the rival systems have been formulated, the only rational control we have on the judgment of adequacy is through the choice of contexts in which applications are attempted. Since we seek as comprehensive an overall system as possible, it seems clear we should try any proposed system in as wide a variety of scientific contexts as is practicable. But this is the most we can do.

[41] This, at least, is my understanding of the discussion in Carnap (1963).

Unfortunately this resolution for the problem of adequacy does not guarantee that any system of inductive reasoning will ever enjoy the kind of widespread approval enjoyed by current systems of deductive reasoning. There is, however, some ground for hope in the fact that there are very few systems that are sufficiently comprehensive and that have been sufficiently well developed even to be active candidates. At the moment the only really serious alternative to something like the N-P logic of testing is a system of Bayesian inference utilizing subjective probabilities. It seems not unreasonable to hope that patient attempts to apply these rival logics in a variety of research situations will make clear which is the more adequate account of scientific inference.

There are those who will feel that by allowing judgments of adequacy to be merely causal outcomes of scientific experience, I have abandoned the justification problem, not resolved it. I would reply that my position simply reflects our true situation regarding inductive reasoning (and probably deductive too!). In this I am in general agreement with Hume for whom inductive reasoning was ultimately a matter of animal habit. This is also the general position of such modern philosophers as Nelson Goodman, who appeals to the simple facts of past linguistic usage, and W. V. O. Quine, who now grounds epistemology in empirical psychology. The main difference between these views and mine is not that one stops trying to give reasons, but where and how one stops.[42]

On my account there are actually two points at which reasons give way to brute fact. One is at the object level, so to speak: some ultimate presuppositions of any test will be unquestioned, and one need not say why. The second is at the meta-level: judgments of adequacy are not justified by reasons. But it is never the case that the acceptance of any specific test hypothesis is itself a matter of brute fact. This remains under rational, objective control, though of course relative to the adequacy of the overall system and the ultimate presuppositions of some particular statistical test. In this way it is possible to insist that inductive

[42] Just as the "new riddle of induction" is only a linguistic version of Hume's old riddle, so Goodman's (1955) entrenchment theory is just a linguistic version of Hume's theory of habits, though more precisely articulated. That Quine (1969) does not make more of the *social* basis of epistemology may be a reflection of his philosophical upbringing at a time when the logic/psychology distinction and the reducibility of sociology to psychology were taken for granted. Then again, knowing has almost always been taken to be a function of the individual "soul." In this the pragmatists are an exception.

inference is a rational process while admitting an essential justificatory role for some brute facts.

9. The Metaphors of Ultimate Justification: Foundations, Nets, and Roots

My overall program for the philosophy of inductive reasoning is far from completed. Much remains to be done in characterizing the logic of statistical hypothesis testing and the logic of theory testing. In addition, the kinds of case studies of actual research that might produce widespread agreement on the adequacy of these methods are practically nonexistent. As it is impossible to pursue these tasks here, I will content myself with some concluding remarks on the metaphor underlying the methodological justification of statistical hypothesis testing presented above. When one reaches the limits of rational discussion, in any subject, the importance of a good metaphor should not be overlooked.

Having rejected the church and reason as guarantors of knowledge, classical empiricists looked to 'experience' as the source and foundation of all knowledge. The implicit ideal was to use only concepts somehow abstracted directly from experience and to admit only hypotheses logically derivable from statements irrefutably certified by experience alone. Knowledge was thus pictured as a solid building rising on unshakable foundations. This metaphor proved remarkably durable. Sustained by Mach's phenomenalism and the logical constructions of Russell's *Principia*, the foundationist picture, in the form of Carnap's *Aufbau*, inspired the Viennese branch of empiricism. Yet even the logical empiricists soon learned that this classical ideal is unattainable. But what is to replace it?

Here we are not concerned with how one obtains a suitable store of singular observation statements, but only with how these may be used to substantiate singular predictions or general laws, especially statistical laws. Once it is admitted that this cannot be done with standard deductive tools alone, there are two major strategies that may be pursued. One is to use the observation statements to assign (inductive) probabilities to all other statements. If, as in Carnap's program, this can be done with a purely semantical probability relation, much of the spirit of the original foundationist program is preserved. If one has doubts about using logical probabilities, a retreat to subjective probabilities is always possible.[43] The second major approach is to construct an inductive logic which contains a bona fide inductive rule of inference, i.e., one that validates empirical statements themselves and not merely probability relations with observation statements. Here one retains the spirit of the original program by insisting that the inductive rule of inference be a *primary* rule, i.e., one whose application presupposes only observation statements. Reichenbach's rule of induction is the best known rule of this type.

These weakened versions of foundationism still have many supporters, both explicit and tacit. Yet there are also many who would like to abandon foundationism altogether. For the latter, Quine has provided the tempting metaphor of knowledge as a "net" or, more recently, as a "web of belief." [44] One obvious possibility, therefore, would be to elucidate my conditions for methodological justification in terms of Quine's net metaphor. This move would be easy because Quine and his followers have never developed a precise characterization of statistical inference. The N-P logic of testing might therefore be viewed as giving the detailed structure of parts of the Quinean net. Tempting as this move might be, it seems to me mistaken.

Quine's net is really Neurath's ship.[45] The elements, i.e., the sea and wind, are experience. Owing to the impact of experience and the lack of a dry dock, the inhabitants of the ship must continually rebuild their craft while at sea. The only necessity is keeping out the wind and the sea. Beyond this the primary principles of shipbuilding are simplicity and elegance, however these may be defined.

My first objection to Quine's picture is merely the intuition that knowledge should be more than something bobbing about on the surface of experience, no matter how internally elegant that something might be. Knowledge should somehow reach into experience. At the very least, Neurath's ship could use an anchor. But perhaps this intuition is merely the product of nostalgia for the foundationist picture.

[43] Jeffrey (1965) is a good example of someone who has made this shift.

[44] The classic source of the net metaphor is Quine's "Two Dogmas of Empiricism" reprinted in Quine (1953) and elsewhere. *The Web of Belief* (1970) is a recent elementary text by Quine and Ullian.

[45] Recall Otto Neurath's famous remark quoted by Quine at the beginning of *Word and Object* (1960, p. viii): "Wie Schiffer sind wir, die ihr Schiff auf offener See umbauen müssen, ohne es jemals in einen Dock zerlegen und aus besten Bestandteilen neu errichten zu können."

My second objection is that the net metaphor is too vague, even for a metaphor. While it is compatible with a methodological justification for N-P testing, it is also compatible with a Carnap-style foundationism. One need only interpret the strings of the net as logical or subjective probability relations. This very natural interpretation virtually eliminates any substantial differences between Quine and Carnap.

The appropriate metaphor is suggested by the epistemological structure of N-P tests — a system of roots. This suggests the organic metaphor of knowledge as a tree with branches growing up and roots growing down into the soil of experience. The roots, however, do not consist solely of observations, but, like the branches, contain statistical hypotheses as well. Moreover, there is no theoretical limit to the size of the tree so long as the roots are extensive enough and deep enough. The principal value of this metaphor is that it makes intuitively clear that there can be an extensive structure of knowledge rising above the ground of experience without there being any foundation at all. Roots are sufficient. In addition, a tree metaphor goes well with a methodological account of ultimate justification. The methodological rules give conditions on the state of the roots and how they must grow if we are to have a genuine "tree of knowledge." [46]

The picture of knowledge as a plant with roots and branches is in keeping with the contemporary passion for things organic. And a pragmatic, methodological account of ultimate justification reflects another current in contemporary thought, an interest in the social aspects of all things. Whether these themes will prove lasting enough to deserve a place in our epistemology is now impossible to judge. If my conception of judgments of adequacy is correct, our task is to develop comprehensive systems and try them in scientific contexts. Only in this way can we decide whether, having abandoned the foundationist picture, knowledge is now to be viewed as a net, a tree, or something completely different.

[46] There is an obvious parallel between a tree of knowledge supported by roots and Popper's metaphor of a building supported by piles driven into the swamp of experience (1959, p. 111). The structure of the N-P inductive logic makes the root metaphor clearly preferable.

REFERENCES

Anscombe, F. J. (1963). "Tests of Goodness of Fit," *Journal of the Royal Statistical Society B*, vol. 25, pp. 81–94.

Arbuthnot, J. (1710). "An Argument for Divine Providence Taken from the Constant Regularity Observed in the Births of Both Sexes," *Philosophical Transactions of the Royal Society of London*, vol. 27, pp. 186–90.

Bar-Hillel (1968). "The Acceptance Syndrome," in I. Lakatos, ed., *The Problem of Inductive Logic*. Amsterdam: North Holland.

Birnbaum, A. (1962). "On the Foundations of Statistical Inference," *Journal of the American Statistics Association*, vol. 57, pp. 269–306.

Birnbaum, A. (1969). "Concepts of Statistical Evidence," in S. Morgenbesser, P. Suppes, and M. White, eds., *Philosophy, Science and Method*. New York: St. Martin's.

Birnbaum, A. (1971). "A Perspective for Strengthening Scholarship in Statistics," *The American Statistician*, vol. 25, no. 3, pp. 14–17.

Black, M. (1971). *Margins of Precision*. Ithaca, N.Y.: Cornell University Press.

Carnap, R. (1950). *Logical Foundations of Probability*. 2nd ed. Chicago: University of Chicago Press, 1962.

Carnap, R. (1963). "Replies and Systematic Expositions," in P. A. Schilpp, ed., *The Philosophy of Rudolf Carnap*. La Salle, Ill.: Open Court.

de Finetti, B. (1937). "Foresight: Its Logical Laws, Its Subjective Sources," in H. E. Kyburg and H. E. Smokler, eds., *Studies in Subjective Probability*. New York: John Wiley, 1964.

Dewey, J. (1939). *Logic: The Theory of Inquiry*. New York: Holt, Rinehart, and Winston.

Earman, J. (1971). "Laplacian Determinism," *Journal of Philosophy*, vol. 68, pp. 729–44.

Edwards, A. W. F. (1971). *Likelihood*. Cambridge: At the University Press.

Feyerabend, P. K. (1962). "Explanation, Reduction and Empiricism," in H. Feigl and G. Maxwell, eds., *Minnesota Studies in the Philosophy of Science*, vol. 3. Minneapolis: University of Minnesota Press. Pp. 28–97.

Fisher, R. A. (1925). *Statistical Methods for Research Workers*. London: Oliver and Boyd.

Fisher, R. A. (1935). *The Design of Experiments*. London: Oliver and Boyd.

Fisher, R. A. (1950). *Contributions to Mathematical Statistics*. New York: John Wiley.

Fisher, R. A. (1956). *Statistical Methods and Scientific Inference*. New York: Hafner.

Giere, R. N. (1973a). "Objective Single Case Probabilities and the Foundations of Statistics," in P. Suppes, L. Henkin, A. Joja, and C. R. Moisil, eds., *Logic, Methodology and Philosophy of Science*, vol. 4: *Proceedings of the 1971 International Congress, Bucharest*. Amsterdam: North Holland.

Giere, R. N. (1973b). "History and Philosophy of Science: Intimate Relationship or Marriage of Convenience?" *British Journal for the Philosophy of Science*, vol. 24, pp. 282–97.

Giere, R. N. (1973c). Review of D. H. Mellor, *The Matter of Chance*, in *Ratio*, vol. 15, pp. 149–55.

Giere, R. N. (1975). "Empirical Probability, Objective Statistical Methods, and Scientific Inquiry," in C. Hooker and W. Harper, eds., *Foundations of Probability and Statistics and Statistical Theories in Science*. Dordrecht: Reidel.

Gilles, D. (1973). *An Objective Theory of Probability*. London: Menthuen.

Glymour, C. (1971). "Determinism, Ignorance, and Quantum Mechanics," *Journal of Philosophy*, vol. 68, pp. 744–51.

Good, I. J. (1950). *Probability and the Weighing of Evidence.* London: Griffin.
Good, I. J. (1965). *The Estimation of Probabilities.* Cambridge, Mass.: M.I.T. Press.
Goodman, N. (1955). *Fact, Fiction and Forecast.* Cambridge, Mass.: Harvard University Press. 2nd ed., 1965.
Hacking, I. (1965). *Logic of Statistical Inference.* Cambridge: At the University Press.
Hanson, N. R. (1958). *Patterns of Discovery.* Cambridge: At the University Press.
Hays, W. L. (1963). *Statistics for Psychologists.* New York: Holt, Rinehart, and Winston.
Hintikka, J. (1966). "A Two-Dimensional Continuum of Inductive Methods," in J. Hintikka and P. Suppes, eds., *Aspects of Inductive Logic.* Amsterdam: North Holland. Pp. 113–32.
Hodges, J. L., and E. L. Lehman (1964). *Basic Concepts of Probability and Statistics.* San Francisco: Holden and Day.
Jeffrey, R. C. (1956). "Valuation and Acceptance of Scientific Hypotheses," *Philosophy of Science*, vol. 23, pp. 237–46.
Jeffrey, R. C. (1965). *The Logic of Decision.* New York: McGraw-Hill.
Kempthorne, O., and Leroy Folks (1971). *Probability, Statistics, and Data Analysis.* Ames: Iowa State Press.
Kendall, M. G., and A. Stuart (1958–66). *The Advanced Theory of Statistics*, 3 vols. London: Griffin.
Kuhn, T. S. (1962). *The Structure of Scientific Revolutions.* Chicago: University of Chicago Press. 2nd ed., 1970.
Kyburg, H. E., Jr. (1961). *Probability and the Logic of Rational Belief.* Middletown, Conn.: Wesleyan University Press.
Kyburg, H. E., Jr. (1964). "Recent Work in Inductive Logic," *American Philosophical Quarterly*, vol. 1, pp. 249–87.
Kyburg, H. E., Jr. (1968). "The Rule of Detachment in Inductive Logic," in I. Lakatos, ed., *The Problem of Inductive Logic.* Amsterdam: North Holland. Pp. 98–119.
Kyburg, H. E., Jr. (1970). "Conjunctivitis," in M. Swain, ed., *Induction, Acceptance and Rational Belief.* Dordrecht: Reidel.
Kyburg, H. E., Jr., and H. E. Smokler, eds. (1964). *Studies in Subjective Probability.* New York: John Wiley.
Laplace, P. Simon Marquis de (1814). *A Philosophical Essay on Probabilities.* Paris.
Laudan, L. (1973). "Peirce and the Trivialization of the Self-Correcting Thesis," in R. N. Giere and R. S. Westfall, eds., *Foundations of Scientific Method: The Nineteenth Century.* Bloomington: Indiana University Press.
Lehman, E. L. (1959). *Testing Statistical Hypotheses.* New York: John Wiley.
Levi, I. (1967). *Gambling with Truth.* New York: Knopf.
Lindley, D. V. (1971). *Bayesian Statistics, a Review.* Philadelphia, Pa.: Society for Industrial and Applied Mathematics.
Luce, R. D., and H. Raiffa (1957). *Games and Decisions.* New York: John Wiley.
Mellor, D. H. (1971). *The Matter of Chance.* Cambridge: At the University Press.
Neyman, J. (1950). *First Course in Probability and Statistics.* New York: Henry Holt.
Neyman, J. (1957). "'Inductive Behavior' as a Basic Concept of Philosophy of Science," *Review of the International Statistical Institute*, vol. 25, pp. 7–22.
Neyman, J. (1967). *A Selection of Early Statistical Papers of J. Neyman.* Berkeley: University of California Press.
Neyman, J., and E. S. Pearson (1967). *Joint Statistical Papers.* Berkeley: University of California Press.

Nicod, J. (1930). *Foundations of Geometry and Induction*, trans. P. P. Wiener. London: Routledge and Kegan Paul. Trans. John Bell and Michael Woods, London, 1969.

Pap, A. (1962). *An Introduction to the Philosophy of Science*. New York: Free Press.

Pearson, E. S. (1966). "The Neyman-Pearson Story: 1926–34," in F. N. David, ed., *Research Papers in Statistics*. New York: John Wiley.

Peirce, C. S. (1934–58). *Collected Papers of Charles Sanders Peirce*, 8 vols. Cambridge, Mass.: Harvard University Press.

Popper, K. R. (1959). *The Logic of Scientific Discovery*. London: Hutchinson.

Popper, K. R. (1962). *Conjectures and Refutations*. New York: Basic Books.

Quine, W. V. O. (1953). *From a Logical Point of View*. Cambridge, Mass.: Harvard University Press.

Quine, W. V. O. (1960). *Word and Object*. Cambridge, Mass.: M.I.T. Press.

Quine, W. V. O. (1969). *Ontological Relativity and Other Essays*. New York: Columbia University Press.

Quine, W. V. O., and J. S. Ullian (1970). *The Web of Belief*. New York: Random House.

Reichenbach, H. (1949). *The Theory of Probability*. Berkeley: University of California Press.

Rogers, B. (1971). "Material Conditions on Tests of Statistical Hypotheses," in R. C. Buck and R. Cohen, eds., *Boston Studies in the Philosophy of Science*. New York: Humanities Press.

Salmon, W. C. (1966). *The Foundations of Scientific Inference*. Pittsburgh: University of Pittsburgh Press.

Savage, L. J. (1954). *The Foundations of Statistics*. New York: John Wiley.

Savage, L. J. (1962). "Bayesian Statistics," in R. E. Machol and P. Gray, eds., *Recent Developments in Decision and Information Processes*. New York: Macmillan.

Savage, L. J. (1967). "Implications of Personal Probability for Induction," *Journal of Philosophy*, vol. 64, pp. 593–607.

Shimony, A. (1970). "Scientific Inference," in R. G. Colodny, ed., *The Nature and Function of Scientific Theories*. Pittsburgh: University of Pittsburgh Press.

Strawson, P. F. (1952). *Introduction to Logical Theory*. New York: John Wiley.

Suppes, P. (1962). "Models of Data," in E. Nagel, P. Suppes, and A. Tarski, eds., *Logic, Methodology and Philosophy of Science*. Stanford: Stanford University Press.

Tukey, J. W. (1961). "The Future of Data Analysis," *Annals of Mathematical Statistics*, vol. 33, pp. 1–67.

van Fraassen, B. C. (1971). "On the Extension of Beth's Semantics of Physical Theories," *Philosophy of Science*, vol. 38, pp. 325–39.

van Fraassen, B. C. (1972). "A Formal Approach to the Philosophy of Science," in R. G. Colodny, ed., *Paradigms and Paradoxes*. Pittsburgh: University of Pittsburgh Press. Pp. 303–66.

von Mises, R. (1957). *Probability, Statistics and Truth*. London: Allen and Unwin.

Wald, A. (1950). *Statistical Decision Functions*. New York: John Wiley.

Watkins, J. (1964). "Confirmation, the Paradoxes and Positivism," in Mario Bunge, ed., *The Critical Approach to Science and Philosophy*. New York: Free Press.

HENRY KYBURG

The Uses of Probability and the Choice of a Reference Class

Many people suppose that there isn't any real problem in arguing from known relative frequencies or measures to epistemological probabilities. I shall argue the contrary — that there is a very serious and real problem in attempting to base probabilities on our knowledge of frequencies — and, what is perhaps more surprising, I shall also argue that this is the only serious problem in the epistemological applications of probability. That is, I shall argue that if we can become clear about the way in which epistemological probabilities can be based on measures or relative frequencies, then we shall see, in broad outlines, how to deal with rational belief, scientific inference of all sorts, induction, ethical problems concerning action and decision, and so on.

Since my main concern is to establish the seriousness of solving certain problems in basing probabilities on frequencies, and since I have no definitive solutions to those problems, I propose to work backward. First I shall characterize the probability relation (actually, it will be defined in terms of the primitive relation 'is a random member of') in a general axiomatic way adequate to the purpose of showing how such a probability concept can provide a framework for discussing various philosophical problems. In the next section of the paper, I shall discuss what several writers of very different persuasions have had to offer on this topic of the relation between measure statements and the probabilities we apply to particular cases, and then I shall attempt to characterize some of the difficulties that seem to me to have been insufficiently considered and offer my own very tentative solution to these difficulties. Finally I shall discuss a number of uses of this probability concept in three aspects of epistemology (observational knowledge,

AUTHOR'S NOTE: Research underlying this paper has been supported by the National Science Foundation by means of grants 1179 and 1962.

knowledge of empirical generalizations, and theoretical knowledge), in connection with the problem of choosing between courses of action whose outcomes are characterized by probabilities and utilities, and in connection with ontological and metaphysical questions.

I

The basic situation we shall consider is that in which we have a set of statements, in a given language *L*, that we regard as practically certain and which we shall take as a body of knowledge or a rational corpus. Let us denote this set of statements by '*K*'. Probability will be defined as a relation between a statement, a body of knowledge *K*, and a pair of real numbers. The language of *K* is assumed to be rich enough for set theory (at least including the ordinary theory of real numbers) and we assume that it contains a finite stock of extralogical predicates. Following Quine's proposal in ML, we shall suppose that there are no proper names in the language. We shall suppose that these extralogical primitive predicates include both relatively observational predicates such as 'is hot' or 'is blue' and relatively theoretical predicates such as 'the magnetic field at x, y, z, is given by the vector *v*'. I shall have more to say about these predicates later; for the moment what is important is that the stock of these predicates is assumed fixed. Next, we assume that the meaning relations that hold among these predicates have been embodied in a set of axioms in the language. One could no doubt find a number of plausible ways of doing this. In the first place, one could find a number of different axiomatizations that yield the same set of theorems in the language. These will be regarded here as different formalizations of the same language. In the second place, there will be a number of axiomatizations, plausible in themselves, that yield different sets of theorems. These will be regarded as formalizations of different languages. Finally, the language must be tied to experience through relatively ostensive rules of interpretation; a language in which 'green' has the meaning which we ordinarily attribute to 'red' would be a different language than English. There will be ostensive rules for relations as well as for one-place predicates; thus 'being hotter than' or 'seeming to be hotter than' is quite as 'directly observable' as 'being green' or 'appearing green'. The general form of an ostensive rule of interpretation is this: "Under circumstances C (described in the psychological part of the metalanguage, or simply exhibited in a general way as 'like this'), ϕ holds (or holds with a certain probability)." There are all kinds of problems here that a serious epistemology would have to grapple with; but for our present purposes it suffices to suppose that there is some set of relatively ostensive rules of interpretation that is characteristic of a language.

Thus the language *L* may be characterized as an ordered quadruple $<V,T,M,O>$, where V is the set of well-formed expressions of the language, *T* the set of theorems of the language that depend only on the meaning of the logical constants 'ϵ', *M* the set of theorems that depend on the meaning of the extralogical constants of the language, and O the set of statements of the languge that can be potentially established (with certainty or with probability) on the basis of "direct observation." We assume that each of V, *T*, *M*, and O admits of a recursive enumeration. Two languages $<V,T,M,O>$ and $<V',T'M',O'>$ are *essentially the same* if and only if there is an isomorphism *m* from V to V′, such that $S \epsilon T$ if and only if $m(S) \epsilon T'$, $S \epsilon M$ if and only if $m(S) \epsilon M'$, and $S \epsilon O$ if and only if $m(S) \epsilon O'$. Thus a language similar to English with the exception that 'red' and 'green' were interchanged in meaning would be essentially the same as English.

Since we shall be concerned with logical relationships both in the object language and in the metalanguage, certain notational conventions will come in handy.

Capital letters, '*S*', '*T*', '*R*', etc., will be used to denote statements of the object language.

If *S* is a statement of the object language, '*nS*' will denote the statement of the object language consisting of a negation sign, followed by *S*. The expression *nS* will be called the negation of *S*. If *S* and *T* are statements of the object language, '*S cd T*' will denote the statement of the object language consisting of *S*, followed by a horseshoe, followed by *T*, the whole enclosed in parentheses; *S cd T* will be called a conditional, *S* is its antecedent, and *T* is its consequent. If *S* and *T* are statements of the object language, '*S cj T*' will denote their conjunction, and '*S al T*' will denote their alternation, understood as above; similarly, '*S b T*' will denote the biconditional whose terms are *S* and *T*.

We shall also be concerned with certain *terms* of the object language. Lowercase and capital letters from the front end of the alphabet will be used to denote terms of the object language; if *a* and *A* are terms of the object language, 'a ε *A*' will be the expression of the object language formed by concatenating a, epsilon, and *A*. In general, terms expressed by lowercase and capital letters from the beginning of the alphabet will be assumed not to contain free variables. Thus, in general, 'a ε *A*' will be a closed sentence of the object language.

Two classes of mathematical terms that are of particular interest to us are those denoting natural numbers and those denoting real numbers. There are, of course, a lot of ways in the object language of denoting a given number; thus the number 9 could be denoted by '9', by 'the number I am thinking of now', 'the number of planets', etc. For our purposes here, however, we wish to restrict our attention to numerical terms expressed in some canonical form, say as an explicit decimal, i.e., a decimal given by a purely mathematical rule, where this is construed as a function whose domain is natural numbers, whose range is the natural numbers between 0 and 9 inclusive, which can be effectively calculated, and which can be expressed by a formula containing no nonlogical constants. Such real number terms in canonical form will be denoted by lowercase letters *p*, *q*, *r*, etc., as will the numbers themselves. Natural number terms in canonical form will be denoted by lowercase letters *m*, *n*, etc. On those occasions on which we may wish to speak of the occurrence of particular number-denoting expressions, like '1', '2', '0.5000 . . . ' etc., we shall make use of single quotation marks.

The ambiguity of the variables *p*, *q*, etc., must be kept in mind. This ambiguity is very natural and useful in the characterization of probability that follows, but it is permissible only because we have committed ourselves to admitting only number expressions in canonical form as values of these variables. Thus we can only say that the probability of something is *p* (the real number *p*) when a certain expression involving *p* (the numerical expression *p* in canonical form) occurs in a certain set of expressions.

If *p* and *q* are real number expressions in canonical form, then '*p* gr *q*' denotes the expression (statement, in fact) consisting of *p*, followed by the inequality sign, followed by *q*. Thus to write *Thm* '*p* gr *q*' is to assert that the statement consisting of *p* followed by the inequality sign '>' followed by *q* is a theorem. Some statements of the form *p* gr *q* will be undecided or undecidable, of course; but such statements will turn out not to be of interest to us. We will never be interested in statements of the following form: the proportion of *A*'s that are *B*'s is *p*, where p is the infinite decimal consisting of 8's if Fermat's theorem is true and of 7's if it is not true.

If A and B are any terms whatever (including natural number expressions and real number expressions in canonical form) then 'A *id* B' will denote the expression in the object language consisting of A, followed by the identity sign '=', followed by B. If A and B are class expressions, A *int* B is the class expression consisting of A, followed by the intersection sign, followed by B; A *un* B is the class expression consisting of A, followed by the union sign, followed by B; and A *min* B is the class expression consisting of A, followed by the difference sign, followed by B. If A and B are class expressions, A *incl* B is the statement consisting of A, followed by the inclusion sign '$\subset$', followed by B.

The most important kind of statement with which we shall be concerned is what I shall call a statistical statement. Statistical statements are expressed in ordinary and scientific language by a variety of locutions: "51 percent of human births result in the birth of boys"; "the evidence shows that half the candidates can't even read English adequately"; "the odds against heads on a toss of this coin are about even"; "the distribution of foot sizes among recruits is approximately normal with a mean of 9 inches and a standard deviation of 1.3 inches." We shall take as a standard form for statistical statements the simple sort: "The proportion of potential or possible A's that are B's lies between the limits p_1 and p_2," or "The measure, among the A's, of the set of things that are B's, lies between p_1 and p_2." For most purposes we need consider only statements of this form, since even when we are dealing with the distribution of a random quantity, such as foot size, or of an original n-tuple of random quantities, when it comes to making use of that knowledge, we will be interested in the probability that a member of the reference class belongs to the set of objects having a certain property; the distribution functions are so much mathematical machinery for arriving at statistical statements of the straightforward form in question.

Such statements will be expressed in the object language with the aid of a new primitive term, '%'; it is a four-place relation, occurring in contexts of the form $\%(x,y,z,w)$, where x and y are classes, z and w real numbers. '$\%(x,y,z,w)$' means that the measure, in the set x, of those objects which belong to y, lies between z and w. This statement will be named in the metalanguage by the expression 'M('x', 'y', 'z', 'w')'. Put otherwise: $M(A,B,p,q)$ denotes that statement in the object language consisting of '%', followed by '(', followed by the term A, the term B, followed by the real number expressions p and q in canonical form. If p and q are not expressions in canonical form denoting real numbers, then we take $M(A,B,p,q)$ to denote "$0 = 1$."

We are now in a position to begin to characterize the set of statements K that we will call a rational corpus or body of knowledge, to the extent that this is necessary for the axiomatic characterization of probability. For reasons which have been discussed at length elsewhere,[1] we shall not suppose that K is deductively closed or consistent in the sense that no contradiction is derivable from it. We will make only the bare minimum of assumptions about K: namely, that it contains no self-contradictory statement and that it contains all the logical consequences of every statement it contains. To be sure, this is already asking a lot in a sense; it is asking, for example, that all logical truths be included in the rational corpus K. But although nobody knows, or could know, all logical truths, in accepting a certain system of logic and a certain language, he is *committing* himself to all these logical truths. Then why not complete logical closure? Because whereas he accepts a language and its logic all at one fell swoop, with a single act or agreement, he accepts empirical statements piecemeal, one at a time. Furthermore, although the logical truths cannot be controverted by experience, the empirical statements he can accept, with every inductive right in the world, can be controverted, and some of them no doubt will be. The man who never has to withdraw his acceptance of a factual statement has to be kept in a padded cell. Thus we state as axioms:

AXIOM I $\quad S \in K \;\&\; Thm\ S\ cd\ T \supset T \in K.$

AXIOM II $\quad (\exists S)(\sim S \in K).$

In stating the weak consistency of K by axioms I and II, we are, of course, presupposing a standard logic. Furthermore, axioms I and II entail that this standard logic is consistent. Since we are supposing not only a first order logic, but a whole set theory, this consistency is not demonstrable. Nevertheless we always do suppose, so long as we know no better, that the language we speak is consistent. And if we come to know better, we will change our language.

The basic fact about probability that I shall stipulate is that every

[1] "Conjunctivitis," in M. Swain, ed., *Induction, Acceptance, and Rational Belief* (Dordrecht: Reidel, 1970), pp. 55–82.

probability is to be based on a known statistical statement. It is, of course, precisely the burden of part of this paper to show that this is plausible. Statements that are not of the simple membership form a e A —the only form that lends itself directly to this statistical analysis— require to be connected in some fashion to statements that are of that form. The obvious way of making this connection is to use the principle that equivalent statements are to have the same probability. This *equivalence principle* has been criticized in recent years by Smokler, Hempel, Goodman, and others; nevertheless the arguments that have been presented against it have focused on its plausibility or implausibility in a system of Carnapian logical probability and on its connections with various natural principles of qualitative confirmation. Since we do not have here a Carnapian form of logical probability, and since we are not at all concerned with qualitative concepts of confirmation, those arguments are beside the point. I shall therefore make the natural assumption that if we have reason to believe that *S* and *T* have the same truth value (i.e., if we have reason to accept the biconditional *S b T*) then *S* and *T* will have the same probability. Furthermore, to say that a statement *S* has the probability (p,q) entails that *S* is known to be equivalent to a statement of the form a e A, that a e *B* is known, and that $M(B,A,p,q)$ is known. It entails some other things, too, in particular that the object denoted by a be a *random member* of the set denoted by *B*.

Of course it follows logically from the fact that the measure of x in y lies between ½ and ¾ that it also lies between 1/5 and 4/5; but this latter fact is not of interest to us. Indeed, it is a nuisance to have to keep track of it. Therefore when we speak of measure statements belonging to *K*, we shall mean the strongest ones in *K* concerning the given subject matter. We shall write '$M(A,B,p,q)\ \epsilon_s K$' to mean that $M(A,B,p,q)$ is the strongest statement in *K* about *B* and *A*—i.e., that statement belongs to *K*, and any measure statement about *A* and *B* in *K* is merely a deductive consequence of that statement.

DEFINITION I $\quad M(A,B,p,q)\ \epsilon_s K \equiv_{df} (M(A,B,p,q)\ \epsilon\ K\ \&\ M(A,B,p',q')\ \epsilon\ K) \supset \text{Thm}\ M(A,B,p,q)\ \text{cd}\ M(A,B,p',q')$.

Randomness is customarily defined in terms of probabilities; here we shall adopt the opposite course and take the probability relation to be definable in terms of randomness. The basic form of a statement of randomness is the following: the object *a* is a random member of the set *A*, with respect to membership in the set *B*, relative to the rational corpus *K*. We shall express the metalinguistic form of this statement by '$\mathrm{Ran}_K(a,A,B)$'.

Two obvious axioms are as follows:

AXIOM III $\quad \mathrm{Ran}_K(a,A,B) \supset a\,e\,A \in K.$

AXIOM IV $\quad \mathrm{Ran}_K(a,A,B) \supset (\exists p)\,(\exists q)\,(M(A,B,p,q) \in_s K).$

Since we have not taken *K* to be deductively closed, it is perfectly possible that we should have the statements $S\,b\,T$ and $T\,b\,R$ in *K*, without having the statement $S\,b\,R$ in *K*. Nevertheless, we want to be able to take account of the relationship between *S* and *R*. We shall say that they are connected by a *biconditional chain* in *K*. In general, if we have a chain of statements, $S\,b\,T_1$, $T_1\,b\,T_2$, . . . , $T_n\,b\,R$, each member of which is a member of *K*, we shall say that *S* and *R* are connected by a biconditional chain in *K*.

DEFINITION II $\quad S'\mathrm{bc}_K\,T' =_{\mathrm{df}} (X)\,((R)\,(S)\,(T)\,((R\,b\,S \in K \supset R\,b\,S \in X)\ \&$
$((R\,b\,S \in X\ \&\ S\,b\,T \in X) \supset R\,b\,T \in X))\supset S'\,b\,T' \in X)$

In order to establish the uniqueness of probabilities (clearly a desideratum), we do not need to stipulate that there be only one set of which a given object is a random member with respect to belonging to another set; it suffices that our statistical knowledge about any of the sets of which *a* is a random member with respect to *B* (relative to the rational corpus *K*) be the same. Thus we have the following axiom characterizing the randomness relation.

AXIOM V $\quad (a'eB'\ \mathrm{bc}_K\ aeB)\ \&\ \mathrm{Ran}_K(a,A,B)\ \&\ \mathrm{Ran}_K(a',A',B')\ \&$
$M(A,B,p,q) \in_s K \supset M(A',B',p,q) \in_s K.$

We shall also stipulate that if *a* is a random member of *A* with respect to *B*, it is also a random member of *A* with respect to the complement of *B*:

AXIOM VI $\quad \mathrm{Ran}_K(a,A,B) \supset (\mathrm{Ran}_K(a,A,A\ \mathrm{min}\ B)\ \&\ \mathrm{Ran}_K(a,A,A)).$

We now define probability in terms of randomness: to say that the probability of *S* is (p,q) is to say that we know that *S* is equivalent to some membership statement $a \mathrm{e} B$, that a is a random member of A with respect to membership in B, and that we know that the measure of B and A is (p,q).

DEFINITION III $\quad \mathrm{Prob}(S,p,q) =_{\mathrm{df}} (\exists a)(\exists B)(\exists A)(S\ bc_K\ a \mathrm{e} B\ \&\ \mathrm{Ran}_K(a,A,B)\ \&\ M(A,B,p,q) \epsilon_s K)$.

We can already prove a number of theorems about probability. Two statements connected by a biconditional chain in K turn out to have the same probability. We have already made use of a part of this principle in the definition of probability, but we wish it to hold generally and not merely for pairs of statements of which one is of the form $a \mathrm{e} B$, where for some A, $\mathrm{Ran}_K(a,A,B)$.

THEOREM I $\quad (S\ bc_K\ T\ \&\ \mathrm{Prob}_K(S,p,q)) \supset \mathrm{Prob}_K(T,p,q)$.
Proof: Immediate from definitions II and III and the fact that if $S\ bc_K\ T$ and $T\ bc_K\ R$, then $S\ bc_K\ R$.

The next theorem states an even stronger version of the same fact, from which theorem III, which asserts the uniqueness of probabilities, follows directly.

THEOREM II $\quad (S\ bc_K\ T\ \&\ \mathrm{Prob}_K(S,p,q)\ \&\ \mathrm{Prob}_K(T,p',q')) \supset \mathrm{Thm}\ p\ id\ p'\ cj\ q\ id\ q'$.
Proof: From theorem I and the hypothesis of the theorem, we have

$$\mathrm{Prob}_K(S,p,q) \text{ and } \mathrm{Prob}_K(S,p',q').$$

From definition III, there must exist terms a, a′, A, A′, B, and B′ such that

$$\mathrm{Ran}_K(a,A,B) \text{ and } \mathrm{Ran}_K(a',A',B')$$

and

$$M(A,B,p,q) \epsilon_s K \text{ and } M(A',B',p',q') \epsilon_s K.$$

From axiom V it therefore follows that

$$M(A,B,p',q') \epsilon_s K.$$

By definition I, we then have

$$\mathrm{Thm}\ M(A,B,p,q)\ cd\ M(A,B,p',q')$$

and

$$\mathrm{Thm}\ M(A,B,p',q')\ cd\ M(A,B,p,q).$$

Since these measure statements belong to K, they are not simply "1 = 0"; thus in order for the conditionals to be theorems, we must have *p id p′* and *q id q′* as theorems, and thus

Thm *p id p′ cj q id p′*.

THEOREM III (Prob$_K$(S,p,q) & Prob$_K$(S,p′,q′)) ⊃ Thm *p id p′ cj q id q′*.

Proof: Clear from the fact that *S* bc$_K$ *S*.

THEOREM IV Prob$_K$(S,p,q) ⊃ Prob$_K$(nS, 1 − q,1 − p), where '1 − *q*' denotes the real number expression consisting of '1' followed by a minus sign, followed by *q*.

Proof: If p and *q* are real number expressions in canonical form, so are 1 − *q* and 1 − p. *M*(A,*B*,p,q) cd *M*(A,A min *B*, 1 − q,1 − p) is a theorem of measure theory. The theorem now follows from axiom VI and definition III.

THEOREM V (*S* bc$_K$ a e *B* & Ran$_K$(a,A,*B*) & (∃X) (*S* cj *T* bc$_K$ a e *B int* X & Ran$_K$(a,A,*B int* X)) & Prob$_K$(S,p,q) & Prob$_K$(*S* cj *T*,r,s)) ⊃ Thm p gr r *al* p *id* r.

Proof: By definition III, and axiom V,

M(A,*B int* X,r,s) ϵ_s K.

By a theorem of measure theory,

Thm *M*(A,*B int* X,r,s) cd *M*(A,B,r′, '1').

By axiom I, therefore,

M(A,B,r, '1') ϵ K.

But by the hypothesis of the theorem and definition III,

M(A,B,p,q) ϵ_s K.

And so by definition I,

Thm *M*(A,B,p,q) cd *M*(A,B,r,'1').

And thus

Thm p gr r *al* p *id* r.

THEOREM VI (Thm S cd *T* & Prob$_K$(S,p,q)) ⊃ Prob$_K$(S cj *T*,p,q).

Proof: Thm *S* cd *T* ⊃ Thm *S b T cj S*; theorem I.

THEOREM VII (Thm *S* cd *T* & $\text{Prob}_K(S,p,q)$ & $\text{Prob}_K(T,r,s)$) ⊃ Thm *r id p al r gr p*.

Proof: Theorems V and VI.

THEOREM VIII (S bc_K a e *B* & $\text{Ran}_K(a,A,B)$ & $(\exists X)$ (S al T bc_K a e *B un* X & $\text{Ran}_K(a,A,B \text{ un } X)$) & $\text{Prob}_K(S,p,q)$ & $\text{Prob}_K(S \text{ al } T,r,s)$) ⊃ Thm *s id q al s gr q*.

Proof: From theorem V.

Under certain special circumstances concerning randomness we can obtain, on the metalinguistic level, a reflection of the usual probability calculus generalized to take account of interval measures. Circumstances of the kind described occur in gambling problems, in genetics, in experiments in microphysics, in the analysis of measurement, etc. So they are not as special as they might at first seem.

THEOREM IX Let $\{S_i\}$ be a set of statements of *L* such that $(S)(S \epsilon \{S_i\} \supset (\exists X)(\text{Ran}_K(a,A,X) \;\&\; (\exists p_X)$ $(\exists q_X)(M(A,X,p_X,q_X) \epsilon_s K) \;\&\; a\, e\, A \epsilon K \;\&\; S\; bc_K$ $a\, e\, X))$.

Then, (i) $(S)(S \epsilon \{S_i\} \supset (\exists p)(\exists q)(\text{Prob}_K$ $(S,p,q))))$.

(ii) $(S)(S \epsilon \{S_i\} \;\&\; \text{Prob}_K(S,p,q) \supset$ $\text{Prob}_K(nS, 1 - q, 1 - p)))$.

(iii) $(S)(T)((S, T \epsilon \{S_i\} \;\&\; S\; bc_K\; a\, e\, X \;\&\; T\; bc_K$ $a\, e\, Y \;\&\; X \text{ int } Y \text{ int } A \text{ id } `\phi' \epsilon K \;\&$ $M(A,X,p_s,q_s) \text{ cj } M(A,Y,p_t,q_t) \epsilon K \;\&$ $\text{Prob}_K(S,p_s q_s) \;\&\; \text{Prob}_K(T,p_t,q_t) \;\&$ $S \text{ al } T \epsilon \{S_i\} \;\&\; \text{Prob}_K(S \text{ al } T,r,s)) \supset$ $\text{Thm}(r \text{ gr } p_s + p_t \text{ al } r \text{ id } p_s + p_t)$ $\text{cj } (q_s + q_t \text{ gr } s \text{ al } q_s + q_t \text{ id } s))))$.

A strengthening of the hypothesis of this theorem will lead to a metalinguistic reflection of the standard calculus of probability.

THEOREM X Let *K* be closed under deduction, and let $\{B_i\}$ be a set of terms; let the sets denoted by the terms B_i be a field of subsets of the set denoted by A. Suppose that

(a) $(X)(X \epsilon \{B_i\} \supset (\exists p_i)(M(A,B_i,p_i,p_i) \epsilon K))$ and
(b) $(S)(S \epsilon \{S_i\} \supset (\exists X)(X \epsilon \{B_i\}$ & Ran_K (a,A,X) & $S\ bc_K\ a\ e\ X$ & $a\ e\ A \epsilon K))$.

Then: (i) $(S)(S \epsilon \{S_i\} \supset (\exists p)(\text{Prob}_K(S,p,p)$.
(ii) $(S)(S \epsilon \{S_i\}$ & $\text{Prob}_K(S,p,p) \supset \text{Prob}_K(nS, 1-p, 1-p))$.
(iii) $(S)(T)(S,T, \epsilon \{S_i\}$ & $S\ bc_K\ a\ e\ B_i$ & $T\ bc_K\ a\ e\ B_j$ & $B_i\ int\ B_j\ int\ A\ id\ `\phi' \epsilon K \supset (\text{Prob}_K(S,p,p)$ & $\text{Prob}_K(T,q,q) \supset \text{Prob}_K(S\ al\ T, p+q, p+q)))$.
(iv) $(S)(S\ bc_K\ a\ e\ A \supset \text{Prob}_K(S,`1',`1'))$.

Two more theorems may come in handy:

THEOREM XI $(\exists a)(\exists A)(\exists B)(\text{Ran}_K(a,A,B)$ & $S \epsilon K) \supset \text{Prob}_K(S,`1',`1')$.

Proof: By axiom III, a e A ϵ K.

Since $S\ cd\ (a\ e\ A\ b\ S\ cj\ a\ e\ A)$ is a theorem, the biconditional a e A *b S cj* a e A belongs to K.

Since a e A *cd* ($S\ b\ S\ cj$ a e A) is a theorem, the biconditional $S\ b\ S\ cj$ a e A belongs to K. Thus

$S\ bc_K$ a e A.

By axiom VI,

$\text{Ran}_K(a,A,A)$.

Since $M(A,A,`1',`1')$ is a theorem, so is $S\ cd\ M(A,A,`1',`1')$, and thus

$M(A,A,`1',`1') \epsilon_s K$.

By definition III,

$\text{Prob}_K(S,`1',`1')$.

THEOREM XII $(\exists a)\ (\exists A)\ (\exists B)\ (\text{Ran}_K(a,A,B))$ & $S \epsilon K) \supset \text{Prob}_K(nS, `0', `0')$.

Proof: Theorem XI and theorem IV.

II

How are we to construe randomness? Most writers on probability and statistics have no difficulty with the concept of randomness, because they think they can define it in terms of probability. In a certain sense this is so: one may say that a is a random member of the class C if a is selected from C by a method which (let us say) will result in the choice of each individual in C equally often in the long run. This is a statistical hypothesis about the method *M*, of course. It might better be expressed by the assertion that *M* is such that it produces each member equally often in the long run. We may or may not, in particular cases, have reason to accept this hypothesis. In any event, it applies only in exceptional cases to applications of our statistical knowledge. There is no way in which we can say that Mr. Jones—the 35-year-old who took out insurance—is selected for survival during the coming year by a method which tends to select each individual with equal frequency. If there is a method, which many of us doubt, it is a method which is presumably not indifferent to the virtues and vices of the individual involved. Furthermore, even if we did know that A was selected from C by an appropriate random method, whether this knowledge, combined with the knowledge of *p* of the *C*'s are *B*'s, would yield a probability of *p* that A is a *B* would depend on what else we knew about A. We could know these things and also know that in point of fact A was a *B*; then we would surely not want to say that the probability of its being a *B* was *p*. Similarly, even under the most ideal circumstances of the application of this notion of randomness, we could encounter information which would be relevant to the probability in question. Thus suppose that the probability of survival of a member of the general class of 35-year-old American males is .998. Let us assign a number to each American male of this age and choose a number by means of a table of random numbers. It is indeed true that this method would tend, in the long run, to choose each individual equally often. But if the method results in the choice of Mr. S. Smith, whom we know to be a wild automobile racer, we would quite properly decline to regard his probability of survival as .998. Finally, the next toss of a certain coin, although it is the toss that it is, and no other, and is chosen by a method which can result only in that particular toss, is surely, so far as we are concerned now, a random member of the set of tosses of that coin; and if we know that on the whole the coin tends to land heads ⅔'s of the time, we will properly attribute a probability of ⅔ to its landing heads on that next toss.

Probability theorists and statisticians have tended to take one of two tacks in relation to this problem; either they deny the significance of probability assertions about individuals, claiming that they are meaningless; or they open the door the whole way, and say that an individual may be regarded as a member of any number of classes, that each of these classes may properly give rise to a probability statement, and that, so far as the theory is concerned, each of these probabilities is as good as another. Thus if we happen to know that John Smith is a 35-year-old hairless male Chicagoan, is a lawyer, and is of Swedish extraction, the first group would say that 'the probability that John Smith is bald is *p*' is an utterly meaningless assertion; and the second group would say that if, for example, we know that 30 percent of the male lawyers in Chicago are bald, and 10 percent of the males, and 5 percent of Chicagoans, and 3 percent of the males of Swedish extraction, then all the probability statements that follow are correct.

The probability that John Smith is bald is 0.30.
The probability that John Smith is bald is 0.10.
The probability that John Smith is bald is 0.05.
The probability that John Smith is bald is 0.03.
The probability that John Smith is bald is 1.00.

In either case, however, we are left with a problem which is in all essentials just the problem I want to focus on. If we say that no probability statement can be made about the individual John Smith, then I still want to know what rate to insure him at. No talk about offering insurance to an infinite number of people will help me, because even if I am the biggest insurance company in the world, I will only be insuring a finite number of people. And no talk about a large finite set of people will do, for the finite group of n people to whom I offer insurance is just as much an *individual* (belonging to the set of n-membered subsets of the set of people in that category) as John Smith himself. So what number will I use to determine his rate? Surely a number reflecting the death rate in some class — or, rather, reflecting what I believe on good evidence to be the death rate in some class. What class? Tell me what class to pick and why, and my problem concerning randomness will have been solved. The same is true for those who would tell me that there are any number of probabilities for John's survival for the coming year. Fine, let there be a lot of them. But still, tell me which one to use in assigning him an insurance rate.

Some philosophers have argued that this is merely a pragmatic problem, rather than a theoretical one. The scientist is through, they say, when he has offered me a number of statistical hypotheses that he regards as acceptable. Choosing which hypothesis to use as a guide to action is not his problem, but mine, not a theoretical problem, but a practical one.

But calling a problem practical or pragmatic is no way to solve it. If I am a baffled insurance company before the christening, I shall be a baffled insurance company after it.

Some of those who regard the problem of choosing a reference class as a practical problem rather than an interesting theoretical one, nevertheless have advice to offer. Some statisticians appear to offer the advice (perhaps in self-interest), 'Choose the reference class that is the most convenient mathematically,' but most suggest something along the lines of Reichenbach's maxim: 'Choose the narrowest reference class about which you have adequate statistical information.' "If we are asked to find the probability holding for an individual future event, we must first incorporate the case in a suitable reference class. An individual thing or event may be incorporated in many reference classes . . . We then proceed by considering *the narrowest reference class for which reliable statistics can be compiled*."[2] Reichenbach recognizes that we may have reliable statistics for a reference class *A* and for a reference class *C*, but none for the intersection of *A* and *C*. If we are concerned about a particular individual who belongs both to *A* and to *C*, "the calculus of probability cannot help in such a case because the probabilities $P(A,B)$ and $P(C,B)$ do not determine the probability $P(A.C,B)$. The logician can only indicate a method by which our knowledge may be improved. This is achieved by the rule: look for a larger number of cases in the narrowest common class at your disposal." This rule, aside from being obscure in application, is beside the point. If we could always follow the rule "investigate further," we could apply it to the very individual who concerns us (wait a year and see if John Smith survives), and we would have no use for probability at all. Probability is of interest to us because we must sometimes act on incomplete information. To be sure, the correct answer to a problem may sometimes be—get more information. But from the point of view of logic this answer can be given to any empirical question. What we expect from logic is rather an indication of what epistemological stance we should adopt with respect to a given proposition, under given cir-

[2] H. Reichenbach, *Theory of Probability* (Berkeley: University of California Press, 1949), p. 374. Italics in the original.

cumstances. Perhaps in the situation outlined by Reichenbach, our answer would be, "Suspend judgment." That would be a proper answer. But since, as we shall see, every case can be put in the form outlined by Reichenbach, that answer, though relevant, seems wrong.

Surprisingly, Carnap does not give us an answer to this question either. The closest he comes to describing the sort of situation I have in mind in which we have a body of statistical information and we want to apply it to an individual is when the proposition *e* consists of a certain structure-description, and *h* is the hypothesis that an individual *a* has the property. *P*. The degree of confirmation of *h* on *e* will be the relative frequency of *P* in the structure-description *e*. This does not apply, however, when the individual *a* is mentioned in the evidence—i.e., when the evidence consists of more than the bare structure description. This is simply not an epistemologically possible state of affairs. Hilpinen, similarly, supposes that if we have encountered *a* at all, we know everything about it. This is the principle of completely known instances.[3] It follows from this principle that if we know that *a* belongs to any proper subset of our universe at all, we know already whether or not it belongs to (say) *B*. In real life, however, this situation never arises. Even Reichenbach's principle is more helpful than this.

To make precise the difficulty we face, let us return to the characterization of a body of knowledge in which we have certain statistical information, i.e., a body of knowledge in which certain statistical statements appear. The exact interpretation of these statements can be left open—perhaps they are best construed as statements of limiting frequencies, as propensities, or as certain characteristics of chance setups, or abstractly as measures of one class in another. In any event, I take it to be the case that we do know some such statements. I shall further suppose that they have the weak form, the measure of *R*'s among *P*'s lies between p_1 and p_2: '$\%(P,R,p_1,p_2)$'.

Three ways in which we can come to know such statements—neither exclusive nor exhaustive—are the following:

(a) Some such statements are analytic: the proportion of 1000-member subsets of a given set that reflect, with a difference of at most .01, the proportion of members of that set that belong to another given set lies between .90 and 1.0. Such statements are set-theoretically true

[3] R. Hilpinen, *Rules of Acceptance and Inductive Logic*, Acta Philosophica Fennica, vol. 7 (Amsterdam: North Holland, 1968), p. 74.

of proportions and limiting frequencies; or true by virtue of the axioms characterizing 'measure' or 'propensity'; in either case derivable from an empty set of empirical premises.

(b) Some such statements are accepted on the basis of direct empirical observation: the proportion of male births among births in general is betwen 0.50 and 0.52. We know this on the basis of established statistics, i.e., by counting.

(c) Since we know that having red or white flowers is a simply inherited Mendelian characteristic of sweet peas, with the gene for red flowers dominant, we know that the proportion of offspring of hybrid parents that will have red flowers is ¾. This statistical statement, like many others, is based on what I think is properly called our theoretical knowledge of genetics.

There are a number of expressions that one would never take to refer to reference classes. One would never say that the next toss of this coin is a random member of the set consisting of the union of the unit class of that toss with the set of tosses resulting in heads. The fact that nearly all or perhaps all of the tosses in that class result in heads will never be of relevance to our assessment of probabilities. Neither is the fact that practically all the members of the union of the set of tosses yielding tails with the set of prime numbered tosses are tosses that yield tails ever relevant to our expectations. Similarly not all properties will be of concern to us: the grueness of emeralds is not something we want to bother taking account of. This may be regarded as a commitment embodied in the choice of a language. We may, for example, stipulate that probabilities must be based on our knowledge of rational classes, where the rational classes may be enumerated by some such device as this: Let L be monadic with primitive predicates 'P_1', 'P_2', ... , 'P_n'. Then the set of atomic rational classes is composed of sets of the form $\{x:P_ix\}$ or of the form $\{x:\sim P_ix\}$. The general set of rational classes is the set of all intersections of any number of these atomic rational classes. An object will be a random member of a class, now, only if the class is a rational class. A similar arbitrary and high-handed technique may be employed to elude the Goodmanesque problems. To be sure, all the problems reappear in the question of choosing a language—e.g., choosing a language in which 'P_1' doesn't denote the union of the unit class of the next toss with the set of tosses yielding tails. But this is a somewhat different problem. It is solved partly by the constraint that the language must be such that it can be learned on the basis of ostensive hints and partly by the constraints which in general govern the acceptance of theoretical structures. In any event, so far as our syntactical reconstruction of the logic of probabilistic statements is concerned, the question of the source of the particular language *L* we are talking about is academic.

Since what we are concerned about here is the choice of a reference class, we may simply start with a large finite class of classes, closed under intersection. To fix our ideas, let us focus on a traditional bag of counters. Let the counters be round (R) and triangular (R^c), Blue (B) and Green (B^c), zinc (Z) and tin (Z^c). Let our statistical knowledge concerning counters drawn from the bag (C) be given by the following table, in which all the rational reference classes appropriate to problems concerning draws from the urn are mentioned.

Reference class	*Subclass*	*Lower and upper bounds of chance*	
C	R	0.48	0.52
C	B	0.60	0.70
C	Z	0.45	0.55
$C \cap R$	B	0.55	0.68
$C \cap R$	Z	0.50	0.65
$C \cap R$	R	1.00	1.00
$C \cap B$	R	0.45	0.65
$C \cap B$	Z	0.30	0.50
$C \cap B$	B	1.00	1.00
$C \cap Z$	B	0.25	0.65
$C \cap Z$	R	0.35	0.70
$C \cap Z$	Z	1.00	1.00
$C \cap R \cap B$	Z	0.50	0.80
$C \cap R \cap Z$	B	0.30	0.80
$C \cap B \cap Z$	R	0.25	0.90

Now let us consider various counters. Suppose that we know that $a_1 \in C$, and, in terms of our basic categories, nothing more about it. What is the probability that $a_1 \in R$? Clearly (0.48,0.52). C is the appropriate reference class, since all we know about a_1 is that it is a member of C.

Suppose we know that $a_2 \in C$ and also that $a_2 \in R$. (Perhaps a_2 is drawn from the bag in the dark, and we can feel that it is round.) Surely knowing that a_2 is round should have an influence on the probability we attribute to its being blue. Since we have statistical information about $R \cap C$, and since $R \cap C$ is 'narrower' than C, again there is no problem: the probability is (0.55,0.68). Note that Carnap and Hilpinen and others who hold a logical view of probability must here abandon the attempt to base this probability strictly on statistical measures—other things enter in. The object a_2 is neither completely known nor completely unknown.

Suppose that we know of a_3 that it belongs to both C and B. What is the probability that it belongs to R? Again we have two possible reference classes to consider: C and $C \cap B$. According to the broader reference class, the measure is (0.48,0.52); according to the narrower one it is (0.45,0.65). It seems reasonable, since there is no *conflict* between these measures, to regard the probability as the narrower interval, even though it is based on the broader reference class.

On the other hand, suppose we ask for the probability that $a_3 \in Z$. There the measure in the broader class is (0.45,0.55), while the measure in the narrower class is (0.30,0.50). Here there does seem to be a conflict, and it seems advisable to use the narrow reference class, even though the information it gives is not so precise; we would tend to say that the probability of '$a_3 \in Z$' is (0.30,0.50).

How about an object drawn from the urn about which we can tell only that it is blue and round? Is it zinc? The probability that this counter (say, a_4) is zinc will have one value or another depending on the statistical knowledge we have. We know that between 0.45 and 0.55 of the counters are zinc, that between 0.50 and 0.65 of the round counters are zinc, that between 0.30 and 0.50 of the blue counters are zinc, and that between 0.50 and 0.80 of the round blue counters are zinc. Here again it seems natural to use the smallest possible class ($C \cap B \cap R$) as the reference class, despite the fact that we have relatively little information about it, i.e., we know the proportion of zinc counters in it only within rather wide limits.

Now take a_5 to be a counter we know to be blue and zinc. Here, as distinct from the previous example, it seems most natural to take the broadest class C as the reference class. It is true that we know something about some of the smaller classes to which we know that a_5 belongs, but what we know is very vague and indefinite, whereas our knowledge about the class C is very precise and in no way conflicts with our knowledge about the narrower class.

Finally, let a_6 be known to belong to $C \cap R \cap Z$. What is the probability of the statement '$a_6 \in B$'? This case is a little more problematic than the preceding one. One might, by an argument analogous to the preceding one, claim that the probability is (0.60,0.70): the smallest rational reference class to which a_6 is known to belong is one about which we have only vague information which seems not to conflict with the very strong information we have about C and B. But there is a difference in this case: a_6 is also known to belong to $C \cap R$, and our information about the measure of B in $C \cap R$ *does* conflict with our information about the measure of B in C. If one interval were included in the other, we could simply suppose that our knowledge in that case was more precise than our knowledge in the other case. But the intervals overlap: it is not merely a case of greater precision in one bit of knowledge than in the other. Thus one might want to argue (as I argue in *Probability and the Logic of Rational Belief*)[4] that the knowledge that a_6 belongs to $C \cap R$ *prevents* a_6 from being a random member of C with respect to belonging to B, i.e., prevents C from being the appropriate reference class.

The various situations described here are summarized in the following table:

We know (altogether)	*We are interested in*	*Possible reference class*	*Measure*	*Probability*
'$a_1 \in C$'	'$a_1 \in R$'	C	(0.48,0.52)	(0.48,0.52)
'$a_2 \in C \cap R$'	'$a_2 \in B$'	C	(0.60,0.70)	
		$C \cap R$	(0.55,0.68)	(0.55,0.68)
'$a_3 \in C \cap B$'	'$a_3 \in R$'	C	(0.48,0.52)	(0.48,0.52)
		$C \cap B$	(0.45,0.65)	
	'$a_3 \in Z$'	C	(0.45,0.55)	
		$C \cap B$	(0.30,0.50)	(0.30,0.50)
'$a_4 \in C \cap B \cap R$'	'$a_4 \in Z$'	C	(0.45,0.55)	
		$C \cap B$	(0.30,0.50)	
		$C \cap R$	(0.50,0.65)	
		$C \cap R \cap B$	(0.50,0.80)	(0.50,0.80)

[4] Middletown, Conn.: Wesleyan University Press, 1961.

'$a_5 \epsilon C \cap B \cap Z$'	'$a_5 \epsilon R$'	C	(0.48,0.52)	(0.48,0.52)
		$C \cap B$	(0.45,0.65)	
		$C \cap Z$	(0.35,0.70)	
		$C \cap B \cap Z$	(0.25,0.90)	
'$a_6 \epsilon C \cap R \cap Z$'	'$a_6 \epsilon B$'	C	(0.60,0.70)	
		$C \cap R$	(0.55,0.68)	
		$C \cap Z$	(0.25,0.65)	
		$C \cap R \cap Z$	(0.30,0.80)	(0.30,0.80)

To formulate the principles implicit in the arguments that have resulted in the foregoing table, we require some auxiliary machinery. We shall say that X *prevents,* in K_1, a from being a random member of Y with respect to membership in Z—in symbols, X $\text{Prev}_{K_1}(a,Y,Z)$; when X is a rational class, a e X ϵ K_1, there are two canonical number expressions p_1 and p_2 such that $M(X,Z,p_1,p_2)$ ϵ_s K_1, for every pair of canonical number expressions q_1 and q_2, if $M(Y,Z,q_1,q_2)$ ϵ_s K_1 then either p_1 *gr* q_1 *cj* p_2 *gr* q_2 is a theorem, or q_1 *gr* p_1 *cj* q_2 *gr* p_2 is a theorem, and finally, provided Y incl X (the statement asserting that Y is included in X) is *not* a member of K_1. Formally:

DEFINITION: $X \text{ Prev}_{K_1}(a,Y,Z) \equiv (X \text{ is a rational class } \& \text{ a e } X \epsilon K_1 \& (\exists p_1)(\exists p_2)(M(X,Z,p_1,p_2) \epsilon_s K_1 \& (q_1)(q_2)(M(Y,Z,q_1,q_2) \epsilon_s K_1 \supset \text{Thm } p_1 \textit{ gr } q_1 \textit{ cj } p_2 \textit{ gr } q_2 \vee \text{Thm } q_1 \textit{ gr } p_1 \textit{ cj } q_2 \textit{ gr } p_2)) \& \sim Y \text{ incl } X \epsilon K_1)$.

In informal terms, the two principles that have guided the construction of the table above are as follows:

Principle I: If a is known to belong to both X and Y, and our statistical knowledge about X and Z differs from that about Y and Z, i.e., if one interval is not included in the other, then, unless we happen to know that X is included in Y, Y prevents X from being an appropriate reference class. If neither is known to be included in the other, then each prevents the other from being an appropriate reference class.

Principle II: Given that we want to avoid conflict of the first sort, we want to use the most precise information we have; therefore among all the classes not ruled out by each other in accordance with principle I, we take as our reference class that class about which we have the most precise information.

These two principles might be embodied in a single explicit definition of randomness:

DEFINITION: $\text{Ran}_{K_1}(a,Y,Z) \equiv X$ is a rational class & $a \in Y \in K_1$ & $(X)(\sim X\,\text{Prev}_{K_1}(a,Y,Z))$ & $(Y')((X)(\sim X\,\text{Prev}_{K_1}(a,Y',Z)) \supset (p_1)(p_2)(q_1)(q_2)((M(Y,Z,p_1,p_2) \in_s K_1$ & $M(Y',Z,q_1,q_2) \in_s K_1) \supset (\text{Thm}\; n(p_1\; gr\; q_1\; cj\; p_2\; gr\; q_2)$ & $\text{Thm}\; n(q_1\; gr\; p_1\; cj\; q_2\; gr\; p_2)))))$.

These principles, though they seem plausible enough in the situations envisaged above, do lead to decisions that are mildly anomalous. For example, suppose our body of knowledge were changed from that described by the first table, so that when we know that '$a_6 \in C \cap B \cap R$', our relevant statistical knowledge is represented by:

$\%(C,Z,0.45,0.50)$
$\%(C \cap B,Z,0.48,0.51)$
$\%(C \cap R,Z,0.49,0.52)$
$\%(C \cap B \cap R,Z,0.20,0.90)$.

Then by the principles set forth, a_6 is a random member of $C \cap B \cap R$ with respect to belonging to Z, and its probability is 0.20,0.90.

Various alternatives might be suggested. One would be to take the intersection of the measure intervals to be the probability in the case of conflict. But the intersection may be empty, in some cases, and even when this is not so, the intersection would give a very deceptive image of our knowledge. How about the union? The union will not in general be an interval, but we might take the smallest interval that covers the union. Thus in the last example, one might base a probability statement on both '$\%(C \cap B,A,0.48,0.51)$' and '$\%(C \cap R,A,0.49,0.52)$', taking the probability of '$a \in Z$' to be $(0.48,0.52)$.

Another suggestion—this one owing to Isaac Levi[5]—is that the set of potential reference classes be ordered with respect to the amount of information they give. In the simple case under consideration, this information is simply the width of the measure interval. We could then proceed down the list, employing principle I. The intention would be to ignore conflict, when it was a class lower in the information content ordering that conflicted with one higher in the list.

[5] I. Levi, oral tradition.

III

What statements go into a rational corpus K? Well, part of what a man is entitled to believe rationally depends on what he has observed with his own two eyes. Or, of course, felt with his own two hands, heard with his own two ears (in at least some senses), or smelled with his own nose. There are problems involved with the characterization of this part of a rational corpus. The line between observation statements and theoretical statements is a very difficult one to give reasons for having drawn in one place rather than in another. It will be part of the burden of this section to show, however, that so far as the contents of K is concerned, it is irrelevant where it is drawn.

Let us distinguish two levels of rational corpus; let K_1 denote the set of statements belonging to the higher level rational corpus, and let K_0 denote the set of statements belonging to the lower level rational corpus. By theorem XI, $K_1 \subseteq K_0$.

A. *Observation Statements.*

We wish sometimes to talk about the body of knowledge of an individual (John's body of justifiable beliefs) and sometimes we wish to talk about the body of knowledge of a group of individuals (organic chemists, physicists, scientists in general).

There are three treatments of observational data that seem to have a rationale of some degree of plausibility. Each of them, combined with an account of epistemological probability of the sort that has been outlined, seems to lead in a reasonable way to bodies of knowledge that look something like the bodies of knowledge that we actually have.

The most complex, least natural, and most pervasive treatment is a phenomenological one in which observations are regarded as indubitable and in which observations are of seemings or appearings. They may not be of appearances: I observe that I am having a tree appearance; I do not observe the appearance of a tree. Let us write 'L_P' for the set of sentences of our original language L that correspond to what we naively call 'observation statements' or, more technically, 'physical object observation statements'. These include such statements as 'the meter reads 3.5', 'there is a dog in the house', 'there are no elephants in the garden', 'the soup is not hot', and the like. Let us write 'L_S' for the set of sentences in L which correspond to what we regard as phenomenological reports: 'I seem to see a tree,' 'I take there to be a meter reading 3.5,' and so on. The statements of this class include (but may not comprise) statements of the form 'S seems to be the case to me', where S is a statement belonging to L_P. This special subclass — which may perfectly well be a proper subclass — of L_S is what we shall take to include (in this view) the foundations of empirical knowledge. Let us take such statements to be of the form $J(I,t,S)$, where I is an individual or a set of individuals and t is a time parameter. Contrary to the impression one gets from most who write in this vein, there seems to be no reason why this sort of view cannot be developed directly on a multipersonal basis. To say that S appears to be the case to a set of individuals I is to say that the individuals in I collectively judge that S seems to be the case (at t). Perhaps they are all looking at the cat on the roof. These statements — statements of the form $J(I,t,S)$ — which belong to L_S, and thus to L, are indubitable, and therefore may properly occur in any rational corpus, say K_1, however rigorous its standards of acceptance.

We must now account for the acceptance of ordinary observation statements concerning physical objects, i.e., sentences belonging to L_P. On the view under consideration, sentences belonging to L_P must be viewed as relatively theoretical statements — as miniature theories — and can have nothing but their probability to recommend them. From statements about physical objects (say, about my body, my state of health, the illumination of the scene, the placement of other physical objects, the direction of my attention, and the like), together with certain statements about what seems to be the case to I (that I seems to be in such and such a position, seems to be in good health, etc.), together with a fragment of psychophysical theory, certain other seeming statements follow. To put the matter slightly more formally, let T be the set of statements about physical objects ($T \subseteq L_P$) and let B be the set of statements about what seems to I to be his circumstances ($B \subseteq L_S$). In order to form the connection between T and B, on the one hand, and another subset O of L_S (the predictions of the theory), on the other, we need to supplement the theory by a baby psychophysical theory PT. From the point now being considered, we must take the whole theory $PT \cup T$, which is a subset of L, and form the conjunction of its axioms (we suppose it finitely axiomatizable). Let us abbreviate that conjunction by $PT \,\&\, T$. From $PT \,\&\, T$, conjoined with B, we obtain the subset O of L_S. The statements in the set O are of the form $J(I,t,S)$; and they may thus be definitively confirmed. Their definitive confirmation provides probabilistic confirmation of the conjunction of axioms of PT and T, and thus also of the axioms of T, and thus also of the elements of T. But T is just the subset of L_P whose elements we were trying to find justification for. The question of probabilistic confirmation of theories is complex and incompletely understood, but it is no more difficult in this case than in others. If we may include elements of B and of O in K_1, thus may provide grounds for the inclusion of the conjunction PT & T in K_0; and, since the elements of T are deductive consequences of the axioms of PT & T, therefore for the inclusion of elements of T in K_0. Furthermore, given the general theory PT in K_0 a single element of L_S in K_1 may suffice for the (probabilistic and tentative) inclusion of the corresponding statement L_P in K_0.

The second treatment of observational knowledge is much like the first. Again, seeming statements belonging to the set of statements L_S are all that can be known directly and indubitably. Again, physical object statements are components of theories. In this version, however, the psychophysical part of the theory, PT, is regarded as analytic. The argument is that to deny that most of the time when I thinks he is seeing a cat would be to deprive the term 'cat' of its conventional meaning. It is a part of the *meaning* of the term 'cat' that a cat is an identifiable physical object; and that consequently most of the time when a person thinks he sees a cat, he *does* see a cat. Under this treatment, both axioms for PT and observations (I thinks, I sees . . .) occur in K_1 and, with statements O, support the inclusion of statements of T in K_0. For example, one of the analytic PT statements might be as follows: when a person feels awake and attentive, and thinks he sees a horse, 99 percent of the time a horse is there. Putting it more precisely, we can say that the measure of the set of times during which I feels awake and thinks he sees a horse, which are also times during which he is seeing a horse, lies close to 0.99. The time t_1 is a time when I feels awake and thinks he is seeing a horse. *If* the time t_1 is a *random member* of such times, with respect to the property in question, then the probability is about 0.99 (relative to the body of knowledge K_1) that I is seeing a horse at t_1. And if this probability is high enough for inclusion in the body of knowledge K_0, there we are. Notice that error is no problem here, any more than for any other kind of probabilistic knowledge. At some other time, say t_2, the body of knowledge K_1 may have changed and become the body of knowledge K_1^*. Relative to the body of knowledge K_1^*, it may *not* be the case that t_1 is a random member of the times during which *I* thinks he is awake and is seeing a horse, with respect to being a time when *I* is in fact seeing a horse. And thus at t_2 it may not be the case that the probability is 0.99 that *I* is seeing a horse at t_1. And thus at t_2 the statement that *I* is (was) seeing a horse at t_1 would no longer, under these circumstances, be included in K_0. Furthermore, it is not always the case, even at t_1, that the time t_1 is random in the appropriate sense. If *I* knows that he has just been given a hallucinogenic drug, or if *I* finds himself in utterly unfamiliar and dreamlike surroundings, or if *I* finds himself simultaneously in the presence of people who have long been dead, or . . . then the time in question will not be a random member of the set of times when *I* feels awake and thinks he sees a horse, with respect to the times when he does see a horse; for these times are members of subsets of those times in which the frequency of horses (or of veridical perceptions in general) is far slighter than it is in the more general reference class.

The third approach to observation statements proceeds more directly and naturally, but leads to essentially the same results. According to this approach, we take observation statements to be physical language observation statements — members of L_P — and take them initially at face value. To be sure, they can be in error: that is, an observation statement may have to be rejected in the face of contrary evidence. To allow for this probabilistic element, we take account of what experience has taught us: that not all our observation judgments are veridical. Which is to say, the measure of the set of times such that *I* judges *S* to hold at *t* ($J(I,t,S)$, where $S \epsilon L_P$), which are times such that *S* does not hold, is not zero. This is something we have *learned*. The statistical statement expressing it is a member of a pretty high level rational corpus, say K_1. Let $J(I,t_1,S)$ also belong to K_1. If t_1 is a random member of the set of times in which judgments of the sort $J(I,t,S)$ are made, with respect to being veridical, then the probability, relative to K_1, that *S* holds at t_1, is very high, and *S* may be included in K_0. Note that we do not have to lump all observation judgments together, but can treat them as being of different sorts, with different frequencies of error.

No reference to any relation between phenomenological judgments (of the form $J(I,t,S)$) and physical object statements, other than the bare statistical, learned, fact that in a small proportion of cases these judgments are in error, is required.

Whichever approach to observational knowledge is adopted, there is a great deal more required for an adequate explication than is indicated by the preceding sketches. But in any event, it seems clear that (a) an epistemological concept of probability is central to any such explication regardless of which approach is adopted and (b) such a concept of probability may perfectly well be one in which probabilities are based on statistical statements, combined with a condition of randomness. The first case, in which physical object statements are construed as parts of a general theory designed to account for phenomena, is the most questionable; but we shall see shortly that this statistical-epistemological account of probability can serve in general to explicate the grounds of acceptance of theories. Anyway, what concerns us here is not the general primordial argument from phenomena to physical objects (which must, on the first view, be accounted for) but the direct and straightforward and generally veridical argument from a given physical appearance to a physical state of affairs, which is essentially the same on each of the approaches considered.

B. *Empirical Laws.*

There is no problem about the acceptance of statistical generalizations in K_0, given the appropriate data in K_1. At the very least in K_1 we will have such analytic measure statements as 'the measure of the set of n-membered subsets of X, in which the proportion of p of Y's is such that the difference between p and the true measure of Y's among the set of all X's is less than ϵ_1, is between $1 - \epsilon_2$ and 1'. Furthermore, we may perfectly well have in K_1 (on observational grounds), 's is an n-membered sample of X's in which the proportion of Y's is p^o.' Under appropriate circumstances it may be the case that s is a *random member* of the set of n-membered subsets of X with respect to membership in the representative set described above, relative to what is known in K_1. If it is, then clearly the probability of the statement 'p differs from p^o by less than ϵ_1' is $(1 - \epsilon_2, 1)$; and if ϵ_2 is small enough, we will be able to include this statement in K_0. But this statement is equivalent to $M(X,Y,p^{\circ} - \epsilon_1,p^{\circ} + \epsilon_1)$. Thus we obtain a measure statement—which is a general empirical statement—in the body of knowledge K_0.

Universal generalizations can be approached by statistical generalizations. For example, if in our sample all the *X*'s turned out to be *Y*'s, we would be entitled (under appropriate circumstances) to include '$M(X,Y,1 - \epsilon_1,1)$' in K_0. The content of this statement is that *practically all* (at least) of the *X*'s are Y's. This is still a far cry from a universal generalization: it is not equivalent, for example, to 'practically all of the non-Y's are non-*X*'s'. (This has a bearing on the traditional problem of the ravens: to examine white shoes is irrelevant to the confirmation of the statistical generalization, 'practically all ravens are black'.) To pass from 'practically all' statements to an 'all' statement is only legitimate under certain conditions and requires a new epistemological axiom.

The content of the axiom is essentially that if it is known that practically all *A*'s are *B*'s, and if our body of knowledge is such that anything known to be an *A* is, relative to that body of knowledge, practically certain to be a *B*, and furthermore such that, at the next lower level, it is not known that there is anything that is not a *B*, though it is an *A*, then, since we will in every case be able to include that any given *A* is a *B* in the lower level rational corpus, we might as well include the universal conditional that all *A*'s are *B*'s in that lower level body of knowledge. This is embodied in the following axiom: (Let us write *eqv* ϕ for the existential quantification with respect to the object language variable v of ϕ, and *uqv* similarly for universal quantification.)

AXIOM G $\quad uqv(v\,e\,X\,cd\,v\,e\,Y) \,\epsilon\, K_0 \equiv [M(X,Y,1 - \epsilon,1) \,\epsilon\, K_1\ \&\ 1 - \epsilon$ is greater than the acceptance level for K_0 & $(Z)(Z\ e\ X\ \epsilon\ K_1 \supset)\ \ Z\,e\,Y\,\epsilon\,K_1 \vee \mathrm{Ran}_{K_1}\ (Z,X,Y))\ \&\ {\sim}eqv(v\,e\,X\,cj\,n\,v\ e\,Y)\,\epsilon\,K_0]$.

It is evident that a universal generalization can only be included in a rational corpus K_0 if it is already the case that every occurrence of an instance of the antecedent of the generalization in K_1 is accompanied by the occurrence of the corresponding instance of its consequent in K_0. If the observational basis of K_1 should change, of course, we may come to reject the universal generalization—we may find a counterexample. For that matter, given an enlargement of the observational basis of K_1, we may come to reject the statistical generalization. There is no offense to fallibilism in axiom G.

C. Theories.

The question of theories is somewhat more complicated than the question of statistical or universal generalizations in our given language *L*. For in general it is the case that the introduction of a theory involves a change, or at least an extension, of our language. As is well known, it is very difficult to draw sharp lines in a theory between sentences which are to be construed as giving a partial analysis of the meaning of the theoretical terms introduced and sentences that are to be construed as embodying part of the empirical content of the theory. In adopting a theory, it is correct to say that we are adopting a particular extension of our language; it is also, appropriately construed, no doubt correct to say that we are adopting a certain conceptual scheme. So far as a body of knowledge K_0 is concerned, however, the distinction between empirical generalizations and meaning postulates is irrelevant. What counts so far as K_0 is concerned is simply whether or not a given statement is acceptable; and both generalizations that are true 'by virtue of the meanings of the term involved' and factual generalizations can perfectly well come to be accepted in K_0. Although there are other criteria for the acceptance of a theory — for example, we prefer a theory such that there is no stronger one or simpler one equally acceptable — one basic criterion is that the theory entail no statements that conflict with statements in K_0. (This harmony may sometimes be achieved by rejecting hitherto accepted members of K_0, as well as by being cautious about accepting theories.) No doubt some of the general statements entailed by the theory will already be members of K_0; others, for which there is no independent evidence, will be members of K_0 simply by virtue of being consequences of an acceptable theory. As to the question of whether the theory as a whole may be accepted (as opposed, that is, to the question of which of several competing theories should be accepted), again we may look at the matter in an essentially statistical manner: the theory is certainly acceptable if all of its consequences are acceptable; but to say of a theory that all its consequences are acceptable is simply a universal generalization (on a metalinguistic level) of the sort we have already considered. To be sure, matters are not so simple in the case of the consequences of theories as they are in the case of crows; it is much easier to obtain a sample of crows that is random in the required statistical sense than a sample of consequences of a theory that is random in the required sense. That is why the emphasis in sampling consequences of theories is on sampling a wide variety of different sorts of consequences. But what is involved here is a question of logical technique which cannot be profitably discussed until we have an epistemological concept of randomness in terms of which to discuss it.

D. *Philosophical Theses.*

If ontological questions are to be answered in probabilistic terms rather than in a priori terms, it seems reasonable to take as a criterion of existence the acceptability of a corresponding existential statement in a body of knowledge. Thus on the basis of the preceding arguments it seems plausible to say that both physical objects and sense data exist, and to characterize these assertions by the number that represents the level of acceptability in K_0, since on either approach, existential statements of both sorts may occur in K_0. As for theoretical entities, why should it not be reasonable to say that, at a given level of acceptance, we are committed to those entities which exist according to the best (most powerful, most simple . . .) theory acceptable at that level?

Metaphysical assertions may be treated much like ontological ones. For example, let us look at the question of universal causation, leaving aside the complicated question of formulating that proposition or identifying particular instances of causation. Let us assume (with most writers on these topics) that we do in fact know of many causal relationships: that cold, under appropriate boundary conditions, causes water to freeze; that virus X, under appropriate conditions, causes a certain cluster of symptoms we call 'flu'; and so on. Let us further suppose, also in accordance with most writers, that we do not know of any event that is, in point of fact, not caused. Starting from these premises, some writers have concluded that the law of universal causation is very probable; others have concluded that all we can say is that some events are caused by others. If we have a frequency based epistemological concept of probability, can we adjudicate this dispute?

The answer is yes. We can reconstruct the argument as follows: in all the cases about which we have knowledge, the law of causation operates.

Regard these cases as a sample of all cases whatever. It is a logical truth (alluded to earlier) that practically all large samples from a class closely reflect the measure of a specified subclass in that class. Let both of these statements belong to that body of knowledge K_1. Assuming that 'practically all' is represented by a large enough number to represent acceptability in the rational corpus K_0, what more is needed? We also need to be able to say that the sample is a random member of the set of samples of that size, from this population, with respect to reflecting the proportion of caused events in the population as a whole. This is precisely where the dispute lies. Is the sample random in the required sense? One cannot answer this question too confidently in the absence of a detailed examination of the concept of randomness, since our intuitions about randomness (construed epistemologically) aren't too strong. But the randomness of the sample seems highly doubtful in view of the fact that a condition of admission into the sample is having the characteristic in question of being caused. Thus the sample belongs to the class of samples with the highly special characteristic of having only caused events in it; and thus such samples can be representative only if the set of all events is such that practically all or all of them are caused. If not practically all events are caused, then we could be sure in advance that our particular sample comes from a subclass of those samples among which representative ones are very rare. The assertion of randomness seems very doubtful, and therefore the probability assertion also seems doubtful. The doubt is not empirical, note; it is purely logical, like doubt about the validity of a sketchy deductive argument. A number of metaphysical questions may be amenable to this sort of treatment, if we have an appropriate concept of randomness.

E. Decisions.

The application of this epistemological concept of probability to decision problems is perfectly straightforward. Let us take utility to be interval measurable (as is customary), and let us define the expectation of an event E, for an agent whose body of knowledge is K_1, to be the utility of E, multiplied by the probability of E, in the following way:

DEFINITION If $\text{Prob}_{K_1}(E,p,q)$, and U is the utility function of the agent, then $ME(E) = (U(E) \cdot p, U(E) \cdot q)$.

The fact that expectations of utilities, on this definition, turn out to be pairs of numbers (or intervals) may provide a plausible explanation of the fact that preference rankings sometimes turn out to be circular.

Suppose that there are three consequences C_1, C_2, and C_3 that an act A might have and that these are exclusive and exhaustive in the sense that the statement "If A, then exactly one of C_1, C_2, and C_3 holds" is a member of K_1. The expectation of the act is as follows:

$$ME_{K_1}(A) = (\Sigma U(C_i) \cdot p_i, \Sigma U(C_i) \cdot q_i), \text{ where } \mathrm{Prob}_{K_1}(C_i, p_i, q_i).$$

In view of the fact that we have intervals here rather than simple numbers, we cannot simply adopt the standard Bayesian maxim: maximize your mathematical expectation in choosing among available acts. We can adopt various versions of it: maximize the minimum expectation $(\Sigma U(C_i) \cdot p_i)$; maximize the maximum expectation $(\Sigma U(C_i) \cdot q_i)$; maximize the average expectation $(1/2 \cdot \Sigma U(C_i) p_i + 1/2\, \Sigma U(C_i) q_i)$, or some weighted average $(r \cdot \Sigma U(C_i) \cdot p_i + s \cdot \Sigma U(C_i) \cdot q_i)$. Or we can adopt a minimax principle of some sort, which may lead to different results. Again since we are dealing with pairs of numbers, we have the freedom to consider a wide variety of possible rules. Although clearly a lot of complications are possible, and there is a fruitful field for research here, there are no special or unfamiliar problems that arise in connection with a theory of rational action based on an epistemological concept of probability.

IV

I have argued that we could achieve plausible answers to a number of problems, philosophical and otherwise, if only we had a clear and useful conception of randomness; and that given such a conception all probability statements could be construed—or rather reconstructed—as statements based on (a) knowledge of a relative frequency or measure and (b) knowledge concerning an individual such that relative to that knowledge the individual is a random member of one class with respect to membership in another class. The relation is properly a metalinguistic one, and probability is a metalinguistic concept, so that in fact what one must define are certain relations holding among terms, sentences, etc. Nevertheless, it is far more convenient to talk as if the relations held between objects and classes.

There are those who say that it is obvious that not all probability statements can be based on statistical statements, but the examples they use to support their claim always seem to admit of some sort of statistical reconstruction. This is not merely my own opinion, but also the opinion of those who, unlike me, want to offer an empirical limiting frequency, or an empirical dispositional, analysis of probability statements. They too tend to say: no measure statement, no probability.

I have also argued, much more hesitantly, for a couple of principles which would, in any situation of a fairly simple sort, pick out one class as a reference class—or, if it allowed more than one, would at least ensure that the relevant measure was the same in all the possible reference classes. Perhaps someone will be able to find either some holes in or some more persuasive support for the principles I have enunciated. Or perhaps someone will find some principles that are yet more persuasive. In any event, I take this to be the most fundamental problem in both inductive logic and epistemology. It is worthy of our most serious efforts.

KEITH LEHRER

Induction, Rational Acceptance, and Minimally Inconsistent Sets

1. *Introduction.* The purpose of this paper is to present a theory of inductive inference and rational acceptance in scientific inquiry. A concept of *relevant* deduction is defined in which the truth of each and every premise of a deductive argument is essential to establishing the truth of the conclusion by deduction from the premises. This definition is based on the completely abstract notion of a minimally inconsistent set of statements. In terms of this same abstract logical concept and the relation of probability, we shall define a concept of inductive inference that is a principle of rationality. This concept of inductive inference is shown to form the basis of a principle of acceptance in which two important epistemic utilities are maximized.

2. *Conditions of Relevant Deduction.* Suppose we are engaged in reasoning concerning some scientific conclusion that is neither contradictory nor logically true. If we intend to establish the truth of the conclusion by deduction from the premises, then it is important that several conditions be met to ensure that the deductive argument is relevant to that end. First the premises of such a relevant deductive argument, *RD*, must be logically consistent:

> CR1. The set of premises *P*1, *P*2, . . . , and *P*n of *RD* is logically consistent.

The need for imposing (CR1) as a condition of relevance arises because an argument with an inconsistent set of premises is valid no matter what conclusion the argument has. An inconsistent set of premises is quite irrelevant to establishing the truth of a conclusion deduced from them. Some philosophers and logicians have maintained that every statement is relevant to itself and hence that a contradiction is relevant to itself. But this concept of relevance is a different one from that under consideration here.[1] I am concerned with a concept of relevance such that the premises of an argument are relevant to establishing the *truth* of the conclusion and nothing is relevant to establishing the truth of a contradiction.

The second condition of relevance is that all the premises be essential to the deduction of the conclusion from the premises. An argument is only as relevant as its premises, and thus if an argument contains an irrelevant or inessential premise it is not completely relevant. We may formulate this condition as follows:

> CR2. The set of premises *P*1, *P*2, . . . , *P*n of *RD* is nonempty and such that no proper subset of the set is sufficient for the deduction of the conclusion *C* of *RD*.

This condition is required to ensure that the argument is relevant, but it is also useful for the clarification, defense, and criticism of arguments in a scientific context. Suppose that a deductive argument is offered in defense of a scientific conclusion when the argument does not satisfy condition (CR2). In that case, the investigator may be misled into thinking that evidential support is required in defense of his conclusion because it is needed for the defense of some inessential premise. When such a premise is inessential for the deduction of the conclusion, the evidence supporting it may be otiose. Similarly, if an argument does not satisfy (CR2), then criticism of the argument directed against some premise of the argument may be fruitless because the premise is inessential to the deduction of the conclusion. Thus, if a deductive argument intended to establish the truth of a scientific conclusion satisfies (CR2), this will help to avoid useless evidence gathering and equally fruitless criticism.

The characterization of relevance contained in condition (CR2) logically entails the result that no argument having a tautological conclusion (and one or more premises) is a relevant deductive argument. This is a consequence of the feature of systems of natural deduction in which any tautology may be derived by conditional proof from the null set of premises. Thus, given (CR2), it follows that any argument having a tautological conclusion and one or more premises will fail to be relevant

[1] See A. R. Anderson and N. D. Belnap, "The Pure Calculus of Entailment," *Journal of Symbolic Logic*, 27 (1962), 19–52, esp. 20, 36, 39, and 42. Various systems by other authors are discussed which postulate that *P* is relevantly deducible from *P* even when *P* is contradictory.

because the null set of premises is sufficient for the deduction of the conclusion. Moreover, no argument with a tautological premise will be relevant either. Again given a system of natural deduction, any set of premises containing a tautology is such that the proper subset of premises containing only those premises which are *not* tautological will suffice for the deduction of the conclusion. Therefore, an argument that is relevant in the sense that the truth of each and every premise is essential to establishing the truth of the conclusion by deduction from premises will have neither tautological premises nor a tautological conclusion. This follows from (CR2). It follows from (CR1) that no premise of a relevant deductive argument may be contradictory, and obviously no contradiction can be the conclusion of such an argument.

The preceding results indicate that a deductive argument is relevant in the sense under consideration only if the premises and the conclusion of the argument are all logically contingent, that is, neither contradictory nor tautological. Though this condition is not independent of the two previous conditions, we shall add it as a third condition.

> CR3. The premises *P*1, *P*2, . . . , *Pn* of *RD* as well as the conclusion *C* of *RD* are neither contradictions nor tautologies.

This condition ensures that relevant deductive arguments are ones in which the truth of contingent premises, and only such, is essential to establishing the truth of a contingent conclusion by deducing the latter from the former.

It might seem to be an objection to the preceding characterization of relevance that an argument can be shown to be irrelevant in a scientific context by virtue of the fact that the conclusion of the argument can be established by deduction from totally different premises. However, it is not this sense of relevance I am attempting to explicate. I am only attempting to explicate that sense in which an argument is relevant because the truth of each and every premise of the argument is essential to establishing the truth of the conclusion *by deduction from those premises*. Thus, I am only concerned with the relevance of the premises within the argument and not with the more general question of whether the argument is required at all.

3. *Analysis of Relevant Deduction.* Having laid down three conditions of relevant deductive argument, I shall now offer an analysis of the concept of relevance under consideration. First, let us recall that a valid

deductive argument is such that the set of statements consisting of the premises of the argument and the denial of the conclusion (or any truth functional equivalent) is logically inconsistent, that is, one may deduce a contradiction from the set of statements. A valid deductive argument that is also relevant is one such that the set of statements having as members the premises and the denial of the conclusion (or any truth functional equivalent) is not only logically inconsistent but *minimally inconsistent*. A set of statements is minimally inconsistent if and only if the set is logically inconsistent and such that no proper subset of the set is logically inconsistent. In short, a minimally inconsistent set is one such that all the statements of the set are required to make it inconsistent.

The preceding considerations yield the following analysis of a relevant deductive argument:

> AR1. An argument *RD* is a relevant deductive argument if and only if *RD* contains a nonempty set of premises *P*1, *P*2, . . . , *P*n and a conclusion C such that a set of statements consisting of just *P*1, *P*2, . . . , *P*n, and ~*C* (or any truth functional equivalent of ~C) is a minimally inconsistent set. A set of statements is a minimally inconsistent set if and only if the set of statements is logically inconsistent and such that every proper subset of the set is logically consistent.

It follows from (AR1) that conditions (CR1), (CR2), and (CR3) are all met. The proof that (CR1) is met is that if the premises of the argument *RD* were inconsistent, then the set of premises would be an inconsistent proper subset of the set of statements consisting of the premises and the denial of the conclusion (or some truth functional equivalent). The proof that (CR2) is met is that if the set of premises of the argument *RD* were such that a proper subset of the set was sufficient for deduction of the conclusion, then the set of statements consisting of that proper subset of the premises and the denial of the conclusion would be an inconsistent proper subset of the set of statements consisting of all the premises of *RD* and the denial of the conclusion. The proof that (CR3) is met is that any two-membered minimally inconsistent set is such that neither member may be a contradiction or a tautology. For if one statement were a tautology, the other would have to be a contradiction to make the set inconsistent, in which case the proper subset containing just the contradiction would be an inconsistent set. If one statement of a two-membered set is a contradiction, then the proper subset of the set containing that contradiction as an only member would be an inconsistent set. Thus conditions (CR1), (CR2), and (CR3) are logical consequences of (AR1).

The preceding discussion characterized and analyzed a concept of a relevant deductive argument. The concept is of some interest simply because any argument in which the truth of a conclusion is established by deduction from premises must always be such that either it is a relevant deductive argument or else some relevant deductive argument could be substituted for it, thereby clarifying which premises of the original argument are essential. By means of such clarification we may avoid ineffective criticism and superfluous defense.

4. *Induction.* The preceding discussion was concerned with relevant deductive arguments in which the truth of each and every premise is required to establish the truth of the conclusion by deduction from the premises. The question to which we shall now turn is that of investigating the role of such arguments in scientific inquiry. A relevant deductive argument may fail to establish the conclusion because the epistemic worth of the argument depends on the value of the premises. If the premises are without merit and would not be accepted in scientific inquiry, then the argument is ineffective for the purpose of gaining acceptance for the conclusion. Consequently, any adequate account of justificatory reasoning must be based on an explication of the rational acceptance of statements as premises.

Philosophers have notoriously disagreed about the status of such premises. Some have argued that premises should be confined to observation reports and others have argued to the contrary that this restriction is unrealistic epistemologically and/or semantically. It is unrealistic epistemologically if it would prohibit us from accepting well-established scientific theories, and it is unrealistic semantically if the distinction between theoretical terms and observation terms is untenable. I shall not here attempt to resolve such disputes. Instead, I shall leave open the question of the exact character of initial scientific premises, if there are such premises, and consider the epistemological situation we confront in any given stage in the history of scientific inquiry. Ultimately, my reason for doing so is that I believe scientific inquiry can proceed quite nicely without any permanent foundation of statements that justify other statements without being so justified themselves. However, even if I am quite mistaken in this assumption, there is some methodological justification for putting the question of ultimate premises to rest and attempting to explicate the character of scientific inference as it proceeds against the background of premises accepted as belonging to the body of scientific knowledge. Thus, I shall assume that we have such a background of accepted statements and examine the character of inductive reasoning.

Though some reasoning may be explicated as reasoning from accepted premises to conclusions by means of relevant deductive argument, other reasoning may not be so explicated and this I shall refer to as inductive reasoning. The aims of inductive reasoning have proven to be controversial. Since an inductive argument is not necessarily truth preserving, we cannot affirm that our only objective in inductive argument is to arrive at the truth. On the contrary, if all we cared about was that the conclusion of our argument be true, then we would not be justified in venturing beyond the deductive consequences of our premises. For even if our premises are true, to reason inductively is to risk error and for such a risk there must be some redeeming benefit.[2] A number of such benefits have been proposed. One is that of explanation, that it is an aim of scientific inquiry to arrive at conclusions that explain and that are explained.[3] Another aim is to arrive at conclusions that give us additional informational content.[4] When we risk error in inductive reasoning, the risk is justified by the fact that the conclusions at which we arrive will, if true, add to our information and facilitate explanation. Of course, the goal of arriving at true conclusions is not to be eschewed, but neither should it exclude other cognitive interests.

This approach to inductive reasoning is the one that I shall adopt.

[2] Cf. Isaac Levi, *Gambling with Truth* (New York: Knopf, 1967), pp. 119–28.

[3] This goal has been stressed by many writers, especially Karl Popper, W. V. O. Quine, W. H. Sellars, Gilbert Harman, and Paul Ziff. In an earlier paper, Keith Lehrer, "Justification, Explanation, and Induction," in M. Swain, ed., *Induction, Acceptance and Rational Belief* (Dordrecht: Reidel, 1970), pp. 119–20, I present a rule designed to achieve this objective.

[4] See Levi, *Gambling with Truth*, pp. 57–90, Karl Popper, *The Logic of Scientific Discovery* (London: Hutchinson, 1959), pp. 119–21, Jaakko Hintikka and Juhani Pietarinen, "Semantic Information and Inductive Logic," in J. Hintikka and P. Suppes eds., *Aspects of Inductive Logic* (Amsterdam: North Holland, 1966), pp. 96–112, and Risto Hilpinen, *Rules of Acceptance and Inductive Logic* (Amsterdam: North Holland, 1968), pp. 103–22. Carl G. Hempel, *Aspects of Scientific Explanation* (New York: Free Press, 1965), pp. 73–78.

However, the importance of arriving at true conclusions, if not the only goal, is in a way a supervening one. For, conclusions that are false do not add to our information nor do they facilitate explanation. No matter how informative a statement would be if it were true, we shall be misled rather than enlightened if we accept it when it is false. Similarly, what explains must be true or the explanation fails, and what is explained must be a fact, something described by a true statement not a false one, or else there is nothing to explain.

A proper concern for truth may be guaranteed by requiring that the conclusions of inductive reasoning should be more probable than the premises of any argument sufficient to demonstrate the falsity of the hypothesis. Thus we shall restrict inductive inference so that there is no acceptable argument demonstrating the falsity of an inferred hypothesis. Grice, Chisholm, and Thomas Reid have suggested that some hypotheses may be regarded as innocent until proven guilty.[5] Others, notably Popper and his followers, have suggested that the strongest hypothesis consistent with the evidence is the one to accept.[6] Both of these theses incorporate the idea that there must not be any acceptable argument demonstrating the falsity of a hypothesis inferred from background knowledge.

A hypothesis may be inductively inferred from background knowledge only if there is no acceptable argument demonstrating the falsity of the hypothesis. Demonstrative arguments must be deductive. Moreover, among deductive arguments we may restrict our consideration to relevant ones for reasons cited above. Thus, an argument is an acceptable demonstrative argument against a hypothesis only if it is a relevant deductive argument whose conclusion contradicts the hypothesis under consideration. But this is not the only condition that an argument must satisfy in order to be an *acceptable* demonstrative argument against a hypothesis. For, there remains the question of the viability of the premises of such an argument. The premises of the argument must each be at least as probable on the basis of the background knowledge as the hypothesis under consideration or else the argument is unacceptable as a demon-

[5] H. P. Grice, "The Causal Theory of Perception," in Robert J. Swartz ed., *Perceiving, Sensing and Knowing* (Garden City, N. Y.: Doubleday, 1965), pp. 470–71. R. M. Chisholm, *Perceiving: A Philosophical Study* (Ithaca, N. Y.: Cornell University Press, 1957), p. 9. Thomas Reid, *The Works of Thomas Reid, D.D.* (Edinburgh: Maclaugh and Steward, 1863), p. 234.

[6] Popper, *The Logic of Scientific Discovery*, 57–90.

stration of the falsity of the hypothesis. If the hypothesis in question is more probable on the background knowledge than one of the premises of the argument, then the argument is not an acceptable argument against the hypothesis. The reason is that a relevant argument is one in which all the premises are essential for the deduction of the conclusion, and consequently such an argument is only as acceptable as its weakest premise.

5. *An Inductive Rule.* On the basis of these considerations, it is contended that a hypothesis may be inductively inferred from background knowledge if and only if the hypothesis is such that any relevant deductive argument whose conclusion is the contradictory of the hypothesis is such that at least one premise of the argument is less probable on the background knowledge than the hypothesis inferred. The rule may be expressed in terms of the notion of a minimally inconsistent set as follows: a hypothesis may be inductively inferred from background knowledge if and only if every minimally inconsistent set to which it belongs has at least one other member which is less probable on the background knowledge than the hypothesis.

This very abstract characterization of an inductive inference has features that make it suitable as a principle of rationality. For example, the set of hypotheses that may be inductively inferred from background knowledge by such a rule must be logically consistent as a set and the set must be consistent with the background knowledge. The proof is that if we assume that the set S of hypotheses inferred from the background knowledge is logically inconsistent, then, since each of the hypotheses belonging to the set is consistent, it follows that some subset of S having at least two members must be a minimally inconsistent set. Since each of the two members was inferred by the rule, each member of the minimally inconsistent set must be more probable than at least one other member. But this is impossible. Hence the original set of inductively inferred hypotheses must be logically consistent.

This consistency feature is lacking in some plausible alternatives. The reason is that a set of hypotheses having any probability less than one on background knowledge can always be so selected that the set is inconsistent with the background knowledge. If someone adopts a rule to the effect that any hypothesis having a probability greater than m/n (m being a smaller integer than n) on background knowledge may be inductively inferred from that knowledge, then, as Kyburg has shown, he will be selecting a rule that will permit him to inductively infer a set of hypotheses from background knowledge that is inconsistent with that knowledge.[7] For example, suppose he has a fair lottery and background knowledge saying that of *n* consecutively numbered tickets, exactly one is the winner. In this case, the probability that it is not the number one ticket is at least m/n, the probability that it is not the number two ticket that wins is at least m/n, and so forth. Thus, the set of inferred hypotheses would be such that one could deduce from the set that no ticket is the winner. This contradicts the background knowledge that one ticket is the winner. Thus the consistency of the inferred hypotheses with the background knowledge is a significant feature of the rule. If one regards it as rational to infer a consistent rather than an inconsistent set of conclusions from what one knows, then the rule fulfills a requirement of rationality.

To formulate other results of applying the rule, it will be useful to have a more formal statement of it. Letting "$I(h,b)$" mean "*h* may be inductively inferred from *b*," the rule is as follows:

I. $I(h,b)$ if and only if for any set *S* of statements of *L*, if *h* is a member of *S* and *S* is minimally inconsistent, then there is a *k*, $k \neq h$, such that *k* is a member of *S* and $P(h,b)$ exceeds $P(k,b)$.

The probability functor is a relation between the statements "*h*" and "*b*" which satisfies the calculus of probability. We may think of the concept of probability as being interpreted in the manner explicated by Carnap or Hintikka, but other analyses would be appropriate as well.[8] Moreover, we shall limit our discussion to finite languages which may be defined in terms of what Levi calls an ultimate partition, a partition being a set of statements that are exclusive in pairs and exhaustive as a set.[9] The ultimate partition may also be thought of as defining the language in the sense that any statement of the language is logically equivalent to a disjunction of members of the partition or to a single member of a partition or to conjunction of a member and its denial. The state-descriptions are, of course, members of a partition in Carnap's

[7] H. E. Kyburg, Jr., *Probability and the Logic of Rational Belief* (Middletown, Conn.: Wesleyan University, 1957), p. 197.

[8] Rudolph Carnap, *The Foundations of Probability* (Chicago: University of Chicago Press, 1950), and Jaako Hintikka, "A Two-Dimensional Continuum of Inductive Methods," in Hintikka and Suppes, eds., *Aspects*, pp. 113–32.

[9] Levi, *Gambling with Truth*, pp. 32–36.

system. Languages without individual constants may be thought of as being defined by a partition, only in this case the partition will not be a set of state-descriptions. Finally, let it be noticed that a partition can be generated from any set of statements. If one has a set of statements, *h*1, *h*2, and so forth, then one can construct a partition by forming maximal conjunctions such that each h_i or its negation (but not both) is contained in order in each maximal conjunction. When one thus obtains a set of such conjunctions containing all possible permutations and then eliminates those maximal conjunctions that are inconsistent, the remaining set of maximal conjunctions will be a partition.

6. *Results of the Inductive Rule.* With the foregoing technical remarks on the board, we may then ask what sort of results (I) yields. The results may be succinctly characterized. A disjunction of members of the ultimate partition may be inductively inferred from background knowledge if and only if the probability of the hypothesis on background knowledge exceeds the probability of the denial of at least one disjunct on the same knowledge.[10] A single member of the ultimate partition may be inductively inferred from background knowledge if and only if it is more probable than its denial on background knowledge.[11] Only (but not all) those hypotheses having a probability greater than 1/2 on background knowledge will be ones that can be inductively inferred from background knowledge.[12]

We obtain a more concrete idea of the results by again considering lotteries. In the fair lottery, we may not inductively infer that the number one ticket will not win, because this hypothesis may be considered as a disjunction of the hypotheses saying that either the number two ticket or the number three ticket and so forth will win. The denial of each of these disjuncts is equal in probability to that of the hypothesis that the number one ticket will not win. Hence the probability of the hypothesis does not exceed the probability of the denial of any disjunct contained in it.

Now let us consider some skewed lotteries where the tickets have differing probabilities of being the winner. In the event that it is more probable that a ticket will win than lose, we may inductively infer the

[10] See T6 in the appendix for a proof of this result.

[11] See T5 in the appendix for a proof of this result.

[12] The proof of this hypothesis follows from the theorem that a hypothesis having a probability no greater than 1/2 has no greater probability than its denial.

hypothesis that the ticket will win from the background knowledge. If there is no ticket such that the probability of its winning exceeds the probability of its losing, then we cannot inductively infer that any specific ticket will win. However, we may inductively infer from the background knowledge that some ticket will not win or that one or the other of some group of tickets will win. We may infer that a ticket will not win if there is any other ticket that has a higher probability of winning on the background knowledge. We may infer a disjunction asserting that one of a specified group of tickets will win if it is more probable that one of the group will win than it is that at least one specific one of the tickets of the group will lose. Thus, for example, one may inductively infer that the winning ticket will be one of the first ten if that hypothesis is more probable than the hypothesis that the number one ticket will lose, or more probable than that the number two ticket will lose, or . . . more probable than that the number ten ticket will lose.

The foregoing results have been expressed in terms of lotteries for the sake of clarity. However, it should not be concluded that the inductive method in (I) is therefore limited to such contexts. On the contrary, whenever hypotheses are under scientific consideration, we may construct a partition from those hypotheses in the manner indicated above, the hypotheses then being equivalent to a single member of the partition or to some disjunction of such members, and then apply the inductive rule. The set of hypotheses must be finite, but this is not as restrictive as it might appear. As Hempel has argued, even such a simple problem as determining the length of some object by measurement allows for an infinite number of different results, and such considerations might appear to indicate that (I) will have very limited applicability.[13] But this conclusion would be precipitous. Though it is mathematically true that an object may have any of an infinity of different lengths, in fact any procedure of measurement will have some least measure, and, thus once we set some limits to the length the object might have, even very rough limits, the measurement problem becomes amenable to the application of (I). If we suppose that m is the least measure of our measurement procedure and find a number mn (n being an integer) such that the probability that the length is at least mn is equal to one, then we may

[13] Carl G. Hempel, "Fundamentals of Concept Formation in Empirical Science," in Otto Neurath ed., *International Encyclopedia of Unified Science*, vol. 2, no. 7 (Chicago: University of Chicago Press, 1952), pp. 29–35.

restrict our consideration to a set of hypotheses affirming that the length of the object is $mn + m$, $mn + 2m$, $mn + 3m$, and so forth until we reach a hypothesis such that the probability of the length exceeding that number is zero. We will then have a partition to which *IR* may be applied.

Another apparent problem dealing with the application of *IR* in a scientific context concerns the inference of general hypotheses and theories. General hypotheses, though often inconsistent with each other, do not seem to form a partition because one rarely has an exhaustive set to consider. However, this difficulty can be easily overcome by adding what Shimony has called a remainder hypothesis, namely, a hypothesis affirming that none of the other hypotheses are correct.[14] If the other hypotheses are incompatible in pairs, then those hypotheses together with the remainder hypothesis will constitute a partition. If the others are not inconsistent in pairs, we may generate a partition by employing the procedure of forming maximal conjunctions of the hypotheses, including the remainder hypothesis, and rejecting the inconsistent conjunctions. The remaining set will be a partition. We can then apply *IR* to decide which hypotheses should be inferred.

It is clear from these remarks that the selection of the partition is of fundamental importance. It is tempting simply to cast this problem into the pit of pragmatics, and say, with Levi, that the selection of an ultimate partition is a pragmatic choice.[15] However, this approach obscures important issues. Certain selection procedures would be arbitrary and unreasonable. Moreover, the selection of two different ultimate partitions, two different languages, both of which are used at the same time, could lead to the inference of inconsistent results by (I) even though the application (I) to a single partition is consistent. Thus, to retain the consistency feature of (I) in application, one must restrict the choice of ultimate partitions. Some philosophers, who stress the relation of scientific advance to conceptual innovation, will regard the selection of ultimate partitions as constituting a basic feature, if not *the* basic feature, of scientific inquiry. At any rate, the problem of selecting a language, an ultimate partition, if pragmatic, must be so restricted as to conform to reasonable standards of scientific procedure. The selection must not be arbitrary, it must not lead to the simultaneous selection of a multi-

[14] Abner Shimony, essay in this volume.
[15] Levi, *Gambling with Truth*, pp. 67–68, 91–95.

plicity of different partitions that yield conflicting results, and it must not fail to meet other standards of cogency, completeness, and simplicity. Thus, the choice of a language or ultimate partition which is required for the application of (I) raises some of the most important issues in the philosophy of science. However, even once a choice has been made concerning the language or ultimate partition and the evidence or background knowledge, there remains the problem of deciding what hypotheses in that language or partition may be inferred from that evidence or knowledge. We may attempt to solve that problem without pretending to resolve those issues concerning what languages to select or what statements to consider as background knowledge.

One final problem of application concerns the assignment of probabilities. It might be thought that it is not feasible to suppose that one can meaningfully assign degrees of probability to all hypotheses in all scientific contexts. The reply to this line of objection is twofold. First, developments in subjective theories of probabilities owing to Savage and Jeffrey suggest that probabilities may be assigned on the basis of empirically determined preferences where initially assignment of probabilities might appear impossible or arbitrary.[16] So, even if the logical theories of Carnap and Hintikka prove inadequate in some cases, there remains the possibility of appealing to subjective theories or appropriate modifications thereof. Second, the rule (I) only requires that we be able to make comparative judgments of probabilities, that we be able to determine whether two hypotheses are such that one is more probable, less probable, or equal in probability to another. Thus, although I have employed a quantitative concept of probability in the formulation of the rule, a comparative concept of probability could replace the quantitative one for the application of *IR* in contexts in which only comparative judgments of probability are feasible. As so modified, the rule would have broad application, because such comparative judgments would be feasible in a wide variety of scientific contexts.

7. *The Rationality of the Inductive Rule.* The next task is to show that these results are rational, that is, that it would be reasonable to infer inductively such hypotheses from background information according to

[16] Leonard J. Savage, "Implications of Personal Probability for Induction," *Journal of Philosophy*, 64 (1967), 593–607, and *The Foundations of Statistics* (New York: John Wiley, 1954). Richard Jeffrey, *The Logic of Decision* (New York: McGraw-Hill, 1965).

(I). We have already noted that the set of statements inductively inferred from background knowledge by (I) will be logically consistent with the background knowledge. Secondly, the rule has the feature of restricted deductive closure. This principle may be formulated as follows:

PRDC. If $I(h,b)$ and k is a deductive consequence of h, then $I(k,b)$.[17]

This principle of restricted deductive closure tells us that for any hypothesis that may be inductively inferred from background knowledge, a deductive consequence of the hypotheses may likewise be inferred. And again this is a desideratum of rationality in that it would be irrational to refuse to infer inductively a deductive consequence of a hypothesis one had inductively inferred.[18]

Thirdly, (I) has a claim to being a principle of rationality because anything that one infers will be such that one can always offer a probabilistic reason for the inference, and for any inference one cannot make one also has a probabilistic reason for not making it. Consider, for example, members of the ultimate partition. If one infers that such a member is false, it will always be the case that some other member was more probable. If one infers that some member is true, then it will always be the case that it was more probable than its denial and also more probable than any other member. And if one cannot infer that a member is true, then it will be at least as probable that some other member is true instead and that the member is false rather than true. With respect to disjunctions, if one infers that a disjunction is true, then it will be more probable that the disjunction is true than it is that at least one of its disjuncts is false and also more probable that it is true than false. If one cannot infer that a disjunction is true, then it will be at least as probable that each of its disjuncts is false as it is that the disjunction is true.

8. *Induction and Explanation.* Moreover, when the rule is applied to the appropriate ultimate partitions, the inferred hypotheses will, if true, explain certain facts contained in the background knowledge and others as well. Hintikka and Hilpinen have shown that partitions of general hypotheses describing the kinds of individuals there are in the universe

[17] See T7 in the appendix for proof of this result.

[18] H. E. Kyburg, Jr., "Probability, Rationality, and a Rule of Detachment," in Y. Bar-Hillel ed., *Proceedings of the 1964 Congress for Logic, Methodology and Philosophy of Science* (Amsterdam: North Holland, 1965), pp. 303–10.

are such that, when probabilities are assigned by Hintikka's methods, the probability that all unobserved individuals are of the same kind as those observed goes up sharply as the number of observed individuals increases.[19] When the probability of one such hypothesis exceeds 1/2 and thus exceeds the probability of any other hypothesis belonging to the ultimate partition, then by (I) the hypothesis may be inferred. Such a hypothesis could be employed to explain why we have observed the kinds of individuals we have and not other kinds, to wit, because all individuals are of the kinds we have observed and there are no other kinds. When we seek to explain why all individuals are of the kinds they are by appealing to higher level theories, we may again apply (I) to enable us to decide which theory we may infer, provided we can make comparative judgments concerning the probability of such theories. At both levels, the application of (I) will facilitate explanation.

We may also consider whether (I) is itself a principle of inductive explanation, that is, whether we may conclude that a hypothesis inductively inferred from background knowledge by (I) is also inductively explained by that knowledge. I have elsewhere argued that a more restrictive inductive rule is required to ensure that the hypothesis is inductively explained by the probabilities.[20] The reason is that by (I) it may turn out to be the case that two hypotheses are inferred, the first more probable than the second, when the first turns out to be false and the second true. For example, suppose that we have a three-membered ultimate partition whose members $h1$, $h2$, and $h3$ have probabilities of 1/6, 2/6, and 3/6 respectively. By (I) we may infer $\sim h1$ and $\sim h2$. However, if $h1$ turns out to be true, it would be peculiar to say that the truth of $\sim h2$ is explained by the probabilities, because $\sim h1$ was even more probable and it turned out to be false. It is possible to formulate a more restrictive inductive rule than (I) having the consequence that an inferred hypothesis will, if it is true, be more probable than any hypothesis that turns out to be false. But (I) lacks this feature. Nevertheless, we may say of hypotheses inductively inferred from background knowledge (and not deducible from such knowledge) by (I) that they are *potentially* explained by the background knowledge, because if all equally

[19] Jaako Hintikka and Risto Hilpinen, "Knowledge, Acceptance, and Inductive Logic," in Hintikka and Suppes, eds., *Aspects*, pp. 1–20; and see Hilpinen, *Rules*, pp. 50–67.

[20] Lehrer, "Justification," pp. 110–27.

probable or more probable hypotheses inferred from such knowledge by (I) turn out to be true, then the truth of such hypotheses together with the background knowledge does inductively explain the truth of the hypotheses inferred. It is explained by the probabilities.

The foregoing considerations imply that if one seeks to reach conclusions in scientific inquiry which explain what needs to be explained and are themselves explained, it is reasonable to apply (I) to obtain these objectives. For, given certain ultimate partitions, the hypotheses inductively inferred from background knowledge by (I) will explain facts contained within the background knowledge as well as other facts, and hypotheses inductively (but not deductively) inferred from background knowledge by (I) are potentially explained by the probabilities.

9. *A Rule of Acceptance.* However, a more direct proof of the rationality of the rule is possible. Some philosophers, Levi and Kyburg for example, have argued that rules of inductive logic may be considered as directives to *accept* hypotheses on the basis of background knowledge. According to this account, inference aims at acceptance. We may regard anything inductively inferred from background knowledge as suitable to be accepted on the basis of that knowledge, but there are other requirements that must be met by a theory of acceptance. One of these, Levi has argued, is a principle of deductive cogency according to which the set of accepted statements includes the deductive consequences of the background knowledge and accepted hypotheses.[21] In short, the set of accepted statements must be deductively closed. The reason for this is that it would be irrational for a man to refuse to accept the deductive consequences of the statements he accepts.

In order to provide a theory of acceptance built upon our inductive rule we shall say that a hypothesis is to be accepted on the background knowledge if it may be inductively inferred from background knowledge or may be deduced from some set of hypotheses that may be inductively inferred from that knowledge. Formally stated, we obtain the following rule where we let "$A(h,b)$" mean "h may be accepted on the basis of b."

AR. $A(h,b)$ if and only if either (i) $I(h,b)$ or (ii) there is some

[21] Levi, *Gambling with Truth*, pp. 26–28. H. E. Kyburg, Jr., has argued against this principle in "Conjunctivitis," in Swain, ed., *Induction, Acceptance, and Rational Belief*, pp. 55–82.

set S of which h is a deductive consequence and such that k is a member of S if and only if $I(k,b)$.

This rule is deductively closed; that is, the set of statements that may be accepted relative to background knowledge includes its own deductive consequences.

The results that we obtain by employing (AR) are very easy to characterize. If there is some member of the ultimate participation that is more probable on background knowledge than any other member of the ultimate partition, then the set of hypotheses we may accept consists of the logical consequences of that hypothesis and the background knowledge. This follows from the fact that we may inductively infer that each of the other members is not true by (I), and the conjunction of those logically implies that the remaining member of the partition is true. If there is some set of members of the ultimate partition that are equally probable and more probable than any member not belonging to the set, then the set of statements we may accept will consist of the logical consequences of the disjunction of those equally probable members and the background knowledge. This follows from similar considerations, namely, that we may inductively infer the falsity of each of the members of the ultimate partition not belonging to the set of members having the highest probability.[22]

10. *The Objectives of Acceptance.* The next question to confront is, How we are to prove that such an acceptance rule as (AR) is a principle of rationality? Levi has argued that a policy of accepting a statement is rational if and only if it can be shown to satisfy the *objectives* of acceptance.[23] To decide whether it is rational to accept a hypothesis we must first determine what aims or goals we hope to achieve by accepting hypotheses. One set of goals is that of accepting hypotheses that are true and informative. If we restrict ourselves to aiming at accepting true hypotheses, it would never be rational to venture beyond accepting the deductive consequences of our background knowledge. For, as we said earlier, there is always some risk of accepting a false hypothesis whenever one accepts any hypothesis that is inductively inferred from such knowledge, and, such a risk would not be rational if our only interest were to accept hypotheses that are true. Crudely put,

[22] See T8 in the appendix for a proof of this result.

[23] Levi, *Gambling with Truth*, pp. 7–19.

it is not just truth that we desire but also the whole truth or as much of the whole truth as we can get. The more content our accepted statements have the more of the truth we get if such statements are true. The more content a statement has the more informative it is. Thus, we may, with Levi, regard truth and content as objectives of acceptance.

11. *Maximizing Expected Utility.* We may now show that AR aims at the satisfaction of these objectives. Following Hempel, Hintikka, and Levi, we formulate some utility function and then show that, where we have the twin objectives of truth and content, a policy of maximizing expected utility yields the same results as (AR). The notion of maximizing expected utility is, informally, the following. First one considers the alternatives before one and the value or utility one attaches to the possible outcomes of adopting the various alternatives. One also has to consider how probable it is that the selection of a given alternative will produce each outcome. The expected utility of selecting a given alternative is the sum that one gets from adding the products of the probability of each outcome on that alternative times the utility of that outcome.

When applying this technique to matters of acceptance, we may restrict consideration to a pair of possible outcomes. When we accept *h* either we accept *h* and *h* is true, that is, we correctly accept *h*, or we accept *h* and *h* is false, that is, we incorrectly accept *h*. To put the matter more formally, we have only to define two utilities, the utility of accepting *h* when *h* is true and the utility of accepting *h* when *h* is false.

To render the preceding considerations more precise, it will be useful to introduce some formal notation derived from Levi. First, letting "$E(h,b)$" mean "the expected utility of accepting *h* on *b*," "$U(h,b)$" mean "the utility of correctly accepting *h* on *b*," and "$u(h,b)$" mean "the utility of incorrectly accepting *h* on *b*," we obtain the following equation:

$$E(h,b) = P(h,b)U(h,b) + P(\sim h,b)u(h,b).$$[24]

Given this formulation of expected utility, and given that we agree that the rational course of acceptance is to maximize expected utilities, all we need to do in order to determine whether our acceptance is rational is to define the utilities and a notion of maximality. Let us first

[24] Cf. *ibid.*, p. 83, and Hempel, *Aspects*, pp. 73–78.

consider the matter of defining the utilities in question.

12. *Maximizing Truth and Content.* Suppose that we are seeking to maximize the utilities of truth and content. In the equation, the probability of h on b and the probability of $\sim h$ on b are given equal weight with the utility. Probability is the probability of *truth*. Thus the objective of truth is given equal weight to that of the utility. Consequently, in defining the utilities in question, we may be satisfied if they represent the objective of content. It might in fact seem that we could simply identify the utility of correctly accepting h on b when h is true with the content of h. Letting "cont(h,b)" mean "the content of h on b," we would then obtain the equation

$$U(h,b) = \text{cont}(h,b)$$

and similarly

$$u(h,b) = -\text{cont}(-h,b).$$ [25]

These equations, a simplification of ones proposed by Levi, embody the idea that if we accept h when h is true, then we obtain the full benefit of the information contained in h, and if we accept h when h is false, then we suffer the full loss of not having the information contained in $\sim h$.

The foregoing is very plausible, but the question of whether it is adequate can only be resolved once we have produced some method for assigning content value. One way of doing this, again suggested by Levi, is to assign equal content values to each of the members of the ultimate partition. Where there are n members in the partition we might accomplish this by defining a regular measure function m on the members $k1$, $k2$, . . . , kn, as follows:

$$\text{m}(k_i b) = 1/n$$

and where d is a disjunction of $k_i s$ each appearing but once and having *r disjuncts*

$$\text{m}(d,b) = r/n.$$

On the basis of this measure function we define content as follows:

$$\text{cont}(h,b) = 1 - \text{m}(h,b).$$ [26]

[25] Cf. Levi, *Gambling with Truth,* p. 80.
[26] Cf. *ibid.,* pp. 69–70.

The rationale behind this assignment is a pragmatic one, to wit, that each of the members of the ultimate partition is equally a good answer to a problem that leads to the choice of the ultimate partition and hence each member of the ultimate partition is equally informative.

There are other ways of assigning content, but on the basis of this assignment we find some reason for rejecting the equation of utilities with content in the manner outlined. I have argued elsewhere that doing so leads to an insensitivity to the differences of probability between members of the ultimate partition and to a certain indifference to accepting hypotheses of high content. The latter can be simply illustrated. If one values high content over low, then the utility of accepting *h* when *h* is true should not be equated with the content of *h*, because that would not show a strong preference for high content over low content. For example, suppose that one has a partition of ten members. In that case, the content of a member of the partition is 9/10 while a disjunction of two different members has a content of 8/10, only slightly less. However, if one is aiming at accepting highly informative conclusions, then one should value accepting a single member twice as much as accepting such a two-membered disjunction, because the latter is only half as informative as the former. Similarly, one should value accepting a member three times as much as a three-membered disjunction of different members, four times as much as a four-membered disjunction, and so forth. In short, one should value the content of a member *n* times as much as a disjunction of *n* different members.

We may obtain such results by defining the utilities not only in terms of content, but also in terms of a number n_h equal to the number of different members of the ultimate partition occurring in a disjunction of members which is equivalent to *h*. If *h* is a member of the ultimate partition, then n_h equals 1. This proposal is expressed in the following equations:

$$U^*(h,b) = \text{cont}(h,b)/n_h.$$
$$u^*(h,b) = -\text{cont}(\sim h,b)/n_h.$$

From these equations we obtain the following equalities:

$$E(h,b) = P(h,b) - \text{cont}(\sim h,b))/n_h.$$
$$E(h,b) = P(h,b) - \text{cont}(\sim h,b)$$

if *h* is a member of the ultimate partition.

$$E(h,b) = (\sum_{i=1}^{j} E(k_i,b))/n_h$$

if h is equivalent to a disjunction of the set of members $k1$, $k2$, . . . , kj of the ultimate partition.[27]

Thus, we may now determine which hypotheses it is rational to accept by defining maximality. With respect to language L, we may, again following Levi, define two notions of maximality:

> $E(h,b)$ is maximal in L if and only if, for every s in L, $E(s,b)$ is no greater than $E(h,b)$.
>
> $E(h,b)$ is strongly maximal in L if and only if $E(h,b)$ is maximal, and for every s, $s \neq h$, such that $E(s,b)$ is maximal in L, cont(h,b) is less than cont(s,b).[28]

The notion of strong maximality is introduced to tell us what to accept in cases of ties. It tells us, among other things, that if the expected utility of each of a set of members of the ultimate partition is maximal, then the disjunction of those members is strongly maximal. We can now formulate a rule for maximizing expected utility having deductive closure as follows:

> (R) $A(h,b)$ if and only if either (i) $E(h,b)$ is strongly maximal or (ii) h is a deductive consequence of the set S which has as its only members b and k where $E(k,b)$ is strongly maximal.

Rule (R) formulated as a principle for maximizing utilities that aim at accepting true and informative statements is equivalent in the results it yields to the earlier principle (AR).[29] Thus (AR) is a principle of rationality for the purpose of obtaining the objectives of truth and content.

13. *Summary*. We began with the concept of a relevant deductive argument analyzed as an argument in which the denial of the conclusion together with the premises constitutes a minimally inconsistent set of statements. In terms of this concept we defined an acceptable argu-

[27] Keith Lehrer, "Induction: A Consistent Gamble," *Nous*, 3 (1969), 290–94.
[28] Levi, *Gambling with Truth*, p. 83.
[29] See T8, T9, and T10 in the appendix for proof of this result.

ment demonstrating the falsity of a hypothesis as a relevant deductive argument in which the conclusion is the contradictory of the hypothesis in question and the premises are each at least as probable as the hypothesis on background knowledge. An inductive rule was then proposed: a hypothesis may be inductively inferred from background knowledge if and only if there is no acceptable argument demonstrating the falsity of the hypothesis. It was shown that this inductive rule meets various conditions of rationality. First, the set of conclusions inductively inferred from background knowledge by the rule must be logically consistent with that knowledge. Second, any deductive consequence of a hypothesis inductively inferred by the rule could itself be inductively inferred by the rule from the same knowledge. Third, any application of the rule is such that there is a probabilistic reason for inferring what is inferred by the rule. Fourth, partitions may be so chosen that hypotheses inductively inferred by the rule will, if true, explain facts contained within background knowledge and others as well. Moreover, a hypothesis that is inductively (and not deductively) inferred from background knowledge is potentially explained by the probabilities. Finally, an acceptance rule based on the inductive rule yields the acceptance of hypotheses equivalent to adopting the policy of maximizing expected utility in the acceptance of hypotheses when what one values is truth and informational content.

The accceptance rule articulated and defended recommends a very adventurous policy of inductive inference. In effect, it tells us that whenever our background knowledge favors a strong hypothesis over others of equal strength, where strength is a measure of informational content, then we should inductively infer that hypothesis. It gives us this direction no matter how slightly the background knowledge favors the hypothesis in question. Some might consider such a policy too risky and advocate some more cautious policy instead. However, in my opinion, the history of science reveals that scientific inquiry has advanced more significantly through rational but adventurous attempts to obtain new information than through safe and cautious attempts to avoid error. It is for that reason that the acceptance rule is proposed as a principle of rational acceptance for the explication of scientific reasoning.[30]

[30] Work on the present article was supported by a National Science Foundation Grant. Other articles written under this grant on the subject of induction by Keith Lehrer include "Induction and Conceptual Change," *Synthese*, 23 (1971), 206–25,

Appendix

The purpose of this appendix is to prove various theorems assumed in the main text. The proofs offered are only intended to constitute a sketch of a completely formalized proof, but one from which the latter could be constructed quite mechanically. All theorems are proven for a finite language L which consists of an ultimate partition, that is, a set of statements $k1, k2, \ldots, kn$ that are disjoint in pairs and exhaustive as a set, and truth functional combinations of such statements. I assume the following calculus of probability for such languages, letting "$p(h,s)$" mean "the probability of h on s," on the condition that the statement to the right of the comma in the probability expression is logically consistent:

P1. $0 \leq p(h,b) \leq 1$.

P2. If b is logically equivalent to b' and h is logically equivalent to h', then $p(h,b) = p(h',b')$.

P3. $p(h,b) = 1$ if and only if h is a deductive consequence of b.

P4. If $\sim(h \,\&\, k)$ is a deductive consequence of b, then $p(h \vee k,b) = p(h,b) + p(k,b)$.

P5. $p(h \,\&\, k,b) = p(h,b) \times p(k,h \,\&\, b)$.

The following theorems are required to establish various results concerning the rules of induction and acceptance. Let $k1, k2, \ldots, kn$ be the n members of the ultimate partition P and let G be a set consisting of the members of P and disjunctions of such members.

T1. If k is a disjunction of $n-1$ different members of P, then k is a consequence of a set S of members of G only if k is a deductive consequence of some member of S. Proof: If k is a disjunction of $n-1$ different members of P, then there is only one member of P that is not a disjunct of k. Let h be that one member. Any disjunction of members of P not containing h as a disjunct will have k as a deductive consequence. Hence if a set of members of G has k as a deductive consequence but no single member of the set has k as a deductive consequence, then all members of the set must either contain h as a disjunct or else be h itself. But such a set of statements would all be true if h were true even though k would then be false. Therefore, h is not a deductive consequence of such a set.

and "Reasonable Acceptance and Explanatory Coherence: Wilfrid Sellars on Induction," *Nous*, 7 (1973), 81–103.

T2. If d is a disjunction of members of P, then any minimally inconsistent set M of which d is a member will be such that once d is deleted from M the remaining set R of statements will have the negations of each disjunct of d as deductive consequences. Proof: Suppose there was some denial of a disjunct of d that was not a deductive consequence of R. In that case, the members of R could all be true (all proper subsets of M are consistent) when d was also true because that disjunct is true whose negation is not a consequence of R. Hence the set M would not be inconsistent.

T3. For any k and b, if k is a member of P, and d is a disjunction of all the remaining members of P, then there is no member m of G of which $\sim k$ is a deductive consequence such that $p(m,b)$ is greater than $p(d,b)$. Proof: The disjunction d is logically equivalent to $\sim k$. Moreover, from P1, P3, and P4 of the calculus of probability it follows that for any statement s that is a deductive consequence of h, $p(s,b)$ is at least as great as $p(h,b)$. Therefore, if d is a deductive consequence of any member m of G, it follows from the preceding result and P2 in calculus of probability that $p(m,b)$ is not greater than $p(d,b)$.

T4. For any disjunction d of two or more members of P, and for any minimally inconsistent set S of statements whose members are also members of G and include d, and for any negation n of a disjunct of d, there is some member m of S such that $p(m,b)$ is no greater than $p(n,b)$. Proof: If d is a disjunction of two or more members of P and k is a disjunct of d, then $\sim k$ must be a deductive consequence of some member m of any minimally inconsistent set S of members of G that includes d by T1 and T2. From T3 it follows that for any such disjunct k and member m of such an S, $p(m,b)$ is not greater than $p(\sim k,b)$.

The next theorems concern the inductive rule (I) which is as follows:

I. $I(h,b)$ if and only if for any set S of statements of L, if h is a member of S and S is minimally inconsistent, there is a k, $k \neq h$, such that k is a member of S and $p(h,b)$ exceeds $p(k,b)$.

It is a theorem of combinatorial logic that every noncontradictory member of L is logically equivalent to some member of G. Since no contradictory statement of L belongs to any minimally inconsistent set of statements having more than one member; since logically equivalent statements have the same probability on the same statement by P2; and since any minimally inconsistent set from which one statement is deleted and a logically equivalent statement added will remain a minimally inconsistent set, the set of statements C such that h is a member of C if and only if $I(h,b)$ by (I) is logically equivalent to the set of statements C′ such that h is a member of C′ if and only if $I'(h,b)$ by (I′) as follows:

> (I′). $I'(h,b)$ if and only if for any set S of statements of G of L, if h is a member of S and S is minimally inconsistent, then there is a k, $k \neq h$, such that k is a member of S and $p(h,b)$ exceeds $p(k,b)$.

Hence, in the subsequent proofs concerning (I) we may in some proofs restrict our consideration to statements of G for the sake of simplicity without any significant lack of generality.

T5. If h is a member of P, then $I(h,b)$ if and only if $p(h,b)$ exceeds $p(\sim h,b)$. Consider a set S of members of G that includes h which is a member of P and is minimally inconsistent. S must have some member m of which $\sim h$ is a deductive consequence. This follows from T1 and the consideration that $\sim h$ is logically equivalent to a disjunction of the $n-1$ remaining members of P. Thus, from T3 it follows that $p(m,b)$ is not greater than $p(\sim h,b)$. Thus, if $p(h,b)$ exceeds $p(\sim h,b)$, then there will be some member m of S such that $p(h,b)$ exceeds $p(m,b)$. This shows that if $p(h,b)$ exceeds $p(\sim h,b)$, then $I(h,b)$. If $p(h,b)$ does not exceed $p(\sim h,b)$, then it is not the case that $I(h,b)$ simply because the set containing just h and $\sim h$ as members will be minimally inconsistent. This shows that $I(h,b)$ only if $p(h,b)$ exceeds $p(\sim h,b)$.

T6. If d is a disjunction of two or more members of P, then $I(d,b)$ if and only if there is a k which is a disjunct of d such that $p(d,b)$ exceeds $p(\sim k,b)$. Consider any set S of members of G that includes a disjunction d of two or more members of P and is minimally inconsistent. For any disjunct k of d, S contains some member m such $p(m,b)$ is not greater than $p(\sim k,b)$ by T4. Hence if k is a disjunct of d, and $p(d,b)$ exceeds $p(\sim k,b)$, there is a member m of S such that $p(d,b)$ exceeds $p(m,b)$. This shows that if d is a disjunction of two or more members of p and there is a k which is a disjunct of d such that $p(d,b)$ exceeds $p(\sim k,b)$, then $I(d,b)$. If there is no disjunct k of d such that $p(d,b)$ exceeds $p(\sim k,b)$, then the set of statements consisting of d and the negations of each disjunct of d will be a minimally inconsistent set such that for no member m of the set will it be the case that $p(d,b)$ exceeds $p(m,b)$. This shows that if d is a disjunction of two or more members of P and there is no k which is a disjunct of d such that $p(d,b)$ exceeds $p(\sim k,b)$, then it is not the case that $I(d,b)$.

T7. If $I(h,b)$ and k is a deductive consequence of h, then $I(k,b)$. Proof: Let h and k be members of G and such that k is a deductive consequence of h and $I(h,b)$. Either h is a member of P or it is not. Suppose h is a member of P. In that case, either k is logically equivalent to h or k is a disjunction of which h is a disjunct. If k is logically equivalent to h, then $I(k,b)$ by P2. If k is a disjunction of which h is a disjunct, $p(k,b)$ is at least as great as $p(h,b)$ by P4. But $p(h,b)$ is greater than $p(\sim h,b)$ by T5. Hence $p(k,b)$ is greater than $p(\sim h,b)$. Therefore, $I(k,b)$ by T6. This establishes the theorem on the supposition that h is a member of P. Suppose h is not a member of P. In this case, h is a disjunction of members of P. Either k is logically equivalent to h or k is a disjunction of members of P having as disjuncts at least all those members of P that are disjuncts of h. If k is logically equivalent to h, then $I(k,b)$ by P2. If k has as disjuncts all those members of P that are disjuncts of h, then $p(k,b)$ is at least as great as $p(h,b)$ by P4. Since $I(h,b)$ by T6 it follows that there is some member m of P which is a disjunct of h such that $p(h,b)$ is greater than $p(\sim m,b)$. But m is also a disjunct of k. Thus $p(k,b)$ is also greater than $p(\sim m,b)$. Therefore $I(k,b)$ by T6. This proves the theorem.

The next theorems concern the following two rules of acceptance:

AR. $A(h,b)$ if and only if either (i) $I(h,b)$ or (ii) there is some set of statements S of which h is a deductive consequence and such that any k is a member of S if and only if $I(k,b)$.

R. $A(h,b)$ if and only if either (i) $E(h,b)$ is strongly maximal or (ii) h is a deductive consequence of the set S which has as its only members b and k where $E(k,b)$ is strongly maximal.

T8. For any s in G, $A(s,b)$ by AR if and only if either (a) s is a member m of P such that for any other member k of P, $p(m,b)$ exceeds $p(k,b)$ or s is a deductive consequence of such a member m and b, or (b) s is a disjunction d of the members of a set S of members of P such that m is a member of S if and only if, for any member k of P, if $p(m,b) \neq p(k,b)$, then $p(m,b)$ exceeds $p(k,b)$ or s is a deductive consequence of such a disjunction d and b. Proof: Either there is some member m of P such that for every other member k of P, $p(m,b)$ exceeds $p(k,b)$ or there is not. Suppose there is such a member m. In that case, by condition (i) of AR and T6 it follows that for every other member k of P, $I(\sim k,b)$, and the set of such negations of other members of P has m as a deductive consequence. Hence by condition (ii) of AR, it follows that $A(m,b)$. Moreover, $I(b,b)$ from the fact that $p(b,b) = 1$ by P3, and thus there must be some k in any minimally inconsistent set to which b belongs such that $p(k,b)$ is less than 1. Hence by condition (i) of AR, $A(b,b)$. By condition (ii) of AR it follows that for any s that is a deductive consequence of m and b, $A(s,b)$. This proves that for any s that satisfies condition (a) of the theorem $A(s,b)$. Suppose that there is no member m of P such that for any other member k of P, $p(m,b)$ exceeds $p(k,b)$. In that case, there will be a disjunction d of members of a set S of members of P such that m is a member of S if and only if for any member k of P, if $p(m,b) \neq p(k,b)$, then $p(m,b)$ exceeds $p(k,b)$. For any member k of P that is not a disjunct of k (if there are any), $I(\sim k,b)$ by T6, and the set of such negations of members of P that are not disjuncts of d has d as a deductive consequence. (If the set is null, d would be a disjunction of all members of P and, hence, would be a deductive consequence of the set by natural deduction.) Hence $A(d,b)$ by condition (ii) of AR. Again, $A(b,b)$ by the argument given above. By condition (ii) of AR it follows that for any s that is a deductive consequence of d and b, $A(s,b)$. This proves that for any s that satisfies condition (b) of the theorem $A(s,b)$ by AR. This shows that if s satisfies either condition (a) or (b) of the theorem, then $A(s,b)$ by AR. We shall now prove that if $A(s,b)$ by AR, then s satisfies either condition (i) or (ii) of the theorem. Again, either there is some member m of P such that, for every other member k of P, $p(m,b)$ exceeds $p(k,b)$, or there is not. Suppose there is such a member. For any s in G, either s is a deductive consequence of m and b or it is not the case that $A(s,b)$ by AR. For consider the set G' of all those members of G that are not deductive consequences of m and b. Any g that is a member of G' is such that the set whose only members are m and g is a minimally inconsistent set, because m is inconsistent with every member of G' and no such member is itself inconsistent. But $p(m,b)$ exceeds $p(g,b)$ by T5, P1, and P4. Hence for any such

g, it is not the case that $I(g,b)$ and thus it cannot be shown that $A(g,b)$ by condition (i) of AR. Thus, only those members h of G which are deductive consequences of m and b are such that $I(h,b)$, and no set of statements all of which are deductive consequences of m and b has any deductive consequence that is not a deductive consequence of m and b by the transitivity of deduction. This proves that $A(s,b)$ by AR only if s satisfies condition (a) of the theorem on the supposition that there is some member m of P such that for any other member k of P, $p(m,b)$ exceeds $p(k,b)$. Suppose now that there is no such member m. In that case, as we have seen, there is a disjunction d of the members of a set S of members of P such that m is a member of S if and only if, for any member k of P, if $p(m,b) \neq p(k,b)$, then $p(m,b)$ exceeds $p(k,b)$. For any s in G, either s is a deductive consequence of d and b or it is not the case that $A(s,b)$ by AR. Assume s is not a deductive consequence of d and b. Then $p(s,b)$ is no greater than $1 - f$ where for any m that is a member of S, $p(m,b) = f$ by P1, P3, and P4. Thus, if s is a member of P, then by the supposition it is not the case that $I(s,b)$ and, if s is not a member of P, then it is a disjunction of members. Thus, if s is a disjunction of members of P, then for any member k that is a disjunct of s, $p(s,b)$ is not greater than $p(\sim k,b)$ because no member k of P is such that $p(\sim k,b)$ is less than f. This follows from P1, P3, and P4. Hence it is not the case that $I(s,b)$ by T6. Moreover, if s is not a disjunction of members of P, then it is a member of P, and, by supposition it is not the case that $I(s,b)$. Since a member s of G is either a disjunction of members of P or a member of P, it follows that any s that is a member of G but not a deductive consequence of d and b is such that it is not the case that $I(s,b)$. Hence for any such s, it cannot be shown that $A(s,b)$ by condition (i) of AR. Thus, only those members h of G which are deductive consequences of d and b are such that $I(h,b)$, and no set of statements all of which are deductive consequences of d and b has any deductive consequence that is not a deductive consequence of d and b. This proves that $A(s,b)$ by AR only if s satisfies condition (b) of the theorem on the supposition that there is no member m of P such that for any other member k of P, $p(m,b)$ exceeds $p(k,b)$. Since there either is or is not such a member, it follows that for any s in G, $A(s,b)$ by AR only if s satisfies either condition (a) or condition (b) of the theorem.

T9. For any s in G, $A(s,b)$ by R if and only if either (a) s is a member m of P such that for any other member k of P, $p(m,b)$ exceeds $p(k,b)$ or s is a deductive consequence of such a member m and b, or (b) s is a disjunction d of the members of a set S of all those members of P such that m is a member of S if and only if, for any member k of P, if $p(m,b) \neq p(k,b)$, then $p(m,b)$ exceeds $p(k,b)$ or s is a deductive consequence of such a disjunction d and b. Proof: By the definitions given in the text, for any member m of P, there is a fixed number q such that $E(m,b) = p(m,b) - q$. For any disjunction d of a set of n different members mi of P,

$$E(d,b) = \sum_{i=1}^{n} E(mi,b)/n.$$

It follows from these equalities by elementary algebra that if there is some member m of P such that for any other member k of P, $p(m,b)$ exceeds $p(k,b)$, then $E(m,b)$ is strongly maximal. If there is no such member, then it follows that for some disjunction d, $E(d,b)$ is strongly maximal, where the members of d belong to a set S of members of P such that m is a member of S if and only if, for any member k of P, if $p(m,b) \neq p(k,b)$, then $p(m,b)$ exceeds $p(k,b)$. Therefore, $A(s,b)$ by R if and only if s satisfies either condition (a) or (b) of the theorem.

T10. $A(s,b)$ by AR if and only if $A(s,b)$ by R for any s in G. Proof: Immediate consequence of T8 and T9.

GEORGE SCHLESINGER

Confirmation and Parsimony

1. Jeffreys' Problem

I shall attempt to deal with the most basic issues of confirmation, inquire what precisely the term 'induction' stands for, what the problem of induction amounts to and what, if anything, can be suggested in a way of solution.

Let us begin by picturing to ourselves Galileo rolling down round objects along an inclined plane in an effort to discover experimentally the law governing the relationship between the distance traveled by the rolling body, subject to the earth's gravity and the time taken to cover that distance. We shall imagine that at the end of seven experiments he has collected the following entirely unlikely perfect results:

t . . .	0	5	10	15	20	25	30
s . . .	0	5	20	45	80	125	180

It seems at first that from this table of results Galileo can at once read off the law correlating time and distance

(G) $\quad s = 1/5\, t^2$

i.e., that distance equals the square of time multiplied by a constant which, for the plane used with its particular inclination, equals 1/5. H. Jeffreys has, however, pointed out that there are infinitely many other equations which fit these seven results just as well as (G):

(J) $\quad s = 1/5t^2 + t(t-5)(t-10)\ \ldots\ (t-30)\, f(t).$

It should be noted that (J) is an infinite set of equations, since $f(t)$ may assume infinitely many different forms. Infinitely many members of the set denoted by (J) will give a different value, for corresponding values of s, to the next t to be tested, although for all the t's so far tested they gave values of s identical with those yielded by (G). But at the same time, (J) contains an infinite subset of equations which will yield the same value of s as does (G) for the next t to be tested. It is because of this last fact that the possibility of a crucial experiment to decide between (G) and any of the (J) equations is ruled out. For suppose we try to adjudicate between the (G) equation and any equation belonging to (J), by testing what will happen when $t = n$. Suppose the result satisfied (G). This, of course, eliminates an infinite number of (J)-type equations, but we would be still left with infinitely many equations competing with (G), namely, with all those (J) equations, for which

$$f(t) = (t - n)\, g(t)$$

where $g(t)$ may assume infinitely many forms.

Now what exactly does our problem amount to? It would be a total misrepresentation of the situation if we said that the difficulty raised was merely that the claims of science, including even those which concern quite elementary relations like Galileo's law of free fall, are not securely founded. The situation seems rather comparable to the one in which I buy one lottery ticket where there are many billions of tickets and only one of them will be drawn to win a prize. Given that each number has an equal chance to win, it would be wrong to say that it is merely uncertain that I shall win; the correct thing to say is that it is virtually certain that I am not going to win. Similarly, given Galileo's experimental results and the principle that the true hypothesis must satisfy them all, together with the fact that there are infinitely many hypothesis capable of doing this, implies not merely that we cannot have full confidence that (G) stands for the law governing the correlation of distance and time but rather that it is virtually certain that (G) does not truly represent this law.

It will, of course, be noted that of all the admissible equations (G) is the simplest and therefore the most convenient of the available equations. This, however, in the absence of further argument carries about as much weight in favor of adopting (G) as would, in the case of the aforementioned lottery, the argument that the most convenient thing for me would be for the ticket in my possession to win and therefore my confidence that it will win is, to some extent, justified.

One must refrain from using a number of faulty arguments which seem to present themselves and which one may be strongly tempted to employ in defending Galileo's choice of (G). Among these the most often heard is: but equation (G) works! We are much more interested in the practical aspects of science than in its capacity to produce ultimate truth. (G) has proven itself to work in the past; therefore we believe it will work in the future as well. For practical scientists this is what really matters.

This argument can be seen as faulty without our questioning the underlying assumption that whatever has worked in the past is going to work in the future as well. At this stage we shall ignore Hume's problem of how we know that the unobserved will be like the observed or the future will be like the past. We shall assume we know. The trouble with the argument is that it overlooks the fact that there are infinitely many other equations which too have worked in the past. The argument cannot be applied to all these since infinitely many of them yield incompatible prediction. What then is there in (G) which justifies our belief, that to it, rather than to its rivals, we should apply the argument that an equation which has worked in the past will work in the future as well?

Some may wish to argue that the credibility of (G) is ultimately based, not on the experimental results listed or on any further evidence of this kind, but rather on its derivability from higher order laws and its having become an integral part of a large interconnected system known as Newtonian mechanics.

This line of approach is wrong on a number of accounts. Nobody will deny that with (G) becoming integrated into Newtonian mechanics the evidence for its truth has undergone an important qualitative change. But to claim that, until this took place, the empirical support for (G) was nonexistent would be a misrepresentation of how scientists view the situation. Galileo is not criticized for having had confidence in (G); in fact, he is admired for his efforts to establish his law of free fall by the only method available at the time, which was to test s against the various values of t. Furthermore, if we disqualified this method we should never reach a situation in which we had any higher order laws from which to derive (G). Higher order laws with their unifying power are only discovered on the basis of our knowledge of at least some of the laws they imply.

There is, of course, a more immediate reason as well why this line of approach is useless: Jeffreys' difficulty, which he happened to illustrate by using Galileo's laws, applies with equal force to any hypothesis, including Newton's laws, and until his problem is solved we cannot be said to have any evidence for any type of law.

Another suggestion which might seem to have merit is that, in view of all the competing equations, we should leave no t untested and thus render the problem for practical purposes irrelevant. From a practical point of view, the range of t is limited. If t exceeds a few minutes we are no longer in the neighborhood of the earth's surface and (G) is not applicable. Between the limited range of zero and a few minutes, t assumes a finite number of different values since, for practical purposes, we are not interested in time intervals less than, say, one-tenth of a second. After we have determined the values of s for all the relevant values of t, the continued existence of infinitely many (J) equations no longer need to trouble us. All the (J) equations left in the field yield identical results for all the tested values of t and the fact that they give different results for all the untested values of t is of no consequence since in practice we are never going to encounter those.

What is wrong with this suggestion is not merely that it may not work everywhere since in general the range of values which are of interest to us is not as limited as in this particular case, but that it goes counter to one of the most central ideas of science. One of the main aims of the scientific enterprises is the construction of hypotheses on the basis of which we may make predictions concerning situations never encountered before. The reason why Galileo's equation is of scientific interest lies in its economy and fruitfulness, and in the fact that it is a very simple equation which has been established on the basis of a few experiments and is nevertheless capable of yielding the values of distances corresponding to a large number of yet untested values of time. Galileo regarded his law as well confirmed long before he had tested it for all the values of t which are of practical interest and, far from condemning this as a deficiency in his scientific judgment, it is taken as a sign of his true insight into the nature of scientific hypotheses.

2. The Problem of Additional Variables

One conspicuous difference, of which we have made no mention yet, exists between equation (G) on the one hand and all the (J) equations which remain unrefuted after a considerable number of experiments. This difference, it has struck many people, cannot be devoid of significance although it is not immediately clear what the significance is. On the one hand, it is only after we have performed all the experiments, the results of which we have tabulated, that we could put forward any particular (J) equation, while on the other hand, the (G) equation was advanced earlier on the basis of fewer experimental results than we have now at our disposal.

Suppose we have on our list the outcome of twenty experiments. It seems unquestionably true that (G) equation was advanced as an equation likely to represent the law we are after on the strength of the first five or six results, and it is quite immaterial which five or six of the twenty values of t we have tabulated had been experimented on first, for the same (G) equation would have been advanced. Not so with (J) equation. The twenty factors preceding $f(t)$ in the second term of the right-hand side of the equation would not have been spelled out by us (or at any rate, it is exceedingly unlikely that they would have been spelled out) before the particular values of t for which the distance fallen is going to be determined were known.

What exactly are we going to make of this acknowledged difference? Would it be reasonable to argue that we select (G) equation because it is much more strongly confirmed since it was advanced right after the fifth experiment and thus was confirmed by every subsequent experiment, the outcome of which could have falsified but did not? And we reject any particular (J) equation which we are setting up as a rival to (G) equation and which has been advanced only now and thus has not yet been exposed to any refutation yet?

I do not believe that many would find such an argument very convincing. It may well be that nobody was likely to have thought of the particular (J) equation we are now setting up as a rival to (G) equation before all the twenty values of t to be tested were known. This, however, does not amount to saying that the (J) equation in question has not in fact been tested just as severely as its rival (G) equation. An equation exists independently of whether anyone thinks of it and every experimental result which does not show that it cannot represent the law of free fall lends it exactly as much support as it does to (G) equation whether or not at the time of the experiment anyone had his mind fixed on that particular equation.

It is of course possible to point out that there is nevertheless a difference between (G) equation advanced fairly early in the course of the series of experiments performed and any particular (J) equation put forward merely at the very completion of the series. Both of course have survived the same number of tests. In the case of the former, however, active attempts were made by the experimenter to falsify it, while in the case of the latter no conscious effort was made to expose it to experimental refutation. If we agree that confirmation consists in putting a hypothesis to severe tests, with the honest intention to see whether it can survive it, then (G) equation has received confirmation whereas (J) equation has not. It is, however, hard to see why anyone should agree to such dogmatic and arbitrary viewpoints of what confirmation consists of.

Another way to try to exploit the difference here, noted between (G) equation and (J) equation, is to stipulate that no equation is likely truly to represent a law of nature if it is not discovered after m experiments where m is some agreed upon number. We therefore perform m experiments and advance the various equations which we think are suggested by the outcome of these experiments. Next we perform another experiment for an additional value of t selected at random. If it should turn out that (G) equation is the only equation compatible with all the $m + 1$ results, then, now that we know the $m + 1$ value of t, we are not permitted to set up the appropriate rival (J) equation containing the extra factor $(t - t_{m+1})$, since this would be contrary to our stipulation.

Once again it would not be easy to justify such a stipulation. But there is also a much more basic reason why no suggestion attempting to exploit the difference between (G) equation and (J) equation, namely that the former could be thought of much before all the experimental results were in, while the latter would be advanced only after all the values of t were known, can succeed. There is a whole group of (J) equations we have so far neglected to consider, each one of the infinitely many members of which can as easily be thought of and advanced as soon as (G) equation.

So far we have unquestioningly assumed that for any tested value of t the value of s obtained on one occasion will be the value obtained on all occasions. Jeffreys has taken for granted that the same value of s corresponds at all times to a given value of t and raised doubts only concerning values of s which correspond to yet untested values of t whether they will turn out as predicted on the basis of (G) equation. One may, however, quite legitimately raise doubts about whether future values of s corresponding to already tested values of t will turn out in accordance with (G) equation. And what is important, one may do so without questioning the assumption that the future resembles the past or that the unobserved will be like the observed. The experimental results we have collected so far are perfectly compatible with the supposition that the law of free fall is of the form:

(K) $s = 1/5\, t^2 + A \sin N!\pi/n$,

where N stands for some physical parameter which assumes integral values only and A and n are constants (the latter an integer). Then it is clear that as long as $N > n$, $N!\pi/n$ is an integer and $A \sin N!\pi/n$ equals zero. Thus the values so far obtained for s, instead of showing that there is no other physical parameter operative, should be taken as indicative of the fact that the third parameter had so far values exceeding n. (K) equation, for instance, may stand for the temperature at the center of the earth measured on a (quantum) scale admitting of integral values only. The temperature at the center of the earth is known constantly to be decreasing; let n be the value of the temperature to be reached in two weeks time at the center of the earth. Clearly (K) equation is not in the least more hard to think of as (G) equation, and every test for the last three centuries and for the next two weeks which supports (G) equation supports (K) equation to the same degree. Both have yielded so far and will yield for another two weeks that for $s = 5$, $t = 5$, but with respect to times beyond two weeks from now $t = 5$ will correspond to some other value of s according to (K) equation.[1]

It could of course be pointed out that (K) equation introduces a third variable into the picture and hence suggests that, as a matter of principle, we do not assume that an extra variable is operative until forced to do so. This suggestion would not be of much use until we justified the alleged principle. It would certainly be wrong to claim that there was anything in the evidence pointing to the absence of a third variable. All that one may say about the evidence is that it is not indicative at all that there is such a variable, but then it fails precisely to the same extent to give any indication that there is no extra variable involved.

[1] The problem raised here affects not only generalizations of a more sophisticated nature where, as in Galileo's law of free fall, an equation relating variables representing physical parameters expresses the law governing the covariation of these two physical parameters. Simpler generalizations like 'All ravens are black' are equally affected. The fact that hitherto all observed ravens have been black can be expressed by

No. of black ravens observed/No. of ravens observed $= 1$

and also

No. of black ravens observed/No. of ravens observed $= 1 - A \sin N!\pi/n$

the two yielding different predictions.

With the introduction of a third variable we may now give our additional reason for rejecting a suggestion made in the previous section for the solution of Jeffreys' problem. It was suggested that we could eliminate all the (J) equations which may from a practical point of view be regarded as rivals of (G) equation by performing enough experiments for every relevant value of t which are after all limited in number. Now, of course, the suggestion is at once seen as useless since we have raised doubts about the future values of s for already tested values of t.

This latter way of rendering our problem has the advantage of clearly bringing out the fact that the difficulty with empirical inference is completely all-pervasive, affecting the most familiar propositions of common sense no less than the sophisticated results of advanced science. It is easily seen that the assumption that past regularities will continue in the future or that the unobserved resembles the observed does not, in the least, imply even that just because the sun has risen every day in the past it is going to rise tomorrow or that because unsupported bodies near the surface of the earth have always fallen to the ground they will continue to do so in the future. In prerelativistic physics, it was asserted that

$$(\text{L}) \qquad \text{mass} = \text{constant},$$

but now our attention has been drawn to the fact that all observations throughout history were compatible with the rival hypothesis that

$$(\text{M}) \qquad \text{mass} = \text{constant} - A \sin N!\pi/n,$$

where N stands for a physical parameter, the value of which drops below n tonight for the first time in history. We have no arguments to support the claim that (L), rather than (M), describes truly the nature of mass. If, however, (M) is the true hypothesis, then overnight masses will assume large negative values, setting up repulsive forces between the earth and the sun, in consequence of which the sun will not rise tomorrow and unsupported bodies will cease to fall to the ground.

I believe that, in order to make any progress toward a solution, it is essential to realize that the notion of doubt with respect to scientific method is very different from the notion of doubt as it applies to some particular claim about nature once we have accepted the validity of a particular method of empirical inference.

Normally, when we entertain serious doubts about the validity of an equation purported to represent a law of nature, the reasonable thing to do is to suspend judgment until more evidence is forthcoming and, in the meantime, refrain from employing or relying on the equation in question. Suppose the questionable equation has been used in the construction of a new type of aircraft; a rational person will avoid traveling on it because of its unproven safety. Sensible people refrain from taking serious risks and, whenever the safety of a given line of action depends on the validity of a putative law of nature, they will not commit themselves to that line of action until the law in question is confirmed beyond reasonable doubt.

Suppose no one finds any solution to the difficulties we have been discussing. How is this to affect the practical conduct of our daily lives? Given, for example, that in the construction of airplanes the validity of all sorts of laws has been assumed without there being any way of justifying why they have been assumed to be true rather than any one of their many rivals, is it advisable to avoid flying by plane? The answer is that there cannot be any point in avoiding air travel when one is just as much exposed to dangers when staying on the ground. We are, after all, unable to justify our belief that the ground beneath our feet will not melt, evaporate, or explode; ought we not therefore get off it and seek the relative safety of the air? It should be clear that since the problem we have been discussing is an all-pervasive one, affecting to the same degree every empirical proposition, there is no escape from it. Avoiding boarding airplanes, or not venturing outside our homes, or even lying immobile in bed will not shield us from the dangers of our environment; as long as we cannot opt out of this world we remain exposed to these dangers. Complete agnosticism with respect to scientific method is an impossibility, for being agnostic with respect to any given hypothesis implies our refusal to commit ourselves to it. Whatever we do or refrain from doing, we commit ourselves to a large number of empirical hypotheses.

The last point has the important consequence that we cannot be called reckless or said to be taking risks when adopting any particular method of hypothesis selection, since there is no alternative we may regard as safer. The labeling of a given action as hazardous carries with it a disapprobation and implicit opinion that a different, less risky action should have been substituted in its place. Similarly, our expression of doubt implies the advocacy of the withholding of judgment and when the efficacy of a given line of action is doubted it is implied that an alternative, more secure line should be taken instead—which again cannot be held with respect to a line of action taken on the basis of any principle of hypothesis selection. Indeed, concepts such as hazard and uncertainty as well as safety, reliability and credibility are not defined until a given scientific method is chosen and in relation to it. To rely on a given hypothesis may be said to be hazardous if and only if it is done not in accordance with the requirements of a scientific method we subscribe to. When, for instance, it is said of an aircraft that it is highly reliable and that it is extremely improbable that it should fail to land safely after take-off, what is meant is that the laws of nature which have been employed in the construction of the vehicle have been adopted as confirmed in accordance with the selection principles laid down by the method universally prevailing in the scientific community and have gone through all the tests to the degree of rigor required by it. On this construal of the notion of safety, reliability, and probability, one just cannot speak of the safety, of the reliability of the selection principles themselves, and of the probability that they will lead to success.

But it by no means follows from this that there is no need or possibility to defend scientific method. Admittedly, now we claim that not only cannot one show that the prevailing scientific method is more assured of success than others, but that the whole notion of likelihood to succeed is inapplicable to scientific method as such. There are, however, other desiderata scientific method has to fulfill. One reasonable demand is that our method should be such that it makes maximum use of past experience in recommending the hypothesis to be selected. In what follows I shall give an outline of a defense, according to which the prevailing method is not the best method but the *only* method which is universally applicable in selecting hypotheses on the basis of past experience.

3. The Only Usable Methodological Principle

It will be recognized at once that a methodological rule requiring that, in any given situation, we choose the least simple of the admissible hypotheses would render us completely paralyzed. Given any hypothesis, it is easy to construct one which is more complex than it and therefore, unlike the most simple, the most complex hypothesis is an undetermined and undeterminable proposition. Thus, if the problem facing us was confined to the question, should we adopt as our universal rule the principle of simplicity or the principle of maximum complexity, we would reject the latter not merely by saying that it is an inferior principle to the principle of simplicity. We would reject it because it is not a principle at all; the rule requiring us to accept the most complex of all the available hypotheses does not really require us to accept anything and is therefore not a rule in the minimal sense of that term.

The next step in my argument is to show that if we were to adopt a rule always to choose the *second* simplest of all the admissible hypotheses, or for that matter the *third, fourth,* or anything but the simplest hypothesis, we would not be provided with any means by which to decide what hypothesis to choose and would, therefore, be left without a rule.

I shall illustrate this claim with an example, which has a rather restricted applicability but which has the advantage of reflecting three important aspects of methodology.

Suppose three nonlinear points are given, lying on the path of a planet known to move along a closed orbit. The simplest of all available hypotheses under these circumstances would appear to be that the planet moves along a circular path. It will be agreed by most that, in the absence of anything but geometrical considerations, the hypothesis with simplicity of rank two is that the orbit of the planet is elliptical.

There are, however, infinitely many ellipses which pass through the three points. Thus if we were to adopt the rule that, in any situation, one is to choose the hypothesis of rank-two simplicity we would not be provided with a definite instruction regarding what specific orbit to postulate. On the other hand, the rule to choose the hypothesis of first-rank simplicity gives us the unique circle determined by the three points.

It is of great interest to note a second feature of the principle to maximize the simplicity which is that the situation remains unchanged with respect to the basic difference between it and the alternative methodological principles we are comparing it with if we wait until more points along the path of the planet are given. Suppose at some later time five points lying on the path of the planet become known and that no circle but an ellipse passes through these points. Under these new circumstances, the hypothesis which postulated an elliptical orbit moves up to rank-one simplicity. But, under these circumstances, it is also our good fortune that no more than one ellipse passes through the points given, since through five points either no ellipse or, at least, one ellipse may be drawn.

This then is a small illustration of what I suggest may be universally the case: While other principles of selection lead us to a whole set of hypotheses of certain kind, the principle to pick the simplest of all hypotheses generally leads us to pick a unique hypothesis. And although what is the simplest hypothesis changes with changing circumstances, it is an invariant aspect of our principle that it is usable.

Before leaving this example, let me raise a very crucial third point which may be made with its aid. On the surface, it may seem that the following question could be posed: even though our data consist of no more than three nonlinear points on the path of the planet under investigation, any ellipse we may draw through these points is just as concrete, specific, and uniquely different from any other curve as is the circle determined by the three points. What then is the basis for saying that this unique curve, which is perfectly compatible with our data, is less probable to represent the orbit of our planet than the circle?

The answer to this question lies in the realization of the essential idea that we are not considering the prior probability of any putative hypothesis. We are approaching the matter from the point of view of the *rules* according to which a hypothesis is adopted. Any curve which passes through our three points is unique, but it is only in the case of the circle that we can name the universal methodological principle which has directed us specifically to postulate it, rather than any other curve, as the shape of the orbit of our planet.

Ultimately, then, our adherence to the rule to maximize simplicity is seen to be caused by the indeterminate nature of all alternative rules. All ellipses which pass through the three points are symmetrical with respect to one another. Why pick this ellipse rather than that? The principle to choose the hypothesis of simplicity rank two seems to operate symmetrically with respect to all the infinitely many ellipses. No other principle which singles out this ellipse and is also universally applicable seems to present itself which could provide grounds for our choice. We can, however, explain why we pick the circle we do pick; sufficient reason for our choice can be provided: It is uniquely given by the rule which instructs us to choose the hypothesis compatible with the data and is the simplest of those which satisfy them.

Now we may return to our previous examples. Until the beginning of the twentieth century, our experiences allowed us to describe mass by

(L) mass = constant

or by

(N) mass = constant + $F(x)$

where $F(x)$ reduces to zero for all past observations. The rule to maximize parsimony bids us to select (L). I should point out that we are not ignoring the fact that there are difficulties in exhaustively characterizing the notion of simplicity. However, no matter how intricate these difficulties may be in general, in the present context the selection of the most parsimonious of all the available equations is a straightforward matter. It is our good fortune that, in all the situations associated with the universal problem we are considering, namely, where in the case of every hypothesis compatible with the data which is advanced, infinitely many others are automatically generated, it is immediately obvious which is the most parsimonious of all these. That equation, the terms of which on the right-hand side form a proper subset of all the terms on the right-hand sides of the other equations, is the most parsimonious of all.

Which of all the equations is of parsimony rank two? Let $F^*(x)$ be the simplest $F(x)$; then it is clear that

(N*) mass = constant + $F^*(x)$

is the equation of parsimony rank two. Suppose we adopted the rule to choose the hypothesis of rank two, then, of course, we would have to face the difficulty of how to identify $F^*(x)$ since we have no generally applicable criterion for the ordering of mathematical expressions with respect to their simplicity. This, however, may not be an insurmountable difficulty since there is no reason to believe that such a criterion may, in principle, never be constructed. The real difficulty is the x may stand for indefinitely many physical parameters and thus (N*) represents not a single equation but an equation form. Here again, then, in this much more general situation we find exactly what we found in our more specific example, that is, if we adopt the principle to maximize parsimony, we are led to the choice of a specific hypothesis, namely, that (L) describes properly the nature of mass. If we were to adopt, however, a rule to choose the hypothesis of parsimony rank two, we would be presented with an indefinitely large set to choose from without being provided with sufficient reason why we should prefer a specific equation of type (N*) over another.

The second point made earlier, concerning what happens when increased knowledge replaces one hypothesis by another, as being the most parsimonious in the context of the newly arisen situation is well illustrated here too. With the acquisition of new knowledge that led to the adoption of special relativity, the most parsimonious expression describing the behavior of mass became

(R) $\text{mass} = \text{constant} + F(v)$

where v is the velocity of the body whose mass is measured relative to the system in which it is measured.

Once again, there are infinitely many equations which are equivalent to (R) with respect to all past observations, namely

(S) $\text{mass} = \text{constant} + F(v) + G(x)$

where $G(x)$ reduces to zero for all those observations. Suppose $G^*(x)$ is the simplest form of $G(x)$, then

(S*) $\text{mass} = \text{constant} + F(v) + G^*(x)$

is the equation of simplicity rank two describing mass in a manner compatible with contemporary knowledge. But, again, although (R) is a single equation (S*) represents an indefinite number of equations, since x may stand for all sorts of physical parameters. Thus the principle of parsimony yields a unique hypothesis, while the principle to select a hypothesis with any degree of parsimony other than the first fails to yield a unique hypothesis. This, of course, is not too surprising. The crucial advantage of the principle of parsimony over all other principles described here is a necessary consequence of the very notion of parsimony. The most parsimonious of all the equations which will fit all that has been observed is the one which is free of all extraneous[2] elements.

[2] It should be clear that the extraneous element is not necessarily, as in the case before us, a term; it may also be a factor or an index. For example the equation

$$\text{mass} = \text{constant} \times \cos T!\pi/n$$

is, as long as $T > n$, equivalent to

$$\text{mass} = \text{constant}$$

and is a less parsimonious rival to the latter, containing the extraneous element $\cos T!\pi/n$.

It is important to realize that each one of the infinitely many less parsimonious rival hypotheses h' of the most parsimonious hypothesis h is parasitic upon the latter. Hypothesis h' is formed from h by adding to it an extra element such that

There is only one way of being free of all extraneous elements; there are infinitely many ways of having such an element—by having *any* one term which is extraneous.

For a moment it might seem perhaps that we could adopt some other principles as well which would always yield a unique hypothesis. It might be suggested in a fanciful way, for instance, that in order to overcome the problem of not being provided with a specific hypothesis whenever we required one, we should appoint a certain person as our "hypothesis chooser" whose task it would be to give clear instructions in every instance what hypothesis we should subscribe to. Alternatively, it might be suggested that whenever presented with a set of hypotheses all compatible with past observations, we should choose the one we think of first.

Now it is unquestionably true that there are uncountably many rules one could adopt to pick out always a particular hypothesis compatible with past observations and which would yield concrete predictions. We could agree, for example, on the rule that from all the competing hypotheses we should choose the one to be adopted by drawing lots or the United Nations should appoint a "universal hypothesis chooser," whose task it would be to pick one hypothesis out of all those which have so far accounted for all the events of the past, for the whole world to adopt. Thus, it would not seem impossible to adopt the convention that the phrase "the future will resemble the past" should be taken to mean that the future will resemble the past described by the statement picked out by any one of the aforementioned rules.

Momentarily it may seem, therefore, that our original problem has shifted to the next level: we now appear to require a meta-rule to help

both h and h' are equally confirmed by any member of a set of evidence e, namely, the evidence so far observed, but there is a set of evidence e' any member of which if it confirms h does not confirm h' and if it confirms h' does not confirm h. Of course, h may sometimes have genuine rivals which are not parasitic upon it, in the way Copernican celestial machanics rivaled Ptolemy's or Young's theory of light rivaled Newton's. Such hypotheses require, however, the services of a creative scientist; they are not mass produced. Such rival hypotheses do not present an important philosophical problem since, because there are never more than a few of them, they are eliminated by crucial evidence which supports one hypothesis and not the other. Of course, there is always specific evidence which will adjudicate between any h' and h and hence, if h is the correct hypothesis, h' will drop out of the competition because of that evidence. But since there are infinitely many h''s, no matter how many are eliminated through the use of crucial evidence, there are still infinitely many left to compete with h.

us to decide which first-order rule for selection of hypotheses to adopt. But it will easily be recognized that the first-order rule which has universally been adopted is different from all other possible rules. For, it is true that no matter which convention we adopted the future would be assumed to resemble the past according to some description of the past. There would be nothing in the past events nor in the types of regularities exhibited by these, which, *in themselves*, determined which particular description we should choose for projecting into the future. Only according to the prevailing convention is it the case that the past is assumed to determine the future in a *strong sense*. The future behavior of masses is predicted on the basis of their past behavior without appeal to anything extraneous (e.g., the drawing of lots or a special person). It entirely depends on the nature of past events what regularities these exhibit and which statement corresponds to the most parsimonious true description of these.[3]

4. A Defense of the Induction Principle

Now we are in the position to say something about the problem of induction in general. What is the problem of induction? There are not many people who, in the twentieth century, have restated the problem of induction more lucidly than Bertrand Russell in a chapter called

[3] Suppose now it is objected: admittedly the selection principle according to which we are to adopt the most parsimonious of the available hypotheses is indeed different from all other such principles inasmuch as it alone relies on nothing extraneous to the listed results of our observations. But the selection principle according to which we are to subscribe to the hypothesis provided for us by the "universal hypothesis chooser" appointed by the U.N. is unique too inasmuch that it alone ensures that we always adopt the hypothesis selected by the special person assigned to the task. Thus what makes us adopt the principle according to which the future is determined by the past in a strong sense, which is unique in one way, rather than another principle such as the one just mentioned which is unique in another way?

The answer to this question is similar to the answer to the question Why, if we do employ the principle to assume that the future is like the past in the strong sense, should we select the most parsimonious rather than the second or third most parsimonious hypothesis, since it is only through the practice we have adopted that we end up with something determinate? If we were permitted to use anything extraneous to our observations, such as a universal hypothesis chooser, then indefinitely many candidates present themselves and there is no way to select among the indefinitely many diverse selection principles which employ these. Only if we adopt the rule not to admit any extraneous elements do we have a unique principle of hypothesis selection which also selects a unique hypothesis, namely the most parsimonious hypothesis.

"On Induction" in his book *The Problem of Philosophy*. He begins his exposition with an example:

Let us take as illustration a matter about which none of us, in fact, feels the slightest doubt. We are all convinced that the sun will rise tomorrow. . . . It is obvious that if we are asked why we believe that the sun will rise tomorrow, we shall naturally answer 'Because it always has risen every day.' We have a firm belief that it will rise in the future because it has risen in the past. If we are challenged as to why we believe that it will continue to rise heretofore, we may appeal to the laws of motion: the earth, we shall say, is a freely rotating body, and such bodies do not cease to rotate unless something interferes from outside, and there is nothing outside to interfere with the earth between now and tomorrow. Of course, it might be doubted whether we are quite certain that there is nothing outside to interfere, but this is not the interesting doubt. The interesting doubt is as to whether the laws of motion will remain in operation until tomorrow (pp. 60–61).

He goes on to explain why this doubt cannot be resolved through any argument based on our knowledge of the past:

It has been argued that we have reason to know that the future will resemble the past, because what was the future has constantly become the past and has always been found to resemble the past, so that we really have experience of the future, namely, of times which were formerly future, which we may call past futures. But such an argument really begs the very question at issue. We have experience of past futures, but not of future futures, and the question is will future futures resemble past futures? (Pp. 64–65.)

It seems to be implied that if the question, "Will future futures resemble past futures?" could be answered in the affirmative, then we could rest assured about tomorrow's sunrise; our only cause for worry is our inability to provide an affirmative answer. Russell appears quite clearly to be maintaining that if it were given that there has been no outside interference to stop the rotation of the earth and yet the sun failed to rise, this would constitute an example that nature cannot be relied upon to remain always unchanged and future events do not necessarily continue to obey the past regularities. But this, as we have seen in section 2 of this paper, is not so. The sun may fail to rise tomorrow because masses have always obeyed and continue to obey equation (M) rather than equation (L). Clearly then the inductive principle does not merely bid us to assume that the future will be like the past, but that the future will be like the past in accordance with the most parsimonious description thereof.

In the light of what we have said in the previous section, however, the Russellian way of describing induction does not necessarily appear faulty. It need not be objectionable to express oneself in the manner of Russell as long as it may be assumed as understood that 'the future will be like the past' is merely brief for 'like the past described in a particular fashion' everyone knowing in which particular fashion, namely, in accordance with the most parsimonious description. The reason why it can be assumed that this is what is meant is that there does not seem to exist any other way of interpreting the principle. As we have seen in the previous section, any other interpretation does not really allow us to make predictions on the assumption of the strict similarity of past and future alone (since one is required to employ extraneous factors beside one's knowledge of past regularities exhibited by the event type we are concerned with) or does not allow us to make any specific predictions at all because it renders one's principle indeterminate.

Now we are in the position to see the outlines of a possible defense of induction in general. This defense presupposes that all agree that no guarantee whatever can be given that any method of predicting the future yields correct results. On the other hand, it is also understood that we cannot refrain from acting altogether and therefore we must subscribe to some method of hypothesis selection. In view of this it may be sufficient to show that the method adopted is superior to any of its alternatives.

Suppose we agree that it is an essential desideratum that all our hypotheses about the unobserved should be constructed on the basis of what is equally accessible to all people. We shall take it for granted that the raw material upon which our principle of empirical reasoning is to operate is the observed facts of the kind about which we are to hypothesize.

It would seem then that the method which arrives at conjectures based on past observations through the principle to assume the resemblance of the future and the past has an edge over all alternative 'methods'; it is the only method capable of being applied to experience.

Suppose it were recommended that we should employ the principle — which sometimes has been called 'counterinductive' — always to assume that the future will be unlike the past. Does the principle bid us to assume that the future will not be like the past as it is described by inductivists? Such assumption leads us nowhere, since it does not yield determinate predictions; there are infinitely many ways in which the future may differ from the past. Thus, this version of counterinduction offers no methodology at all. But we would not be better off if we were to assume that the future will not be like the past, as described in any other particular way.

Is it perhaps feasible to employ the principle to assume that the unobserved will not be like the past as described in any but one particular way? Which particular way are we to take as being singled out by this principle? Surely not the most complex one, since, as we have already said, that simply does not exist. But descriptions of parsimony of rank two as well as of rank three and, in fact, of any other rank except one exist in infinitely large numbers, and we would be provided with no means of choosing a particular hypothesis among all the hypotheses of equal rank. Only the description which contains no redundant elements is a unique description of the observed in the sense that every other description differs from it in the rank of the parsimony it possesses. But, of course, if the principle to assume that the future will not be like the past is to be interpreted to mean that it will not be like the past as it is described in any but the most parsimonious way, then the principle is equivalent to the inductive principle. The latter, therefore, appears to be the only principle which leads to the choice of a determinate hypothesis among the infinitely many hypotheses which are automatically generated by the addition of terms extraneous with respect to the past and provides us with a methodology of coping with the future with the aid of a rule applicable to past observations.[4]

[4] It is being assumed throughout that the mathematics we work with and the language in which we formulate our statements are the ones in common use. We do not, for example, admit artificially constructed predicates such as grue and bleen. Apart from all the obvious objections to such predicates from considerations of simplicity and teachability, there is the objection that there are infinitely many possible artificially created words and there are infinitely many diverse predictions we are led to make through the use of the various artificial predicates. But there is no criterion to guide us how to choose among these predicates; thus there is nothing to tell us which of the contrary predictions to accept. If we stick, however, to natural languages, we have no problem since their use yields a unique prediction in all cases. Thus the principle to use natural language and it alone produces determinate predictions.

PAUL TELLER

Comments on "Confirmation and Parsimony"

Professor Schlesinger presents us with an argument which he hopes will justify use of his methodological principle, that of picking among competing hypotheses the simplest one which is not contradicted by the evidence. The argument seems to take several different forms none of which I find convincing. Let me try to explain briefly what seems to go wrong.

Professor Schlesinger remarks (p. 331) that, at least when our behavior is concerned, we cannot remain agnostic with respect to many propositions about the unobserved. And (though this is not explicitly said) if we are not to form our beliefs about the unobserved arbitrarily or randomly, we presumably must form them by rules, or at least in a rule-governed way which admits of description by lawlike regularities. If such a rule or regularity picks out general hypotheses that cover observed cases as well as unobserved, clearly it must be required that the rule not pick out any hypothesis which actually conflicts with what has been observed. Professor Schlesinger's argument now seems to proceed like this. There is only one rule which meets these conditions, that of picking the simplest hypothesis compatible with the evidence. Thus, if we are to use any rule at all, it must be this one.

That only the "principle of simplicity" can pick determinate general hypotheses is initially argued by example: Suppose we know that something moves on a closed orbit, and we know only three (nonlinear) points of the orbit. Professor Schlesinger claims that only the rule of picking the simplest curve satisfying these conditions uniquely determines a specific hypothesis describing the way the curve might look (p. 334). It is then suggested that in general only "the principle to pick the simplest of all hypotheses generally leads us to pick a unique hypothesis" (p. 335).

The initial example is disquieting, because it suggests that Professor Schlesinger means to require of a methodological rule that, given the data, it literally *determine* the favored hypothesis; that, given the data and the rule, one should be able to determine the favored hypothesis mechanically. But no methodology, much less the suggested rule of simplicity, will enable an investigator to crank out from the data Newton's mechanics or Einstein's general theory of relativity. We may hope that Professor Schlesinger had some weaker interpretation of "determine" in mind. He might have wished to say that the principle together with the data must make a unique selection from among all actually proposed hypotheses.

Much more to the point, the central claim, that only the rule of simplicity picks determinate hypotheses, is plainly false. Professor Schlesinger himself recognizes this (p. 338), observing that the rules directing us to choose by lots or to choose the hypothesis thought of first would both have the property claimed for the rule of simplicity, of yielding a determinate selection. Let me add to the counterexamples one which may help to make later points more clear. My proposed alternative to the rule of simplicity is the lexicographical rule: pick the hypothesis which occurs lexicographically first among those compatible with the data. The lexicographical ordering of hypotheses is determined by rewriting all mathematical expressions occurring in hypotheses as their verbal equivalents and then ordering as one does in a dictionary. If one is of a mind to identify certain syntactically distinct forms as representing the same hypothesis, a hypothesis' place in the ordering is determined by the first occurrence of one of its representations.

Armed with this further example, let us see what Professor Schlesinger says about the multiplicity of uniquely selecting rules. He claims that we have reason for preferring the rule of simplicity because any alternative rule will rely on something "extraneous" to the observations in yielding predictions about the unobserved (p. 339). In such cases, he seems to believe, past observations would not "*in themselves*" determine "which particular description we should choose for projecting into the future" (p. 339). These remarks leave me puzzled on many accounts. For example, what is to count as genuine evidence and what as extraneous to the evidence? Why should not the outcome of drawing lots or the word of a specially designated person (Professor Schlesinger's examples of extraneous information (p. 339)) simply count as part of the evidence to be used in evaluating hypotheses? I am also wholly unclear about what is meant by saying that past observations *in themselves* do or do not determine choice of hypotheses. Consequently, I am at a loss over what the connection is supposed to be between such determination and the presence or absence of things "extraneous." Perhaps most troubling is that we are given no reason to opt for a rule which makes no use of extraneous information and which enables past observations in themselves to determine selection of hypotheses. Hence, even if something were to be said about the unclear terms, we would still have been given no reason to use the rule of simplicity rather than some competing rule.

It seems to me unlikely that these difficulties can be surmounted. No appeal to what is and is not extraneous and what it is for data to function in itself in confirmation, however these terms are plausibly understood, can justify selection of Professor Schlesinger's rule of simplicity over all others. To be convinced of this, compare the rule of simplicity with the lexicographical rule. In any plausible sense of the term, the one rule is as free from reliance on "extraneous" information as the other. Both rules rely only on the evidence (however "evidence" is to be analyzed) and on a certain ordering of hypotheses. These orderings seem, from a preanalytic point of view, to be entirely comparable in every general relevant respect which we know how to describe. Both are relative to the same language. Both are completely determined by formal characteristics of the syntactic objects which represent the hypotheses in the language. (This last claim can, of course, be known to hold for the ordering according to simplicity only when we know what that ordering is. But Professor Schlesinger's examples indicate quite clearly that he has such a formally described ordering in mind.) Similarly, what ever plausible (and nonquestion-begging) sense is given to the expression "determine in themselves," both rules seem equally to allow observations in themselves to determine selection of hypotheses. The two rules simply do not seem to be differentiated by these two characteristics. Hence, even if these characteristics could be argued to be relevant, it does not seem that they could provide a basis for preferring the rule of simplicity to the lexicographical rule.

In footnote 3 Professor Schlesinger takes up the objection that no reason has been given to prefer the method distinguished by freedom from "extraneous" elements. He replies by saying that unless we settle on the principle of simplicity we are faced with indefinitely many principles to choose from and no way to choose among them.

This is (as Professor Schlesinger recognized) just a meta-level version of the original argument for the principle of simplicity; and reapplication of the argument at a meta-level fares as poorly as did the original application. The point is, very simply, that there are a great many meta-rules of methodology selection which one could adopt. Some of these will select the principle of simplicity; many others will select other principles. For example, the meta-rule of selecting the first counterexample to Professor Schlesinger's thesis which came to Paul Teller's mind yields the lexicographical rule. We have been given no reason for selecting one meta-rule rather than another (or, for that matter, a reason for not selecting a determinate methodology arbitrarily), and it is hard to see how, a priori, we could be given any such reason. Consequently, despite Professor Schlesinger's forthright efforts, I do not see that we have been given any reason in support of his methodology of simplicity.

GEORGE SCHLESINGER

Rejoinder to Professor Teller

At the outset of his comments Professor Teller expresses disquiet at what seems to him my claiming to have a rule which is capable of automatically generating the hypothesis to be adopted when given nothing but the experimental data. Such a claim would indeed be quite preposterous since everybody knows that there are no clear-cut principles for the production of hypotheses, but that these are suggested by the imaginative insights of a high talent which conjectures without the aid of any rules of scientific discovery. Subsequently, he expresses the hope that I may not have had such a claim in mind.

I have, of course, explicitly explained at considerable length that I wish to make a sharply different claim. I have been dealing with a situation where a hypothesis to account for all the observations had already been suggested by a scientist. But as soon as this is done, one can mechanically generate out of the suggested hypothesis infinitely many parasitic rival hypotheses by the simple method I have described. The principle of maximum parsimony determines which one to adopt among these hypotheses.

But let me come to the truly surprising part of Professor Teller's paper, which is his expression of puzzlement by what is meant by extraneous evidence. He says: "[W]hat is to count as genuine evidence and what extraneous to the evidence? Why should not the outcome of drawing lots or the word of a specially designated person (Professor Schlesinger's examples of extraneous information) simply count as part of the evidence to be used in evaluating hypotheses?" (p. 344) Although I find it almost impossible to believe that Professor Teller should not himself be able to answer this question, let me briefly reply: When we are, for example, searching for a hypothesis correlating the time traveled by a particle and the distance covered during that time, then the fact that in three seconds the particle traveled four feet I take as genuine evidence supporting the hypothesis which shall be admitted as a candidate for adoption, because such a hypothesis logically implies that such a fact obtains. On the other hand, the word of a specially designated person is not logically implied by that hypothesis. The same goes for the results of drawing lots.

He then continues by saying, "I am also wholly unclear about what is meant by saying that past observations *in themselves* do or do not determine choice of hypotheses. Consequently, I am at a loss over what the connection is supposed to be between such determination and the presence or absence of things 'extraneous'" (p. 345). Once more the situation does not seem to call for elaborate explanations. By past observations I mean, of course, past observations of what is genuine evidence. When we have used nothing beside genuine evidence but the fact that it is genuine evidence (which is determined logically by seeing that the hypothesis concerning the law governing the phenomena under investigation logically implies it) and the principle of maximum simplicity (which is once more a purely logical operation) in choosing our hypothesis, then the genuine evidence alone, unaided by any extraneous fact, determines which hypothesis we adopt.

Why is such a practice to be preferred? I think a brief answer is simply this: if there is merit in basing hypotheses on evidence, then the more we base them on the evidence the better. The best thing is to base a hypothesis maximally on the evidence, that is, base the hypothesis on the evidence and nothing else.

Lastly, I should like to make a point or two about Professor Teller's lexicographical example. First, it is to be noted that the construction of the lexicon involves insurmountable difficulties. Given any hypothesis, there are infinitely many ways in which it can be worded. But perhaps what Teller had in mind is that we restrict the number of letters to be used. In this case, though there is a vast number, there is still only a finite number of formulations for each hypothesis. But there are infinitely many hypotheses. We cannot begin ordering these hypotheses into our dictionary until we have formulated them all!

Thus we cannot have the required dictionary. But we shall ignore this matter. Suppose there were only a finite number of hypotheses and that these were listed in some order. Using this lexicon, we would be resorting to the extraneous empirical information in what of the many possible ways the hypotheses have actually been ordered.

There is also another point. The principle of maximum parsimony is unique among all the principles which select the hypothesis to be adopted by taking account of its complexity: it, and it alone, leads to a determinate hypothesis. The principle to select our hypothesis through Professor Teller's lexicon is one among many comparable principles, each of which will yield determinate results; every dictionary listing all the hypotheses according to some order will choose for us a definite hypothesis. We would then need a higher order principle to help us to decide which lexicon to use. 'Choose the lexicon devised by Teller' is such a principle. It, however, raises the same problems once more on a second level. First, we would have to be furnished with the extra empirical information of which dictionary has been compiled by Teller. Secondly, we would need a third-order principle which would make us choose Teller's dictionary rather than some other.

DAVID MILLER

The Measure of All Things

In this paper[1] I present no results, only questions; and, I hope, difficulties.

1. Central to all theories of induction is an insistence on instances. Instances of a universal theory are needed to confirm it. The problem is: how many?

Hume of course showed that no quantity of instances could verify a universal generalization. In his *Abstract* (1740, p. 15) he argued further that no number of instances could make a universal generalization even probable. The former result is not seriously contested these days; if we can judge by the continued interest in probabilistic inductive logic, the latter is. Since the argument is as good as the same in each case, this is somewhat puzzling.

Perhaps one difference is that it actually seems to be possible, for very simple languages, to give probability measures according to which universal generalizations receive high probabilities (or 'degrees of confirmation') relative to finite evidence. Certainty, it appears, cannot be so achieved.

This is really rather a rough way of putting the matter; shortly I shall smooth it out a little. But for the moment I just want to say that I think the difference is no difference. Just as there is not a certain number of instances that will make a hypothesis certain, so there is not a reasonable number that will make it reasonable.

[1] The present paper has little in common — beyond sameness of spirit — with the paper read at Minnesota in 1968. Most of what I said there either was unoriginal or is no longer my opinion. In the latter category I include all the doubts I expressed on whether Popper's noninductive methodology could handle problems of pragmatic or practical preference. I should now like unreservedly to withdraw all such doubts; the problems seem to me now susceptible of an elegant and simple solution. See Popper, ch. 1 (1972) and especially sec. 14 (1974).

2. It has undoubtedly been acknowledged that, so abundant is the supply of probability measures, there is no real difficulty in locating one which makes any hypothesis, unrefuted by particular evidence, as probable as one likes thereon. The trouble comes in making the measure even tolerably plausible. For it has been appreciated also that the choice of one probability measure over all others is hardly to be justified by pure logic alone. (This is obvious in the case of subjectivists and has been made explicit by such writers as Carnap (1952, sec. 18) and Hintikka (1970, pp. 23–24).) To this extent, anyway, Hume's extension of Hume's argument to the probabilistic case has not been utterly disregarded. Levi sums up a fairly popular position well when he says, "The skeptical despair that threatens quests for global justification need not infect efforts at local justification" (1967, p. 6). Likewise, Hacking remarks, "Objective *local* inductive logic is already a reality. It is practiced by statisticians" (1971, p. 610).

Such admissions are surely sufficient to refute inductive logic's claim (if it makes the claim) to be a viable theory of knowledge. But never mind. Let us grant local justification, or local inductive logic, as a possible activity (without identifying it with statistics); grant, that is, that we often make a (nonlogical) choice of 'logical' probability measure and in terms of it perform inductive inferences. Even so, I maintain, inductive methods are next to useless. Hume's arguments cannot be so readily dismissed.

In most cases to date high probabilities for universal theories have been taken as a goal of inductive methods. (Kyburg, 1961, Hintikka and Hilpinen, 1966, and Hilpinen, 1968, are examples.) Sometimes, however, acceptance according to some rule or other is the primary goal, and high probabilities, though valued, enter only indirectly (Levi, 1967). In apparent opposition to both groups has been Carnap, who has on occasion been less than a champion either of universal laws or of acceptance rules. But it is clear that he attaches some significance to both, especially if purely theoretical issues alone are at stake (Carnap, 1966, pp. 258–60). If high probability means nothing for a theory, it is indeed somewhat hard to see the point of probabilistic inductive logic. And this holds despite the fact that very often high probability is hardly deemed on its own a satisfactory understudy for certainty. After all, to the extent that the measure involved is not 'logical', to that extent what Carnap calls the 'logical factor' may need overruling by the 'empirical factor'. Repetition or enumeration of instances is supposed to be as valuable in strengthening weight of evidence, in making probability judgments reliable, as in boosting probability; though admittedly in some cases the two jobs seem to merge into one (Levi, 1967, ch. 9; Hilpinen, 1968). In accordance with all this I shall hereafter suppose that high probabilities are esteemed, without inquiring into the reason for such treatment.

3. The case against inductive inferences which aim at certainty can be split in two. (1) To render certain a universal statement we would have to examine every object in the universe. (2) Even if we did this, it would not be enough; we would still have to ensure that we *had* examined every object — and this involves another universal statement.

To put it round the other way: even were we to know exactly what the universe consisted of, we would still have to investigate every object in it in order to confer certainty on generalizations like "All *A*'s are *B*'s." It is not surprising, therefore, that if we permit ourselves the luxury of concluding "All *A*'s are *B*'s" after observing just $a_0, \ldots, a_{n-1}$ as 'positive instances', then we can conclude also that $a_0, \ldots, a_{n-1}$ exhaust the universe of discourse. More precisely, if the rule "x_0 is *A* and *B*; x_1 is *A* and *B*; . . . ; x_{n-1} is *A* and *B*; *therefore* all *A*'s are *B*'s" is allowed to be inductively valid for sufficiently large *n*, then, by substituting a tautological property for *A* and "observed" for *B*, we can conclude that beyond the observed objects $a_0, \ldots, a_{n-1}$ the universe is quite empty. This argument is basically due to Goodman.[2]

If our conclusions are supposed to be certain, the point of the argument is indeed obvious. For if we cannot be certain that all the universe has been inspected, neither can we be certain that there does not exist an *A* which is not a *B*. Thus it is just not true that deduction is the special case of induction that obtains when all objects in the universe have been observed. I would contend that there is, likewise, no process of 'near-deduction' that obtains when nearly all objects have

[2] It is usually presented as a criticism of the rule "All observed *A*'s are *B*'s; *therefore* all *A*'s are *B*'s." But this rule — having, as it has, a universal premise — can only be applied after application of a rule like that given in the text. It is therefore less fundamental. The unconvincing form of the rule in the text — symmetric in A and *B* in the premises and asymmetric in the conclusion — is a symptom of the nonexistence of such things as positive instances. This objection is perhaps too well known to need repetition; see, however, Popper's reply to Quine (Popper, 1974).

been observed — we need to know the size of the universe in this case too. On the other hand, suppose we permit ourselves the near-luxury of concluding with near-certainty "All *A*'s are *B*'s" after observing just $a_0, \ldots, a_{n-1}$ as 'positive instances'. Can we conclude also (with near-certainty) that $a_0, \ldots, a_{n-1}$ exhaust the universe of discourse?

Probabilistic replicas of the argument I have ascribed to Goodman will be the main concern of this paper. In this section we deal with some informal replicas; in later sections with formal ones.

No assumption is made about the size of the universe except that it is no smaller than the number *n* of individuals $a_0, \ldots, a_{n-1}$ we have observed. All are 'positive instances'. We conclude, "Probably all *A*'s are *B*'s"; or " 'All *A*'s are *B*'s' is probably true"; or "The probability of 'All *A*'s are *B*'s' relative to the evidence = r." By the same token, may we not conclude also that probably everything has been observed; or that "All things have been observed" is probably true; or that the probability of "All things have been observed," relative to the evidence, is equal to r?

In the first two cases it is hard to know how to reply. "Probably all *A*'s are *B*'s" and its metalinguistic analogue " 'All *A*'s are *B*'s' is probably true" are often cited, in a typically vague manner, as typical inductive conclusions. There is no obvious reason why the conclusion that everything has been observed should be inadmissible, except that it is unexpected. Indeed, it is disastrous; for as soon as we can conclude that probably all *A*'s are *B*'s, after observing some, we are entitled to conclude that probably our conclusion is vacuous anyway — that is, it probably tells us no more than we know already. But I shall not say more here about these informal inductions; there is indeed very little more to say.

It is with examples such as the metric example that much current work in induction is concerned. Here, at least, something can be said. Remember, we have a conclusion, "The probability of 'All *A*'s are *B*'s' relative to the evidence = r." Can we hope that the probability of "All things have been observed" is likewise r, or near r?

In most systems on the market it would appear that we cannot, for reasons that I shall sketch in the next section. But the question still remains: If high probabilities can be obtained reliably only after a large number of instances have been observed, are we not in danger of making it also highly probable that the universe has been exhausted? (In this case the induction is hardly worth making.) Or, more generally: How many instances are needed before we can make reliable inferences? Too many?

4. Henceforth I shall restrict myself to the systems of Carnap and Hintikka, and to related systems (such as Levi's treatment of universal generalization, 1967, ch. 13). My detailing will be extremely sketchy.

We are concerned with elementary languages containing the usual logical trappings, identity, individual constants, and monadic predicate symbols. (The systems discussed in any detail by Hintikka do not include the sign of identity; but there is a generalization developed by Hilpinen (1966) which does.) Identity does not in fact have much to do in most of the systems under consideration, since no two individual constants (names) are allowed to name the same object. Indeed, the only permissible interpretation for a language with N names is a domain of N individuals, each named by exactly one name. (The case of an infinite universe is handled by considering what happens when N grows without limit.) Consequently universal statements are, within the language, deemed logically equivalent to N-fold conjunctions of instantial statements.

Evidence statements are intended to be singular statements, containing predicate symbols and individual constants. Though many different ways have been invented of supplying an initial probability distribution, all of them lead to some relative probabilities' depending on the number of names in the language. Certainly that number is fixed in advance; so the size of the universe is known — assumed, that is to say — in advance.

In formal systems like these such an assumption is hardly enough, as it was in our more informal treatment above, to make the sentence "All things have been observed" in any way acceptable. For normally there is no predicate "observed." And if it does exist as primitive it can expect no special (formal) relationship with evidential statements. "All things have been observed" can just as well attain, or fail to attain, a high probability as can "All things are blue," given the right sort of evidence. On the other hand, if we impose the (methodological) requirement of total evidence any actual evidence will be weighed down with positive occurrences of "observed," so that "All things have been observed" will be as well confirmed as any universal sentence. This,

I suppose, is why workers in inductive logic have chosen not to feature "observed" as a primitive predicate (see also p. 358 below).

Given the sign of identity we can of course get round this lack of expressibility, more or less. If $a_0, \ldots, a_{n-1}$ are the objects that have been observed, the sentence $(x)(x = a_0 \vee \ldots \vee x = a_{n-1})$ can be used to say that everything has been observed (though it is not equivalent to it). But this statement will stick at probability zero so long as $n < N$, the preordained cardinality of the universe, whatever the evidence. For given the interpretation of universal statements it states a contradiction.

It would look, then, as if the simple version of Goodman's paradox outlined above cannot be revived in most influential systems of inductive logic. This is hardly because the systems are too cleverly designed; they can scarcely be said to *surmount* the criticism, only to *avoid* it. Is it possible to make these systems more susceptible to this line of attack, whether or not the attack succeeds?

Crucial, of course, is the assumption that the size of the universe, and even its composition, can be known in advance. It must be tempting for an inductivist to pigeonhole this assumption among the 'global' ones that are agreed to be necessary and to treat inquiries as 'local' only when such an assumption has already been made. But this, I suspect, is asking a bit much. To be sure, the universe whose elements we are supposed to know about in advance is not in general the universe. It may be an urnful of balls; it may be a pack of cards. (In this case we shall surely be prepared to acknowledge a previous acquaintance.) But the universe is quite often something like the set of future tosses with a coin, a set whose cardinality seems far from fixed.[3] It does seem that an inductive logic that cannot cope with such universes may be an inductive logic without a future.

5. I have yet to offer any details of the way in which probability is distributed in the systems of Carnap and Hintikka. Carnap calls a *state-description* a sentence which is consistent and otherwise maximally strong; it tells of every object exactly what properties it has. (Quantifiers can be ignored, since all quantified sentences are equivalent to unquantified ones.) We also ignore, of course, differences in order of conjuncts in state-descriptions. Call two state-descriptions isomorphic

[3] This oversimplifies. See my "What's in a Numeral?" (unpublished).

if each can be obtained from the other by a permutation of individual constants. Given a state-description, we call the *structure-description* to which it belongs the disjunction of all state-descriptions isomorphic to the given state-description. Structure-descriptions of course also contain individual constants, and Carnap initially distributes his probability measure over structure-descriptions (c* does it equally); it is then subdivided, within each structure-description, among state-descriptions. Right from the start, then, many probabilities depend considerably on the size of the universe. Carnap succeeds in softening somewhat this dependence by his requirement C6 (1952, p. 13) that probabilities of singular sentences on singular evidence be the same, whatever the size of the universe. But as Hintikka notes, "The main drawback of . . . Carnap's procedure seems . . . to be its dependence on the domain of individuals which the language in question presupposes. . . . In most applications, however, this domain is largely unknown. In general it seems . . . perverse to start one's inductive logic from the assumption that one is in some important respect already familiar with the whole of one's world . . ." (1965, p. 279). Hintikka's systems avoid this dependence in the early stages, but not for long. For his initial handout of probability is over what he calls *constituents*, which are the strongest consistent sentences of the language which contain no individual constants. Each structure-description can be treated as a disjunct in just one constituent, so structure-descriptions (and thereafter state-descriptions) receive their absolute probabilities according to some subdistribution (usually a uniform one in each case). Thus it is far from true that the absolute probabilities are in general independent of *N*. What is true is that the 'possible states of the world' over which we initially distribute probabilities are independent of *N*; this Hintikka acknowledges (see 1965, p. 282, n. 19).

Like Hintikka I feel that no dependence of this sort is particularly desirable. How, however, can it be avoided in systems akin to Carnap's and Hintikka's? For, first of all, if we don't agree on some set of individual names then we have no determinate set of state- or structure-descriptions. In Carnap's system this means that no probabilities can be assigned at all; in Hintikka's no probabilities on singular evidence. (I exclude, of course, the trivial cases.) Or, rather, probabilities can be assigned (using "*N*" as a variable for the power of the universe) but there is no way of comparing them with one another. What is more, if our set of names is not finite, state- and structure-descriptions cease to be well formed — because infinitely long. And secondly, if the names are in any sense 'proper names' we cannot suppose that we use them without being familiar with (or perhaps 'acquainted with') their nominata. Thus the very use of names seems to presuppose familiarity with members of the universe; so is it worth trying to fight against it?[4]

Both these points can, in my opinion, be answered. It is possible to develop systems, similar to those already discussed, in which the size of the universe is in no sense 'fixed' in advance. It is possible, therefore, to produce systems in which the query "What is the probability that the universe has been exhausted, given the evidence?" does not always, when it hasn't been exhausted, give the answer zero.

To start with suppose that the universe is finite, though its cardinality is otherwise secret. A way around the first point is then almost obvious. In both Carnap's and Hintikka's approaches the crucial handout of probability is among structure-descriptions. So suppose that we have examined two objects. We consider to start with the set S_2 of all structure-descriptions in a universe with two names, a_0, a_1; call p_2 the measure we choose to put on S_2. Now add to S_2 all structure-descriptions involving those names and one other, a_2. This gives us a larger set of structure-descriptions, S_3, but one over which we can distribute probabilities as before. In terms of such a distribution, p_3, it makes good sense to ask for the probability, given the evidence, of the sentence $(x)(x = a_0 \vee x = a_1)$. For this is just asking for the measure, according to p_3, of the state-descriptions, consistent with the evidence, which contain only the names a_0 and a_1. It seems not unreasonable to demand that p_3, the distribution over S_3, be a genuine extension of p_2, so that the latter probabilities are obtainable from the former by relativization with respect to the sentence $(x)(x = a_0 \vee x = a_1)$; for any h not containing the name a_2 we have $p_2(h) = p_3(h, (x)(x = a_0 \vee x = a_1))$. This means that when, more generally, we add further names a_3, a_4, . . . , the probability of $(x)(x = a_0 \vee x = a_1)$ with respect to the successive measures p_3, p_4, . . . , will, if anything, decrease. As the number of individual names increases indefinitely, the probability of $(x)(x =$

[4] This paragraph is a trifle naive. Hintikka's later system (1966) enables probabilities on singular evidence to be assigned in an infinite universe. There is, naturally, no need to mention the universe's cardinality. But what the state- and structure-descriptions are is left a little vague. See the discussion by Hilpinen (1968, pp. 53–58).

$a_0 \vee x = a_1$) will therefore tend to a limit from above. Further conditions may be required on the sequence of measures to ensure that other probabilities have limits as the number of names increases. Then we can take the probability of a sentence as the limit of the probabilities it receives under the sequence of measures p_2, . . . outlined above.

What this amounts to is a distribution over all possible finite cardinalities for the universe. That is, we assign to each sentence of the form $(x)(x = a_0 \vee \ldots \vee x = a_{n-1})$ a positive probability. Should we wish to allow the possibility of an infinite universe, we need only take care that the limit of the probabilities above is not 1. Of course, the infinitude of the universe cannot be stated in the language, so this gambit will not assist us in calculating probabilities on the assumption that the universe is infinite. But this is a separate problem.

A method such as this for avoiding any decision on the size N of the universe is intuitively worthless as long as the names a_0, . . . are imagined to be proper names of the usual sort. For it is quite obvious that no account has been taken above of any structure-descriptions that contain, for example, just the names a_0 and a_2. We have required that the names featured in any structure-description form *an initial segment of the list of all names*. An argument due to Popper, however, has indicated that most systems of inductive logic are anyway unsatisfactory if they employ proper names like "John" and "Mary" (Popper, 1967). And a reply to Popper by Jeffrey suggested what can be done about it. Popper's point can be put as follows. Suppose that there are two buttons in a box, Udolpho and Warren. If we observe that one is amber, without 'observing' its name, we have to formulate our evidence as "Udolpho is amber or Warren is amber." Probabilities calculated relative to evidence so formulated will differ in general from probabilities calculated relative to "Udolpho is amber" or to "Warren is amber." Yet the information that the selected button is in fact Udolpho (or that it is in fact Warren) should hardly make any difference to the probabilities (for example, to the probability that both buttons are amber). Jeffrey responds (1968) by remarking that in such circumstances to choose a name like "Udolpho" is to ask for trouble; instead we should make use of names like "the observed button," "the other button." Our evidence can then read "The observed button is amber"; there is no need to consider the weaker sentence "Either the observed button is amber or the other button is amber." Thus no problem arises.

In our more general case we must therefore take a_0, . . . not as orthodox proper names, but as some such names as "the first object observed," "the second object observed," and so on. The supposition that objects are observed one at a time is perhaps slightly fanciful, but hardly outrageous compared with the other idealizations that inductive logic asks us to concede. But once we allow it, most of the problems caused above by proper names are at an end.

For example, there seems no longer much to be said for the assumption that every object is endowed with a name. Indeed, since the universe is not known, neither are the objects in it. Names become attached to particular objects ('fixed' or 'known' objects) only when these objects swing into sight and become observed (but see fn. 7 below) — or anyway when they become identified; up to then, the most that we can say of, for example, the name "the tenth observed object" is that it is the name of the tenth observed object (if there is or will be one). And since we are hardly going to be persuaded a priori that all objects have been or will be observed, we must concede that some of them may never succeed in getting named in this way. (Of course, persuasion a posteriori that all objects have been observed is just what I want to argue that inductive logic ought to lead to.) Note, however, that such names as "the first unobserved object" are also possible and indeed are needed for singular prediction. But not all things can be named even in this manner, unless *the unobserved part of the universe can be well ordered* — a postulate that I would not particularly want to make. (See my "What's in a Numeral?") It must be admitted, however, that structure-descriptions and state-descriptions lose much of their importance in a scheme like this. For if some objects are anonymous there is not much of a sense in which a structure-description expresses a possible world or type of world. The same is of course not true of Hintikka's constituents. One way to accommodate this deficiency is to distribute whatever probability is allotted to a constituent *not* over the structure-descriptions it involves (it doesn't involve structure-descriptions), but over 'truncated structure-descriptions'. By a truncated structure-description, I hasten to explain, I mean a sentence lying half way between a constituent and a structure-description: a sentence which attributes a suitable galaxy of properties to the named objects, but for the rest of the universe merely says which properties are instantiated (*if any*). Since all observed objects have names, relative probabilities on evidence are hereby determined. (Note that there is no obvious way of solving this problem in a system like Carnap's which probes no deeper than structure-descriptions.)

Let us take a very simple example. We have three individual names, a_0, a_1, a_2. Two monadic predicates P_0, P_1 produce four Q-predicates Q_0, Q_1, Q_2, Q_3. If the sign of identity is absent from the language then the following is a fairly typical constituent:

$$(\exists x)Q_1x \,\&\, (\exists x)Q_2x \,\&\, (x)(Q_1x \vee Q_2x).$$

It asserts that Q_0 and Q_3 are the Q-predicates that the universe omits. Now consider the sentence

$$Q_1a_0 \,\&\, Q_2a_1 \,\&\, Q_1a_2 \,\&\, (\exists x)Q_3x \,\&\, (x)(Q_1x \vee Q_2x \vee Q_3x);$$

this tells us which Q-predicates the named objects a_0, a_1, a_2 subscribe to, and of the rest of the universe is no more particular than is a constituent. And in fact it is actually less particular, for it does not tell us whether or not there is any object besides a_1 with the property Q_2. To get what we want, therefore, we shall have to introduce the sign of identity. (Of course, we need it for other reasons too; and I excluded it above only for simplicity's sake.) What would be a fairly typical constituent when we enrich this miniature language with the identity sign?

Hilpinen (1966, p. 134) has pointed out that we need to limit the maximum number of layers of quantifiers in a sentence (its *depth*) if we are to solve this problem. Suppose that it is limited to 2. Then a constituent might be something like

$$(\exists x)(Q_1x \,\&\, (\exists y)(y \neq x \,\&\, Q_1y)) \,\&\, (\exists x)(Q_2x \,\&\, (y)(Q_2y \rightarrow y = x)) \,\&\, (x)(Q_1x \vee Q_2x).$$

This tells us that Q_1 and Q_2 are still the only Q-predicates admitted by the universe, that at least two things satisfy Q_1, and that one thing only satisfies Q_2. Now a truncated state-description would be some sentence like the following:

$$Q_1a_0 \,\&\, Q_2a_1 \,\&\, Q_1a_2 \,\&\, (\exists x)(x \neq a_0 \,\&\, x \neq a_2 \,\&\, Q_1x \,\&\, (\exists y)(y \neq x \,\&\, Q_1y)) \,\&\, (\exists x)(x \neq a_1 \,\&\, Q_2x \,\&\, (y)(Q_2y \rightarrow y = x)) \,\&\, (x)(Q_1x \vee Q_2x).$$

It tells us the Q-predicates of the observed objects a_0, a_1, a_2, and of the rest of the universe it tells us how the Q-predicates are distributed, as well as is possible with two levels of quantifiers only.

This is a truncated state-description. A truncated structure-description is the disjunction of all truncated state-descriptions that are identical up to the permutation of individual names. It is over these sentences that I would propose distributing the measure already accorded (by some procedure akin to Hilpinen's) to the constituent cited just above. We can then make a further, presumably uniform, subdistribution among truncated state-descriptions. In this way we will generate a sequence of measures p_0, . . . Since all observed objects have names, this is enough to determine all the relative probabilities that we could ever need. Of course there are problems; for example, what depth can we allow our constituents to achieve? It would be reasonable, I think, to have this increase with the number of names in the language. But then what further restrictions would be needed to ensure that our sequence of measures has the required limiting properties? And so on.

I shall not discuss these problems, since we have already taken ourselves rather far from conventional systems of inductive logic; and it is not my task, even less my desire, to develop a new system. What I want to return to is the original problem of this paper, the probabilistic replication of Goodman's argument stated on p. 352 above.

6. We have, therefore, a sequence of measures defined over an ever expanding set of structure-descriptions — those involving just a_0 and a_1, those involving just these two or alternatively a_0, a_1, and a_2, and so on. So suppose that we have observed n objects. Whatever objects they are, their names will be $a_0, \ldots, a_{n-1}$. What now is the probability, given this evidence, of the sentence $(x)(x = a_0 \vee \ldots \vee x = a_{n-1})$; the probability, that is, that every object is among those observed? It is clear that this will depend a good deal on the sequence of measures $p_0, \ldots$ employed; and it will be small, if not zero, unless the structure-descriptions with few individual names (at least n, of course, for the evidence demands that) are very heavily weighted in comparison with the others. The critical point is, To what extent will universal statements suffer in the same way? Can we ensure that most of them will not have negli-

gible probabilities too? To be sure, $(x)(x = a_0 \vee \ldots \vee x = a_{n-1})$ will be stronger, so not more probable, than any other universal statement; but it may be that it will be only marginally less probable.

This is all too vague to amount to anything like a direct criticism. Yet what it seems must be true is this. Suppose that we attach no credence at all to the view that the universe is infinite. Then if $a_0, \ldots, a_{n-1}$ have been observed, for any $\epsilon > 0$ there is some finite number k such that, on the evidence, the sentence "There are no more than k objects other than $a_0, \ldots, a_{n-1}$" has probability in excess of $1 - \epsilon$. If, therefore, high probability means anything, it must mean something for this hypothesis. If, for example, we have an acceptance rule under which sufficiently probable hypotheses are acceptable (subject perhaps to other conditions), then this hypothesis is acceptable. If the universe is assumed to be finite, it turns out that we can put a bound on it. On the other hand, if we consider the possibility of an infinite universe — as we surely should — two cases can arise, depending on how seriously we take the possibility. If it is an outside one, so that the limit as m grows of the probabilities of the sentences $(x)(x = a_0 \vee \ldots \vee x = a_{m-1})$ is greater than $1 - \epsilon$, then what has just been said goes through in just the same way; we can put a finite bound on the size of the universe. If the opposite obtains, this cannot be done. But then, can we be sure that any universal statement can reach a probability as great as $1 - \epsilon$? It seems to me that we cannot. For since the infinitude of the universe is not statable in the language, how are we to calculate the probability of any universal statement relative to the assumption that the universe is of infinite power? And without this contribution we are lost. (Note, incidentally, that the talk of probability in this paragraph can be interpreted as either absolute or relative probability; and that when there is clearly evidence available I take it for granted that the probability is relative. It is easily seen that it hardly matters at all what the evidence is.)

Thus if an inductive logic of this sort is to be of any use, it must allow us to put a finite bound on the universe. I don't see inductivists relishing this conclusion. But they will of course point out that there is no inductive logic of this sort in existence and perhaps that I have made plain the unattractiveness of such a system. But if we bar from the object language statements about the size of the universe, how are we to judge whether we might not have observed too many instances? It is to this question that we finally turn.[5]

7. Against the view that inductive methods are next to useless enthusiasts may cite the following partial successes. Suppose that we assume in advance (a global assumption) that the universe is of size N. Then it does seem possible for universal hypotheses to receive gratifyingly high probabilities when the number of observed objects n is a small fraction of N.[6] Thus informative conclusions can be reached despite the fact that only a small part of the universe has been inspected. There is no reason to suppose that this is not true when the size of the universe is known to lie between certain limits, though not known exactly.

My criticism here — it is really the same criticism over again — is that if the conditions envisaged really obtain, then the n objects observed are not a 'small part' of the universe. One hundred buttons need not be a 'small sample' from a box of ten million buttons; they may amount to almost all of them. Likewise, one hundred buttons may make up almost all of even a countable infinity of buttons. *It all depends on how we measure the universe.* To be sure, if the universe is finite, size N say, the most natural way of measuring a subset of size n is to allot it n/N. This uniform distribution is the only one under which every subset of any given cardinality is accorded the same measure; in particular, the only one in which all individuals in the universe are treated alike. *In an infinite universe no measure satisfies these conditions.* But this is not the only way of measuring the universe; and though it seems to me as good a way as any other, I don't think that an inductivist ought to think that.

For *it follows from the inductivist point of view that the individuals in the universe should not be treated alike.* In particular, induction treats an *observed* object as in some way of deeper significance than an *unobserved* one. If only one object has been observed the believer in induction thinks that he has already learned something about the rest of the universe; consequently, the first observed object ought to

[5] The lack of precise — or anyway precisely stated — results in this section is no doubt deplorable. But I hope that I have given sufficient indication of the hazards any system developed along the lines suggested would encounter.

[6] For doubts, and discussions, see my "The Acceptance World."

attract greater measure than any other does (though not necessarily more than do all the others put together). The second object observed deserves for the same reason to be the next most heavily weighted, and so on down the line.[7] It seems reasonable that the rate of decrease should satisfy the following: when we reach a number n of observations after which we feel entitled to ignore the rest of the universe, and come to a conclusion, then the sum of the measures of the first n objects should be substantial. (This can only be rough, since n is likely to vary for hypotheses of different strength; this happens in Hilpinen (1968), for example. But it may be possible to make a refinement by taking into account utility considerations.)

In other words it seems to me reasonable that if a theory manages to attain so high a probability as to be acceptable, then a very considerable part of the universe should have been examined. Naturally, I do not expect an inductivist to feel the same. But what measure for the universe *will* he use? Some independent argument for using a uniform measure has to be provided. Certainly it should not be uncritically assumed that the uniform measure is the 'natural' way of measuring the universe. There is no more a 'natural' (or 'correct') way of measuring the universe than there is a 'natural' or 'correct' way of measuring probability. Indeed, if the universe is assumed to be infinite, then the uniform distribution — anyway not countably additive — would assign zero measure to each element. Thus in this case *every finite sample would constitute a negligible part of the universe*. In such a situation it is difficult to see why we should take its message so seriously.

If anything, I should have thought, our probability measure and our measure of the universe should harmonize in the way suggested. Then

[7] It may be said that this is no proper measure, since we do not know which object the first one will be. So which object is it that gets the largest measure? The answer is: the first one observed. We have to get used to the idea that any other names owned by the individuals in the universe are incidental and that the objects are identifiable by the names suggested on p. 358. (For a fuller defense see my "What's in a Numeral?") But this is not especially bizarre. Think of "the first card drawn," "the first throw of the coin," and so on.

That not all individuals are treated alike in inductive logic may seem to be contradicted by Carnap's condition C7 (1952, p. 14). But if we use the sort of names suggested here, perhaps we will be more inclined to give up this condition of Carnap's. I cannot discuss here the complications that this would lead to. (It is interesting that in the Kolmogorov/Martin-Löf theory of randomness an attribute with a very low frequency or probability in a Bernoulli sequence *cannot* occur at the first trial, though it can of course occur at others. See Martin-Löf, 1966, p. 616.)

inductive methods would be more or less fruitless. But if there is no such harmony, there is certainly a need for an argument why there is not. Yet even the recognition that such an argument is lacking is sadly lacking itself. This is the current state of probabilistic inductive logic.

In conclusion I say just this. I have attempted to spell out the implications of the view, acknowledged but not apparently taken sufficiently seriously by Carnap (1966, pp. 250–51) and Hintikka (1970, p. 9), that *high relative probability means low relative content*; that a universal statement that is rendered highly probable by evidence has, relative to that evidence, almost nothing of interest to say. I have suggested, tentatively enough, I hope, that this lack of content is revealed in the paucity of the unobserved part of the universe — as it should appear to anyone who assigns such high probabilities. And that if one wants to treat what has not been observed as still of interest, one must give up after all the craving for high probabilities.

REFERENCES

Carnap, R. (1952). *The Continuum of Inductive Methods.* Chicago: University of Chicago Press.

Carnap, R. (1966). "Probability and Content Measure," in P. K. Feyerabend and G. Maxwell, eds., *Mind, Matter, and Method.* Minneapolis: University of Minnesota Press. Pp. 248–60.

Hacking, I. (1971). "The Leibniz-Carnap Program for Inductive Logic," *Journal of Philosophy*, vol. 68, no. 19, pp. 597–610.

Hilpinen, R. (1966). "On Inductive Generalization in Monadic First-Order Logic with Identity," in J. Hintikka and P. Suppes, eds., *Aspects of Inductive Logic.* Amsterdam: North Holland. Pp. 133–54.

Hilpinen, R. (1968). *Rules of Acceptance and Inductive Logic*, in *Acta Philosophica Fennica*, vol. 22. Amsterdam: North Holland.

Hintikka, J. (1965). "Towards a Theory of Inductive Generalization," in Y. Bar-Hillel, ed., *Logic, Methodology and Philosophy of Science.* Amsterdam: North Holland. Pp. 274–88.

Hintikka, J. (1966). "A Two-Dimensional Continuum of Inductive Methods," in J. Hintikka and P. Suppes, eds., *Aspects of Inductive Logic.* Amsterdam: North Holland. Pp. 113–32.

Hintikka, J. (1970). "On Semantic Information," in J. Hintikka and P. Suppes, eds., *Information and Inference.* Dordrecht: Reidel. Pp. 3–27.

Hintikka, J., and R. Hilpinen (1966). "Knowledge, Acceptance, and Inductive Logic," in J. Hintikka and P. Suppes, eds., *Aspects of Inductive Logic.* Amsterdam: North Holland. Pp. 1–20.

[Hume, D.] (1740). *An Abstract of a Book lately Published entituled A Treatise of Human Nature, &c.* London: C. Borbet.

Jeffrey, R. C. (1968). "The Whole Truth," *Synthese*, vol. 18, pp. 24–27.

Kyburg, H. E., Jr. (1961). *Probability and the Logic of Rational Belief.* Middletown, Conn.: Wesleyan University Press.

Levi, I. (1967). *Gambling with Truth*. New York: Knopf.
Martin-Löf, P. (1966). "The Definition of Random Sequences," *Information and Control*, vol. 9, pp. 602–19.
Miller, D. (forthcoming). "The Acceptance World," *Theoria*.
Miller, D. (unpublished). "What's in a Numeral?"
Popper, K. R. (1967). "The Mysteries of Udolpho," *Mind*, n.s. vol. 76, no. 301, pp. 103–10.
Popper, K. R. (1972). *Objective Knowledge*. London: Oxford University Press.
Popper, K. R. (1974). *Replies*, in P. A. Schilpp, ed., *The Philosophy of Karl Popper*. La Salle, Ill.: Open Court.

WILLIAM H. HANSON

Names, Random Samples, and Carnap

An important feature of Carnap's approach to the problems of probability and induction is his sharp distinction between inductive logic proper and the methodology of induction.[1] His formula for the singular predictive inference, for example, belongs to the former, and his requirement of total evidence to the latter. Under the methodology of induction he subsumes, among other things, all matters concerning random samples. His view seems to be that a methodological rule requires samples to be chosen by random methods, and hence the theorems of inductive logic itself need not concern themselves with considerations of randomness.[2]

A less-noticed and seemingly unrelated feature of Carnap's inductive logic is its heavy reliance on individual constants (i.e., names). For example, his formula for the singular predictive inference (which can be used to give a complete characterization of an inductive method) is stated in such a way that names of all the individuals involved must appear in the evidence and hypothesis statements.[3]

The purpose of this paper is to show that these two features of Carnap's approach — the relegation of considerations of randomness to methodology and the prominence of individual names — are related in an important way. More particularly, it will be demonstrated that at least in certain cases the use of names in probability problems gives exactly the same result as would be obtained in an intuitively more natural approach if evidence statements were allowed to contain the information that the samples they describe had been selected by random

[1] *Logical Foundations of Probability*, 2nd ed. (Chicago: University of Chicago Press, 1962), sec. 44. Hereafter cited as *Foundations*.

[2] *Ibid.*, p. 494.

[3] See *The Continuum of Inductive Methods* (Chicago: University of Chicago Press, 1952), secs. 4, 9, and 10. Hereafter cited as *Continuum*. See also *Foundations*, pp. 567–68.

methods. It will also be shown that use of this more natural approach would avoid certain little-noticed[4] but serious difficulties that arise at a very elementary level in Carnap's inductive logic.

Section I presents the basic difficulty in question through the use of a simple example. Section II then discusses and rejects two seemingly plausible suggestions for resolving this difficulty. In section III it is shown that the difficulty extends far beyond the simple example discussed in I and II and hence is really a serious problem for Carnap's inductive logic. Finally section IV makes some tentative suggestions for resolving the difficulty.

I

The problem I wish to consider can be illustrated by a very simple example. Let us suppose that we have a set of boxes, each of which contains two buttons, and that each button is either amber or black. Suppose also that the set contains an equal number of boxes with no amber buttons, with one amber button, and with two amber buttons. Suppose finally that in each box one of the buttons has the letter 'a' engraved on it and the other the letter 'b', and that these identifying letters are independent of the colors of the buttons (i.e., among those boxes with one button of each color there are just as many boxes in which the button marked 'a' is amber and the one marked 'b' is black as there are boxes with the reverse combination of colors and letters). Let a box (call it 'α') be drawn at random from the set. There are now two probability problems concerning α that I wish to consider.

Problem 1. The box α is given to an honest and competent person who examines its entire contents without letting us see it. We then ask this person whether or not α contains at least one amber button and he answers, "Yes." What is the probability that both of the buttons in α are amber?

[4] Some of the matters considered in this paper have been discussed in the literature by Popper, Jeffrey, and Bar-Hillel. But the discussion is not widely known, and none of the discussants seems to have noticed the crucial point — namely, the relation between the use of names and random sampling. The papers in question are the following: Sir Karl Popper, "On Carnap's Version of Laplace's Rule of Succession," *Mind*, 71 (1962), 69–73, and "The Mysteries of Udolpho: A Reply to Professors Jeffrey and Bar-Hillel," *Mind*, 76 (1967), 103–110; Richard C. Jeffrey, "Popper on the Rule of Succession," *Mind*, 73 (1964), 129, and "The Whole Truth," *Synthese*, 18 (1968), 24–27; Yehosua Bar-Hillel, "On an Alleged Contradiction in Carnap's Theory of Inductive Logic," *Mind*, 73 (1964), 265–67.

Problem 2. We randomly select a button from α for observation, and we discover that it is amber. What is the probability that the unexamined button is amber and hence that both of the buttons in α are amber?

I think that everyone who knows a little about the mathematical theory of probability will agree that the answers to these problems are ½ and ⅔, respectively. For in the first problem our observer's report merely tells us that the range of equally probable alternatives has been narrowed from three to two — hence the answer ½. In the second problem, however, our seeing an amber button gives us all the information contained in the observer's report of the first problem and more besides. For if α contains just one amber button the likelihood that it will be selected by a random draw is only ½. Hence for the second problem it seems intuitively that the answer should be higher than ½, and a simple calculation shows that it is ⅔.

In spite of their different answers, the two problems are quite similar. For the hypothesis ('Both buttons are amber') is the same in each, and what we know of the composition of α is also the same ('At least one button is amber'). The only difference between the two is in the way in which we learn what we know of the composition of α. In Problem 1 we are merely told that at least one button is amber by a person who has examined the entire contents of α and who is therefore omniscient for purposes of the problem. In Problem 2, however, we find this out by random sampling of α. Clearly it is this difference in the way we obtain our incomplete knowledge of the composition of α that makes the problems different. I shall call problems of the first sort *omniscient-observer problems* (hereafter OO problems) and those of the second sort *random-sampling problems* (hereafter RS problems). Notice, incidentally, that the identifying marks that appear on the buttons play no role in either problem.

Let us now try to express and solve in a Carnapian system the probability problems that we have been considering. We can use the language L_2^1, which contains just one primitive predicate, A, and two individual constants, *a* and *b*. This language has just two *Q*-predicates, A and —A, which I take to be abbreviations for 'amber' and 'black', respectively. Each individual constant is taken to be the name of the correspondingly engraved button in box α. The confirmation function c* (which at one time, at least, was Carnap's preferred confirmation function)[5] will be used to determine probabilities in L_2^1 since it conforms exactly to the conditions of our problems.[6]

It is easy to obtain the correct answer to Problem 1 in L_2^1. Clearly the hypothesis is *Aa* & *Ab* and the evidence is *Aa* v *Ab*. Values are assigned by m* to the state-descriptions of L_2^1 in accordance with Table 1. Hence it is easy to show that c* (*Aa* & *Ab*, *Aa* v *Ab*) = ½.

State-descriptions of L_2^1	m*	Structure-Descriptions of L_2^1 Are the Following Sets of State-Descriptions
1. *Aa* & *Ab*	1/3	{1}
2. *Aa* & —*Ab*	1/6	{2, 3}
3. —*Aa* & *Ab*	1/6	{4}
4. —*Aa* & —*Ab*	1/3	

Table 1

The situation with Problem 2, however, is quite different. It will be remembered that the crucial difference between the two problems was that in Problem 2 our knowledge that at least one button is amber was obtained by random sampling. Hence the evidence statement for Problem 2 should include this information. But how do we express this information in L_2^1? There seems to be no way to do so, since L_2^1 contains only the single color predicate A. And besides, Carnap apparently thought that this sort of information need never appear in the evidence statement of a probability problem.[7] Yet if our evidence statement fails to specify the way in which we obtained our incomplete knowledge of the composition of α and states only what we know about that composition — namely, *Aa* v *Ab* — then c* will give us the same answer for Problem 2 as it did for Problem 1. Hence it looks as though Carnap's c* is unable to accommodate correctly Problem 2.

There is, however, a way to make c* yield the correct answer to Problem 2. All we need to do is look closely at the amber button our random draw produced and note its identifying mark. Suppose that it is 'a'. Then we can express the evidence as Aa, and, since c* (Aa & Ab,

[5] *Foundations*, pp. 562–67.

[6] The assumption that α is selected at random from a set of boxes in which each statistically different distribution occurs as often as any other entails that all the possible structures of α are equally probable. Analogously, c* assigns equal probabilities to each of the structure-descriptions of L_2^1. It also assigns equal probabilities to each state-description within a given structure-description, and this corresponds to our assumption that there are just as many boxes in which the button marked 'a' is amber and the one marked 'b' is black as there are boxes with the reverse combination of letters and colors.

[7] See *Foundations*, p. 494. Cf. fn. 20.

Aa) = ⅔, Problem 2 receives its correct answer. Of course we get the same answer if the identifying mark is 'b', since c* (Aa & *Ab*, *Ab*) = ⅔. So Carnap's c* is able to solve Problem 2 after all. But there is something very strange about the solution. When we first considered Problems 1 and 2 we noted that the identifying marks played no role at all in determining the correct answers. Hence for purposes of these problems the buttons need have no identifying marks. Yet without them the solution to Problem 2 for c* that I have just given is impossible.

The bizarreness of this situation is perhaps more apparent if we suppose that the marks exist but are so small that they can be read only with a magnifying glass. Then if we randomly select a button from α, find it to be amber, but don't use the glass, the only evidence we have that can be stated in L_2^1 is *Aa* v *Ab*. Hence c* will give ½ as the answer to Problem 2. But if we use the glass on the selected button, then no matter what we see, 'a' or 'b', c* will give the correct answer to Problem 2. But surely the odds on a fair bet based on Problem 2 do not change because of the use of a magnifying glass.

Summarizing, it seems that Carnap's c* gives the correct answer to an RS problem like Problem 2 only if the evidence statement contains names of the individuals in the sample. If the evidence does not contain names, then c* treats an RS problem as if it were an OO problem (like Problem 1) and may give an incorrect answer. In section III, I shall prove rigorously that this result holds not only for c* but for all the c-functions that satisfy Carnap's conditions C1–C11.[8] But first I shall discuss informally, in terms of the simple examples given above, two proposals for resolving the difficulty.

II

As was pointed out earlier (in footnote 4), the difficulty under consideration has been discussed in the literature by several authors. In "The Whole Truth" Jeffrey proposes a solution to the problem that at first seems simple and natural. He suggests that we treat names (i.e., the individual constants of Carnap's languages) like disguised descriptions rather than tags.[9] Thus in Problem 2 instead of using *a* and *b* for the

[8] See *Continuum*, pp. 12–29.

[9] I am not here suggesting that Carnap thought of individual names as mere tags,

correspondingly engraved buttons, we might let *a* stand for 'the button I will see' and *b* for 'the other button'. As long as we look at one and only one button it is clear that such an approach avoids the difficulty that arises in Problem 2 when names are treated like tags. For if the uniqueness conditions of the descriptions abbreviated by a and *b* are satisfied, then we will be able to state the evidence without using a disjunction. The difficulty with this approach is, of course, that the uniqueness conditions of the descriptions may not always be satisfied. Suppose that we are dealing with an *N*-membered population and that we are using the individual constants a_1, a_2, . . . , a_N. We might then think of letting a_i abbreviate 'the ith thing I see'. But suppose that in making my third observation I inadvertently see two things at once. Then a_3 does not denote anything, and familiar problems arise for sentences that contain it. Similar problems arise for $a_j (j \geqq 5)$ if I never see more than four things. Or suppose that on my sixth observation I inadvertently see the same thing I saw on the second observation. Then a_6 and a_2 denote the same thing, contrary to one of the basic assumptions of Carnap's semantics, and more than *N* individual constants will be needed to provide a name for each object.

Of course all these difficulties can be avoided if just the right descriptions are used. But there is no assurance that we will always be able to construct just the right descriptions, and this for two reasons. First we simply may lack the ingenuity or the vocabulary, and second (as the previous paragraph shows) the problem may change in ways that we do not anticipate. In addition to these difficulties, Jeffrey's approach has the drawback that the semantics of the language has to be modified constantly to accommodate different probability problems, even different problems dealing with the same set of objects. Suppose I intend to make several estimates of the composition of the same population of *N* objects based on several series of random draws. Then a_i can no longer abbreviate 'the ith thing I see', but will have to be relativized to the set of draws and become something like 'the ith thing I see in the first set of draws'. But then when the second set of draws begins it will have to be modified in an appropriate way. For all these reasons it seems to me

that is, as signs having denotation but no sense or meaning. It is clear from his talk about individual concepts in *Meaning and Necessity*, for example, that he did not. The point is rather that he did not think of them as disguised descriptions either, and, furthermore, for purposes of his inductive logic they could just as well be thought of as tags. On this latter point see *Foundations*, p. 59.

that Jeffrey's approach is a desperate expedient and should be avoided if at all possible. Surely Carnap himself would not have endorsed it since he did not think of names as disguised descriptions.

Another solution to the problem under consideration has been suggested by several statisticians with whom I have discussed it. They typically find my distinction between RS and OO problems a bit arbitrary since, they say, the former can be transformed into the latter by merely enlarging the sample space and building the assumption of randomness into the distribution. With Problem 2 this can be done by adding a predicate O for 'is observed', yielding the eight possibilities of Table 2.

1. *Aa* & *Ab* & *Oa*	1/6
2. *Aa* & *Ab* & *Ob*	1/6
3. *Aa* & —*Ab* & *Oa*	1/12
4. *Aa* & —*Ab* & *Ob*	1/12
5. —*Aa* & *Ab* & *Oa*	1/12
6. —*Aa* & *Ab* & *Ob*	1/12
7. —*Aa* & —*Ab* & *Oa*	1/6
8. —*Aa* & —*Ab* & *Ob*	1/6

Table 2

Following the constraints on Problem 2 we have that each of the first two and the last two sentences in Table 2 has a probability of 1/6, while each of the remaining four has a probability of 1/12. We can now consider the evidence in Problem 2 to be merely that an amber button is observed, and express this as $(Aa \,\&\, Oa) \vee (Ab \,\&\, Ob)$. Problem 2 can now be handled correctly without having to notice the mark engraved on the observed button since

$$P(Aa \,\&\, Ab/(Aa \,\&\, Oa) \vee (Ab \,\&\, Ob)) = 2/3.$$

Since the statement of evidence in the equation above contains no explicit mention of random sampling, we are here treating an RS problem in the way we previously treated OO problems. The reason this approach works is that we have divided each of the four state-descriptions of Table 1 into two parts by adding either *Oa* or *Ob*, and then divided the value previously assigned to the state-description *equally* between its two parts. For example, state-description 1, which previously was assigned the value 1/3, has been replaced by the first two sentences of Table 2, each of which is assigned the value of 1/6. It is this equal division of the old state-description values between their new parts that amounts to asserting that the observations were made at random. This is what my statistician friends mean by building the assumption of randomness into the distribution. It should also be noticed that we continue to get the correct answer to Problem 1 on this approach since

$$P(Aa \,\&\, Ab/Aa \vee Ab) = ½.$$

The success of the method sketched in the preceding paragraph may lead one to believe that the difficulty involved in handling RS problems in Carnap's inductive logic has been solved. That such a belief is premature can be seen by noticing that the sentences in Table 2, unlike those in Table 1, are not state-descriptions. For in adding the new predicate O we have transformed the language L_2^1 into L_2^2, and the state-descriptions of L_2^2 all contain four conjuncts. Hence there are sixteen state-descriptions of L_2^2; they are given in Table 3. In order to handle Problem 2 in L_2^2 without giving favored status to either *a* or *b*, we have to express the evidence, e, as $(Aa \,\&\, Oa \,\&\, -Ob) \vee (Ab \,\&\, Ob \,\&\, -Oa)$. The hypothesis, *h*, continues to be *Aa* & *Ab*. It now turns out, unfortunately, that

$$c^*(h, e) = ½.$$

We also get the wrong answer for Problem 1, since

$$c^*(h, Aa \vee Ab) = 3/7.$$

The reason for the discrepancy between the c*-values in L_2^2 and the values based on Table 2 is that L_2^2 has *ten* structure-descriptions, each of which is weighted equally by m*, while Table 2 is just a refinement of Table 1 and hence preserves the effect of that table in which m* assigns equal values to the *three* structure-descriptions of L_2^1. In order to get the right answers to Problems 1 and 2 in L_2^2 we would have to construct a special measure function that preserves the effect of Table 2. The function m** given in Table 3 is the obvious choice. It preserves all the values assigned by Table 2 and also makes explicit what was implicit in that table, namely, that cases in which neither or both of the objects are observed simply are not being considered. Unfortunately there are several difficulties with m**. One is that it violates both conditions C9 and C10 of *Continuum* and hence is unavailable within a strictly Carnapian approach.[10] This is not so serious, since if there were

[10] $c^{**}(Aa \,\&\, Oa, Ab \,\&\, -Ob) = 2/3$ *while* $c^{**}(Aa \,\&\, Oa, -(Ab \,\&\, Ob)) = 1/3$. According to C9 these two values should be the same, and according to C10 the latter should fall within the interval 0–1/4.

State-Descriptions of L_2^2	m*	m**	Structure-Descriptions of L_2^2 Are the Following Sets of State-Descriptions
1. Aa & Ab & Oa & Ob	1/10	0	[1]
2. Aa & Ab & Oa & —Ob	1/20	1/6	[2, 3]
3. Aa & Ab & —Oa & Ob	1/20	1/6	[4]
4. Aa & Ab & —Oa & —Ob	1/10	0	[5, 9]
5. Aa & —Ab & Oa & Ob	1/20	0	[6, 11]
6. Aa & —Ab & Oa & —Ob	1/20	1/12	[7, 10]
7. Aa & —Ab & —Oa & Ob	1/20	1/12	[8, 12]
8. Aa & —Ab & —Oa & —Ob	1/20	0	[13]
9. —Aa & Ab & Oa & Ob	1/20	0	[14, 15]
10. —Aa & Ab & Oa & —Ob	1/20	1/12	[16]
11. —Aa & Ab & —Oa & Ob	1/20	1/12	
12. —Aa & Ab & —Oa & —Ob	1/20	0	
13. —Aa & —Ab & Oa & Ob	1/10	0	
14. —Aa & —Ab & Oa & —Ob	1/20	1/6	
15. —Aa & —Ab & —Oa & Ob	1/20	1/6	
16. —Aa & —Ab & —Oa & —Ob	1/10	0	

Table 3

good reasons to adopt m** Carnap's conditions could be modified to accommodate it. A more serious difficulty with m** (and indeed with the entire approach being discussed here) is that use of a special predicate for 'is observed' forces us to revise constantly the semantics of the language. For *Oa* does not mean simply that *a* is observed at some time by someone. If this were the case then the probability of *Oa* would be much higher than ½, and the probability of *Oa* & *Ob* would not be zero. Clearly *Oa* is to be interpreted as saying that *a* is observed within a certain span of time or within the context of a certain specified set of procedures. But then if the drawing of an object from the set containing just *a* and *b* is to be repeated at a later time, the interpretation of *O* must be changed. Surely it would be better if this sort of frequent modification could be avoided.

The most serious difficulty with m**, and with similar functions that might be defined to accommodate the predicate O, is that for a given domain of individuals we will need many such probability functions, one for each possible sample size. To illustrate this, suppose that we have a set of ten objects each of which is either amber or black. Suppose also that we intend to draw five of them at random, observe their color, and on the basis of our observation assess the probability of predictions about the frequency of the two colors in the entire set. Then we will need a probability function that assigns zero to each state-description in which some number of individuals other than five is observed. But if we next decide to make similar assessments on the basis of a six-membered sample, the probability function must change. Now we will need one that assigns zero to each state-description in which some number of individuals other than six is observed. It hardly needs to be said that changing the probability function as the problem changes, even though the number of properties and individuals remains the same, is not the sort of thing Carnap had in mind. And independently of Carnap's own preferences I think it is desirable to avoid constant modification of the probability function, just as in the previous case it seemed desirable to avoid constant modification of the semantics of the language.

The upshot of all this is that neither Jeffrey's approach nor the approach suggested by my statistician friends is very satisfactory for handling the difficulty that I have been considering. In section IV I will sketch the beginnings of a solution that avoids the kinds of problems encountered by these two approaches. Unfortunately, this solution will require the formal languages and the probability functions to be more complicated than any considered by Carnap.

III

The conclusions of section I regarding difficulties and limitations inherent in Carnap's inductive logic were based on the single example that gave rise to Problems 1 and 2. Hence it is legitimate to ask if similar examples give rise to similar difficulties. If not, then Carnap's inductive logic will merely have been shown to contain a single anomaly, and this is probably not of much interest. But if similar examples do give rise to similar difficulties, then the criticisms of the previous sections will have been shown to be broadly based and serious. In the present section I will show that the latter is indeed the case. Specifically, the difficulties involved in handling Problem 2 in Carnap's inductive logic will be shown to be independent of the number of individual constants, the number of predicates, and the c-function.

I shall begin by giving Carnap's formula for the singular predictive inference as it appears in *Continuum*. This will serve as a point of departure for the results to be presented in this section, since it is Carnap's general formula for a large class of RS problems. But first some notation must be introduced. In *Continuum* Carnap restricts his attention to languages with a finite number of primitive predicates (π) and a finite number of individual constants (N). These languages are designated L_N^π. The strongest noncontradictory predicates that can be formed in L_N^π are called Q-predicates. L_N^π contains $\kappa = 2^\pi$ Q-predicates, and they are designated as $Q_1, Q_2, \ldots, Q_\kappa$. A Q-sentence is a sentence that results from applying a Q-predicate to an individual constant. The function $\lambda(\kappa)$ takes as values the non-negative real numbers and ∞ and gives a complete characterization of the confirmation function c_λ. In the formula for the singular predictive inference to be given below the evidence statement, e_Q, says that s_1 individuals have property Q_1, s_2 have $Q_2, \ldots, s_\kappa$ have Q_κ. $s_1 + s_2 + \ldots + s_\kappa = s$. The hypothesis, h_i, says that some individual not mentioned in the evidence has Q_i. More precisely, e_Q is a conjunction of s distinct Q-sentences each of which contains a distinct individual constant, where s_1 of these constants occur in Q-sentences with Q_1, s_2 with $Q_2, \ldots, s_\kappa$ with Q_κ. For each i, $1 \leq i \leq \kappa$, h_i is a Q-sentence made up of Q_i and an individual constant that does not appear in e_Q. Carnap's formula for the singular predictive inference involving h_i and e_Q is as follows:[11]

$$(1) \qquad c_\lambda(h_i, e_Q) = \frac{s_i + (\lambda(\kappa)/\kappa)}{s + \lambda(\kappa)}$$

To see how (1) applies to an RS problem, recall that Problem 2 used the language L_2^1 and the c-function c^*. So we have $\pi = 1$, $N = 2$, $\kappa = 2^\pi = 2$, $Q_1 = A$, $Q_2 = -A$, and $\lambda(\kappa) = \kappa = 2$.[12] In Problem 2, $s_1 = s = 1$ and $s_2 = 0$, so $e_Q = Aa$ and $h_1 = Ab$. Hence (1) gives

$$c^*(Ab, Aa) = (1 + 1)/(1 + 2) = 2/3.$$

And this is the same result that we got in the previous section after we had noticed the identifying mark on the observed button.[13]

[11] Equation (1) follows from (9-6) and (4-4) of *Continuum*. The evidence statement e_Q is what Carnap calls an individual distribution. Equation (1) also apparently holds if the evidence is expressed as a statistical distribution. (See *Foundations*, pp. 111, 567–68.) Replacing individual distributions with statistical distributions does not, however, obviate the criticisms of Carnap's inductive logic made in this paper since a statistical distribution is always relative to a fixed set of individual constants.

[12] Function c^* is that c-function for which $\lambda(\kappa) = \kappa$. See *Continuum*, sec. 15.

[13] Actually in section I our hypothesis was Aa & Ab rather than Ab. But as long as the evidence is Aa in both cases the probabilities must be the same. Similarly, in (1) the hypothesis could be h_i & e_Q rather than h_i without affecting the probability. This fact should be kept in mind when comparing (1) with (2) below.

In order to generalize the difficulty that was encountered in section I in connection with Problem 2, I shall have to produce and prove an equation that is similar to (1), but that contains instead a general formula for OO problems.[14] If it turns out that the general formulas for RS and OO problems give different results in a large number of cases, then I will have shown that the difficulty of section I is widespread. For the difficulty was just that sometimes one is forced to treat an RS problem like an OO problem (i.e., when the individual constants that have been assigned to the individuals in the sample are unknown). In giving the general result about OO problems it will be convenient yet sufficient to restrict attention to those cases in which $s = N - 1$. With this restriction in mind I introduce some additional notation.

For any non-negative integers $s_1, s_2, \ldots, s_\kappa$ such that $s_1 + s_2 + \ldots + s_\kappa = s = N - 1$, let h_i' $(1 \leq i \leq \kappa)$ be the structure-description with Q-numbers $s_i + 1$ and s_j (for all $j \neq i$).[15] Then let e_Q' be the disjunction $h_1' \vee h_2' \vee \ldots \vee h_\kappa'$. This has the effect of making e_Q' a sentence which says that at least s_1 individuals have Q_1, at least s_2 individuals have $Q_2, \ldots,$ at least s_κ individuals have Q_κ. Yet e_Q' doesn't say which individuals have which Q-properties. And each h_i' gives a complete yet purely structural description of the domain with which we are concerned. Hence e_Q' and h_i' are, in a certain obvious sense, structural analogues of e_Q and h_i, respectively, for the case $s = N - 1$.[16] As an example consider how this notation applies to Problem 1. The language is again L_2^1. Since $s_1 = s = N - 1 = 1$, h_1' is $Aa \,\&\, Ab$, h_2' is $(Aa \,\&\, -Ab) \vee (-Aa \,\&\, Ab)$, and e_Q' is $h_1' \vee h_2'$ which is logically equivalent to $Aa \vee Ab$. Hence h_1' and e_Q' correspond precisely to the hypothesis and evidence statements of Problem 1.

In view of the interpretation of h_i' and e_Q', it should be obvious that a formula for the value of $c_\lambda(h_i', e_Q')$ will give us the answers to a large

[14] In most of this section I simply assume that (1) is a correct formula for the singular predictive inference in those cases where the initial probability distribution is given by m_λ and the sample is selected at random. Hence I am assuming that (1) is a correct formula for a large class of RS problems. This assumption seems unproblematic. Nevertheless equation (3), which appears at the end of this section, supports (1) since it gives the same result for the case in which $s = N - 1$.

[15] That is, h_i' is a structure-description in which s_j individuals have Q_j, for all $j \neq i$, and $s_i + 1$ individuals have Q_i.

[16] Actually, h_i' is more closely analogous to $h_i \,\&\, e_Q$ than to h_i, since the first sentence is a structure-description and the second is a state-description, while the third is just a single Q-sentence. But this is an inessential difference for my purposes since $c_\lambda(h_i, e_Q) = c_\lambda(h_i \,\&\, e_Q, e_Q)$. Cf. fn. 13.

class of OO problems, just as (1) gives us the answers to a large class of RS problems. The formula is contained in equation (2) below.

For any h_i' $(1 \leq i \leq \kappa)$,[17]

$$(2) \qquad c_\lambda(h_i', e_Q') = \frac{(s_i + \lambda/\kappa)/(s_i + 1)}{[(s_1 + \lambda/\kappa)/(s_1 + 1)] + [(s_2 + \lambda/\kappa)/(s_2 + 1)] + \ldots + [(s_\kappa + \lambda/\kappa)/(s_\kappa + 1)]}$$

Equation (2) holds for any c-function that satisfies conditions C1–C11 of *Continuum* and any choice of non-negative integers $s_1, s_2, \ldots, s_\kappa$ such that their sum $s = N - 1$. The proof of (2) is given in the appendix.

As a formula for OO problems (2) is correct and unobjectionable. But if one were to attempt to use (2) rather than (1) for RS problems the results would be disastrous. To illustrate, suppose that we have a domain of N objects for which we distinguish π primitive predicates, and that we have applied L_N^π by way of the usual semantical rules to this domain. Part of the semantics of L_N^π will involve specifying a one–one mapping from the N individual constants of L_N^π to the objects of the domain. Suppose also that we randomly select $N-1$ objects and determine that s_1 of them have the property Q_1, s_2 have $Q_2, \ldots, s_\kappa$ have Q_κ. We now want to know what the probability is that the remaining object has the property Q_i (when $1 \leq i \leq \kappa$) according to our favorite c-function, c_λ. If we know which individual constants have been assigned to the objects we have observed, then we can apply (1) and get an answer that is correct, given the initial assignment of probabilities made by m_λ. If, however, we do not know which individual constants have been assigned to the observed objects, then we may be tempted to state the evidence and hypothesis as strongly as we can (i.e., with e_Q' and h_i') and apply (2). But an analysis of the consequences of (2) will show that this approach is unthinkable for an inductivist like Carnap. For Carnap thought that any c-function worth serious consideration should, when applied to problems of the kind we are considering, satisfy at least two conditions: (a) The value of the c-function should vary directly with s_i/s; and (b) for a sufficiently large sample it should be close to s_i/s.[18] These conditions are often summarized by saying that a

[17] In (2) I use 'λ', as Carnap often does, as an abbreviation of the functional notation 'λ(κ)'.

[18] *Continuum*, p. 24.

satisfactory c-function should allow us to learn from experience. It is now easy to show that if (2) were applied to RS problems, learning from experience would be sadly deficient.

If $0 \leqq \lambda < \kappa$, then each of the terms $\frac{s_j + \lambda/\kappa}{s_j + 1}(1 \leqq j \leqq \kappa)$ of (2) varies directly with s_j. Hence (2) allows some learning from experience, but usually very little. For even when $\lambda = 0$ (2) becomes

$$c_0(h_i', e_Q') = \frac{(s_i)/(s_i+1)}{[(s_1)/(s_1+1)] + [(s_2)/(s_2+1)] + \ldots + [(s_\kappa)/(s_\kappa+1)]}$$

which (although it gives the same results as the "straight rule" for $s_i = 0$ and $s_i = s$) gives results quite close to $1/\kappa$ as long as each of the numbers $s_1, s_2, \ldots, s_\kappa$ is fairly large, even though these numbers may differ widely among themselves. For example, let $\kappa = 2$, $N = 500$, $s_1 = 100$, and $s_2 = 399$. Then

$$c_0(h_1', e_Q') = .498$$

and

$$c_0(h_2', e_Q') = .502.$$

It should be remembered that when $\lambda = 0$ learning from experience is supposed to be at its maximum in Carnap's systems.

As λ moves from 0 toward κ learning from experience declines, and when $\lambda = \kappa$, which gives us the function c^*,[19] we have

$$c^* (h_i', e_Q') = 1/\kappa.$$

For $\lambda > \kappa$ each of the terms $(s_j + \lambda/\kappa)/(s_j + 1)$ $(1 \leqq j \leqq \kappa)$ of (2) varies inversely with s_j. Hence if (2) is applied to an RS problem it here yields counterinductive methods. The worst of these is $c\infty(h_i', e_Q')$ which according to Carnap's conventions is defined as

$$\lim_{\lambda \to \infty} [c_\lambda (h_i', e_Q')].$$

It is not difficult to show that

$$c\infty(h_i', e_Q') = \frac{(1)/(s_i+1)}{[(1)/(s_1+1)] + [(1)/(s_2+1)] + \ldots + [(1)/(s_\kappa+1)]}$$

[19] See fn. 12.

As an example of the strongly counterinductive nature of (2) when $\lambda = \infty$ let $\kappa = 2$, $N = 100$, $s_1 = 10$, and $s_2 = 89$. Then

$$c_\infty (h_1', e_Q') = .891$$

and

$$c_\infty (h_2', e_Q') = .109.$$

It should now be clear that the discrepancy between equations (1) and (2) is widespread, and hence that (2) cannot reasonably be used for RS problems. So the criticisms of Carnap's inductive logic made in section I turn out to be broadly based.

Up until now I have simply assumed that (1) gives a correct formula for a large class of RS problems. I still see no reason to doubt this assumption, but it will be instructive to show that for the case in which $s = N - 1$ it is possible to state the problems in such a way that no individual constant is given favored status and yet to obtain the same formula. In order to obtain the desired result I retain all the notation introduced in connection with (1) and (2), and in addition I introduce e′ as an abbreviation of the sentence 'A sample selected at random contained s_1 objects with Q_1, s_2 objects with Q_2, . . . , s_κ objects with Q_κ'. It is fairly clear that Carnap did not intend any of the predicates of his languages L_N^π to be interpreted as 'is (was) selected by a random method',[20] so e′ should not be thought of as a sentence of L_N^π. But suppose that we have probability functions P_λ that assign values to sentences like e′ as well as to the sentences of L_N^π, and suppose further that if A and B are any sentences of L_N^π

$$P_\lambda(A,B) = c_\lambda(A,B).$$

If we now take $s_1, s_2, \ldots, s_\kappa$ to be any non-negative integers whose sum $s = N - 1$, we will find that equation (3) gives the result we seek. The proof of (3) is given in the appendix.

For any h_i' $(1 \leqq i \leqq \kappa)$,

[20] In *Foundations*, p. 494, Carnap argues that no theorem of inductive logic proper need contain the requirement that a sample be random (or even that it not be known not to be random) since this will be taken care of by the requirement of total evidence — a methodological principle. Thus he avoids bringing considerations of randomness into inductive logic proper by saying, in effect, that whenever we seriously *apply* inductive logic to a practical problem we should, as a matter of correct methodology, make sure that we have no evidence indicating that our samples are not random.

$$(3) \qquad P_\lambda(h_i', e') = \frac{s_i + (\lambda(\kappa)/\kappa)}{s + \lambda(\kappa)}.$$

Notice that the right side of (3) is identical with that of (1), Carnap's equation for the singular predictive inference.

Equation (3) shows that it is possible (at least when $s = N - 1$) to understand the singular predictive inference in such a way that individual constants are entirely irrelevant to it, and yet to agree with Carnap's formula for it as given in (1). More precisely, we can take the only essential evidence to be that $s = N - 1$ individuals have been randomly selected and have been found to be distributed among κ properties in a certain specified way. This is what e′ expresses. We can then construe any prediction about the remaining individual as a statement about the complete structure of the population. The sentences h_i' $(1 \leq i \leq \kappa)$ express all such hypotheses that are compatible with e′. Finally (as the proof of (3) in the appendix shows), if we take Carnap's m_λ to assign the correct probability value to each h_i', and if we assign values to each of the likelihoods $P_\lambda(e', h_i')$ in the only reasonable way, then the value for the singular predictive inference is the same as that given by (1).

With (3) we have come full circle, for (3) is the formal embodiment of the informal approach that was initially used to solve Problem 2 in section I. And the fact that (3) and (1) give the same result can be thought of as vindication of Carnap's formula for the singular predictive inference, at least for $s = N - 1$. It also shows that given a probability distribution m_λ, there is a trade-off between framing the evidence and hypothesis statements in terms of the preassigned names of the objects involved and stating the evidence and hypothesis in purely structural terms but including in the evidence an explicit statement that the sample was obtained by random methods. Thus the use of names can be thought of as a device for coding the information that the sample was chosen at random. Whether or not Carnap would have been better advised to make use of sentences like e′ instead of the coding device provided by names will be discussed in the next section.

IV

I think the foregoing results show that giving an adequate formulation of the logical approach to probability is much more difficult than Carnap and his sympathizers thought. Serious problems arise at an entirely elementary level, and they do not seem to have any simple solutions. All this, I fear, bodes ill for any approach that attempts to make comprehensive assignments of probabilities to the sentences of a formalized language. Nevertheless, in closing I will attempt to sketch the first step or two of an adequate solution to the problem presented in sections I and III. It is not surprising that this solution will be much more complicated than either of those considered in section II.

As I pointed out at the end of the previous section, the difficulty that gave rise to this paper can be traced to Carnap's requirement that the evidence and hypothesis statements of RS problems be framed in terms of preassigned names of the objects involved. But as equation (3) showed, this requirement could be dropped (at least where $s = N - 1$) if we had (a) some way of including in the evidence statement of RS problems the information that the sample was selected by a random method, and (b) some procedure for assigning probabilities to statements that contain talk of random selection. I shall here concern myself only with a sketch of a solution to (a).

It should be fairly easy to see that (a) cannot be solved simply by adding to Carnap's languages a special one-place predicate that is always interpreted as 'was selected by a random method'. For all the difficulties that arose for the special predicate O ('is observed') discussed in section II would also arise for this predicate. But something like this can be done. Suppose that we introduce a special set of individual constants $t_1, t_2, \ldots$ that designate moments of time. We could then introduce a special two-place predicate R such that $R\alpha\beta$ is always interpreted 'Object α is selected by a random method at time β'. The fact that R takes a temporal argument eliminates some of the problems that arose for O in section II. And it is possible to use R and the t_i to give plausible statements of the evidence and hypotheses in RS problems. These statements will not contain preassigned names of particular objects, but will express what is important in such problems — namely, the frequency of the observed properties and the fact that the sample was selected by a random method.[21] As an example, consider a

[21] It might be objected at this point that the only important difference between R and O is that the former is a two-place predicate that takes a temporal argument whereas the latter is a one-place predicate. Hence, the objection might go as follows:

typical case of the singular predictive inference in which the evidence is that two objects have been selected at random from some population and found to be amber (A), and the hypothesis is that the next object to be selected at random will be amber. I suggest that this problem can be understood in terms of e, *i*, and *h*, below, as the probability of *h* given e & *i*. (Here e and *h* are the evidence and hypothesis, respectively, and *i* is background information that specifies the nature of the singular predictive inference. $E\alpha\beta$ is to be interpreted as 'α is earlier than β'.)

$$e = (\exists x)(\exists y)(x \neq y \,\&\, Rxt_1 \,\&\, Ryt_2 \,\&\, (z)(Rzt_1 \supset z = x) \,\&\, (z)(Rzt_2 \supset z = y) \,\&\, Ax \,\&\, Ay)$$

$$i = Et_1t_3 \,\&\, Et_2t_3 \,\&\, (\exists x)(Rxt_3 \,\&\, (y)(Ryt_3 \supset y = x)) \,\&\, (x)((Rxt_1 \vee Rxt_2) \supset -Rxt_3)$$

$$h = (\exists x)(Rxt_3 \,\&\, (y)(Ryt_3 \supset y = x) \,\&\, Ax)$$

Sentence e should require no explanation. Sentence *i* assures that the thing mentioned in the hypothesis is distinct from and is selected later than the things mentioned in the evidence. It also says that exactly one thing is selected at random at t_3. This clause is included because without it P(*h* given e & *i*) would be the probability that a third thing is amber *and* is selected at t_3, given e. But in the singular predictive inference it is assumed as part of the problem that this additional thing will be selected. What we wish to know is the probability that this thing will be amber given e *and* that it will be selected.[22]

if in section II I had allowed O to be a two-place predicate with a temporal argument, then the problems that this paper raises for Carnap's inductive logic could have been solved along the lines suggested there without eliminating preassigned names. But this objection overlooks three important points. The first is that the intended interpretation of R involves random selection while that of O involves merely selection or observation, not necessarily by a random method. Hence O might be interpreted in ways that do not reflect a crucial feature of the problems under consideration. The second point overlooked by this objection is that since the use of preassigned names is at best a misleading way of coding other information (namely, that the sample was selected at random) their use should be avoided if at all possible. The third point is that whichever approach one chooses for *expressing* the evidence and hypothesis in an RS problem (i.e., whether one chooses to use R and eliminate names as in the present section or to use O and retain names as in section II) the problem of determining how to *assign probabilities to* those statements remains. This is the problem referred to as (b) in the present section. We encountered this same problem in section II when we noticed that functions like m** were relative to the size of the sample referred to in the evidence statement.

[22] It might be objected that the uniqueness claims of e and *h* are too strong since

Notice that no individual constants designating objects appear in e, *i*, or *h*. Hence the sort of difficulty that faces Carnap's inductive logic in handling RS problems cannot arise. Individual constants for times do appear, but I do not think they create any special difficulties. So I suggest that anyone interested in pursuing Carnap's program of assigning probability measures to formal languages begin by using something like the techniques sketched in the example above for expressing RS problems. I leave to him the much more difficult task of determining how the measure functions are to be specified in order to yield the desired values. The only thing about this latter task that is clear to me is that *R* will have to be treated differently from other predicates, and this will probably mean substantial revision of Carnap's approach.

V

The lesson to be learned from this paper, if there is just one such lesson, is that Carnap's move in banishing considerations of random sampling from inductive logic proper to the methodology of induction[23] was a mistake. Undoubtedly this move greatly simplified matters and facilitated Carnap's impressive technical achievements. But it ultimately led to confusion and absurdity, and if my general line of argument is correct it must be abandoned by the serious logical probability theorist. I have tried to indicate above how such a theorist might begin making the necessary repairs. Their completion will very likely be a long, difficult, and thankless task which I gladly leave to him.[24]

Appendix

Proof of (2). Since e_Q' is defined as $h_1' \vee h_2' \vee \ldots \vee h_K'$ and the h_j' are pairwise mutually exclusive, we have by Bayes's theorem that

it is highly implausible, for example, that just one thing in the entire world is selected at random at t_3. This could be remedied by making *R* a three or more place predicate that takes arguments for places and/or observers as well at times.

[23] See Foundations, pp. 493–94.

[24] This paper was written before I became aware of the work on inductive logic done by Hintikka and his followers. (See, for example, several of the articles in J. Hintikka and P. Suppes, eds., *Aspects of Inductive Logic*, Amsterdam: North Holland, 1966.) As far as I am able to tell, however, these authors have neither discovered nor attempted to resolve the difficulties raised in this paper.

$$(4)\ c_\lambda(h_i', e_Q') = \frac{m_\lambda(h_i') \cdot c_\lambda(e_Q', h_i')}{m_\lambda(h_1') \cdot c_\lambda(e_Q', h_1') + \ldots + m_\lambda(h_\kappa') \cdot c_\lambda(e_Q', h_\kappa')}.$$

But clearly each h_j' entails e_Q', so (4) reduces to

$$(5) \qquad c_\lambda(h_i', e_Q') = \frac{m_\lambda(h_i')}{m_\lambda(h_1') + \ldots + m_\lambda(h_\kappa')}.$$

Hence the problem reduces to finding the m_λ-values of structure-descriptions. Since all the state-descriptions in a structure-description have the same Q-numbers, and since the m_λ-value of a state-description is a function only of λ, κ, and its Q-numbers, we will be able to calculate the m_λ-value of any structure-description if we know the Q-numbers associated with it and the number of state-descriptions it contains.[25] In view of the fact that $s = N - 1$, the Q-numbers associated with the structure-description h_j' are $s_j + 1$ and s_i, for all $i \neq j$. Hence by (10-7) of *Continuum* each state-description in h_j' has as its m_λ-value,[26]

$$(6) \qquad \frac{(\lambda/\kappa + s_j) \prod_i [\lambda/\kappa(\lambda/\kappa + 1)(\lambda/\kappa + 2) \ldots (\lambda/\kappa + s_i - 1)]}{\lambda(\lambda + 1)(\lambda + 2) \ldots (\lambda + N - 1)}.$$

The number of state-descriptions in h_j' is

$$(7) \qquad \frac{N!}{(s_j + 1)! \prod_{i \neq j} s_i!}$$

since this is just the number of those statistically equal distributions of N individuals among κ properties which assign s_i individuals to the ith property, for all $i \neq j$, and $s_j + 1$ individuals to the jth property, where

$$s_j + 1 + \sum_{i \neq j} s_i = N.$$

Hence the m_λ-value of h_j' is the product of (6) and (7). By substitut-

[25] The Q-numbers of a state-description are the numbers of individuals assigned to each Q-property by that state-description. Equation (10-7) of *Continuum* gives the m_λ-value of a state-description as a function only of λ, κ, and its Q-numbers.

[26] It is important to notice that in the iterated product in the numerator of (6) i takes all values from 1 to κ, including j.

ing an appropriate instance of this product for the numerator and each addend of the denominator of (5) and then removing from each of these terms the common factor

$$(8)\qquad \frac{\prod_i [\lambda/\kappa(\lambda/\kappa+1)(\lambda/\kappa+2)\ \ldots\ (\lambda/\kappa+s_i-1)]N!}{\lambda(\lambda+1)(\lambda+2)\ \ldots\ (\lambda+N-1)\ s_1!\ s_2!\ \ldots\ s_\kappa!}$$

we arrive at (2).

Proof of (3). Since the h_j $(1 \leqq j \leqq \kappa)$ are the only structure-descriptions compatible with e′ and are pairwise mutually exclusive, we have by Bayes's theorem that

$$(9)\qquad P_\lambda(h_i', e') = \frac{P_\lambda(h_i') \cdot P_\lambda(e', h_i')}{P_\lambda(h_1') \cdot P_\lambda(e', h_1') + \ \ldots\ + P_\lambda(h_\kappa') \cdot P(e', h_\kappa')}.$$

Each h_j' is a sentence of L_N^π, so $P_\lambda(h_j') = m_\lambda(h_j')$. Hence $P_\lambda(h_j')$ is the product of (6) and (7) as in the proof of (2). The value of $P_\lambda(e', h_j')$ is $(s_j+1)/(N)$, since in randomly selecting the $(N-1)$-membered sample described by e′ from the N-membered population described by h_j' there are just s_j+1 different individuals that might remain unselected. We now have a value to substitute for each term of (9). Under this substitution an appropriate instance of the product of (6) and (7) appears as a factor of the numerator and each addend of the denominator, so we can remove from each of these terms the common factor (8). This substitution and simplification reduce (9) to

$$(10)\qquad P_\lambda(h_i', e') = \frac{[(s_i+\lambda/\kappa)/(s_i+1)] \cdot [(s_i+1)/(N)]}{[(s_1+\lambda/\kappa)/(s_1+1)] \cdot [(s_1+1)/(N)] + \ \ldots\ + [s_\kappa+\lambda/\kappa)/(s_\kappa+1)] \cdot [(s_\kappa+1)/(N)]}$$

(Comparison of (10) and (2) shows that the only difference between them is the likelihoods. In (2) the likelihoods have dropped out, but not so in (10).) Further obvious simplifications of (10) reduce it to (3).

TOM SETTLE

Presuppositions of Propensity Theories of Probability

Introduction

A fully satisfactory account of a propensity theory has yet to be given even by proponents. Much less satisfactory have been the accounts given of propensity theory by its opponents. This paper is aimed to contribute to the debate over propensity theory by drawing attention to its metaphysical presuppositions. Although I do endorse a propensity theory of probability, this paper aims only secondarily to defend propensity theory against attack. Nonetheless a good starting point in my exposition might be an initial appraisal of a number of recent writings, in the hope that creative criticism may aid a more satisfactory characterization of probability construed as propensity.

Propensity theories of probability are relatively new in the field and are not yet widely endorsed. The still fairly widespread antipathy among scientists and philosophers of science to holding metaphysical beliefs militates against acceptance of a theory which is firmly rooted in ontology, and may even hinder understanding it.

The more popular views (frequency theories, logical theories, subjective theories of probability) are consistent with a positivist philosophy which emphasizes what may be experienced or observed more or less directly. Not so propensity theories. Frequency theories of probability focus upon the empirical outcomes of trials on chance setups. Logical theories focus upon the support some statements afford others. Subjective theories appear to presuppose that the root of probabilities lies in human ignorance — as Joseph Butler put it: "Nothing which is the possible object of knowledge . . . can be probable to an infinite intelligence; since it cannot but be discerned absolutely as it is in itself —

AUTHOR'S NOTE: A portion of this paper, in only a slightly amended form, has appeared in the *British Journal for the Philosophy of Science*, 23 (1972), 331–35.

certainly true, or certainly false. But for us, probability is the very guide of life" (1736, introduction).

By contrast with these theories, propensity theories dig deeply below the phenomenal surface, resembling in this respect what Bunge calls representative theories (1964). Indeed, propensities may be thought of as rooted in just those structures, characteristics and relations to which deep explanatory theories draw attention. This hints at the main ontological presuppositions of propensity theories and noticing it helps us to avoid certain confusions. For instance, failing to recognize propensities as deep, even occult, hence only indirectly testable — resembling in these respects field theories in physics — may mislead thinkers into confusing propensities with their chance outcomes (cf. Hacking, 1965), or into confusing the sense of propensity with how propensities might be measured or tested (cf. Braithwaite, 1953, Sklar, 1970), or into supposing that a syntactical or semantical rather than an ontological analysis will suffice to expose the sense of 'propensity' (cf. Levi, 1965).

Two distinct propensity theories of probability were introduced independently by C. S. Peirce (1910) and Sir Karl Popper (1957). Unfortunately neither philosopher has expressed his view sufficiently fully for the sense of propensity to be unambiguously clear. It is not my chief concern here to do exegesis on the original writings — although a little of that is needed to disentangle propensity theories from various misrepresentations. My aim is rather to give a fuller account of propensity in general and of a version of Popper's propensity theory of probability in particular.

I have discussed the preferability of Popper's view to Peirce's elsewhere (1974) and need not repeat that discussion at this point. Some of the arguments will in any case need to be repeated in my criticisms of recent writings by Mellor (1971) and Giere (1973) who both seem to prefer Peirce's theory where this differs from Popper's.

I

1. Adversus *Sklar: Are Some Propensities Probabilistic?*

The importance of metaphysical presuppositions is well brought out by a consideration of Sklar's attack on Popper's propensity theory of probability. The attack quite misses the mark because Sklar's metaphysical assumptions appear not to allow the formulation of Popper's theory.

For example, Sklar assumes determinism when discussing classical statistical situations (e.g., coin-tossing) (1970, pp. 361, 363, 366) and thus leaves room for only subjectivism or an objectivist theory related to the frequencies of actual outcomes. Thus, if propensity theory of probability were inconsistent with determinism, as I think it is, it would be ruled out at the start. Further, Sklar assumes a questionable version of positivism in seeking to analyze propensity in terms of the outcomes of trials on chance setups, thus taking concrete, individual phenomena as analytically primitive. If propensity were an unanalyzable primitive (as field of force is), no propensity theory could survive such (operationalist) analysis. Consistent with this positivist assumption is Sklar's confusion of the meaning of a term with the means of testing the truth of assertions employing the term (1970, p. 360), a confusion shared by Braithwaite.

Furthermore, it seems to me Sklar is asking the wrong question. My question, "Are some propensities probabilistic?" deliberately inverts the usual kind of question of which Sklar's title "Is Probability a Dispositional Property?" is an example. This type of question invites thinkers to contrast theories on the mistaken tacit assumption that there is only one proper theory of probability — this in spite of all of Carnap's influence, though it was probably Popper who first pointed out that there may be many theories, some corresponding to various interpretations of the calculus of probability, some not. Perhaps probability is not so homogeneous a concept that it must admit of only one sense. Perhaps some propensities could be probabilistic (or be the causes of chance outcomes) without all propensities having to be probabilistic or all statistical probabilities having to be rooted in probabilistic propensities. By contrast, the question I ask invites thinkers to scrutinize the concept of propensity and to ask whether some propensities might not be probabilistic. Thus my question calls for a general theory of propensity: it asks not simply whether certain predicates are dispositional, but rather what is it for an object or an object-environment complex to have a disposition, tendency, bent, proclivity, or propensity. This question will be taken up more forthrightly in later sections.

The divergence of Popper's views from those adversely criticized by Sklar is great. For example, Sklar writes "that probability is a dispositional property of 'the world,' a property that can be fully analyzed by reference to objective properties of things or of states of affairs, but only if the analysis makes use of the subjunctive mood . . . the subjunctive analysis needed is straight-forward, probability being analyzed in terms of what would be the limits of relative frequencies of outcomes in possible sequences of experiments. This is, I believe, Popper's view" (1970, p. 355). Sklar offers telling criticisms against this frequentist (mis)representation of Popper's views, leaving Popper's actual views quite unscathed. It is almost pointless for Popper to respond to Sklar's arguments; a simple denial that his view is captured by Sklar's characterization of it might suffice.

Nonetheless, one or two interesting points are raised by Sklar's analysis, aside from the point that there are important metaphysical presuppositions to a full-fledged propensity theory, for instance, Sklar's claim related to the distribution of values of parameters specifying initial conditions during a sequence of trials. Sklar claims that "what this distribution would be is completely unconstrained by any law-like features of the actual world whatever!" (1970, p. 363.) According to Popper, as he expressed himself to me in a private communication, there is a law of nature, that unless they are constrained, initial conditions have a ("natural") propensity to scatter over the interval left open to them by the (constraining) experimental conditions. Of course, that there is such a law of nature is a matter of conjecture. It is precisely this conjecture which, according to Popper, he and Lande have tried to formulate.

Secondly, Sklar raises the problem of the relativity of estimates of probability to specifications of the trial setup. He raises it in a curious context, namely, embedded in the problem of the utility of probability assertions, but it is an important problem in its own right. Indeed the problem was crucial to Popper's giving up a frequency interpretation of probability. In his 1959 article Popper presents this problem in a sharp fashion and adduces arguments which go to show in advance that Sklar's own theory of probability (compare Sklar, 1970, pp. 364–65) is at fault. Incidentally, Sklar does not cite Popper's 1959 article and thus may be supposed to have missed that presentation of a refutation of his own position. I shall discuss the problem of relativity to specification in section 3, where I discuss Giere's views. Giere considers it interestingly within the context of the problem of the single case, a problem Sklar does not seem to notice.

2. Adversus *White: Propensity Is Not Liability*

Alan White's article (1972) contains a number of errors of exegesis of a note by C. S. Peirce (1910), which should not be left uncorrected, as well as a number of other errors which render his criticisms of attempting to analyze probability in terms of liability beside the point. First the exegetical errors.

In calling Popper's propensity theory of probability (see Popper, 1957, 1959, and 1967) a "resuscitation" of Peirce's (1972, p. 35), White failed to take proper account of two significant differences between the theories. Peirce ascribes his "would-be" 's as a property of objects regardless of environment — in this he is followed by Mellor (see his 1969, 1971) — while Popper explicitly makes them relative to environment. This is not simply a matter of the conventions of language use, as Mellor suggests, but rather, and more importantly, a matter of what we want to assert, as scientists, about the world. Do we wish to say that the disposition of falling to the ground is a disposition of an object, as such, regardless of its environment? Would we not rather say it was a disposition relative to a gravitational field? Secondly, Popper's theory, but not Peirce's, is aimed to make sense of single-case probabilities.

White's second exegetical mistake is to ascribe to Peirce the intention to *analyze* the notion of probability (1972, p. 35). As far as I can tell, Peirce (1910) attempts no such analysis. Certainly Peirce does not describe what he is doing as analysis. Rather he says he is aiming to *define* probability (1910, pp. 404, 405, 409), a word "that we, all of us, use . . . with a degree of laxity which corrupts and rots our reasoning" (p. 405). Peirce's method is first to distinguish probability from likelihood and from plausibility by characterizing the latter terms (pp. 406–8), and then to define the meanings of probability statements by giving synonyms, one of which he coined for the purpose. This is not analysis. Indeed it is questionable whether we are called upon to analyze probability, construed as a propensity, any more than we are called upon to analyze gravitation, construed as a field of force. Some positivists attempt analyses of probability in terms of the relative frequencies of different kinds of outcomes of trials on a chance setup, but their analyses, one suspects, are aimed to eliminate probability as an unanalyzed primitive, in favor of discrete events or inert properties of particular objects (being head uppermost, say) as primitive. What Peirce's theory aimed to do, by contrast, was to suggest probability had roots in the qualities of the relevant physical objects.

Thirdly, White misrepresents what Peirce calls a "habit." White claims that a habit differs from a liability in the same way "what does happen when" differs from "what would happen if" (p. 35), which is an infelicitous contrast in view of Peirce's identification of habit with "would-be." Indeed Peirce is very careful not to confuse habit in his sense with habit in White's sense, that is, repetitious behavior (Peirce, 1910, pp. 409–10, 412).

Perhaps a more serious shortcoming is White's failure to appreciate which property propensity theories ascribe to what objects. White appears to be unable to shake off frequentist predilections sufficiently to think probability could be anything but the chance of certain outcomes of trial runs on chance setups. He seems not to appreciate that, as Popper puts it, "the word 'probability' can be used perfectly properly and legitimately in dozens of senses, many of which, incidentally, are incompatible with the formal calculus of probability" (1967, p. 31). (See also Popper, 1935; Carnap, 1945; and Bunge, 1967a, 1967b.) White writes, "Unfortunately for the propensity theory . . . probability is not *in fact* attributed to such things as coins, dice, . . . etc. It is not the probability of the coin, but of the coin's falling heads, that is in question" (p. 38, my italics). Thus with one sweep of his bold "in fact," White slays all propensity theories of probability! One must protest! If White is right on current usage (which I doubt), propensity theory challenges current usage — as do all novel theories. Propensity proponents, whatever else they disagree about, tend to agree that the probability of a coin's falling heads, *construed as a propensity*, is *not* a property of the coin's falling heads nor even a property of the set of possible outcomes of the coin's being tossed, but rather is a property of the physical object in the game — according to Peirce and Mellor, a property of the coin; according to Popper, Hacking, and myself, a relational property of the complex of tossing device and coin. Of course, probability construed as a frequency may be a property of the coin's falling heads relative to other possible outcomes. (For example see, Von Mises, 1919a, 1919b.) But that is a different theory.

White's most serious mistake, serious because it ruins his argument, is to identify propensity with liability. To be sure, Popper is to blame for introducing the phrase 'is liable to' as a partial explication of propensity (1957, p. 67). But White is to blame for taking the synonymy of liability and propensity seriously, because liability is referred to nowhere else in Popper's writings on propensity theory. Usually Popper uses 'tendency' or 'disposition' as synonyms. (See, for example, Popper, 1957, pp. 67, 68, 70; 1959, pp. 30, 31, 35, 36, 37; 1967, p. 32).

White is correct to notice the categorical difference between liability and likelihood and quite correct to affirm "one cannot analyse the notion of being likely in terms of the notion of *being liable*" (p. 43). But the declared purpose of his paper was to show the propensity theory of probability to be mistaken (p. 35). Remarks about liability are not at all to that purpose. White's criticisms concerning liability do not hold for tendency or disposition, save that likelihood is not to be analyzed in terms of tendency or disposition either. But then propensity proponents do not use these concepts to analyze 'probability' but to interpret it. Propensity theory of probability remains unscathed from White's attack, since precious little remains of his argument if propensity theory does not imply either that probability in the sense of chance is to be analyzed in terms of propensity or that propensity is to be understood as liability.

One final remark. It should not be surprising if proponents of propensity theory of probability differ. There is not simply one theory, as White's title suggests: there are at least eight significantly different theories, varying in their acceptability, ascribable severally to Peirce (1910), Braithwaite (1953), Popper (1957), Hacking (1965), Bunge (1968a, 1969), Mellor (1969, 1971), Giere (1973), and myself (1973a, 1974), aside from a "qualified propensity theory" such as Levi's (1967) and a caricature of a propensity theory such as Sklar (1970) criticized, but which does not belong recognizably to any of the authors mentioned. Incidentally, in none of these theories (save Sklar's caricature) is chance *analyzed* in terms of propensity and in none (save for Popper's slip of the pen) is propensity identified with liability.

3. *Giere versus Popper: The Problem of the Single Case*

Let us suppose, with Popper, that the frequency theory of probability (or the frequency interpretation of the calculus of probabilities) offers a consistent account of probabilities with respect to statistical sequences. There remains the problem of the single case. From a frequentist position, "the probability of an *event of a certain kind* . . . can be *nothing but* the relative frequency of this kind of event in an extremely long (perhaps infinite) sequence of events" (Popper, 1959, p. 29). Now, if the *kind* of event is defined in terms of the sequence, decisive objections can be raised. If, on the other hand, the kind is defined in terms of the generating conditions, there has been a shift from frequency theory to propensity theory. This, at least, is how Popper argues; and I endorse the argument. The crucial point about propensity theory, as I understand it in Peirce and Popper, is this: the probability of this or that outcome's happening given certain initial conditions of a chance setup is determined by the physical characteristics of the setup and not by any sequence to which the outcome might belong.

Giere thinks Popper's account of propensity runs into the same difficulties as Popper thought frequency interpretations did. Specifically, Giere argues that "relativity to specification" (1967, p. 39) plus defining propensity as a disposition "to produce sequences whose frequencies are equal to the probabilities" (1959, p. 35) leads to a relativity to sequences, since, as Popper asserts, "any particular experimental arrangement may be regarded as an instance of more than one specification for 'its' repetition" (1967, p. 38). Let me quote at length from Giere (1973):

> But what counts as a repetition of the same experiment? For example should the maximum height of the coin be specified? If it is not, then the toss in question may be a member of many virtual sequences, e.g. one in which the height is ten feet and one in which it is nine feet. But if the limiting frequency of heads is different in these two virtual sequences, as it might be, then, following the line of argument Popper himself employs against the limiting frequency interpretation (1959), it would be wrong to assign to this single case the virtual frequency of the sequence of ten foot tosses if it was in fact a nine foot toss. Thus Popper fails to solve the problem of the single case for the old-fashioned reason that he provides no solution to the problem of the reference class.

Let me say immediately that Giere is quite right to question what appears to be Popper's view that propensities are essentially tendencies to produce sequences. I say "appears to be" since it is not unambiguous in Popper's writings that this is his view. Giere is right to challenge such a view's capability to solve the problem of the single case. I shall return to this point shortly. First, however, I wish to clear up a confusion in Giere's remarks.

Giere appears to confuse specifying generating conditions with specifying sequences. I have difficulty understanding the relevance of the remarks about maximum height of toss of a coin unless either Giere has confused those two types of specification or Giere wishes us to believe that maximum height of toss is part of the generating conditions. This last I do not accept, though I am quite prepared to accept that the generating conditions may include a factor which *results* in different ranges of height tossed. What, we may ask, is the import of Giere's remarks about different *sequences* to which a nine-foot toss may belong? Certainly we should not be distressed if certain (virtual and virtually infinite) sub-sequences of the (virtual and virtually infinite) sequence of possible trials of any "well-characterized" chance setup had different limiting frequencies. This is perhaps to be expected; it is a component of the problem into which frequentists run when attempting specification of sequences. It fails to be a problem for the propensity theorist, however, since the more or less well-done specification of *generating conditions* gives no grounds for differentiating (virtual) sub-sequences of the (virtual) sequence of possible trials on the setup. Related, though quite different problems which arise for the propensity theorist, but not for the frequentist, are as follows: the relations between the virtual sequence picked out by the specified generating conditions and any actual sequence obtained by trials on a setup governed by those conditions; and the problem of interpreting the results of a statistical test. Giere's remarks about different heights of toss bear on this latter problem: if the factor resulting in different ranges of height tossed is omitted from the specification of generating conditions, actual sequences of tosses so generated may vary, some including, some not including, tosses above nine feet, thus increasing the range of possible members and thus the difficulty of interpreting the observations. It is, of course, open to the experimenter to improve the characterization of the setup, if this should occur to him, to include the missing factor and to try again. Nonetheless Giere's remarks do not hit Popper's propensity view, their intended target. Especially, Giere's claim that Popper has failed to solve the problem of the single case because he has failed to solve the problem of the reference class is beside the mark. The problem of the reference class does not arise in the same manner for propensity theorists as it does for frequentists nor is it the same kind of problem. For frequentists, as Popper has shown, the problem bites deeply into their interpretation of probability since it touches the difficulty of specifying sequences without specifying generating conditions. For propensity theorists, it raises merely the practical problem of interpreting tests, since the virtual sequences picked out by generating conditions will be more or less narrow in range of possible members accordingly as the generating conditions are more or less restrictively specified, the experimental setup more or less well characterized.

Let us now turn to the question whether a propensity theory which interprets propensities as tendencies to produce sequences is to be preferred to one which eschews such interpretation.

Giere clearly prefers to characterize propensity with no reference to a sequence. Popper has done this in referring to weighted possibilities. But Popper has taken a second step in interpreting the "weights" as "measures of a tendency of a possibility to realize itself upon repetition" (1967, p. 32). What end is served by taking this second step? At a first glance, there are two ends that might be served: first, unless propensities are linked to frequencies of chance outcomes, the estimates of probability interpreted as propensity cannot be tested by the most obvious method of comparing those estimates with the empirical measures of probability interpreted as a frequency; secondly, single-case propensity theory could illuminate the sense of probability as used in statistical laws in science only if single-case propensities are linked to the virtual frequencies of various possible outcomes of repeated trials. Nonetheless, those two advantages should not lead us to insist on the propensity *to produce sequences* being essential to the sense of the propensity to produce a single event, else the claim to give an account of single-case propensities is not adequately fulfilled. Let me put it this way: we may take the propensity to produce single events as prior and hence explain propensity to produce sequences. We could also, perhaps, explain the randomization within certain limits, of the values of measures of initial or generating conditions. But if we do so, there is a certain circularity in demanding that the sense of single-case propensities depends upon the sense of propensity to produce sequences. The circularity is not vicious, just disquieting. Perhaps it can be dispensed with. Perhaps we can accept that the sense of single-case propensities is illuminated, where repetitions are appropriate, by considering the virtual distribution of various outcomes on a virtual long run of trials, without its depending on such consideration.

In contrast to the frequency-linked sense of propensity, a thorough-going single-case propensity estimate might be a weighted possibility estimate without regard to repetitions of any trial. Common use of 'probable' illustrates such repetition-free estimates. For example, in the troublesome case of the race horse. Several philosophers have found it difficult to say how propensity theory might apply to horse racing, and this is so, if applying propensity theory means furnishing a method for calculating rational odds. Nonetheless, in estimating odds, a sensible gambler pays attention to various factors related to the physical characteristics of the various horses in the race — the identity of the horse's sire and dam; who his trainer is; news from the stable about his health and how well he moves on different going; and so on. But it is inappropriate here to speak of virtual sequences: each race, composed, as it is, of so many possible factors (horses, jockeys, the going) has such a strong air of uniqueness. The probability of a particular horse's winning a particular race, construed as a propensity, may be envisaged as a weight of a possibility among several variously possible outcomes.

One further question concerning relativity to specificity. Popper has asserted that propensities are properties of repeatable experimental arrangements rather than any particular experimental arrangement. Is this assertion primarily practical (methodological), making an assertion about experiments, especially regarding limitations in how they may be specified, assuming they must be specified before being performed and assuming they are *repeatable* with reference to the terms of the specification for their first performance? Or is the assertion epistemological, making an assertion about our knowledge or possible knowledge of propensities, especially drawing attention to the fallible, corrigible, and simple character statements of propensity share with theories? Or is the assertion ontological, asserting that propensities as properties of a physical setup are not independent of human definition (specification) of a subset of the many other properties characteristic of the setup? I find it hard to accept that Popper's remark is ontological, since, if it were, it would imply that a ghost ("the observer") had crept back into the system: that real propensities are specification-dependent is tantamount to the abandonment of belief in *reality* as opposed to a scientific conjectural reconstruction of reality. I am prepared to say that in the case of conjectured reconstructions of reality *conjectured* propensities are specification-relative, but not that the *real* propensities are so relative. I think the propensity of such-and-such an outcome of a trial or a particular chance setup — say the propensity of the actual setup to yield a six uppermost on the throw of a biased die — has a value, let us say, r. Differing ideas about what constitutes that particular chance setup, before or after its being set up, may lead to varying conjectures regarding the propensity in question, more or less close to r. But the varying specifications and points of view do not alter r; they merely make it more or less difficult to estimate r correctly. Whereas the actual propensity to yield a six uppermost depends upon the gravitational field, the resilience of the surface onto which the die is cast, the location of the bias and whether it is fixed, the shape of the earth's magnetic field at the place of throw if the bias is magnetic, etc., an *estimate* of the value of the propensity will depend upon the values and the weights allowed to each factor *thought relevant*, thus displaying a subjective component.

4. Mellor versus Popper: Of What Is Propensity a Property?

Mellor sides with Peirce, and with common usage, in ascribing propensities to things. As Popper noticed, this has been common in science since Aristotle, who ascribed potentialities to things: "Newton's was the first *relational* theory of physical dispositions and his gravitational theory led, almost inevitably, to a theory of fields of forces" (Popper, 1957, p. 70). I shall assume that common usage shall not be an arbiter over propriety of scientific usage, since at least one aim of science is to reform knowledge and hence to reform common sense. Therefore one result of successful science may be a reform of common usage. This leaves the question of the ascription of propensities open. Let us consider a few examples, say, Mellor's. He cites solubility and fragility, and gives an extended discussion of dying and of spontaneous disintegration of atomic nuclei.

First, solubility. Popularly sugar dissolving in water is a paradigm. But in science textbooks, solubility is considered more generally. It is a relational property defined on at least two substances which mix. Van't Hoft, who pioneered much interesting work on the properties of solutions, defined a solution as a homogeneous mixture of two or more substances. His definition is generally accepted. It is a matter of convention which of two miscible substances is to be called the solvent and which the solute, if these terms are to be applied. If a solid is mixed with a liquid, it is conventional to call the solid the solute even when it furnishes more than half the weight of the solution. Moreover, theories of chemical composition and of dissociation aid understanding of the process of dissolving, thus pointing up more sharply the relational nature of solubility. It distorts our understanding of dissolving to say solubility is a property of the solute. To claim that solubility is a property of a dyad, albeit an ordered dyad, is not to claim that neither member of the dyad has peculiar properties relevant to solubility. On the contrary, chemical and physical theory points not only to the action of one substance relative to another, but also to the structure and properties of each substance relevant to such relative action. For example, work on atomic structure, which had light shed upon it by the theory of dissociation arising from a study of conductivity of solutions, continued alongside other work on solutions, osmosis, and so on.

Second, fragility. To uncover the sense in which being fragile is a property of interest in scientific inquiry (as opposed to a property of interest to normative ethics) we need to explore theories which deal with the behavior of substances or objects of particular shapes, under stress of varying kinds (torque, impact, sheering, etc.). Fracture may then be seen as a limit phenomenon related to elasticity. Clearly, elasticity is dyadic, since it relates things or substances to stresses of various kinds. Fragility is hardly a scientific concept at all, though the propensity to fracture under certain loads is. There are root relevant properties properly ascribable to an individual object, such as the strength of connectivity of its parts (which incidentally is a relational property, also, but it is a property the parts of the whole share). Nonetheless, the propensity to break is a propensity relative to conditions which overwhelm the forces connecting parts together; hence it is not a property an object has in (conceptual) isolation. Common usage confuses the issue, to an extent, since fragility usually is ascribed only to those objects which are disposed to break under *ordinary* impacts, such as may result from falling or from being struck by hand, but such breaking points occupy a mere short span of the range of possible breaking points.

Spontaneous disintegration of radioactive nuclei is one of the few phenomena under current scientific investigation (I forgo predicting what other similar properties may or may not be the subject of future inquiry) for which the relevant propensity is possessed by objects in isolation. (Even so, deeper theories may probe the characteristics of nuclear connectivity and expose how the parts come to disconnect — but that is another matter.) However, examples such as this will not bear the burden of interpretation of the sense of propensity, which more often than not in scientific inquiry is a relational property.

I do not wish to go deeply into Mellor's discussion of death risk since it raises no further questions germane to my purpose in this paper. But I cannot forbear to remark how Mellor comes so close to a dyadic theory of propensity without actually endorsing it. He writes, "With a variety of people in similar environments we take their different chances of death to display different propensities in them. With similar people in a variety of environments we take their different chances of death to display different propensities in the environment" (1970, p. 89). It seems curious of Mellor to fail to connect the various relevant properties of individuals (say, the weakness of their respiratory organs) with relevant properties of the environment (say, the air pollution index) in order to arrive at death risk as displaying (to use his term) a propensity *jointly* of the person and his environment to induce his death or as displaying a propensity of *an individual relative to an environment* to die. I find the notion of a person's propensity to die in isolation from any environment somewhat incoherent, although the propensity of a radioactive nucleus to disintegrate in isolation is not similarly incoherent.

My point is not that all dispositions are relative, but that dispositions of isolated systems should be considered as special cases, with relative dispositions (dispositions of systems relative to environment) as the general and basic case (cf. my discussion of this point, 1974).

II

5. *Variants of Propensity Theories*

If my aim is to expose presuppositions of propensity theory, it behooves me to say whether the exposure is intended to relate to all possible propensity theories or only to some. If the latter, I should say which. Let me first recall the two significant differences between Peirce's view and Popper's to which I referred in section 2: whether propensities are to be ascribed to systems (à la Peirce) or system-environment complexes (à la Popper); and whether the theory is intended to explicate the single case (à la Popper). These differences supply a table into which the eight theories I have referred to fit more or less neatly:

	Frequentist	*Single Case*
Systems	Peirce	Giere Mellor
Complexes . . .	Braithwaite Hacking	Bunge Popper Settle

Of course, we can readily distinguish Braithwaite's views from Hacking's by virtue of Braithwaite's insistence that the meaning of a probability assertion is to be found at least in part in the rules adopted for the acceptance or rejection of attributions of particular degrees of probability to particular outcomes. And we can distinguish Mellor's views from those of others by virtue of his curious dissociation of propensities from statistical probability. For Peirce and Popper, propensity theory was intended to interpret statistical probability; for Mellor, statistical probability is to be grounded on personalist theories while things having dispositional properties is what makes some chance statements true. The differences between the views of Bunge, Popper, and myself are much more slight and may even be illusory: clarification of the various expositions may reveal identity of view. I do not propose to expose these differences now, since I have done so in my (1974) to an extent, except to note that whereas Popper and I talk about the propensity of system-environment complexes to realize certain events, Bunge several times has talked about the propensity of events to realize themselves (e.g., 1967a, 1967b). This may simply be a difference in manners of speaking or it may more seriously be a difference in theories of causality of events. I think it the former and hence I do not wish to pursue it further.

Initially, I shall aim to expose some philosophical presuppositions of my own theory of propensity, which closely resembles Popper's and Bunge's. I shall argue that theories which do not share these presuppositions are less satisfactory on that account. Although I shall not give any detailed exegesis of the literature to pick out which thinkers appear to share my presuppositions and which not, I shall give some cursory indications.

6. *Skeptical Realism*

The first group of presuppositions to which I wish to draw attention cluster together to form a view I have dubbed 'skeptical realism'. Let me explain them one by one.

Realism. In the first place, and perhaps trivially, I should say I do not want to fight over word usage; hence I shall not object to a solipsist or anyone else using the term 'propensity' in any sense he cares. Nonetheless I think the sense of propensity in a metaphysics which failed to assign reality to things and to their properties would be vacuous.

In the second place, and more importantly, the realism I have in mind is opposed to various versions of instrumentalism or operationalism which rule out assigning reality to the possible referents of theoretical terms. I have a number of grounds for thinking false the philosophical theory which assigns existence only to those objects or characteristics of objects which are the matter of immediate perception, but I do not wish to go deeply into the problems of perception at this point. It perhaps suffices here to remark that a propensity theory embedded in such a view would be vacuous. On such a strict phenomenological view, it would be possible to give a construal to the term 'habit' in White's sense (see section 2 above) but not to distinguish between habit in that sense and habit in Peirce's sense of "would-be." Since that distinction is crucial to Peirce's (and Popper's) concept of propensity, a strict phenomenology is inimical to propensity theory.

In the third place, I should remark that the realism I have in mind does not imply either that sense data are reliably veridical (naive realism) or that there is a referent for every theoretical term in the sciences. Rather it implies that theories in the sciences are intended to model the real world more or less closely.

Realism is not peculiar to propensity theories of probability, being a presupposition also of frequency theories. It is not, however, presupposed by subjectivist or personalist theories of probability, since those theories are compatible with ontic determinism. Ontic determinism rules out objective chance and hence evacuates probability statements of any reference to reality except superficially, as mere reports of observed frequencies: probability statements refer instead to ideas, to private states of partial belief. This is not to say, however, that ontic indeterminism may not be wedded to a personalist theory of probability so that reality may be assigned to some referents of some probability statements (see Mellor, 1971).

Depth. Propensities generally do not lie on the surface of things, hence propensity theory presupposes depth. Depth has at least two senses worth distinguishing here. In the first place, depth refers to theoretical sophistication in the sense of maturity, as discussed by Bunge (1968b). In the second place, depth refers to distance from sight, to hiddenness. These are not the same concepts, although their extensions overlap. For example, the classic "dormitive virtue" is an *occult* power but displays no theoretical depth. Some propensities are deep in both senses, for example, the propensity of radioactive nuclei to disintegrate.

Although realism is not peculiar to propensity theory of probability, the presupposition of depth is. Both frequentist and subjectivist theories are compatible with a strict phenomenology, with what Bunge calls "black-boxism" (1964). Indeed, the development of frequency theories of probability in this century as the way to obtain an *objective* interpretation of probability went hand in hand with the development of a fairly strict positivism, which sought to reduce deep concepts (theoretical concepts) to observational concepts or even to operations.

A propensity theory which did not presuppose propensities to be occult properties (if not theoretical) and thus deep would reduce a propensity statement to a shorthand form for a set of statements predicting outcomes of possible trials on various setups.

Skepticism. I deliberately emphasize propensity in general as an occult property to draw attention to a deep and far-reaching tacit error in modern philosophy of science and in modern epistemology, namely the assumption that the propriety of knowledge claims depends upon their being justified, or at least justifiable. This assumption, coupled with the traditional empiricist assumption that experience is the only legitimate source of warrant for knowledge claims about the world, led to a widespread denunciation of serious speculation about what lies below the surface of things. Scientific knowledge was presented as a more or less satisfactory ordering of reports of experience. Where scientists' speculations gave deep theories, thinkers in this empirical justificationist tradition reinterpreted those theories, to their ontological impoverishment. Propensity theory clashes with this tradition, since propensity statements share with deep theoretical statements an incapacity to meet the demands of empirical justificationism without evacuative reinterpretation. Hence propensity theory may be said to presuppose the falsity of empirical justificationism.

Let me explain this point a little more fully. Empirical justificationism initially presupposes two types of scientific (or, simply, cognitive) claim — particular (singular) and general (universal); and two types of scientific (nonlogical) term — observational and theoretical (or nonobservational). Putting these two together leads to a further division of universal claims into empirical laws, using only observational terms; theories, using only nonobservational terms; and bridging laws (or correspondence rules), using both sorts. There are difficulties in carrying through all these distinctions in a thorough fashion (see, for example, Suppe, 1972), but let us not consider these just now. Next, empirical justificationism presupposes that singular statements employing observational terms can in principle be checked for truth or falsity by inspection of the real world by a competent observer who can report sensibly what he sees. Again there are serious problems here, some of which spill over from the difficulty of carrying through the distinctions above in a thorough fashion — for example, the problem that observation reports are theory-impregnated and hence are remote from raw observation (Popper, 1935; Agassi, 1966a). Others are fresh problems, such as Descartes's objection to experience as a final and authoritative arbiter on matters of fact: namely, that any particular putative observer at any particular time may be deluded, dreaming, drunk, drugged, or simply mistaken (Descartes, 1741). Descartes's problem is usually met by some such device as Popper's suggestion that intersubjectivity replace objectivity as a desideratum for scientific claims to satisfy. Of course, that quite dethrones experience as arbiter and places the solution of problems of perception, etc., into the domain of social conventions of justification. (For further discussion of this point see my 1969b, 1971a, 1974, and Agassi, 1971.) Goodman (1955) shows it may be impossible to justify the use of one predicate rather than another similarly justified yet startlingly different predicate in a true description of an object. Goodman's riddle is usually presented as a riddle of induction, but it has repercussions at the simple level of reporting experience. I do not believe that any of the attempts to solve Goodman's riddle within the framework of a logic of confirmation have had success: the riddle seems capable of solution only in the context of a socially acceptable standard for justifiable (excusable) behavior.

Assuming, for the moment, that all those problems can be taken care of, empirical justificationism asserts that for a claim to know a universal statement (law, theory) to be properly made, the statement must enjoy support from other justified statements, principally singular observation reports. Now it is quite clear that a universal statement is not proved by an accumulation of particular statements which report instances of it — except where the instances exhaust the possibilities. This point was first made by Aristotle (*Posterior Analytics*). Hume is famous for his refutation of induction (see his 1748) in which he appears to me to confuse Aristotle's objection with an objection raised by Maimonides (1194): experience cannot tell us that what we take to be the laws of nature will not alter in the future (say, tonight). Coupled with Goodman's riddle (which is a fresh amalgam of Aristotle's and Maimonides' objections) these objections are decisive against any strong form of empirical justificationism (see my 1969b, 1971b). If it is difficult for empirical laws to satisfy the demands of empirical justificationism, it is even more difficult for theories to do so, since they are further removed from raw observation. The same holds for propensities. But for propensities, one further difficulty arises: the problem of potency. Many propensity claims pick out potencies or potentialities of systems-in-contexts. The tough-minded version of empirical justificationism has always disallowed potencies as real, since potency is a metaphysical notion with no counterpart among the concepts of the mechanics of inert matter (compare Wisdom, 1971). This point will be taken up in the next section.

Popper has suggested giving up justificationism, thus allowing once more deliberate speculation about what the world is made of fundamentally. Speculation would not of itself lead afresh to essentialism or intellectualism, since these metaphysical viewpoints are themselves justificationary, though not empiricist. Empiricists had hoped to curb these metaphysical excesses by demanding *empirical justification,* but the demand itself has proved excessive. It is clear from the history of the most creative moments in science that at least part of the task of scientists is freely to propose hypotheses concerning the hidden structure and properties of physical systems and environments which may explain their observable structure and properties — and perhaps even help us to control their future behavior in some respects. Perhaps excesses would be sufficiently curbed and rationality — hence intellectual respectability — sufficiently safeguarded, if all hypotheses were to be held open to criticism, including empirical criticism where any can be devised. (For the background of this theory of rationality and for further discussion of the role of criticism in rationality see Popper, 1945; Bartley, 1962, 1964;

Agassi, 1966b, 1969; Jarvie and Agassi, 1967; Agassi, Jarvie, and Settle, 1970; Settle, 1971a, 1974; Settle, Jarvie, and Agassi, 1974.)

7. *Dynamism*

Propensity theory presupposes propensities to be causes or at least partial causes of the events which the carrier of the propensity is disposed to bring about. It seems merely vacuously a property of a system relative to an environment that if the system be placed in such an environment, certain events follow. The constant conjunction theory of causation evacuates all causal links of their active or dynamic component. Moreover, it does so on empirical justificationist grounds. (One more reason for rejecting empirical justificationism?) Customarily modern philosophers of science eschew use of the ancient concept of efficient cause, replacing causal explanation (in Aristotle's sense) with deductive-nomological explanation (in Hempel's sense) (Hempel and Oppenheim, 1948; Hempel, 1961, 1964). There are some exceptions (Bunge, 1959; Agassi, 1964; Settle, 1973a; Brody, 1972), but not many, considering the trouble proponents of the covering law model have had in trying to block undesirable maneuvers. (For detailed discussion see Nagel, 1961; Eberle, Kaplan, and Montague, 1961; Kaplan, 1961; Kim, 1963; Hempel, 1964; Ackermann, 1965; Ackermann and Stenner, 1966; Omer, 1970; Morgan, 1972.)

The most serious objection to the deductive-nomological theory of scientific explanation is that it appears to provide no adequate grounds for distinguishing between those generalizations which we wish to consider as natural laws (or at least as approximations to natural laws) and those which we think to be only superficial laws (phenomenological laws), or for distinguishing both of those from accidental generalities. I follow Agassi (1964) in thinking that theories regarding what is naturally necessary (natural laws) are properly judged relative to a metaphysics or cosmology. This propriety arises not only because what is to be announced as physically necessary will first, as a matter of scientific custom (compare Bunge, 1967c, 1970c), have to pass metaphysical tests, but also because the very idea of natural necessity is metaphysical, hence already rooted in a metaphysics, or world view. Of course, it is not absurd to deny that there is a natural necessity, though the world view such a denial presupposes might be hard to live with, psychologically speaking.

In spite of the strength of the empiricist tradition stemming from

Hume, the idea of dynamic cause remains a presupposition of much discourse both in the arts — for example, law, medicine, and historiography — and in the sciences, not to mention common speech (!). In physics, for example, we have not been able quite to rid ourselves of the notion of *force* (gravitational, electromagnetic, nuclear, etc.) despite some energetic, not to say forceful and influential, analytic attempts — notably Mach's (1883). For example, the dynamic notion of gravitational collapse is used to explain both the genesis and the nemesis of giant stars or stellar systems (collapsars or "blackholes"). Of course, by no means all laws in science make use of the concepts of force or power or some other version of dynamism or efficient causation; and even in those which do the idea of drive or cause is often masked by the symbolic representations of the theories, which might help to explain why scientific explanation is so widely held to be the subsumption of instances under generalities. I do not mean to deny that some explanations in science are merely matters of such subsumption, as, for example, explanations afforded by phenomenological ("black-box") laws. Nonetheless, explanations, to be fully satisfactory, appear to need some causal or dynamic assumption in the explanations. It is not my intention here to discuss theory of scientific explanation in detail: I merely wish to stress that the causal or dynamic character of propensities brings propensity theory into conflict with the received view of theories and of scientific explanation.

In my view, the idea of propensity is closely linked to the idea of natural law: natural laws imply propensities. A statement of a propensity may perhaps be regarded as a conjectured ontological hypothesis (or at least an hypothesis rooted in an ontology) concerning some physical property of a system-environment complex, declaring or hinting at a natural dynamical relation between the system and its environment, or parts thereof, in certain, perhaps virtual, states, which would lead to the complex entering other states.

8. *Indeterminism*

Nonetheless, the gap between deep theories and reality is great, and perhaps unbridgeable. The Parmenidean rift between reality and appearances has not been bridged by any theory so far proposed, if by bridging it we mean showing how human speculations concerning the constitution and mode of behavior of the real world (of, say, a thing-in-itself) have the right to be correct. In my view, this rift may be bridged, but only after a fashion: our speculation concerning the structure and dynamism of the real world is our attempt to bridge it; and our criticisms of our speculations are efforts to eliminate poor bridges. Even our best bridges may be poor ones. In particular we may notice scientific knowledge is built up piecemeal and not even cumulatively: law statements refer simply to one level of analysis of reality (or to one degree of resolution or focus). Programs for the *integration* of bodies of theory referring to varying domains of reality remain incomplete despite a few successes, such as the integration of theories in the electrical and magnetic domains. I follow Bunge (1967c, 1970b) in regarding the reduction of theories at some levels or in some domains to theories at other levels or in other domains as still largely a program, despite Nagel's (1961) optimism, except where one theory turns out on axiomatization to be a subtheory of another. The contrary would be surprising. It would be surprising if we had so successfully grasped the core of reality that we could integrate all laws into one axiom system.

Although it may be said that all systems of theories intended to illuminate or explain how the world appears to us model the real world to an extent or after a fashion, including existentialist theories of man, the claim of any one particular theory system (say elementary particle physics or existentialism-cum-phenomenology) to sovereignty over all is to be resisted in view of the illumination afforded our comprehension of reality by a variety of systems arising from a variety of types of analysis and comprising a variety of law statements, some qualitative, some quantitative, some deterministic, some stochastic.

In particular, the claim made by some philosophers, oblivious perhaps of the fragmented, many-layered quality of the body of scientific knowledge, the claim namely that there are only two metaphysical options regarding orderliness — strict determinism and the doctrine of sheer chance — is to be resisted.

Both strict determinism and the doctrine of sheer chance are extreme and happen to be oversimplifications: both of them oversimplify the relation between law statements belonging to different levels of analysis by supposing them all reducible to one rock-bottom level which is taken to be either deterministic or stochastic. In my view, both strict determinism, which contradicts ontic probability, and the doctrine of sheer chance, which undermines the lawfulness propensity theory relies upon, are inimical to probabilities as propensities. Hence I say that propensity theory of probability, though not propensity theory in general, presupposes the falsity of both those ontological doctrines as well as the falsity of the reductionism which they presuppose. I shall return to this point shortly.

Of course, even if all natural laws were deterministic, there might still be statistical laws in science, owing to our ignorance or to our habit of referring to just one level of analysis or of focus at once. Such laws would be superficial: to uncover probabilities as propensities, we should have to dig more deeply. The question whether some propensities, some natural hidden potencies of some systems in some environments might be probabilities is equivalent to the ontological question whether the universe is thoroughgoingly deterministic or is at least in part stochastic. This is not perhaps the place to defend my view that the universe is partially indeterministic (see my 1973b), but at least I should distinguish several senses of determinism in order to present the view more sharply. Let us call 'deterministic' a law or theory which yields a unique value for each variable in its formalism when the values of the other variables are fixed. Then we could distinguish determinism as a regulative maxim — "Look for deterministic laws or theories!" — from the epistemological claim that all the laws we know in science are deterministic; and we could distinguish both of these from ontic determinism — "All true laws of nature are deterministic."

Is propensity theory of probability consistent with ontic determinism? Of course, epistemic indeterminism — the view that as far as we know in science at the moment some laws are stochastic — is consistent with determinism both as a maxim and as a metaphysical claim. Hence, on the face of it, theories of probability should also be consistent with determinism. In my view, the realism presupposed by propensity theory and the presupposition of depth commit proponents of a propensity theory of probability to ontic indeterminism. No other theories of probability presuppose indeterminism, although none is inconsistent with it.

9. *The Crux of the Matter*

I began part II by announcing that I aimed to expose the presuppositions of that propensity theory which Popper, Bunge, and I appear to share, after having first differentiated on various grounds between that theory and several others. It is now clear that a further, perhaps more important, differentia could have been used. We may distinguish between, on the one hand, a theory of propensity which is little more than a frequentist theory embellished with an abbreviative concept — disposition — and, on the other hand, a theory which asserts propensity in the ontologically full sense I have sketched. Put another way, we may have a theory in which the propensity concept summarizes possible outcomes, contrasted with a theory in which the propensity is regarded as at least part cause of the outcomes. Of course, viewing probability as a frequency is not incompatible with viewing probability as a propensity. On the contrary, the two may go well together. Bunge has gone further and suggested that what he calls "physical probability theory" should be viewed as a union of three models:

"M_1: Propensity interpretation (quantified objective possibility) . . .
M_2: Randomness interpretation (objective chance) . . .
M_3: Statistical interpretation (relative observed frequency)" (1967a, pp. 90–91).

Not all thinkers wish to enhance a frequency interpretation (a union of Bunge's M_2 and M_3?) by the addition of a causally interpreted propensity theory, no doubt feeling that to allow that probability construed as a frequency is a dispositional property of the objects or generating conditions concerned goes far enough. Perhaps here is the crux of the matter: *Is the introduction of propensity theory of probability to be allowed to reintroduce (or at least to emphasize) an active or dynamic ontology at the root of science?* For my part, I hope so.

There is, of course, much clarificatory work to be done on such an ontology and on its bearing on probability. I could try here only to draw attention to it and to its bearing on propensity theory of probability. I hope I have done enough to provoke more discussion. My view that philosophy may be creatively relevant to science (see my 1971b) gives me optimism that further work may shed yet more light on probabilistic theories in physics and elsewhere as well as on ordinary language discussions of chance.

REFERENCES

Ackermann, R. (1965). "Deductive Scientific Explanation," *Philosophy of Science*, vol. 32, pp. 155–67.

Ackermann, R., and A. Stenner (1966). "A Corrected Model of Explanation," *Philosophy of Science*, vol. 33, pp. 168–71.

Agassi, J. (1964). "The Nature of Scientific Problems and Their Roots in Metaphysics," in M. Bunge, ed., *The Critical Approach to Science and Philosophy*. New York: Free Press. Pp. 189–211.

Agassi, J. (1966a). "Sensationalism," *Mind*, vol. 75, pp. 1–24.

Agassi, J. (1966b). "Science in Flux: Footnotes to Popper," in R. S. Cohen and M. W. Wartofsky, eds., *Boston Studies in the Philosophy of Science*, vol. 3. Dordrecht: Reidel. Pp. 293–323.

Agassi, J. (1969). "Unity and Diversity in Science," in R. S. Cohen and M. W. Wartofsky, eds., *Boston Studies in Philosophy of Science*, vol. 4. Dordrecht: Reidel. Pp. 463–522.

Agassi, J. (1971). "Positive Evidence as a Social Institution," *Philosophia*, vol. 1, pp. 143–57.

Agassi, J., I. C. Jarvie, and T. W. Settle (1970). "Grounds of Reason," *Philosophy*, vol. 45, pp. 43–49.

Aristotle. *Posterior Analytics*.

Bartley, W. W. (1962). *The Retreat to Commitment*. New York: Knopf.

Bartley, W. W. (1964). "Rationality versus the Theory of Rationality," in M. Bunge, ed., *The Critical Approach to Science and Philosophy*. New York: Free Press. Pp. 3–31.

Braithwaite, R. B. (1953). *Scientific Explanation*. Cambridge: At the University Press.

Brody, B. A. (1972). "Towards an Aristotelian Theory of Scientific Explanation," *Philosophy of Science*, vol. 39, pp. 20–31.

Butler, J. (1736). *Analogy of Religion*. New York: Ungar, 1961.

Bunge, M. (1959). *Causality*. Cambridge, Mass.: Harvard University Press.

Bunge, M. (1964). "Phenomenological Theories," in M. Bunge, ed., *The Critical Approach to Science and Philosophy*. New York: Free Press. Pp. 234–55.

Bunge, M. (1967a). *Foundations of Physics*. New York: Springer-Verlag.

Bunge, M. (1967b). *Scientific Research 1*. New York: Springer-Verlag.

Bunge, M. (1967c). *Scientific Research 2*. New York: Springer-Verlag.

Bunge, M. (1968a). "Philosophy and Physics," in R. Klibansky, ed., *Contemporary Philosophy, a Survey*. Florence: La Nuova Italia Editrice.

Bunge, M. (1968b). "The Maturation of Science," in I. Lakatos and A. Musgrave, eds., *Problems in the Philosophy of Science*. Amsterdam: North Holland.

Bunge, M. (1969). "What Are Physical Theories About?" *American Philosophical Quarterly*, monograph no. 3, pp. 61–99.

Bunge, M. (1970a). "Theory Meets Experience," in *Mind, Science, and History*. Albany: State University of New York Press. Pp. 138–65.

Bunge, M. (1970b). "Problems Concerning Intertheory Relations," in P. Weingartner and G. Zecha, eds., *Induction, Physics, and Ethics*. Dordrecht: Reidel. Pp. 285–315.

Carnap, R. (1945). "Two Concepts of Probability," *Philosophy and Phenomenological Research*, vol. 5. Reprinted in H. Feigl and W. Sellars, eds., *Readings in Philosophical Analysis*. New York: Appleton, 1949. Pp. 330–48.

Descartes, R. (1641). *Meditations*.

Eberle, R., D. Kaplan, and R. Montague (1961). "Hempel and Oppenheim on Explanation," *Philosophy of Science*, vol. 28, pp. 418–28.

Giere, R. N. (1973). "Objective Single Case Propensities and the Foundations of Statistics," in P. Suppes, L. Henkin, A. Joja, and Cr. C. Moisil, eds., *Logic, Methodology and Philosophy of Science*, vol. 4, *Proceedings of 1971 International Congress, Bucharest*. Amsterdam, North Holland.

Goodman, N. (1955). *Fact, Fiction and Forecast*. London: University of London Press.

Hacking, I. (1965). *Logic of Statistical Inference*. Cambridge: At the University Press.

Hempel, C. G. (1961). "Deductive-Nomological vs. Statistical Explanation," in H. Feigl and G. Maxwell, eds., *Minnesota Studies in the Philosophy of Science*, vol. 3. Minneapolis: University of Minnesota Press. Pp. 98–169.

Hempel, C. G. (1964). *Aspects of Scientific Explanation*. New York: Free Press.

Hempel, C. G., and P. Oppenheim (1948). "Studies in the Logic of Explanation," *Philosophy of Science*, vol. 15, pp. 135–75.

Hume, D. (1748). *Inquiry Concerning Human Understanding*.

Jarvie, I. C., and J. Agassi (1967). "The Problem of the Rationality of Magic," *British Journal of Sociology*, vol. 18, pp. 55–74.

Kaplan, D. (1961). "Explanation Revisited," *Philosophy of Science*, vol. 28, pp. 429–36.

Kim, J. (1963). "On the Logical Conditions of Deductive Explanation," *Philosophy of Science*, vol. 30, pp. 286–91.

Levi, I. (1967). *Gambling with Truth*. New York: Knopf.

Mach, E. (1883). *Die Mechanik*.

Maimonides, M. (1194). *Guide for the Perplexed*. Eng. trans. Friedlander, 1881.

Mellor, D. H. (1969). "Chance," *Aristotelian Society Supplementary Volume*, vol. 43, pp. 11–36.

Mellor, D. H. (1971). *The Matter of Chance*. Cambridge: At the University Press.

Morgan, C. G. (1972). "On Two Proposed Models of Explanation," *Philosophy of Science*, vol. 39, pp. 74–81.

Nagel, E. (1961). *The Structure of Science*. New York: Harcourt, Brace.

Omer, I. (1970). "On the D-N Model of Scientific Explanation," *Philosophy of Science*, vol. 37, pp. 417–33.

Peirce, C. S. (1910). "Notes on the Doctrine of Chances," in C. Hartshorne and P. Weiss, eds., *Collected Papers of Charles Sanders Peirce*, vol. 2 (1931), pp. 405–14.

Popper, Sir K. R. (1935). *Logik der Forschung*. Vienna: Julius Springer.

Popper, Sir K. R. (1945). *The Open Society and Its Enemies*. London: Routledge.

Popper, Sir K. R. (1957). "The Propensity Interpretation of the Calculus of Probability and the Quantum Theory," in S. Korner, ed., *Observation and Interpretation: A Symposium of Philosophers and Physicists*. London: Butterworth. 2nd ed. entitled *Observation and Interpretation in the Philosophy of Physics*. London: Dover, 1962.

Popper, Sir K. R. (1959). "The Propensity Interpretation of Probability," *British Journal for the Philosophy of Science*, vol. 10, pp. 125–42.

Popper, Sir K. R. (1967). "Quantum Mechanics without 'The Observer,'" in M. Bunge, ed., *Quantum Theory and Reality*. New York: Springer-Verlag. Pp. 7–47.

Settle, T. W. (1969a). "The Point of Positive Evidence — Reply to Professor Feyerabend," *British Journal for the Philosophy of Science*, vol. 20, pp. 352–55.

Settle, T. W. (1969b). "Scientists: Priests of Pseudo-Certainty or Prophets of Enquiry?" *Science Forum*, vol. 23, pp. 22–24.

Settle, T. W. (1971a). "The Rationality of Science *versus* the Rationality of Magic," *Philosophy of the Social Sciences*, vol. 1, pp. 173–94.

Settle, T. W. (1971b). "The Relevance of Philosophy to Physics," in M. Bunge, ed., *Problems in the Foundations of Physics*. New York: Springer-Verlag. Pp. 145–62.

Settle, T. W. (1973a). "Are Some Propensities Probabilities?" in R. J. Bogdan and I. Niiniluoto, eds., *Logic, Language, and Probability*. Dordrecht: Reidel. Pp. 115–20.

Settle, T. W. (1973b). "Human Freedom and 1568 Versions of Determinism and Indeterminism," in M. Bunge, ed., *The Methodological Unity of Science*. Dordrecht: Reidel.

Settle, T. W. (1974). "Induction and Probability Unfused," in P. A. Schilpp, ed., *The Philosophy of Karl R. Popper*. La Salle, Ill.: Open Court.

Settle, T. W., I. C. Jarvie, and J. Agassi (1974). "Towards a Theory of Openness to Criticism," *Philosophy of the Social Sciences*, vol. 4, pp. 83–90.

Sklar, L. (1970). "Is Probability a Dispositional Property?" *Journal of Philosophy*, vol. 67, pp. 355–66.

Suppe, F. (1972). "What's Wrong with the Received View on the Structure of Scientific Theories?" *Philosophy of Science*, vol. 39, pp. 1–19.

von Mises, R. (1919a). "Fundamentalsatze der Wahrscheinlichkeitsrechnung," *Mathematische Zeitschrift*, vol. 4, pp. 1ff.

von Mises, R. (1919b). "Grundlagen der Wahrscheinlichkeitsrechnung," *Mathematische Zeitschrift*, vol. 5, pp. 52–99.

White, A. R. (1972). "The Propensity Theory of Probability," *British Journal for the Philosophy of Science*, vol. 23, pp. 35–43.

Wisdom, J. O. (1971). "Science versus the Scientific Revolution," *Philosophy of the Social Sciences*, vol. 1, pp. 123–44.

JEFFREY BUB

Popper's Propensity Interpretation of Probability and Quantum Mechanics

This paper is a critique of Popper's interpretation of quantum mechanics and the claim that the propensity interpretation of probability resolves the foundational problems of the theory. The first two sections are purely expository. In section 1, I sketch Popper's conception of a formal theory of probability and outline the propensity interpretation. Section 2 is a brief description of the gist of Popper's critique of the Copenhagen interpretation of quantum mechanics and his proposed solution to the measurement problem (the problem of the "reduction of the wave packet"). In section 3, I show that the propensity interpretation of probability cannot resolve the foundational problems of quantum mechanics, and I argue for a thesis concerning the significance of the transition from classical to quantum mechanics, which I propose as a complete solution to these problems. Finally, section 4 contains some general critical remarks on Popper's approach to the axiomatization and interpretation of formal theories of probability.

1

In Appendix *iv of *The Logic of Scientific Discovery,*[1] Popper criticizes Kolmogorov[2] for failing to carry out his program of constructing a purely *formal* theory of probability: Kolmogorov assumes that in an equation like

$$p(a,b) = r$$

AUTHOR'S NOTE: Research was supported by the National Science Foundation.

[1] K. R. Popper, *The Logic of Scientific Discovery* (New York: Harper Torchbooks, 1965).

[2] A. N. Kolmogorov, *Foundations of the Theory of Probability* (New York: Chelsea, 1950).

where $p(a,b)$ is the conditional probability of a given b, the elements a and b are *sets*. Popper's proposal is that nothing should be assumed concerning the nature of the elements $a,b,c, \ldots$ in the set S of admissible elements on which the probability functions $p(a,b)$ are defined, beyond the closure of the set S under the two operations: *product* (or meet, or conjunction) and *complement*. For the axioms of the formal theory of probability determine an algebraic structure of a certain kind for the set S, if we define equivalence in S by

$$a = b \text{ if and only if } p(a,c) = p(b,c) \text{ for every } c \text{ in } S.$$

With Popper's axioms, S is a *Boolean algebra*. (This is demonstrated at the end of Appendix *v.) Thus, the elements of S may be interpreted as sets, or predicates, or sentences, etc.

Popper distinguishes the task of finding one or more suitable *axiomatizations* of the formal calculus of probability from the task of finding one or more suitable *interpretations*. These he classifies as subjective or objective. Among the objective interpretations are the *classical* interpretation of de Moivre and Laplace (in which $p(a,b)$ is interpreted as the proportion of equally possible cases compatible with the event b which are also favorable to the event a), the *frequency* interpretation of Venn and von Mises (where $p(a,b)$ is the relative frequency of the events a among the events b), and Popper's own *propensity* interpretation.

The propensity interpretation may be understood as a generalization of the classical interpretation. Popper drops the restriction to "equally possible cases," assigning "weights" to the possibilities as "*measures of the propensity, or tendency, of a possibility to realize itself upon repetition.*"[3] He distinguishes *probability statements* from *statistical statements*. Probability statements refer to frequencies in *virtual* (infinite) sequences of well-defined experiments, and statistical statements refer to frequencies in *actual* (finite) sequences of experiments. Thus, the weights assigned to the possibilities are measures of conjectural virtual frequencies to be tested by actual statistical frequencies: "In proposing the propensity interpretation I propose to look upon *probability statements* as statements about some measure of a property (a physical property, comparable to symmetry or asymmetry) of *the whole experimental arrangement*; a measure, more precisely, of a *virtual frequency*; and I

[3] K. R. Popper, "Quantum Mechanics without 'The Observer,'" in M. Bunge, ed., *Quantum Theory and Reality* (New York: Springer-Verlag, 1967), p. 32.

propose to look upon the corresponding *statistical statements* as statements about the corresponding *actual frequency*."[4]

2

In his article "Quantum Mechanics without 'The Observer,'"[5] Popper characterizes the Copenhagen interpretation as involving what he calls "the great quantum muddle."

Now what I call the great quantum muddle consists in taking a distribution function, i.e. a statistical measure function characterizing some *sample space* (or perhaps some "population" of events), and treating it as a *physical property of the elements of the population*. . . . Now my thesis is that this muddle is widely prevalent in quantum theory, as is shown by those who speak of a "duality of particle and wave" or of "wavicles." For the so-called "wave" — the ψ-function — may be identified with the mathematical form of a function, $f(P, dP/dt)$, which is a *function of a probabilistic distribution function P*, where $f = \psi = \psi(q,t)$, and $P = |\psi|^2$ is a density distribution function. . . . On the other hand, the *element* in question has the properties of a particle. The wave shape (in configuration space) of the ψ-function is a kind of accident which poses a problem for probability theory, but which has nothing to do with the physical properties of the particles.[6]

Thus, it is a confusion to interpret Heisenberg's uncertainty relations as setting limits to the precision of our measurements. Rather, these relations are statistical scatter relations: they relate the dispersions of conjugate magnitudes such as position and momentum in quantum ensembles.

With this clarification, Popper proposes that the measurement problem of quantum mechanics be resolved in the following way: He considers the example of a symmetrical pin board, constructed so that a number of balls rolling down the board will ideally form a normal distribution curve at the bottom, representing the probability distribution for each single experiment with each single ball of reaching a certain final position. The probability distribution of reaching the various final positions for those balls which actually hit a certain pin will be different from the original distribution, and this conditional probability can be calculated from the probability calculus. Now Popper's thesis is this:

[4] *Ibid.*, pp. 32, 33.
[5] *Ibid.*
[6] *Ibid.*, pp. 19, 20.

"In the case of the pin board, the transition from the original distribution to one which assumes a 'position measurement' . . . is not merely analogous, *but identical with the famous 'reduction of the wave packet'*. Accordingly, this is not an effect characteristic of quantum theory but of probability theory in general."[7] Again:

Assume that we have tossed a penny. The probability of each of its possible states equals ½. As long as we don't look at the result of our toss, we *can* still say that the probability will be ½. If we bend down and look, it suddenly "changes": one probability becomes 1, the other 0. Was there a quantum jump, owing to our looking? Was the penny influenced by our observation? Obviously not. (The penny is a 'classical' particle.) Not even the probability (or propensity) was influenced. There is no more involved here, or in *any* reduction of the wave packet, than the trivial principle: if our information contains the result of an experiment, then the probability of this result, relative to this information (regarded as part of the experiment's specification), will always trivially be $p(a,a) = 1$.[8]

3

For the purposes of this analysis, I propose to characterize a statistical theory as involving a set of physical magnitudes forming an algebraic structure of a certain kind, together with an algorithm for assigning probabilities to ranges of possible values of these magnitudes. The idempotent magnitudes, whose possible values are 0 or 1, represent the propositions of the theory, i.e., the idempotent magnitudes are associated with the properties of the systems described by the theory. I shall refer to the given algebraic structure of the idempotent magnitudes as the *logical space* L_1 of a statistical theory. Roughly, L_1 represents all possible ways in which the properties of the systems described by the theory can hang together, or the possible events open to the systems. It is not required that L_1 be Boolean.

I shall say that two physical magnitudes, A and B, are *statistically equivalent* in a statistical theory just in case

$$p_W(\text{val}(A) \in S) = p_W(\text{val}(B) \in S)$$

for every statistical state W (i.e., for every probability assignment generated by the statistical algorithm of the theory) and every Borel set S of

[7] *Ibid.*, p. 36.
[8] *Ibid.*, p. 37.

real numbers. (For "val(A) ϵ S", read: The value of the magnitude A lies in the set S of real numbers.) So if

$$p_W(\text{val}(X) \epsilon S) = p_W(\text{val}(A) \epsilon g^{-1}(S))$$

for every statistical state W and every Borel set S, where $g:R \rightarrow R$ is a real-valued Borel function on the real line, then X and g(A) are statistically equivalent. (The magnitude g(A) will satisfy the condition

$$p_W(\text{val}(g(A)) \epsilon S) = p_W(\text{val}(A) \epsilon g^{-1}(S))$$

for every possible probability assignment, not only those generated by the statistical states of the theory.) With respect to *this* equivalence relation, I shall say that two magnitudes A and A′ are *compatible* if and only if there exists a magnitude, B, and Borel functions $g:R \rightarrow R$ and $g':R \rightarrow R$ such that[9]

$$A = g(B)$$
$$A' = g'(B).$$

This definition of the compatibility of two magnitudes is due to Kochen and Specker.[10] A linear combination of two compatible magnitudes may be defined as the linear combination of associated functions:

$$aA + a'A' = (ag + a'g')(B)$$

where a, a′ are real numbers, and similarly the product may be defined as

$$AA' = (gg')(B).$$

With linear combinations and products of compatible magnitudes defined in this way, the set of magnitudes of a statistical theory forms a *partial algebra*,[11] and the set of idempotent magnitudes forms a *partial Boolean algebra*. I shall refer to the partial Boolean algebra of idempotents defined in this way as the *logical space* L_2 of a statistical theory.

Evidently, the two algebraic structures, L_1 and L_2, are different. I

[9] I.e., such that

$$p_W(\text{val}(A) \epsilon S) = p_W(\text{val}(B) \epsilon g^{-1}(S))$$
$$p_W(\text{val}(A') \epsilon S) = p_W(\text{val}(B) \epsilon g'^{-1}(S))$$

for every statistical state W and every Borel set S.

[10] S. Kochen and E. P. Specker, *Journal of Mathematics and Mechanics*, 17 (1967), 59. Kochen and Specker use the term "commeasurability" instead of "compatibility." I avoid their term because it suggests a particular and, I think, misleading interpretation of the relation.

[11] The term is due to Kochen and Specker, who have investigated the properties of these algebraic systems.

shall say that a statistical theory is *complete* if and only if L_1 and L_2 are isomorphic. For this is the case if and only if the statistical states of the theory generate all possible probability measures on the logical space L_1.

A partial Boolean algebra is essentially a partially ordered set with a reflexive and symmetric (but not necessarily transitive) relation — compatibility — such that each maximal compatible subset is a Boolean algebra. Thus, a partial Boolean algebra may be pictured as "pasted together" in a certain well-defined way from its maximal Boolean subalgebras. By a probability measure on a partial Boolean algebra, L, I mean any assignment of values between 0 and 1 to the elements of L, which satisfies the usual conditions for a probability measure (as defined, say, by Kolmogorov's axioms) on each maximal compatible subset of L.

Now, the statistical algorithm of quantum mechanics involves the representation of the statistical states of a mechanical system by a certain class of operators in a Hilbert space (the statistical operators), and the physical magnitudes of the system by hypermaximal Hermitian operators in the Hilbert space. Each statistical operator W assigns a probability to the range S of the magnitude A according to the rule

$$p_W(\text{val}(A) \in S) = \text{Tr}(WP_A(S))$$ [12]

where $P_A(S)$ is the projection operator associated with the Borel set S by the spectral measure of A. The peculiarity of quantum mechanics as a statistical theory lies in the fact that the logical space L_1 of a quantum mechanical system is not Boolean. The idempotent magnitudes are represented by the projection operators, and the structure of the logical space L_1 is given by the algebra of projection operators in Hilbert space. There is a 1–1 correspondence between projection operators and subspaces in Hilbert space, each projection operator corresponding to the subspace which is its range.

The completeness problem for quantum mechanics was first solved by Gleason. Gleason's theorem[13] states that in a Hilbert space of three or more dimensions, *all possible* probability measures on the partial Boolean

[12] The right-hand side of this equation is the trace of the product of two Hilbert space operators: the statistical opeartor W and the projection operator $P_A(S)$, corresponding to the range S of the magnitude A. The trace of an operator is an invariant. In the case of an n-dimensional Hilbert space, the trace is the sum of the elements along the diagonal of the matrix of the operator.

[13] A. M. Gleason, *Journal of Mathematics and Mechanics*, 6 (1957), 885.

algebra of subspaces may be generated by the statistical operators W, according to the algorithm

$$p_W(K) = \mathrm{Tr}(WP)$$

where P is the projection operator onto the subspace K. It follows immediately that the logical spaces L_1 and L_2 are isomorphic.

Kochen and Specker pointed out a further corollary to Gleason's theorem: Because there are no dispersion-free probability measures (two-valued measures) on L_1, except in the case of a two-dimensional Hilbert space, there are no two-valued homomorphisms on L_1. The nonexistence of two-valued homomorphisms on L_1 means that it is impossible to imbed L_1 in a Boolean algebra. It follows that it is impossible to represent the statistical states of quantum mechanics by measures on a classical probability space in such a way that the algebraic structure of the magnitudes of the theory is preserved.[14] In other words, it is impossible to introduce a "phase space," X, and represent each physical magnitude by a real-valued Borel function on X, in such a way that each maximal compatible set of magnitudes is represented by a set of phase space functions which preserve the functional relationships between the magnitudes. What cannot be achieved for magnitudes A, A′, . . . which are all functions of the magnitude B, i.e., $A = g(B)$, $A' = g'(B)$, . . . , is that if B is represented by the phase space function f_B, then A is represented by the phase space function $f_{g(B)} = g(f_B)$, A' is represented by the phase space function $f_{g'(B)} = g'(f_B)$, etc. Thus, there is no phase space[15] reconstruction of the quantum statistics which preserves the functional relations between compatible sets of physical magnitudes.

[14] A classical probability space is a triple (X, F, μ), where X is a set, F is a σ-field of subsets of X (the measurable sets), and μ is a probability measure on X. Classical statistical mechanics is explicitly formulated as a statistical theory on a classical probability space, with X the phase space of classical mechanics. The axiomatization of the classical theory of probability along these lines is due to Kolmogorov, *Foundations of the Theory of Probability*. What is fundamental is the notion of probability as a measure function on an event structure or propositional structure represented as the algebra generated by the subsets of a set under the operations of union, intersection, and complement, i.e., a Boolean algebra. It is *always* possible (trivially) to represent the statistical states of an arbitrary theory by measures on a classical probability space, if the algebraic structure of the magnitudes of the theory (or the logical space L_1) is not preserved. See Kochen and Specker, *Journal of Mathematics and Mechanics*.

[15] Of course, the phase space X need not be a phase space in the sense that it is parameterized by generalized position and momentum coordinates, as in classical mechanics. It is a phase space in the sense that the points of this space define two-

The algebra of idempotent magnitudes (propositions) of a classical mechanical system is isomorphic to the Boolean algebra of Borel subsets of the phase space of the system. The work of Gleason, and of Kochen and Specker, shows that the transition from classical to quantum mechanics involves a generalization of the Boolean propositional structures of classical mechanics to a particular class of non-Boolean structures: the propositional structure of a quantum mechanical system is a partial Boolean algebra isomorphic to the partial Boolean algebra of subspaces of a Hilbert space. This may be understood as a generalization of the classical notion of validity — the class of models over which validity is defined is extended to include partial Boolean algebras which are not imbeddable into Boolean algebras.[16]

Popper's position is an implicit commitment to a phase space reconstruction of the quantum statistics, i.e., the representation of the statistical states of quantum mechanics by measures on a classical probability space. His claim is that there is *no more* involved in the quantum mechanical "reduction of the wave packet" than the "trivial principle" according to which additional information transforms an initial probability distribution into a new conditional probability distribution in a Boolean probability calculus. But there *is* more involved here, because the transformation of probabilities is *non-Boolean.*

The "trivial principle" to which Popper refers is the Boolean rule for conditional probabilities. The points in a classical probability space (phase space) correspond to ultrafilters in the logical space, i.e., maximal consistent sets of propositions. The representation of the states of classical mechanics as the points of a space X, which functions in the theory like the Stone space of a Boolean logic, shows that the logical space is Boolean. The sense in which the specification of a point in X is a *state*-description is just this: A point in X corresponds to a maximal totality of propositions in logical space — maximal in the sense that

valued probability measures in the generalized sense, i.e., two-valued homomorphisms on L_1. That is to say, it is a phase space in the sense that the points of this space specify state-descriptions in an analogous sense to the points of a classical mechanical phase space: assignments of values to the magnitudes, satisfying the functional relationships given above.

[16] See Kochen and Specker, *Journal of Mathematics and Mechanics,* and J. Bub and W. Demopoulos, "The Interpretation of Quantum Mechanics," in R. S. Cohen and M. Wartofsky, eds., *Boston Studies in the Philosophy of Science,* vol. 13 (Dordrecht: Reidel, 1974). Also, J. Bub, *The Interpretation of Quantum Mechanics* (Dordrecht: Reidel, 1974).

these propositions are related by the logical relations of a Boolean ultrafilter. The Boolean rule for conditional probabilities is the rule

$$p(s,t) = \frac{\mu(\Phi_s \cap \Phi_t)}{\mu(\Phi_t)} = \frac{P(s \wedge t)}{p(t)}$$

where the initial probabilities are given by the measure μ, and Φ_s, Φ_t are the subsets of points in X which satisfy the propositions s, t respectively. This rule derives directly from the Boolean semantics, i.e., essentially Stone's representation theorem for Boolean algebras. (See, for example, the classic paper by Los.)[17] Thus, the conditional probability $p(s,t)$ is determined by a new measure μ_t, defined by

$$\mu_t(\Phi) = \frac{\mu(\Phi \cap \Phi_t)}{\mu(\Phi_t)}$$

for every measurable set $\Phi \subseteq X$.

The measure μ_t is the initial measure μ "renormalized" to the set Φ_t, i.e., μ_t satisfies the conditions:

(i) $\mu_t(\Phi_t) = 1$

(ii) if $\Phi_u \subseteq \Phi_t$ and $\Phi_s \subseteq \Phi_t$
then

$$\frac{\mu_t(\Phi_u)}{\mu_t(\Phi_s)} = \frac{\mu(\Phi_u)}{\mu(\Phi_s)}.$$

Condition (ii) ensures that μ_t preserves the relative measures of subsets in Φ_t defined by μ.

The theorems of Gleason, and of Kochen and Specker, do not exclude the representation of the statistical states of quantum mechanics by measures on a classical probability space. They exclude a representation which preserves the algebraic structure of the magnitudes of the theory, or the structure of the logical space L_1. It follows that if μ is a measure corresponding to some quantum mechanical statistical state in a phase space reconstruction of the quantum statistics, then the conditional probability μ_t will not in general be a measure corresponding to a quantum mechanical state.

To see that this is so, it suffices to consider "pure" statistical states, represented by statistical operators which are projection operators onto

[17] J. Los, *Proceedings of the International Congress of Mathematics*, 1962, pp. 225–29.

one-dimensional subspaces or vectors in Hilbert space. Suppose μ is the initial measure corresponding to the pure statistical state represented by the Hilbert space vector ψ. Let τ be the eigenvector corresponding to the eigenvalue of t of T, and suppose τ does not coincide with ψ and is not orthogonal to ψ (i.e., τ and ψ do not form part of any orthogonal set of basis vectors in the Hilbert space and so do not correspond to the eigenvectors of any common magnitude). I use the same symbol — t — for the value of the magnitude T and for the proposition corresponding to the property of the system represented by the value t of the magnitude T, i.e., the proposition represented by the idempotent magnitude associated with the projection operator onto the one-dimensional subspace spanned by the vector τ. The conditional probability μ_t, as computed by the Boolean rule, preserves the relative measures of subsets in Φ_t defined by μ, and so depends on ψ. But then μ_t cannot represent a statistical ensemble of quantum mechanical systems each of which has the property t, for such an ensemble is represented by the measure μ_T, corresponding to the statistical state τ, and μ_T is independent of ψ and depends only on τ. Both measures satisfy condition (i), i.e.,

$$\mu_t(\Phi_t) = \mu_T(\Phi_t) = 1$$

but they cannot define the same relative measures on subsets of Φ_t.

According to the Boolean rule for conditional probabilities, the relative probabilities of subsets in Φ_t are unchanged by the additional information that t is true. But in a phase space reconstruction of the quantum statistics, we must assume that any initial information concerning the relative probabilities of subsets in Φ_t is somehow invalidated by the additional information that t is true (or false), if the magnitude T is incompatible with any magnitude of which ψ is an eigenvector. The fact that an initial probability measure is reduced or "collapses" to the set Φ_t, given the additional information that t is true, is not problematic here. What is problematic is that this reduction is apparently accompanied by a randomization process.

Popper writes:

Let us call our original specification of the pin board experiment "e_1" and let us call the new specification (according to which we consider or select only those balls which have hit a certain pin, q_2, say, as repetitions of the new experiment) "e_2." Then it is obvious that the two probabilities of landing at a, $p(a,e_1)$ and $p(a,e_2)$, will not in general be equal, because the two experiments described by e_1 and e_2 are not the same. But this does not mean that the new information which tells us that the conditions e_2 are realized in any way changes $p(a,e_1)$: from the very beginning we could calculate $p(a,e_1)$ for the various a's, and also $p(a,e_2)$; and we knew that

$$p(a,e_1) \neq p(a,e_2)$$

Nothing has changed if we are informed that the ball has actually hit the pin q_2, except that we are now free, *if we so wish,* to apply $p(a,e_2)$ *to this case*; or in other words, we are free to look upon the case as an instance of the experiment e_2 instead of the experiment e_1. But we can, of course, continue to look upon it as an instance of the experiment e_1, and thus continue to work with $p(a,e_1)$: the probabilities (and also the probability packets, that is, the distribution for the various a's) are *relative probabilities*: they are *relative to what we are going to regard as a repetition of our experiment*; or in other words, they are relative to what experiments are, or are not, *regarded as relevant to our statistical test.*[18]

Now this is unobjectionable, but quite irrelevant to the quantum mechanical problem. It is not the mere change from an initial probability assignment to a new probability assignment, conditional on certain information, that is problematic in quantum mechanics — i.e., it is not the fact that probabilities change if we change the reference class that is problematic. What is problematic is the *kind* of change. To refer to Popper's example: Let Φ_b be the set of trajectories which hit the pin q_2 and another pin q_3, and let Φ_c be the set of trajectories which hit the pin q_2 and q_4. Then, of course,

$$p(b,e_1) \neq p(b,e_2)$$
$$p(c,e_1) \neq p(c,e_2).$$

This is not at all puzzling. But in a Boolean theory of the pin board, it would be quite incomprehensible if

$$\frac{p(b,e_1)}{p(c,e_1)} \neq \frac{p(b,e_2)}{p(c,e_2)}.$$

And this inequality is characteristic of the quantum mechanical "reduction of the wave packet."

Feyerabend has made a similar point, although his argument does not explicitly bring out the non-Boolean character of the quantum statistics: "[Popper] pleads with us not to be surprised when a change of

[18] K. R. Popper, "Quantum Mechanics without 'The Observer,' " pp. 35, 36.

experimental conditions also changes the probabilities. This, he says, is a 'quite trivial feature of probability theory.' . . . Quite correct — but irrelevant. For what surprises us (and what led to the Copenhagen interpretation) is not the fact that there is *some change*; what surprises us is the *kind of change* encountered."[19]

I have argued elsewhere[20] that there are two possible interpretations of the role of the Hilbert space in quantum mechanics. The "great quantum muddle" is in effect the confusion of these two interpretations. The first interpretation takes the partial Boolean algebra of subspaces of Hilbert space as the logical space L_1 of a statistical theory. Vectors in Hilbert space (or, rather, one-dimensional subspaces) then represent atomic propositions or atomic events. The problem of specifying all possible probability measures on such propositional structures has been solved by Gleason: the probability calculus is generated by the set of statistical operators (in the technical sense) in Hilbert space. On this interpretation of Hilbert space — call it the *logical interpretation* — the transition from classical to quantum mechanics is understood as the generalization of the Boolean propositional structures of classical mechanics to a class of propositional structures that are essentially non-Boolean. Thus, classical and quantum mechanics are interpreted as theories of logical structure, in the sense that they introduce different constraints on the possible ways in which the properties of physical systems are structured. Just as the significance of the transition from classical to relativistic mechanics lies in the proposal that geometry can play the role of an explanatory principle in physics, that the geometry of events is not a priori, and that it makes sense to ask whether the world geometry is Euclidean or non-Euclidean, so the significance of the quantum revolution lies in the proposal that logic can play the role of an explanatory principle, that logic is similarly not a priori.

The second interpretation — call it the *statistical interpretation* — presupposes that the logical space L_1 of a statistical theory is necessarily Boolean, and takes the Hilbert space as the space of statistical states of a statistical theory, with each unit vector representing a statistical state for the algorithm

[19] P. K. Feyerabend, "On a Recent Critique of Complementarity: Part I," *Philosophy of Science*, 35 (1968), 309. Quotation from p. 326.

[20] "On the Completeness of Quantum Mechanics," in C. A. Hooker, ed., *Contemporary Research in the Foundations and Philosophy of Quantum Theory* (Dordrecht: Reidel, 1973).

$$p_{\psi}(\text{val}(A) \epsilon S) = || P_A(S) \psi ||^2.$$

"Mixtures" of such "pure states" may be defined as more general statistical states, specifying probability assignments representable as weighted sums of probability assignments generated by pure ψ-states. Thus, the Hilbert space is interpreted as specifying the logical space L_2 of a statistical theory, from which it follows that quantum mechanics is incomplete, since L_1 is Boolean and L_2 is non-Boolean. The measurement problem — the problem of the "reduction of the wave packet" or the inexplicable randomization in the transformation rule for conditional probabilities — is characteristic of the statistical interpretation. The logical interpretation avoids this problem by properly characterizing the category of algebraic structures underlying the statistical relations of the theory.

4

In the article "Quantum Mechanics without 'The Observer,'" Popper distinguishes between *theories* and *concepts*. As scientists, we seek true theories, "true descriptions of certain structural properties of the world we live in."[21] A conceptual system is merely an instrument used in formulating a theory, providing a language for the theory. Concepts, then, are "comparatively unimportant."[22] In fact, "two theories, T_1 and T_2, should be regarded as one if they are logically equivalent, even though they may use two totally different 'conceptual systems' (C_1 and C_2) or are conceived in two totally different 'conceptual frameworks.'"[23]

Now it seems to me that the relation between *concepts* and *theories* (as understood by Popper) is precisely analogous to the relation between *interpretations* of probability and *formal theories* of probability (as outlined in section 1). The measurement problem which arises in quantum mechanics represents the inapplicability of Boolean probability theories to a certain domain of events. The problem cannot be avoided by using the language of propensities rather than the language of frequencies, since propensities and frequencies are both proposed as interpretations of a Boolean probability calculus. This is not to say that

[21] K. R. Popper, "Quantum Mechanics without 'The Observer,'" p. 11.
[22] *Ibid.*, p. 14.
[23] *Ibid.*, p. 12.

there is in principle no difference between propensity-talk and frequency-talk, only that whatever difference there is can have no relevance for the measurement problem, just because Popper has explicitly proposed the propensity interpretation as an interpretation of a Boolean theory of probability.

Popper says, "We understand a theory if we understand the problem which it is designed to solve, and the way in which it solves it better, or worse, than its competitors."[24] *With respect to the measurement problem of quantum mechanics,* Popper's informal comments about propensities are quite empty. There cannot be a difference which makes a difference between propensity-talk and frequency-talk, so long as propensities and frequencies are proposed as interpretations of a Boolean probability calculus. For Popper to say something new in terms of propensities, it should be possible to formalize the difference between propensity-talk and frequency-talk in such a way that the new formal theory of probability solves those foundational problems of quantum mechanics which arise because of the inadequacy of the old theory.

It is clear, then, that Popper's propensity interpretation of probability adds nothing to the solution of the measurement problem. What of Popper's critical comments on Kolmogorov's axiomatization of the probability calculus and his own conception of a formal theory of probability? Again, there is no difference which makes a difference between Popper's axiomatization and Kolmogorov's. What the difference amounts to is this: Kolmogorov proposes axioms for a probability calculus on a Boolean logical space L_1. For completeness, the logical space L_2 ought to be Boolean, and this is achieved by Kolmogorov's axiomatization. This seems to me the natural way to proceed. Popper does the same thing backward. He proposes axioms which characterize a Boolean logical space L_2. For completeness, again, the logical space L_1 ought to be Boolean, and this is achieved by Popper's axiomatization. There is nothing to choose between the two approaches.

[24] *Ibid.*, p. 14.

WALTER WEIMER

The Psychology of Inference and Expectation: Some Preliminary Remarks

The problems of inference are manifold: the first task facing the theorist concerned with them is to attempt an ordering of priorities, to decide which of them to tackle. One must assay not only the relative importance of the various problems but also the possibilities of achieving their successful solution; only then may the individual problems profitably be addressed. Historically, the majority of philosophers have concluded that *the* problem of inference is that of induction, and more specifically the problem of the *justification* of induction. According to the point of view from which they operate, the prime task of the philosophy of science is to show how inductive inferences can result in genuine knowledge, or, what is to say the same thing, to show that science is a rational endeavor. Inductive inference is taken by almost everyone to be the method by which science operates: and *if* it is the method of science, *then* it must be shown to be a rational source of knowledge. Hence the philosophical or 'logical' justification of induction and the demonstration that one or another 'confirmation theory' or 'inductive logic' is actually a *rational* source of knowledge loom large in the literature — as the contributions to this volume amply indicate.

Plausible though they may seem, I reject not only the identification of the problems of inference with the justification of induction, but also the conceptual framework from which that identification stems. Neither that point of view nor the problems it suggests will increase our understanding of nondemonstrative reasoning. Of the many legitimate and worthwhile problems surrounding inference, I wish to concentrate upon two that I believe to be among the most important. They may be formulated in two straightforward questions: "*What* do

AUTHOR'S NOTE: This essay was supported in part by a visiting appointment at the Minnesota Center for Philosophy of Science during the summer of 1971. I am grateful to the editors for facilities and constructive criticism provided.

we learn from facts (and experience)?" and "*How* do we learn from them?" Instead of induction I wish to discuss *learning*, or better, the nature of knowledge and its acquisition: my thesis will be that psychology, rather than philosophy, is the domain of inquiry that will ultimately provide answers to the problems of inference. These problems cannot be successfully addressed in a framework that denies their inherently psychological nature; nor can the nature of man, the psychological organism, be divorced from the nature of our knowledge. Epistemology, as the theory of knowledge *and its acquisition*, is properly one of the psychological sciences. Put another way, the problems of inference and expectation are the problems of learning and knowledge; learning and knowledge (and thus epistemology) are problems for the psychological sciences. Hume was correct both in asserting that the senses in combination with logic cannot justify nondemonstrative inference and in relegating such reasoning to "mere animal belief."

The Dilemmas of Justificationism

The burden of this part of the essay is to lead us to an alternative point of view from which to conceive of science and philosophy *that does not lead to despair when the truth of both Humean claims is acknowledged*. To do this we must first explore the point of view from which the majority of philosophers operate and then expose several of the structural fusions and confusions that it engenders. For only if we abandon the justificationist's quest can the problems of inference be cast in such a form that insoluble dilemmas do not result at every turn. Unless it is clear that what I have elsewhere[1] called the justificationist metatheory of science *must* be abandoned, it will make little sense to conceive of the problems of nondemonstrative inference as problems for psychology.

Justificationism as a metatheory of science. Justificationism as a point of view or metatheory of scientific rationality is a "rule system" or "grammar" that specifies how our concepts can be formed. It enshrines

[1] See my *Psychology and the Conceptual Foundations of Science* (Englewood Cilffs, N.J.: Prentice-Hall, in press). The term 'justificational philosophies of criticism' was introduced by W. W. Bartley III in his brilliant and neglected book *The Retreat to Commitment* (New York: Knopf, 1962). By referring to justificationism as a *metatheory* of the scientific endeavor I have, I believe, only developed consistently ideas already implicit in Bartley's presentation.

a number of definitional fusions and confusions that are characteristic of its outlook, and these fusions and confusions, which *are* its rules of conceptual classification, are the defining traits of this point of view. By fusing certain concepts together it defines *in advance* what can be taken as appropriate answers to the fundamental questions of the nature of science, the nature of our knowledge and its acquisition, and even the nature of rationality (and rational inquiry). By listing *some* of these major structural fusions and confusions we can see how a justificationist *must* approach the questions and issues concerning our knowledge and its acquisition that are our concern in this essay.[2]

The justificationist conception of knowledge requires the identification of knowledge with both *proof* and *authority*. A putative knowledge claim cannot be accepted as genuine knowledge unless it can be proven, and it cannot be proven except by submission to the appropriate epistemological authority. For the empiricist such as Locke, *the* epistemological authority is the deliverance of sense: all genuine knowledge must be founded upon the authority of sense experience. For the intellectualist such as Descartes, the supreme epistemological authority needed to certify a knowledge claim as genuine is rational intuition. Regardless of the particular philosophy endorsed, the justificationist philosopher will not accept as genuine knowledge any claim that cannot be validated by whatever ultimate epistemological authority he accepts. To repeat: justificationism fuses and confuses knowledge with proof and with authority.

A third fusion follows from the identification of knowledge with proven assertion: the conception of 'eternally valid' knowledge gradually accumulating (through inductive inference) into the body of scientific knowledge. If knowledge is proven, then once certified it remains so forever. Thus scientific progress *must* be the accumulation of more and more eternally valid truths. Justificationist historiography, enshrining this 'cumulative record' position, rewrites history to guarantee the continuity (and therefore the rationality) of scientific progress.

The goal of science for classic justificationism was to establish that all the propositions of science have the same probability of being true:

[2] The few points we can mention, although crucial to an understanding of justificationism as a metatheory of science, are certainly not exhaustive of the position. A far more comprehensive account is found in my *Psychology and the Conceptual Foundations of Science*.

the probability value 1 or certainty. For every proposition *h* of science, given the relevant evidence e, the probability *p* of that proposition must, according to the demands of justificationist rationality and intellectual integrity, be shown to be expressed as $P(h,e) = 1$. One reads this formula as the probability of hypothesis *h* being true on the basis of evidence e is 1 (or certainty). To establish this, justificationism had to develop a logic of scientific assessment: an *inductive* logic.

By defining its concepts in this manner the justificationist approach *demands* that certain things obtain in science and its methodology. Half a dozen of the most important requirements are the following: First, there must be an 'empirical basis' of facts that are *known for certain.* This is the *foundation of empirical knowledge* (the basis for inductive inference). Second, 'theories' are second-class citizens, being 'derivative' from facts and accumulated generalizations (i.e., they are inductions based upon inductions). Third, science must be cumulative and gradual: facts must be piled upon facts to construct the edifice of scientific knowledge. Fourth, factual *meaning* must be fixed for once and for all *independently* of theory, and must remain invariant. Fifth, explanation consists in showing that a 'proven' factual proposition follows deductively (i.e., logically) from a theoretical proposition. Sixth, science evaluates *one* theory at a time, never two or more. Justificationism enshrines a *monotheoretical* model of assessment.

We could elaborate numerous other fusions and confusions at the heart of justificationism, but there is only one that is essential to the problems of inference: we must see how *certainty* (*proven* assertion) in the definition of knowledge becomes '*probability*'. Classic justificationism gave way to *neo*justificationism when it was gradually realized that *no* scientific proposition is provable, or can be "known for certain." (Fries showed this in 1828, as a special case of the logical thesis that the logical relations, such as provability, consistency, etc., can refer only to propositions. And propositions can be derived *only* from other propositions, never from 'facts'.) Thus *all theories are equally unprovable,* and $P(h,e) = 0$, always. Thus failed the method of justification for knowledge claims. The basis of knowledge failed also when Pierre Duhem noted that science is fact-correcting in its nature rather than fact-preserving. New theories often refute the facts of older ones. There is no eternally valid 'factual basis.' Popper and his students[3] have been calling attention to the failure of both the method and the basis of justificationism for several decades. But their criticism has in the main been ignored, because it has been assumed that with the relaxation of the definition of knowledge to allow probable assertions as genuine knowledge claims, inductive 'logic' could be rescued, i.e., *justified*, as a legitimate *source* of knowledge and simultaneously shown to be the *method* of its acquisition.

So in order to save the rationality of science, justificationists watered down 'proof' to 'probable' in the definition of knowledge. The development of *probabilistic inductive logic* (from the pioneering work of W. E. Johnson at Cambridge) has been the result. Intellectual honesty now demands far less of the 'formula' for the assessment of scientific warrant: today the goal is a "confirmation theory" that will assign a probability ranging from $0 < 1$ in the formula $P(h,e) = c$ (where c is interpreted as *degree of confirmation*). Practically nothing else has changed in the switch from the classic to the 'neo' form of justificationism. The chief problem is still to justify knowledge, but now it is the justification of probable rather than certain truth claims. Induction, as the means by which we acquire and certify scientific knowledge, must be rescued from the skeptical doubts of Hume. The epistemological authority must be *relocated* in the probabilistic inductive procedure: otherwise there is no authority, hence no proof, and therefore no rationality to science. At best, it will be where Hume left it—and animal belief, for the justificationist, is merely psychological and *therefore* irrational. Skepticism, for the justificationist, means that no *genuine* knowledge is possible at all and that our illusion of knowledge is a purely psychological phenomenon.

The major fusion of concepts introduced by neojustificationism is the equation of *induction* with *probability*. Having dropped the impossible

[3] The locus classicus of Popper's argument is *The Logic of Scientific Discovery* (New York: Harper & Row, 1959), especially sec. 1 and ch. 10. See also *Conjectures and Refutations* (New York: Harper & Row, 1963), especially ch. 1. The most telling single presentation thus far is Lakatos' essay "Changes in the Problem of Inductive Logic," in Lakatos, ed., *The Problem of Inductive Logic* (Amsterdam: North-Holland, 1968), pp. 315–417. See also Lakatos's "Falsification and the Methodology of Scientific Research Programmes," in Lakatos and Musgrave, eds., *Criticism and the Growth of Knowledge* (Cambridge: At the University Press, 1970); J. Agassi, *Towards an Historiography of Science*, Beiheft 2 of *History and Theory* (Middletown, Conn.: Wesleyan University Press, 1963); and J. Agassi, "Sensationalism," *Mind*, n.s. 75 (1966), 1–24.

quest for certainty, the sophisticated justificationist now aims at the next best thing: *near*-certainty. And the 'near certain' has been identified, with no reflection at all, with the 'probable.' Probabilistic inductive logic is the result: induction as the means of both knowledge acquisition and assessment has become fused to the calculus of probability. This new logic of scientific method is to be an algorithmic assessment procedure for the acquisition of probable knowledge rather than certain knowledge. That is, this 'logic' is supposed to *prove* putative scientific statements to be probable (rather than prove them true).[4]

The failure of justificationism. In a sense, the real failure of justificationism is that it creates for its adherents insoluble dilemmas. By fusing together concepts that do not belong together, it creates a metatheoretical structure that must inevitably lead even its most brilliant practitioners to an insurmountable wall of dilemmas, infinite regresses, and problems that admit no solutions. The justificationist quest, by its very nature, is impossibly difficult: the goal cannot be reached, no matter how ingenious its pursuer may be. The superiority of nonjustificational approaches to science and its growth is that they show, from alternative points of view, either how these dilemmas do not arise or how the problems they pose are soluble in a consistent manner.

Considering only the problem of inference (or knowledge acquisi-

[4] Logical positivism and empiricism, as the dominant 'received view' philosophies of science, are easily shown to be justificationist at heart. All the "theories" of cognitive significance (principles of verifiability, etc.) are explicit statements of the equation of knowledge with proof (or probability). *Meaning* is likewise assimilated to proof. The appeal is to a rational authority to say "What do you mean?" and "How do you know?" and science, which is the meaningful and the sensible, is demarcated from nonscience by the fact that the latter is unverifiable, i.e., meaningless, and also non-sense. The entire unity of science movement stems from the notion of a common, intersubjective, fixed-for-certain-for-all-time basis of "facts." All science must be Kuhnian normal science "fact accumulation." There can be no revolutionary science since science never overturns prior certified facts. There is no real problem in explaining the growth of scientific knowledge or of our acquisition of knowledge. There remains only the task of justifying induction as the means by which knowledge is achieved (i.e., the task of *proving* that induction yields *valid* knowledge). The construction of a theory of the scientific merit of a proposition is the prime task for philosophy, and that theory is universally assumed to be inductive logic or 'confirmation theory.' The growth of knowledge remains the gradual (inductive) accumulation of (probable) truths (because a theory of scientific merit is *automatically* one of growth). And the guidance of future research, i.e., the problems of pragmatic action in actual scientific practice, provides no problem at all: since induction is the rational means of knowledge acquisition, all who are rational will employ it.

tion),[5] it is obvious that probabilistic inductive logic and the neojustificationist conception of probable knowledge are no better off than the classic positions they are assumed to be such a significant improvement upon. If one understands Fries's argument that propositions cannot be proven by facts, then it is obvious that *they cannot be 'probabilified' either*. It will not do to substitute near-certainty, in the guise of the calculus of probability, for certainty. Fries's argument shows why neither can be obtained: inductive logic cannot assess the 'merit' of propositions where 'merit' is taken to be either truth or probability value. Nor, for that matter, can they be proven false. Within the justificationist framework the skeptic, who holds that no informative knowledge is possible, always triumphs over the 'positive' justificationist. The most that the positive justificationist can do is delude himself into thinking that it is a worthwhile task to render the skeptic's acknowledged *victory* as bloodless as possible.[6] No wonder that contemporary empiricists are the best representatives of existentialist despair and dread — by the consistent application of their own criteria of rationality, they have shown that their trust in empiricism was not justified — and have retreated to a *faith* in empiricism rather than a defense of it.[7] This is, as Bartley

[5] The futility of the other facets of the metatheory is becoming painfully obvious from both internal and external criticism. For example, the 'foundations of knowledge' (which justificationism requires as the basis of inference) conception has been demolished by K. R. Popper in *The Logic of Scientific Discovery*; by W. Sellars in *Science, Perception and Reality* (London: Routledge & Kegan Paul, 1963), especially chs. 3, 4, and 5); and by B. Aune in *Knowledge, Mind and Nature* (New York: Random House, 1967), especially ch. 3. Justificationist 'cumulative record' historiography has been severely criticized by T. S. Kuhn in *The Structure of Scientific Revolutions*, 2nd ed. (Chicago: University of Chicago Press, 1970); by Agassi, *Towards an Historiography of Science*; and by Lakatos, "Changes in the Problem of Inductive Logic," and "Falsification and the Methodology of Scientific Research Programs." The 'meaning invariance' thesis and the conception of explanation as logical deduction have been shattered by P. K. Feyerabend's criticism, especially in "Explanation, Reduction, and Empiricism," in H. Feigl and G. Maxwell, eds., *Minnesota Studies in the Philosophy of Science*, vol. 3 (Minneapolis: University of Minnesota Press, 1962), pp. 28–97. Feyerabend's philosophy of proliferation (see "Reply to Criticism" in R. S. Cohen and M. W. Wartofsky, eds., *Boston Studies in the Philosophy of Science*, vol. 2 (New York: Humanities Press, 1965), pp. 223–61) also demolishes the monotheoretical assessment model proposed by justificationism. Bartley, *The Retreat to Commitment*, pointed out the flaws in the justificationist conception of rationality and showed how the justificationist fuses the concept of criticism to that of proof.

[6] A representative example is A. J. Ayer in *The Problem of Knowledge* (London: Penguin Books, 1956), especially pp. 68–75.

[7] For example, Bertrand Russell in the last paragraph of *Human Knowledge: Its Scope and Limits* (New York: Simon and Schuster, 1948): "Empiricism as a theory

(1962) aptly remarked, a *retreat to commitment*, and it is no different, and no less *irrational*, when practiced by the 'scientific' empiricist than by the Protestant irrationalist. Whatever rationality there is to science must forever lie *outside* the confines of justificationism.

The promise of nonjustificational metatheories. For the justificationist, informative theoretical knowledge is an illusion — a mere 'animal' belief. But what is 'animal belief' if knowledge is *not* identified with proven assertion? If knowledge is instead identified with *warranted* assertion — such that a claim constitutes *genuine* knowledge *if* it is warrantedly assertible — then mere 'animal belief' *can* constitute knowledge. An animal belief could constitute a knowledge claim if there were 'good reasons' that could be adduced in its defense. Those 'good reasons' would never *prove* a claim to be warrantedly assertible: knowledge is always conjectural, always subject to revision and reformulation. What could constitute 'good reasons' for the defense of a conjectural knowledge claim? Why do we believe that science provides 'good reasons' for its claims?

One of the reasons we believe in science is that it advances *theories* (tentative, fallible, conjectural ones) that (attempt to) *explain* the phenomena with which they deal. Human knowledge, be it scientific or commonsensical, proceeds by what Russell called the method of analysis: given data, we attempt to construct theories that most satisfactorily explain that data. The 'good reasons' with which we defend our knowledge claims are *always* theoretical reasons: that is, we defend our knowledge claims by showing that they are theoretically motivated, that they follow from a theory that is sufficiently well corroborated that we deem it worthy of continued exploration and scrutiny. The hypothetico-inferential method of explanation is the characteristic pattern of support for all knowledge claims, scientific or commonsensical.

of knowledge has proved inadequate, though less so than any other previous theory of knowledge. Indeed, such inadequacies as we have seemed to find in empiricism have been discovered by strict adherence to a doctrine by which empiricist philosophy has been inspired: that all human knowledge is uncertain, inexact, and partial. To this doctrine we have not found any limitation whatever" (p. 507).

Compare with M. B. Turner, *Philosophy and the Science of Behavior* (New York: Appleton-Century-Crofts, 1967): "It is always a bit ironical when a house which professes to virtue topples under censure by its own precept. . . . Empiricism itself is culpable, yet we have found no reliable substitute for a knowledge supported by the fact of its public communicability. For the empiricist, the alternative to absolute skepticism is the wistful embrace of a principle of convergence" (p. 7).

But how do we arrive at tentative knowledge claims in the first place? How do our 'animal beliefs' arise? Why have we inferred that reality has the characteristics that common sense and science have attributed to it, and how and why do these accounts differ in their portrayal of reality? And exactly what sort of 'knowledge' is provided by our scientific theories? When and why are our inferences rational? These and other questions loom large in the discussion that follows, because they are central to an understanding of inference and expectation. The promise of addressing these issues from a nonjustificational point of view lies in the fact that answers — albeit tentative, conjectural ones in the form of theories that will explain the phenomena involved — *can* be adduced which are not doomed to incorporate insurmountable difficulties. That is, the psychology of inference and expectation can provide principled explanations of the nature of our knowledge and its acquisition that do not bog down in the impossible philosophical quest for a justification of induction. Hume's conclusions lead to skepticism *only* when the justificationist conceptions of knowledge and rationality are taken to be the only ones available. By abandoning the justificationist quest one can at least formulate problems of inference for which contingent theories can provide answers. And one can even reformulate old questions, such as the one about what guides scientific life, in a manner in which the incorrect answer "induction" does not arise. Consider how this can be so.

Inductive inference and the guidance of life. Because of the manner in which he fuses his concepts, inductive methods may be considered to be indispensable to the justificationist for two main problems. The first role for 'inductive logic' is as a theory of the instant *assessment* of the scientific warrant of a putative hypothesis. This leads to the quest for a computational formula, schematized as $P(h,e) = c$, where c or degree of confirmation stands for the rationality or warrantability of the hypothesis in question. That is, the probability of the hypothesis is equated with its scientific merit (both of which are assumed to be equivalent to its truth-value). I wish to ignore the assessment problem as beyond the scope of our present inquiry. Suffice it to say that I find the Popperian arguments against 'inductivism' as a theory of assessment quite convincing: so-called 'inductive method' *plays no role in the assessment of the scientific merit of theoretical statements*.[8] But there is another role for induction that the justificationist has up his sleeve, and it seems to be even more fundamental than the role of assessment. That is, there are justificationists who will concede that Popper's criticism of inductivism as a means of assessment *is* devastating, yet still hold that inductive methodology is indispensable to the guidance of scientific life. Thus the second major problem for which inductive method is deemed essential concerns knowledge acquisition: it is deemed indispensable in the quest for a guide to future scientific research. From the justificationist point of view, the problem of acceptance merges indistinguishably into that of providing a guide to scientific life. Inductive logic is not only a logic of assessment but it is supposed to be a logic of scientific discovery: the inductive judge (the confirmation formula) is also to guide scientific life. That is, it is assumed that information on the scientific warrant of propositions is somehow capable of guiding future research: that 'degree of confirmation' statements provided by the inductive judge can tell the researcher where to go in subsequent research.[9]

[8] See Popper, *The Logic of Scientific Discovery*, sec. 1 and ch. 10, and *Conjectures and Refutations*, chs. 1 and 11; and also Lakatos, "Changes in the Problem of Inductive Logic," and "Falsification and the Methodology of Scientific Research Programs"; also the contributions of Popper, Watkins, and Carnap in I. Lakatos, ed., *The Problem of Inductive Logic*.

[9] A representative argument in favor of the necessity of inductive method for guidance to future research is due to Wesley Salmon. See especially his book *The Foundations of Scientific Inference* (Pittsburgh: University of Pittsburgh Press, 1967) and such essays as "The Justification of Inductive Rules of Inference," in Lakatos, ed., *The Problem of Inductive Logic*, pp. 94–97.

Salmon's contention is that as long as *any* nondemonstrative inference (such as Salmon sees Popper's concept of corroboration to be) is deemed necessary for science, then the problem of induction is being smuggled in, and since *the* problem is that of justification, one must now justify this form of inference. But why must science necessarily contain some form of 'nondemonstrative' inference? Salmon feels that this is obvious because a science based upon observation statements (a factual basis) and deductive logic *alone* would be impotent: it could tell us nothing new, that is, could not tell us how knowledge is acquired and therefore could not guide future research. Since science "obviously" does tell us new and informative things, deductivism plus corroboration must be an ampliative inference procedure. Salmon writes: "Popper furnishes a method for selecting hypotheses whose content exceeds that of the relevant available observation statements. Demonstrative inference alone cannot accomplish this task, for valid deductions are nonampliative — i.e., their conclusions cannot exceed their premises in content. Furthermore, Popper's theory does not claim that basic statements plus deduction can give us scientific theory; instead, corroboration is introduced. Corroboration is, I think, a nondemonstrative kind of inference. It is a way of providing for the acceptance of hypotheses even though the content of these hypotheses goes far beyond that of the basic statements. *Modus tollens* without corroboration is empty; *modus tollens* with corroboration is induction" (1967, p. 28).

It is, of course, easy to see how a justificationist could be led to think that induction is a guide to scientific life. He would think that information gained by inductive methods could *enrich* the premises from which one could reason one's way to future conduct. For example, if I know that one horse running in a given race has won four out of his last five starts, while no other horse in the field has won more than one out of his last five starts, it would be "natural" (so the argument goes) to bet on that horse rather than one of the others. By similar reasoning, if a certain theory is highly 'inductively' established (i.e., has 'won' a large number of its experimental 'races' in the past) it would be "natural" not only to conclude that it is probably true (the assessment problem) but also to continue research based upon it. Information gathered by 'inductive inference' in the past is supposed to be relevant to the determination of future behavior.

But again, the appraisal of performance in the past is being confused with guidance in the future. It is obvious that the mere confirmation of an expectancy provides no basis for inference at all. Indeed, that *is* the classical problem of induction! The classic problem is that information about one event or series of events cannot provide any information about another event or series of events *unless* one can justify an inductive inference from the one to the other. (The temporal factor is irrelevant: it is the *inference* from past to present that is in question, not the change in time.) So the justificationist who thinks that induction is a guide to life must also think he has solved the classic problem of induction. Since no one *has* solved the justificationist problem of induction, anyone who believes induction is a guide to life is

He then proceeds to rehash the "necessity" for justification of nondemonstrative inference: "With the same force and logic with which Hume raised questions about the justification of induction, we may and must raise problems about the justification of any kind of demonstrative inference" (p. 28).

Notice two things about this argument. First, it is actually directed to corroboration as a problem of nondemonstrative 'inference' in the *appraisal* of theories. It ignores the repeated Popperian reply that corroboration is analytic and hence not nondemonstrative (see Watkins, "Non-inductive Corroboration," in Lakatos, ed., *The Problem of Inductive Logic*, especially p. 63), and therefore not an instrument of prediction or guidance. Second, since it is cast within the justificationist framework, it concentrates upon justification as the only significant problem. That is, it never gets to the problem of the guidance of scientific life at all — it merely states that if science is not to be totally empty there must be such a guide, and it assumes that that guide must be 'inductive inference'. There is no positive argument in favor of induction as a guide to scientific life at all, only the truism that science is not sterile.

simply being intellectually dishonest. To repeat: within this framework, we never get to guidance at all — since *all* the uses of inductive logic hinge upon its justification, *nothing*[10] gets off the ground unless a successful justification can be found.

Like the confirmation theorist, I take very seriously the problem of guidance in scientific research: indeed these problems of *expectation* are intimately related to what and how we learn from scientific "data." But I deny that the problem of guidance is in any simple way connected to the problem of acceptance: the acceptance of either a given proposition or an entire body of scientific knowledge claims says absolutely nothing about the course that future research should take. One could think that these two orthogonal problems were aspects of the same problem only if the justificationist point of view, which identifies both problems with 'induction,' is presupposed. To examine the problems of inference we must hold the problems of acceptance, which are the chief problems of methodology in scientific research, in abeyance. Instead of asking "What guides scientific life?" and providing yet another twist to the incorrect answer "inductive inference," we should ask "What is the nature of scientific concept formation?" and the further question "How is concept formation related to data?"

One way of restating the so-called logical problem of induction within the psychological framework is to note that the difficulty is that there are an infinitude of inferences (or theories, or models) that are equally *not supported* (i.e., neither determined to be true nor shown

[10] Assuming that the *only* information available was an appraisal of past success, *would* it actually be rational to guide conduct (to "bet") on that basis? Certainly not. This is because there is *always* an infinitude of "theories" that are consonant with any "data" whatsoever, and unless one has other, *independent* reasons for believing one of them, it is not rational to support (i.e., to make a commitment, such as placing a bet) *any* of them. The race track analogy disguises this because everyone brings their "intuitive" knowledge of the racing situation to bear upon the information at hand. For example, the "theory" that the horse which has won in the past will win the "present" race is certainly compatible with the "data" as given, i.e., his past performance. But equally compatible a priori are the "theories" that he will be sick or break a leg, have a bad jockey, be frightened by a colorful spectator, etc., etc., and *lose*. The "good reasons" that can be adduced for betting on a horse that has won in the past, which render such behavior rational, always involve knowledge of reality that transcends empirical and logical considerations. And this nondemonstrative knowledge will in its turn have to be justified before it is rational to accept one out of an infinitude of potential theories of future performance. Once again: the justificationist cannot put inductive inference to *any* use until it has been justified as a rational source of knowledge. And it is precisely this that he has never succeeded in doing.

to be trustworthy guides) by any given data. The transition from the justificationist's problem of sanctioning inference to the nonjustificational one of learning and concept formation is made by asking what *constrains* our patterns of inference in such a manner that we sometimes hit upon a 'good' concept or theory. We *do* learn informative things with the aid of (some) data in the process of scientific concept formation. It will not suffice to explain how and why this is so by vague reference to "insight" or "creative imagination." And the problem cannot be avoided by sharply distinguishing between the contexts of discovery and justification. Nor will the neo-Popperian ploy of distinguishing pure science from applied technology avoid the issue.[11] Sooner or later we must face the question of how we learn to form scientific concepts and how scientific concept formation is related to the guidance of scientific practice. The problems of inference and expectation require a psychological solution. Hume was correct in dropping these problems into the (according to the philosopher, obviously disreputable) hands of the psychologist. But before we despair of this fate, let us see what the psychologist must account for, and then ask what theories he has available to explain the phenomena in question. Like Russell, I think that such explanations are all that we, as prudent men, can hope for from a contingent theory.

The Nature of Scientific Knowledge

As far as logic is concerned, *all* scientific and 'common-sense' knowledge is on a par: it is all equally nondemonstrative in character and,

[11] Popperians freely admit that the logic of science is deductive only and that the corroboration of a theory, as an analytic appraisal of past success, says nothing about the future success that can be expected for the theory. Indeed, they often claim that science has no guidance at all (other than creative imagination, which they do not even try to explain) and that only (mere) technology requires *reliable* theories that can guide future practice (i.e., provide *safe* bets). Both J. Agassi, "The Confusion between Science and Technology in Standard Philosophies of Science," *Technology and Culture*, 7 (1966), 348–66, and "Positive Evidence in Science and Technology," *Philosophy of Science*, 37 (1970), 261–70, and T. W. Settle, "Induction and Probability Unfused," in P. A. Schilpp, ed., *The Philosophy of Karl Popper* (La Salle, Ill.: Open Court, 1974), pp. 697–749, have taken this approach. In the first place, the problem of guidance is not *just* the problem of finding reliable or technologically safe theories: everyone (including especially the Popperians) wants to *increase our knowledge* by exploring more adequate theories. Second, the problem of creativity, which is at the heart of the matter, is not addressed at all by this approach. Scientific concept formation is beyond the pale of methodology for the Popperian.

hence, equally incapable of justification. The *assumption* that human beings are in possession of any knowledge *whatsoever* about the world can only be defended by adducing two classes of further assumptions in its support. The first class of assumptions that is necessary to defend the claim that we have knowledge (*any* knowledge) concerns the actual structure of reality — the world we live in as opposed to the possible ones. The second class of assumptions concerns the nature of the organism that *has* knowledge — we must be structured in such a manner that we can come to know the environment in which, as a matter of (contingent) fact, we happen to exist. Indeed, we could not have *knowledge* in (or from) the 'trivialities' of immediate perception (as the empiricist claims) unless our theories of how the knowing organism is structured are correct. The so-called 'direct' knowledge disclosed by perception is as inferential in character as our most speculative conjectures. (That is, the assumption that we have even 'trivial' knowledge by perceptual experience depends upon our theories of cognizing organisms. It is the further assumption that we have other, *non*trivial contingent knowledge that is dependent upon the structure of reality.) Although we cannot justify the belief that we actually possess knowledge, we most certainly can adduce good reasons for thinking that we do have (and can gain more) knowledge. Those 'good reasons' are our contingent theories about the nature of our knowledge and the nature of knowledge acquisition. Although we cannot *prove* that we have knowledge (by this line of defense), there is nothing whatsoever irrational about seeking it and continuing to subject every theory that we deem worthy of entertaining to critical scrutiny. Constructing contingent theories of the nature of ourselves and our knowledge would be irrational *only* within the justificationist framework, where rationality remains equated with proof and certainty and justification.

The justificationist is literally a prisoner within the confines of his own framework: because of the structural fusions and confusions inherent in it, that framework prevents many crucial problems from ever being satisfactorily resolved. The nature of our knowledge and the means of its acquisition are problems that it has done a particularly fine job of botching up, and it will take the remainder of this essay to begin to locate the directions in which genuine solutions to these problems should be sought. To begin this task of "relocating" the problems posed by inference and expectation we shall spend the rest of this part of the essay exploring the character of so-called 'scientific' knowledge, to see what the end product is that our theory of inference must produce. In the subsequent part the nature of the organism that makes inferences will be located within a biological, evolutionary framework. Only after these preliminaries are completed may we address the psychology of inference and expectation.

Theories as structural representations of reality. The classic conception of scientific theories as an amalgam of postulates, syntactic calculus, correspondence rules, and a model, while perhaps not too incorrect as a description, is totally incapable of explaining what theories *do,* i.e., how they function in embodying and generating scientific knowledge. In order to understand the role of inference and expectation in the acquisition of knowledge we must know both the nature of our knowledge and how it is embodied. Only after we are clear on what knowledge is and how it is embodied may we ask how it is transmitted and how we acquire new knowledge. Thus we must begin by asking how theories, as embodiments of scientific knowledge, function in the scientist's head. That is, how does the conceptual structure of a theory make 'understanding' possible?

The thesis I wish to defend is that theories are nothing more, nothing less, than conceptual *points of view* that organize 'observations' or 'factual data' into meaningful patterns. *Theories are a way of seeing reality.* They render their subject matter intelligible by exhibiting the *structure* of the phenomena with which they deal. Theories are structured representations that constitute a way of 'seeing' or perceiving reality. Theories function as instruments of understanding *because* they are structural representations of a domain.

The conception of theories *as a way of seeing* is very old: *theoria* for the ancient Greeks meant a 'vision,' and metaphysicians have long been known as 'visionaries.' This way of 'picturing' theories has the advantage of bringing to the fore their inseparable relation to perception: we perceive reality by means of the structural representations which are our theories. Theories let us see the structure of reality. And the key feature of theories as 'points of view' or 'ways of perceiving' is that they are *generative* or *productive* (in the transformational grammarian's sense): they enable us to *infer* to indefinitely extended instances of unobserved cases. Theories allow us to represent an indefinitely large number of data points. States of affairs are represented by theories, and that representation very literally *is* our understanding of them.

The thesis that theories are structural representations of reality was admirably defended by N. R. Hanson.[12] Consider this comment by Hanson:

> Scientifically understanding phenomena x, y, and z consists in perceiving what kinds of phenomena they are — how they relate one to the other within some larger epistemic context; how they are dependent upon, or interfere with, one another. Insights into such relations "out there" are generable within our perceptions of the structures of theories; these theoretical structures function vis-a-vis our linguistic references to x, y, and z in a way analogous to how the scene stands to the tree and hill "out there" and also to the painted patches on canvas. Thus, I suggest that in contrast to the delineation of theories as "ideal languages" or "Euclidean hypothetico-deductive structures," the important function of scientific theory is to provide such structural representations of phenomena that to understand how the elements in the theoretical representation "hang together" is to discover a way in which the facts of the world "hang together." In short, scientific theories do not always argue us into the truth; they do not always demonstrate deductively and forcefully what is the case. Often they show what could be the case with perplexing phenomena, by relating representations of those phenomena in ways which are themselves possible representations of relationships obtaining "out there." Theories provide patterns for ordering phenomena. They do this, just as much as they provide inference channels through which to argue toward descriptions of phenomena (1970, p. 240).

Scientific theories represent the phenomena of their domains by being structural patterns that are isomorphic to the structural relations that obtain in the phenomena themselves.

What is it that allows one to say that an artist has captured "the same" scene as the actual landscape "out there"? Subject matters and their representations must share something in common: *structure*. Consider a song being sung, its recording as a record, and its musical score on note paper: it is knowledge of their common structure that enables us to say that these disparate manifestations all represent the same (abstract) entity. Likewise, scientific theories represent external reality. But representation is not just (iconic) picturing: theories are not pho-

[12] See his *Patterns of Discovery* (Cambridge: At the University Press, 1958), and "A Picture Theory of Theory Meaning," in R. Colodny, ed., *The Nature and Function of Scientific Theories* (Pittsburgh: University of Pittsburgh Press, 1970), pp. 233–74.

tographs. Scientific theories are *abstract* representations — they are like graphs or flow charts rather than pictures. The correlation between theoretical terms and external objects is conventional, as is the correlation between the symbols of a map and the terrain it represents. But a map is informative because it shares relational structures with the terrain in question. Charts, graphs, etc., enable one to see the dynamical structural characteristics of the phenomena they represent. "They provide a pattern through which the multiform and chaotic manifestations of the original appear as correlated parameters. These patterns provide conceptual gestalts which allow inferences from one parameter to another parameter throughout a charted system of data lines" (*ibid.*, p. 250).

How far is it from graphs and flow charts to our sophisticated mathematico-physical theories? Not far at all: mathematics is simply a tool for transforming the *perceptual* structural representation provided by a graph or flow chart into an *equation* that *shares that same structure.* "Scientific theories enable us to understand perplexing phenomena precisely because they enable us to see on the page some of the same structures which are there in the phenomena themselves. The theory allows us to comprehend what makes things "go" — and to work our ways into the phenomena, along the dynamical structures (as it were) by way of inferences through the algebra which itself has the same structure as the phenomena, or at least a structure compatible with the phenomena" (*ibid.*, p. 273). All our 'perceptual' knowledge of the structure of reality is (in principle) capable of transformation by mathematics into equations that equally represent the structure of reality. Our understanding of reality is nothing more, nothing less, than our ability to perceive such structures (i.e., to infer it from our phenomenal experience). The problem for a psychology of inference and expectation is to characterize the organism that does this 'perceiving,' in order to understand how our knowledge is acquired.

Structural realism and the nature of scientific knowledge. What scientific theories do is to represent, in concise and manageable form, the structure of their domains: they are structural representations of reality. But what is 'structure,' and why is it that when we represent reality we represent only the structure that is definitive of it? Why is our knowledge of the external world structural? The answer, simply put, is that the only nonstructural or intrinsic properties of existents of which we have any comprehension whatsoever are the properties of our own phenomenal experience. Our knowledge by acquaintance, our phenomenal experience, is qualitatively different from our knowledge of the nonmental realm. What we know of 'external' (external to the mental realm of 'direct awareness') reality is knowledge *by description*: we are not acquainted with the intrinsic properties of external objects at all.

The genesis of scientific knowledge of reality, insofar as it is knowledge by description only, may be compared to the unfolding plot of a detective story. Assuming that a crime has been committed, the detective must track down the culprit. Now if no one saw the commission of the crime, if it was not an 'ingredient' (to use Russell's term) in the perceptual knowledge by acquaintance of anyone, then the detective is in exactly the same situation as the scientist. Although the detective is in a sense at a loss because no one saw the culprit, he is not entirely hamstrung: he can gain sufficient knowledge to enable the guilty person to be "brought to justice." That is, the detective can come to *know* the culprit, by a (usually quite involved) process of inferential reasoning, without ever becoming directly acquainted with him. All the "clues" regarding the nature and whereabouts of the culprit, from those present at the scene of the crime through to the inevitable chase scene, can provide the detective with knowledge by description, knowledge of structural properties, of the guilty party. He can come to learn, for instance, that he is male, between 6′ and 6′2″ tall, that he weighs approximately 175 lbs., has a gold ring on a certain finger, walks with a limp, etc., etc. In short, the detective can come to identify positively an individual without ever having 'seen' or become 'acquainted' with him. The guilty party can be known in exactly the same way that an abstract and unobservable scientific entity, such as a proton or a phoneme, can be 'known.'

But there is one major difference between the fictional detective story and the scientific detective story: the analogy is incorrect in one crucial respect. At the end of the novel, the culprit is identified — somebody shouts, "I know him, that's Joe Smith!" or some such. What they mean by identifying the culprit by his proper name is that they are "acquainted" with him, and hence their knowledge of Joe Smith is not only knowledge by description but *also* knowledge by acquaintance. But proper names do not occur in scientific investigation: the scientist never concludes by saying, "And that's how I met Joe Neutrino." This is because the scientist is never directly acquainted with the intrinsic properties of *any* 'external object': *all* the properties which we normally attribute to nonmental objects exist *wholly* within the mind. As Berkeley pointed out, Locke's distinction of primary and secondary qualities is invalid: *all* the properties of objects, including what Locke thought were primary or intrinsic properties, exist only within the mental realm of the perceiver. All such properties and qualities are "in" the mental world rather than "in" the objects themselves. Can science conceive of *any* properties of the entities of the nonmental world? According to the view I am proposing, which is really Bertrand Russell's view,[13] what we can know about external or nonmental objects (including even our physical bodies in the latter class) is only their structural properties and relations, never their intrinsic properties.

Commenting on the question of what science knows of the nonmental realm, Maxwell[14] has said:

> [T]he only aspects of the nonmental world of which we can have any knowledge or any conception are purely structural (or, in other words, purely formal). The *details* of the answer consist of an explication of "structural" or "formal" adequate to the task at hand. . . . The notion of *form* or *structure* needed here may accurately be said to be logical (and/or mathematical) and, in a sense, abstract; *characterizations* of instances of it will be in terms of logic alone, i.e., the logical connectives, quantifiers, and variables — they will contain no descriptive terms. . . . *Structure,* in the sense we require, *must be factual and contingent and,* at least in its exemplifications, *concrete.* Furthermore, it cannot be emphasized too strongly — what should already be obvious — that structure is *not* linguistic nor, even, conceptual in character; it is an objective feature of the real world (p. 153).

But *structure* should not be identified *solely* with the notion of form: rather "it is form plus causal connections with experience" (*ibid.,* p. 154).

In overview, structural realism is *structural* in the sense that our knowledge of the entire nonmental world, from our bodies to physical objects, is of the structural characteristics of that world rather than

[13] As first presented in *The Analysis of Matter* (London: Allen and Unwin, 1927), and developed consistently in *Human Knowledge: Its Scope and Limits.*

[14] G. Maxwell, "Scientific Methodology and the Causal Theory of Perception," in Lakatos and Musgrave, eds., *Problems in the Philosophy of Science* (Amsterdam: North-Holland, 1968), pp. 148–77.

of the *intrinsic properties* of those objects composing it. Structural realism is a *realism* in that physical objects *do* causally influence our perceptions: properly stated, the causal theory of perception is true.

Structural realism may be stated in terms reflecting (and related to) Kant's distinction of *phenomena* from *noumena*:

> On the one hand there is the realm of phenomena. These are wholly *in the mind* (in our sense). Of the phenomena and only of the phenomena do we have *direct knowledge*. On the other hand, there are the things in themselves, and here our divergence from the views of Kant is great: although we have no *direct* knowledge of the latter, the bulk of our common sense knowledge and our scientific knowledge *is* of them. Among them are not only electrons, protons, forces, and fields but also tables, chairs, and human bodies. All of our knowledge of these is, of course, indirect and may be generally characterized as hypothetico-deductive (or, better, hypothetico-inferential). . . . The implications of this are manifold and profound. In fact, it requires considerable time and thought — for most of us at least — to realize what a drastic revision of our usual conceptions, including our scientific conceptions is required (*ibid.*, pp. 154–55).

One of the concepts most in need of revision (actually reinterpretation) is that of *observation*. Both common sense and unreflective scientific thought speak commonly of "observing physical objects." But, strictly speaking, there can be no observation of public objects at all:

> Observation, as usually conceived, is a naive realist concept through and through. Therefore, if structural realism is true, then, in any usual sense of "observation," we observe neither public objects nor entities in our minds: we never observe anything at all. For example, if I have a dream, no matter how vivid, of seeing a white dog, I cannot be said to have observed a dog or, indeed, anything, since there existed nothing corresponding in any straightforward way to the ostensibly observed object. On the other hand, if I do what ordinarily could be called "actually seeing a white dog," then . . . I do not actually observe anything (or even *see* in the usual sense), for there *is* nothing external to me which is white in the usual, qualitative sense, or etc., etc. (*ibid.*, p. 167).

Talk of the observation of objects, as is common in science and common sense, must be taken as a shorthand formulation, couched in terms of naive realism, for discourse about sense impressions and their structural relation to the external objects which are their causes.

Observation and the deductive unification of experience. The ordinary concept of observation, as structural realism makes clear, is a naive realist relic in scientific discourse. But we are supposed to observe facts, and facts are alleged to be the indispensable basis upon which the edifice of scientific knowledge is erected (by so-called inductive inference). Although there is increasing acceptance of the inherent theoretical nature (or contamination) of facts, i.e., acceptance of the doctrine of factual relativity,[15] even among staunch justificationist empiricists, there has been almost unanimous assent to the claim that the ultimate foundation of science is the perceptual experience of the scientist. There has been no general acknowledgment that science is *not based upon perceptual experience at all,* nor has the import of the idealized and abstract nature of scientific entities been acknowledged. But the total *disconnection* of theory from experience, as well as the abstract nature of scientific entities, can easily be demonstrated. When this is done, the nature and role of inference in scientific thought are seen to be drastically different from what the received view doctrines have proposed: that is, the basis of inference must be radically relocated from generalization from facts to the functioning of the nervous system. Let us follow up our presentation of structural realism with a look at the abstract and nonperceptual nature of scientific knowledge in order to relocate the real problems that they pose for inference and expectation.

[15] Factual relativity, or the conceptual nature of facts, is the doctrine that factual propositions exist only within a given conceptual framework rather than independently of theories and conceptual schemes. The data of sensory experience do not come with tags proclaiming their factual status. Observation is more than merely becoming aware of sensory input: it is the assimilation of input into a classificatory scheme that is logically prior to that input. Karl Popper is largely responsible for forcing the admission that the 'basic statements' of science are theoretically determined upon the members of the Vienna Circle; see his *Logic of Scientific Discovery,* especially section 30. Thomas Kuhn has continually emphasized factual relativity in scientific revolutions in *The Structure of Scientific Revolutions.* The pragmatist C. I. Lewis also ably defended the doctrine under the guise of the pragmatic conception of the a priori; see especially chs. 5 and 8 of *Mind and the World Order* (New York: Scribner's, 1929). N. R. Hanson also drove home the relativity of facts, as in this passage from his *Patterns of Discovery*: "If in the brilliant disc of which he is visually aware Tycho sees only the sun, then he cannot but see that it is a body which will behave in characteristically "Tychonic" ways. These serve as the foundation for Tycho's general geocentric-geostatic theories about the sun. . . . Tycho sees the sun beginning its journey from horizon to horizon. He sees that from some celestial vantage point the sun (carrying with it the moon and planets) could be watched circling our fixed earth. Watching the sun at dawn through Tychonic spectacles would be to see it in something like this way. Kepler's visual field, however, has a different conceptual organization. Yet a drawing of what he sees at dawn could be a drawing of exactly what Tycho saw, and could be recognized as such by Tycho. But Kepler will see the horizon dipping, or turning away, from our fixed local star" (1958, p. 23).

We must support two claims: first, there are no "perceptual experiences" whatsoever at the "basis" of science; second, the fundamental propositions of scientific explanatory discourse deal with abstract and ideal entities. Let us consider the latter contention first. It may be restated as the claim that science can never deal directly with particulars or individual things *at all*, but only with "abstract entities" or thing-*kinds*. Most contemporary "philosophy of science," insofar as it deals with substantive theory construction in science, is concerned with the purely formal structure of the subject matter theories it examines (in particular, those that are or can be axiomatized). Thus every textbook has its chapter on "theories in general" which treats the nature of formal systems and their utilization in theory construction and scientific explanation. But one factor that has been quite neglected is what Stephen Körner has examined:[16] the restriction which the logical structure of scientific axiomatic systems imposes upon the subject matter of such theoretical systems.

The empiricist "foundations" view of knowledge holds that that which is given in perceptual experience is to be taken as particular. But one may pose an embarrassing question at this point: how does the nominalist-empiricist *know* (or come to recognize, be acquainted with, or recollect, etc.) his given particulars *as* singular instances or *particulars*? That is, if a "thing" is proffered to me as "an X," where X is any description (i.e., classification) of it *whatsoever*, how can I claim to know that it *is* an X without first knowing what it is to be an instance of a thing-*kind*, namely, of kind X? In short, in order to recognize a thing (or a fact, etc.) mustn't one presuppose knowledge of, or operation within, a framework (theory) of thing-kinds? Reflection on the problem indicates that one must acknowledge that *classification is fundamentally a process of abstraction*: that is, the process of abstraction is a sine qua non for determination of concreta. Let us develop this point in another way by considering what is involved in hypothetico-deductive "scientific" explanation.

The question we must ask is this: to what extent does employment of the hypothetico-deductive method *idealize* the domain to which it is applied? In other words, what does the process of deduction require

[16] See pt. II of his *Experience and Theory* (London: Routledge & Kegan Paul, 1966).

of the empirical predicates of a science before it may legitimately be applied?

Körner considers two classes of such constraints. The first constraint requires the elimination of inexactness and indefiniteness of all the predicates. The logic of the H-D framework is an unmodified classical two-valued logic. Thus, strictly speaking, the H-D framework admits no inexact predicates whatsoever. So the H-D framework must, strictly speaking, be the logic of the finished science report, i.e., admit no inexact or 'open concepts' as Pap[17] termed the concepts of a growing science. The point is simply this: the H-D system is *not* connected to experience directly — the empirical predicates with which it deals *must* be idealizations or abstractions. Raw perceptual experience is rendered into "concrete" categories by the process of abstraction. The abstraction distinguishes between the relevant and irrelevant determinable characteristics and discards the latter. It creates new abstract determinables that *replace* the original perceptual ones at the "basis" of scientific explanatory systems. "This type of abstraction which, in order to distinguish it from other kinds, I shall call 'deductive abstraction,' and which replaces perceptual by abstract determinables, reinforces the general effect of the restrictions which the logico-mathematical framework of every theory imposes upon perceptual characteristics" (Körner, 1966, pp. 166–67).

But even this is not yet sufficient to render a so-called 'empirical' predicate fit for a position in a deductively unified system. Further idealization effectively removes the predicate from perceptual experience *entirely*:

> The disconnection of the theory from its perceptual subject-matter can now be also expressed by saying that no perceptual proposition and no perceptual predicate occurs in any deductive sequence. That this must be so is clear. All inexact perceptual predicates are precluded from occurring in any sequence, and the exact — though internally inexact — determinables have been replaced by non-perceptual predicates through abstraction — to say nothing of the further replacements due to the conditions of measurement, general and special.
>
> No perceptual proposition will be a last term in a deductive sequence. The theory will be linked to perception not by deduction but by identification (*ibid.*, pp. 168–69).

[17] See ch. 11, "Reduction and Open Concepts," in his *Semantics and Necessary Truth* (New Haven, Conn.: Yale University Press, 1958), pp. 302–60.

This means that the reference of empirical concepts cannot be given, but is rather idealized, which is to say, *constructed*, by the active abstraction of our conceptual frameworks. The concept of an ideal straight line within an axiomatic system is a far cry from *any* empirical line, such as a light ray, the surface of a straight edge, etc. But our theories of reality deal with the abstract, idealized elements such as "straight lines" and not with the lines drawn with rulers or physical objects. With that admission, *our theories become nonempirical* in the sense that their reference is to ideal, nonperceptual entities rather than concrete physical ones. The empirical becomes identified with the ideal.[18]

This effectively complicates the traditional usage and interpretation of correspondence rules or bridge laws between 'theory' and 'observation.' It is commonly assumed that correspondence rules somehow link or "translate" empirical predicates into theoretical ones. But clearly there is a "linkage" that precedes the typical notion of correspondence rule, which is required to identify an empirical predicate in the first instance. As Körner notes, most philosophers who talk of correspondence rules "are not aware that one of the fundamental transactions consists in the transposition of internally inexact, empirical predicates from a modified into an unmodified two-value logic and in their consequent replacement by internally exact ones" (*ibid.*, p. 90).

Körner's claim, then, is that the traditional usage of correspondence rules (to link empirical predicates to theoretical propositions) covers only half the problem of reference for theoretical terms: there must be additional "correspondence" rules linking the phenomena of experience, sense data if you like, with empirical predicates. And this linkage is not unambiguous or in any sense "direct": constructing an empirical predicate out of the deliverances of sense requires a theory of idealization and abstraction. That "theory of abstraction" means that cognition is not a passive register of "objective" sensations, as Locke claimed, but rather an active, indeed constructive, process is another way of stating the abstract, conceptual nature of facts. But the "theory" determining the facts in *this* instance is actually the genetic, physiological, and psychological structure of man. We must now turn to a study

[18] It is the *identification* of the empirical with the ideal that strikes at the heart of the matter: "A hypothetico-deductive system is . . . not directly connected with experience. In linking it to experience by 'identifying' some of its predicates and propositions with internally inexact empirical ones one is not ascertaining that they are identical, only treating them as if they were" (Körner, 1966, pp. 89–90).

of the a priori constraints upon human knowledge that are imposed by our physiological and psychological nature. What constrains the input of the potential environmental flux into human consciousness is actually "the way the mind works." And the way the mind works is as an inference machine. As we shall now see, this inference machine not only makes inferences in accordance with the data it has available but also constructs the very data from which it infers.

The Biological Basis of Cognition: The Organism as a Theory of Its Environment

Hume relegated nondemonstrative inference to "animal belief." Justificationist philosophers, while not denying such base accompaniments to this pattern of reasoning, have been trying to salvage a respectable *logical* problem of 'induction' ever since. Typically their resultant 'respectable' problem utilizes two tacit assumptions (among others): first, that we have 'facts' or 'observational statements' as unproblematically given, and that induction leaps beyond the firm foundation of evidence to risky theoretical formulations; second, that the psychological nature of the organism who utilizes nondemonstrative inference to obtain knowledge is all but totally irrelevant to the problems of philosophy. In a sense, the burden of this essay is to argue that these two assumptions are grossly mistaken and that the very nature of the legitimate problems of inference cannot be discerned until these assumptions are exposed as totally indefensible and then their pernicious effects eliminated. Only by purging these inadequate conceptions of knowledge and human nature from our thinking can we understand the indispensable role of inference in cognition and the relation of inference to knowledge and its acquisition. Not surprisingly, these two assumptions of traditional philosophy are intimately related: they are based on an incredibly inadequate psychology of knowledge acquisition (to say nothing of a chimerical conception of contingent yet somehow certain knowledge).

To counter the justificationist approach to the problems of inference we must get clear on the nature of human knowledge and the nature of the organism that acquires that knowledge (as well as indicate *how* knowledge is acquired). We have begun to counter the justificationist conception of knowledge and its acquisition by examining the function of scientific theories and the structural nature of our knowledge of the nonmental realm. We have also indicated the abstract and conjectural nature of observational statements, and discussed the nonperceptual nature of scientific propositions. Now we must turn even more in the direction of theoretical psychology, to examine the nature of the 'inferring' organism. In so doing we will be examining the a priori constraints upon human knowledge and its acquisition that are imposed by our psychological nature.

The fundamental thesis I wish to defend, and ultimately to elaborate, is that the problems of inference can only be understood in the context of a psychology of learning and concept formation which takes as its central tenet the thesis that *organisms* (more properly, the central nervous systems of organisms) *are theories of their environment*. The problems of nondemonstrative inference are the problems of understanding how the nervous system,[19] from its preconscious determination of the orders of sensory experience to the higher mental processes such as human thought, attempts to model, with ever increasing degrees of adequacy, the environment that the organism confronts. There is no difference in kind between the scientist inferring the most esoteric theory of reality, on the one hand, and the simplest organism's inferring the presence of food or danger in its environment. In both cases the fundamental *activity* of the nervous system is *classification* (or abstraction) and the fundamental *function* the nervous system performs is *modeling* (of the environment). Nondemonstrative inference, the process by which we gain our contingent knowledge of reality, is not different from the psychological phenomena of concept formation and learning. And we shall see subsequently that the key to these latter phenomena is found in the psychology of perception and memory.

The proposal that organisms are theories of their environments is to be understood within an adaptational or evolutionary perspective. If an organism is to survive in an uncertain world (such as the one we find ourselves inhabiting), it must be able to operate as a biological mechanism within that world. The mere fact of survival implies that

[19] Although it is more correct to say that the entire organism is a manifestation of a theory of its environment, the nervous system is the key component in the acquisition and generation of knowledge. The central nervous system (CNS) is also a theory of the environment of the CNS, which *includes* the remainder of the organism. Thus we may talk of the CNS as "the organism" and as a theory of the organism's environment without oversimplification or distortion.

an organism has been effective in maintaining an appropriate commerce with its environment. The struggle for survival is the struggle to adapt to, or to learn to utilize effectively, the environment in which the organism is situated. The survival of a species *implies* that the species is efficient in dealing with the contingencies of its environment. It is the nervous system that is ultimately responsible for the organism's perception and knowledge of, and its commerce with, the environment. The question then arises, "How does the nervous system come to have knowledge of the environment and its contingencies?" that is, "How does the CNS function?" The answer is that the nervous system functions as a theory of the environment of the organism: *its job is to make inferences* (about environmental contingencies). To the extent that the theory that the nervous system instantiates is adequate to its task and the inferences are successful, to that extent the species survives. If the theory is inadequate, the species may be expected to die out. Evolution, from this point of view, is a mechanism which allows nervous systems to construct more and more adequate theories of their environment. Man, the conscious, thinking (and sometimes) rational animal, is an emergent phenomenon in that only his CNS has developed to the extent that it can consciously create and reflect upon, as well as operate according to, theories of the environment.

The question now arises *how* organisms (such as scientists) construct their theories of the environment. The answer to this question lies in a proper unpacking of the psychological phenomena of "learning" and "concept formation." That is, this question really asks, "How do organisms form their concepts of reality?" and it is at this point that the psychology of inference and expectation enters, as an indispensable ingredient, into epistemology. The questions that must be asked are "What are 'facts' or 'data'?" and "What and how do we learn from data?" These questions take on a very strange light once it is realized that the reality which not only preconscious sensory experience but scientific thought attempts to model is known *only* as a result of the classifications imposed by the central nervous system.[20] Once (at least rudimentary) answers to these questions of perception and learning are at hand, the problems that "induction" has been presumed to be

[20] This is not, of course, a claim that the nervous system is a *homonculus* whose job is that of a taxonomist; rather it is the case that the nervous system classifies information, a task which homonculi are often postulated to perform.

necessary to solve, such as the guidance of scientific life, can then be addressed.

But first we must explore the ramifications of considering organisms as theories. There would seem at first blush to be little reason to construe either the activity of preconscious nervous processes as instantiating a theory or the conscious activity of all but trained professionals as actually theory construction. After some time is spent developing the approach with the higher mental processes, the manner in which the thesis applies to the functioning of the nervous system as a whole will be considered. But the point of all this discussion remains: unless we can understand the nature and functioning of the organism that makes inferences, we shall never understand either what or how we learn from 'facts'.

Thought and symbolism. So far as I am aware, there has been only one serious hypothesis on the nature of thought offered within the context of an adaptational or evolutionary approach to the problem. This is an hypothesis due to Kenneth Craik.[21] Craik's hypothesis is simply "that thought models, or parallels, reality — that its essential feature is not 'the mind', 'the self', 'sense-data', nor propositions but symbolism, and that this symbolism is largely of the same kind as that which is familiar to us in mechanical devices which aid thought and calculation" (p. 57). The fundamental feature of our neural machinery is its power to parallel or model external events. In terms familiar from the discussion of Hanson's conception of theories as structural representations, thought literally *is* a representation of reality. Man, the conscious organism, as a theory of its environment, models reality by means of thought. The knowledge that resides "in the head" of the scientist resides in his internal model — thinking about the environment in which we find ourselves is ipso facto to model that environment. That is, modeling (as an activity of CNS's) is a richer concept, or a more generic one, than 'representation' (which in turn is richer or more inclusive than iconic 'picturing'). And modeling, as Russell and Maxwell have been at pains to emphasize, is a structural means of representation. The models which constitute scientific theories are (to greater or lesser degree) structurally isomorphic to the reality to which they pertain. To repeat, our scientific knowledge of reality, which is embodied in our theories, which

[21] In his neglected monograph *The Nature of Explanation* (Cambridge: At the University Press, 1943).

are in turn representations, which are in turn the product of thought, which is in turn modeling, is purely of the structural rather than the intrinsic properties of reality. Craik was very clear on the structural nature of models: "By a model we thus mean any physical or chemical system which has a similar relation-structure to that of the process it imitates. By 'relation-structure' I do not mean some obscure non-physical entity which attends the model, but the fact that it is a physical working model which works in the same way as the process it parallels, in the aspects under consideration at any moment" (*ibid.*, p. 51).

Thinking, if Craik's hypothesis is tenable, is always *symbolic*: the model which is the thought is only structurally related to the reality which it imitates. The activity of neural excitation per se is totally unlike, say, the patterns of stress in a bridge, yet the *patterns* of excitation which constitute the thinking about (or calculating) that stress are isomorphic (in a structural sense) to the stress itself. "It is likely then that the nervous system is in a fortunate position, as far as modelling physical processes is concerned, in that it has only to produce combinations of excited arcs, not physical objects; its 'answer' need only be a combination of consistent patterns of excitation — not a new object that is physically and chemically stable" (*ibid.*, p. 56).

If thinking *is* modeling, then 'the organism is a theory of its environment' follows automatically. If the organism carries around in its CNS an internal model of external reality (and, of course, a model of itself and its capabilities) then it will be able to adapt to, to survive within, that environment. It will be able to try out and assess various alternative actions, react to future situations before they arise, and in general *anticipate* the environment in which it finds itself. (Our nervous systems are instruments of adaptation to our environment *because* they permit trial of alternatives for future conduct in an economical manner: thought models *potential* realities.) Modeling has survival value — it enables an organism to anticipate the future course of events *and* to *act* in accordance with that information. As shoud be obvious by now, the function of thought, as modeling, is identical to that which is attributed to inductive or nondemonstrative inference: thought-as-modeling is the vehicle by which we gain our contingent knowledge of reality.

What is the relation of thought (as an obviously "higher" form of evolutionary adaptation) to the other activity of the CNS, specifically to the unconscious and unverbalized aspects of human (and animal) neural functioning? Is 'conscious' thought, as it occurs in words, the only symbolic activity of the nervous system which may be said to model reality, or is there other evidence for modeling? If symbolism is defined in a very general way, as the ability of processes to parallel or imitate each other, then it becomes clear that other aspects of neural activity are equally as symbolic as thought. Indeed, *modeling is the fundamental function of the central nervous system* (of all species possessing such nervous systems), and it is exemplified in every instance of assimilation and accommodation that occurs in the life of an organism. The non-human cases are of only academic interest (for our purposes) and need not be pursued. But the case of *unconscious* modeling is worthy of note, for it makes the point that virtually all neural activity is symbolic and that words and conscious awareness are inessential to this mode of functioning.

One of the simplest appearing neural activities is the phenomenon of habituation. When an organism is exposed to stimulation for an extended duration, its reaction (in terms of its neural responsivity) diminishes. At first it was natural to assume that this phenomenon of habituation to stimulation was the result of "inhibition" that simply made the nervous system less sensitive to input: i.e., stimulation was said to decrease sensitivity if it endured over time, as if the system had a refractory period to overcome. But the lack of reaction to continuing *constant* stimulation is only part of the story of habituation. If the stimulus *changes* slightly, all the alerting and orienting reactions that accompany the onset of *novel* stimulation occur. The work on physiological correlates to stimulation is primarily Russian, stemming from the continued interest in Pavlovian conditioning. The demonstration of the paradoxical nature of habituation, that it *cannot be due to decreased sensitivity*, is largely the work of E. N. Sokolov.[22] Sokolov's findings indicate that what is going on in the cortex during habituation is that the present stimulus input is being matched to a standard that represents prior stimulation. That is, the current input is compared with an internal standard, or model, that is the result of prior stimulation. If the match of sample to standard obtains, "habituation" results; if the match does not obtain, if the new input does not

[22] See R. Lynn, *Attention, Arousal and the Orientation Reaction* (New York: Pergamon Press, 1966); E. Sokolov, "Neuronal Models and the Orienting Reflex," in M. A. B. Brazier, ed., *The Central Nervous System and Behavior* (New York: Josiah Macy, Jr., Foundation, 1960), pp. 197–276; also the references in Karl Pribram, "The Brain," *Contemporary Psychology*, September 1971.

match the stored or *expected* representation, then orienting and investigatory reactions occur (unless the stimulation is extremely strong: then a defense reaction occurs). But the point to note is that habituation, one of the simplest and most ubiquitous phenomena of neural activity, is not just a passive phenomenon. Rather, as Karl Pribram[23] has said, "habituation thus does not indicate loss of sensitivity by the nervous system; it shows that the brain develops a neural model of the environment, a representation, an expectancy against which inputs are constantly matched" (p. 46). Modeling, then, is not just the fundamental function of thought — it appears to be the fundamental function of the nervous system in general.

Now this is not an overstatement: modeling *is* really *the* fundamental function of the CNS. What the CNS "does" is to pick up information and then act upon it. And it is easy to see that all "picking up" of information requires the system to make a choice, to effect a classification, to determine whether a present level of sensory stimulation is to count as input or not. This is because the living organism never is or inhabits a static system in which no "stimulation" occurs — all input occurs against a background of ongoing neural activity. The sensory apparatus of living organisms has neural activity occurring *all* the time. As long as this is so, then the habituation situation is always present: in order to code *anything* as an input, the nervous system must compare it to an internalized model of *what input is*. Surprisingly, the nervous system cannot know what in the flux of its milieu is an input unless it has some internal standard or criterion of what constitutes an input. Thus there are two types of classification inherent in each of our sensory impressions: first, the determination that an ongoing level of activity constitutes an input; second, simultaneous to that determination the input 'event' is classified as instantiating a particular *kind* of sensory event. Pribram's comment cannot be ignored: the brain develops a neural model of the environment, a representation, an expectancy against which inputs are constantly matched. But to this picture must be added the startling rider that the nervous system *creates* its own inputs.

The sensory order. The fact that it is the structure and functioning of the nervous system that create the richness of sensory experience, rather than the environment external to the nervous system, is not

[23] In his article "The Brain." See also his book *Languages of the Brain* (Englewood Cliffs, N.J.: Prentice-Hall, 1971).

often recognized. That this must be so was clearly stated in a fascinating and neglected monograph by F. A. Hayek.[24] Hayek's psychological theory is concerned with sensory *perception* and the physiological correlates of our psychological abilities. The fundamental thesis Hayek advances is that no sensory input is "perceived" (i.e., inputed through the active CNS) *at all* unless it is perceived as one of the kinds of input accepted by the (innate or learned) classes of sensory order. As noted in the discussion of habituation, sensory perception is always an act of classification; the input signal is "processed" (matched to a standard) by any member (to which it "keys") of the sensory "orders" which impart to the phenomenal event the intrinsic properties that we experience. No sensory input can be perceived unless it can be isomorphically accepted as a match by the classes of sensory order. That is, no objects or constructions of phenomenal existents are possible except in terms of the (prior) apparatus of classification inherent in the operation of the functional nervous system. In other words, unless an environmental "event" gives rise to a pattern of sensory input that fits the organism's preexisting (either innate or learned) system of natural-kind classification, it is not perceived *at all. Perception is thus never of the intrinsic properties or attributes of "objects" in the real world*: instead, they (objects) are the results, the *abstractions*, of the actual organization and *memory* of the central nervous system. According to Hayek,

> The point on which [this] theory of the determination of mental qualities . . . differs from the position taken by practically all current psychological theories is thus the contention that the sensory (or other mental) qualities are not in some manner originally attached to, or an original attribute of, the individual physiological impulses, but that the whole of these qualities is determined by the system of connections by which the impulses can be transmitted from neuron to neuron: that it is thus the position of the individual impulse or group of impulses in the whole system of such connexions which gives it its distinctive quality; that this system of connexions is acquired in the course of the development of the species and the individual by a kind of "experience" or "learning"; and that it reproduces therefore at every stage of its development certain relationships existing in the physical environment between the stimuli evoking the impulses. . . . This central conten-

[24] Hayek's psychological theorizing is found primarily in *The Sensory Order* (London: Routledge & Kegan Paul, 1952; reprint, 1963, Phoenix Science Series). A more recent statement is found in "The Primacy of the Abstract," in A. Koestler and J. R. Smythies, eds., *Beyond Reductionism* (London: Macmillan, 1969), pp. 309–33.

tion may also be expressed more briefly by saying that "we do not first have sensations which are then preserved by memory, but it is as a result of physiological memory that the physiological impulses are converted into sensations. The connexions between the physiological elements are thus the primary phenomenon which creates the mental phenomena" (1952, p. 53).

Thus Hayek's thesis is that an "event" (an external physical energy source) is not coded at all unless it is assimilated to a system of classification that already exists within the CNS. In a later paper Hayek makes clear the implications of this thesis for the organismic basis of our knowledge: "What this amounts to is that all the 'knowledge' of the external world which . . . an organism possesses consists in the action patterns which the stimuli tend to evoke, or, with special reference to the human mind, that what we call knowledge is primarily a system of rules of action assisted and modified by rules indicating equivalences or differences or various combinations of stimuli" (1969, p. 316).

Thus there is a circular interdetermination at work in the functioning of the CNS as it gains its knowledge of the environment. The primary mode of functioning of the nervous system is modeling: the input to the system is compared to an internalized standard, and the action patterns that result have their basis in this information. But not only does the nervous system compare its inputs to a standard, it also creates them. We have no *direct* commerce with external reality at all: every input that (causally) results from an external energy source reflects the intrinsic properties of the CNS rather than the intrinsic properties of the external source. This is, of course, the physiological correlate to the distinction of knowledge by description as opposed to acquaintance. As Russell's doctrine so clearly states, all our knowledge of phenomena external to the CNS is of the structural properties of the phenomena: we have no knowledge of the intrinsic or "first-order" properties of external objects at all. Our knowledge by acquaintance, which is knowledge of the intrinsic properties of the nervous system, although causally related to external events, is the result of the nature and functioning of that nervous system rather than the external events themselves. The only modeling of external, nonmental reality that thought can accomplish is of its structural properties.[25]

[25] But is the structure of reality, as it is disclosed by our best contingent (i.e.,

Paradoxically, the upshot of this is that we cannot regard the phenomenal world of immediate experiences as in any sense more 'real' or 'fundamental' than the world of science. This is so because *both* the phenomenal world and the scientific world are *equally* constructions in the mind of man. Indeed, both 'worlds' are the result of the primary activity of the nervous system: classification. But no purely 'phenomenalistic' interpretation of science can suffice: this is because the scientific image constantly contradicts the manifest phenomenal image. Phenomena per se are not subject to completely regular (i.e., invariant) classification — "Our knowledge of the phenomenal world raises problems which can be answered only by altering the picture which our senses give us of that world" (Hayek, 1952, p. 173). The primacy of what Sellars[26] has called the 'scientific image' over the 'manifest image'

scientific) theories, *really* a reflection of the structure of reality, or is it merely the result of our cognitive structure? Considering only the nature and functioning of the nervous system per se, the answer is very likely the latter alternative. For the nervous system is essentially a highly skilled *instrument of classification*: as noted in the discussion of habituation, neural activity must *judge* whether or not other neural activity (which we can only *assume* has its source in 'external' origins) is to count as input. That is, the classes of sensory order determine what counts as stimulation. It follows that if there were a stimulus which was not regular, i.e., not an instance of one of the appropriate kinds, we could not 'know' anything about it. The key concept here is *regularity*: what sensory perception can accept as input can never be unique properties of individual objects, but must always be properties which the objects have in common with other objects. The fundamental activity of the nervous system is classification. Perception is *always* classification, which is to say interpretation: interpretation of something (either 'object' or 'pattern of neural activity') as belonging to one or another class (of 'objects' or 'patterns of neural activity'). Thus, to reinforce again Russell's point, the qualities which we attribute to experienced external objects are not properties of those objects at all, but rather a set of *relations* by which our nervous system effects their classification. This is what knowledge by description means: all we can perceive of external events is their structural relation to each other and to our experience. We truly are theories of our environment: *all* we can know about the world is inherently theoretical (there is no direct awareness or factual bedrock), and all our 'experience' can do is modify our theories!

Return now to regularity. The line of reasoning just outlined indicates that we can 'know' (i.e., classify) only those kinds of events which show a degree of regularity in their occurrence in relation to other events. We could not know (classify) events which occurred in a completely irregular manner. The world *as we know it* remains a construction of the mind of man: "The fact that the world which we know seems wholly an orderly world may thus be merely a result of the method by which we perceive it" (Hayek, 1952, p. 176). This much of the thesis of idealism, it seems to me, is incontestable (if the arguments upon which they are based, which are really contingent theories, are sound). Empiricist theories of the mind are rendered so absurd that only fools or philosophers could have propounded them.

[26] See his essay "Philosophy and the Scientific Image of Man," in *Science, Perception, and Reality*, and *Science and Metaphysics* (London: Routledge & Kegan Paul, 1968). See also B. Aune, *Knowledge, Mind and Nature.*

is that science is better at the task of modeling reality — of approaching more closely the *reproduction* of the objective order of events — than is sensory experience. The phenomenal order, which reflects the (human) organism's initial theorizing or modeling of his environment, is at best a (very) rough approximation of the order of reality. The constructions of scientific thought are, *all* things considered, far superior in their degree of adequacy of representation of reality.

Parenthetically, this casts further light on the relation of consciousness to preconscious sensory perception. In both cases, the function of the nervous system is to model reality. But the model formed by the preconscious senses will often prove to be inadequate in that it will lead to falsified expectations. Thus, although the conscious, self-reflecting mind can know external reality only in terms of the classes that preconscious sensory perception has created, the 'data' of sensory experience form a basis for their own revision (i.e., reclassification). The conscious mind will reclassify the initial sensory experiences in order to model the structure of reality more adequately. "The experience that objects which individually appear as alike to our sense will not always behave in the same manner in relation to other classes of apparently similar objects, and that objects which to our senses appear to be different may in all other respects prove to behave in the same manner, will thus lead to the formation of new classes which will be determined by explicitly (consciously) known relations between their respective elements" (*ibid.*, p. 145). *If* anything can represent reality adequately, it will be the conscious mind, through the construction of scientific theories, rather than the preconscious sensory order with its initial classification of inputs.[27]

So the argument of classical idealism, that the world is a construction in the mind of man, leads from idealism to scientific or representational realism (specifically, to structural realism). If anything is going

[27] That is, ideally completed, Utopian science will be the most adequate representation of reality that human beings can aspire to. Such an ideal scientific account will, of course, have to make explicit the enormous amount of implicit knowledge (which Polanyi and Kuhn aptly term "tacit knowledge") that we all possess as highly evolved theories of our environment. The conscious activities of scientists are based upon almost wholly unconscious processes and principles of determination, and "science" must make explicit those processes and principles before its claim to adequacy can be taken seriously. But this is tangential to the main point in the text above: reflective thought refines and corrects the initial classifications effected by the preconscious sensory order.

to give us an adequate model of reality, it will be the scientific model rather than the manifest, sensory-based model. Since both images are on a par with equally 'infirm' foundations, i.e., since neither is in more direct contact with reality than the other, the choice between them must be made upon their relative efficiency and adequacy of performance upon their common task: modeling reality. The scientific image wins this contest hands down: science can beat the manifest image *at its own game*.

But the skeptic's question may now be repeated: "How do we know that there *is* an objective world, i.e., a reality external to the constructions of our own minds?" The answer depends upon the point of view adopted. Within the justificationist framework, the answer is that we *cannot* know there is an objective world, for we cannot *prove* its existence (or even render it probable). Further, it is not a question for logic alone to arbitrate. Nor can sense experience, the ultimate epistemological authority of empiricism, be other than impotent with regard to this issue.[28] Intellectualism fares no better: there are no a priori arguments that are compelling of a choice for realism or idealism. Solipsism, as Russell aptly remarked, is logically impeccable; it is only *psychologically* intolerable.

The skeptic's question can be answered by reflecting upon how we 'know' anything (other than formal systems) in a nonjustificational metatheoretical framework. For the nonjustificationist, knowledge is a matter of having warranted assertions, i.e., conjectural claims that can be defended by 'good reasons'. And 'good reasons' will consist of other knowledge claims that have been corroborated (as opposed to 'verified' or 'confirmed'). Such knowledge will be rational insofar as it is sub-

[28] Empiricism, pursued conscientiously to its inevitable conclusion, leads to *conceptual a priorism*. Hayek puts the self-stultifying nature of sensationalistic empiricism beautifully: "Precisely because all our knowledge, including the initial order of our different sensory experiences of the world, is due to experience, it must contain elements which cannot be contradicted by experience. It must always refer to classes of elements which are defined by certain relations to other elements, and it is valid only on the assumption that these relations actually exist. Generalization based on experience must refer to classes of objects or events and can have relevance to the world only in so far as these classes are regarded as given irrespective of the statement itself. Sensory experience presupposes, therefore, an order of experienced objects which precedes that experience and which cannot be contradicted by it, though it is itself due to other, earlier experience" (*ibid.*, p. 172).

Note that this reasoning becomes the strongest of possible *contingent* supports for realism rather than idealism: but note also that realism is *not* empiricism!

ject to criticism, where criticism is no longer fused with the attempt to prove or justify. Clearly there can be no justification of such knowledge claims, for there is no ultimate epistemological authority upon which to found them, i.e., ground them in certainty. Within a nonjustificational philosophy the problem of the justification of nondemonstrative inference (as genuine or rational knowledge) does not arise: since *all* knowledge claims are equally inferential rather than certain, equally nondemonstrative, the quest of justification is seen to be misguided. Such a quest can arise only if the ideal of Euclidean systems is taken to constitute genuine knowledge, and all other forms of knowledge (such as those of contingent science) are considered pale approximations to this genuine form. The dilemma the justificationist has created for himself, by defining his concepts in such a manner that genuinely rational contingent knowledge cannot be shown to exist, can be avoided only by stepping outside the justificationist metatheoretical framework.

The existence of an external nonmental world that is causally responsible for our perceptions of it in the mental world of direct acquaintance is a scientific claim, and it must be defended by adducing 'good reasons' for its truth (in exactly the manner that any other contingent proposition(s) must be defended). The good reasons that must be delivered (sooner or later) are an adequate theory of how the mind works which indicates that the mind could not work the way it in fact does unless there was a world external to our senses, a world having the properties that Utopian science attributes to it. To repeat, solipsism is only *psychologically* intolerable. That is, psychology must ultimately deliver a theory of the mind which delimits the a priori constraints upon human mental functioning (the scientifically determined Kantian categories of the understanding, if you will), and an indispensable ingredient of that theory must be a specification of how the mind could not function as it does indeed function without an external world. Nothing less will suffice for determining the warranted assertability, i.e., the *truth,* of this particular empirical claim. But one must never forget that explanations, even if true, are never justifications.

The a priori constraints upon human knowledge. Thus far we have been sketching (synthetic or 'scientific') a priori constraints upon the nature of our knowledge: by considering the nature of our knowledge and our biological nature, we have been adducing facts that a psychological theory of knowledge acquisition must be consonant with. That is, we have described facets of our knowledge and our thought that *any* theory of the acquisition of knowledge must acknowledge. There is one further such constraint that is important enough to warrant special mention.

Since our knowledge of reality is a product of the classificatory ability of the CNS, it is clear that a principle of causality is a presupposition of (the *acquisition of*) human knowledge. Classification must presuppose causal regularity. Further, perception requires for its very definition a concept of causality. Russell was very clear on this: "The conception of 'causal lines' is involved not only in the quasi-permanence of things and persons but also in the definition of 'perception'. When I see a number of stars, each produces its separate effect on my retina, which it can only do by means of a causal line extending over the intermediate space. . . . Generally, what is said to be perceived, in the kind of experience called a 'perception', is the first term in a causal line that ends at a sense organ" (1948, pp. 458–59). Science presupposes the causal theory of perception: human knowledge could not be what it in fact is unless the existence of what Russell aptly called "causal lines" were presupposed. "But for causal interconnectedness, what happens in one place would afford no indication of what has happened in another, and my experiences would tell me nothing of events outside my own biography" (*ibid.*, p. 162).

However, neither experience nor logic can prove the necessity or even the existence of causal lines. Yet without them, human knowledge and science as we know them to be would be impossible to achieve. For the 'scientific' justificationist, this is most perplexing. Russell had to make a *postulate of scientific inference* out of his notion of causal lines. He saw it as the only way to account for our knowledge transcending empirical particulars: such knowledge cannot be wholly based on experience. Speaking of the universal knowledge that science claims to know, Russell the justificationist asks *the* question: "But we most certainly do need *some* universal proposition or propositions, whether the five canons suggested in an earlier chapter or something different. And whatever these principles of inference may be, they certainly cannot be logically deduced from facts of experience. Either, therefore, we know something independently of experience, or science is moonshine" (*ibid.*, p. 505). In this he is caught up in the self-stultifying justificationist quest, and even his "postulates" cannot stave off the verdict that science is moonshine *on justificationist criteria*. But Russell had the right approach to inference and expectation — he considered them as biologically based mechanisms of adaptation *given* the world as it is disclosed by science. "The forming of inferential habits which lead to true expectations is part of the adaptation to the environment upon which biological survival depends" (*ibid.*, p. 507).

What we shall achieve, by diligently studying such inferences, is not knowledge of any postulates of scientific inference that are beyond justification, but rather a theory of the a priori constraints that give human knowledge the character that it does in fact have. What we shall wind up with is a synthetic or scientific theory — a fallible, conjectural one — about the nature, scope, and *limits* of human knowledge. What we need to know about scientific (and other) inference is not how to justify it but rather how it occurs and what constrains it. And it is precisely this sort of knowledge that is, at least in principle, available from the study of the inferring organism — from the psychology of inference and expectation.

The Acquisition of Human Knowledge

To the question of "How do we model reality?" the obvious answer is, "By forming concepts (of its structural properties)." To the question of "How is our knowledge represented?" the obvious answer is, "By our concepts." To the question of "What constrains the patterning of our inferences?" the obvious answer is, "The concepts that constitute our present knowledge." To the question "What is the end product of inference?" the obvious answer is, "Concepts." For these and similar reasons, explicating the nature of concept formation is perhaps the most important task faced by psychology. Although our knowledge of concept formation is quite sketchy and incomplete, we do know a number of things that cast light upon the problems of inference and expectation.

The most important thing we know about concepts is that their acquisition involves the interplay between experienced particulars and abstract rules. The concept literally is the interplay between instances and their rules of determination. All our knowledge consists in the repeated restructuring of classifications effected by the CNS: concepts are these classifications and their patterns of restructuring. The experienced particulars, i.e., the initial sensory classifications effected by the CNS, are alone not our concepts — our experience is not (the same as) our knowledge. *All* knowledge transcends experience: our concepts are a means by which we generate potentially infinite domains of experiences. The near infinite richness of content of the experienced particular is not the starting point from which the mind forms its concepts, but rather the *product* of a fantastic range of restructurings or classifications that the CNS effects. Human knowledge consists of rules of determination by means of which we construct the richness of the particular — all our knowledge is, as Hayek said, a system of rules of determination (or action patterns) which is evoked by the input to the CNS. What we learn in forming a new concept is essentially a new pattern or configuration (of experienced particulars) according to new rules of determination. Thus to *know* is always to be in a position to *infer*: to know is to generate an indefinitely extended domain of potential data points. Our concepts are inevitably inference tickets: to know a particular is always to be prepared to infer to an indefinite number of other particulars. What we know is inextricably bound up with what we expect. This being so, the answer to the question "What guides scientific life?" is obviously "What we (think we) know." But what *do* we know, and *how* do we know it?

Conceptual abstraction. The central problem for the 'psychology of inference' is to characterize *what is learned* and *how our memory utilizes prior knowledge*. If we can get clear on what is learned, or what is to say the same thing, *what our knowledge is*, then we will be in a position to discuss the role of experience in the determination of human knowledge and to see why induction-from-experienced-particulars is an inadequate theory of the nature of knowledge and its acquisition. Furthermore, if we can get clear on *how memory utilizes knowledge* in the formation of new concepts, then the 'guidance of life' problem will automatically be answered.

Most simply stated, the problem is to explicate how organisms classify particular instances as instances of thing-kind categories. For example, how do you know that △ is a triangle? What is involved in learning that particular triangles are all instances of 'triangularity'? The most crucial thing to note is that it is impossible to exhaust the meaning of 'triangularity' with any list, no matter how long, of physical (referential) attributes: there are an infinitude of things that are all equally triangles. No induction-by-enumeration of referents theory can teach an organism the concept. The number of experienced particulars required to learn "what triangles are" (that is, that certain figures instantiate triangularity) would be indefinitely large. Generic concept formation is essentially 'productive' or 'creative' as the linguist uses the term: to know a concept is to be able to apply it appropriately to a totally novel instance, one that need not ever have occurred before in the history of the organism, or indeed the history of the world. The question comes to this: How can an organism recognize all the potential instances, on the basis of *no* prior exposure to them, as instances of *the same* concept? How can the organism construct a theory of what proper subset of its sensory presentations are instances of triangles, *and* then extrapolate from that given corpus of presentations the defining characteristics of 'triangularity' which would enable it to recognize *any* instance as a triangle?

One thing that is very clear is that an organism which has learned the concept of 'triangularity' has learned the *rules of determination* that enable the construction of triangles. Equally clear is the fact that those rules of determination must range over abstract entities (i.e., structures not found in the experienced surface structures of particular triangles). The only way known for organisms to "make infinite use of finite means" as they do in generic concept formation is to employ "grammars" of perception (or cognition, or behavior) that make use of abstract entities in their rules. And only grammars that allow indefinite recursion and which employ nonterminal symbols (for abstract entities) in their derivations of surface structures from underlying or 'deep' structures can do this. What these grammars show is how an abstract, underlying deep structural "meaning" (intension) can be mapped into indefinitely many distinct surface structure representations. A generic concept is the same as an underlying deep structure that can be characterized *only by the rules of determination which are its grammar*, rather than by listing the attributes or "experienced features" that are present in its potentially infinite number of surface structure instantiations. Indeed, your knowledge of the concept of triangularity is independent of any surface structure experiential representation: the concept is defined by purely "verbal" statements that characterize its rules of determination.

Experience per se, as in 'seeing' a triangle, is not the meaning of the concept.[29]

What is learned in concept formation, our knowledge of concepts, is the rules of determination that constitute the invariant relations in a *group*. Concepts are not copies or representations of particulars, they are the rules by which we construct particulars as instances of thing-kind classes. Our perceptual knowledge of reality seems to be based upon the possession of "rules of seeing" that determine the invariants in our environment and the group of transformations applicable to them. But no sensory experience alone is identical with *any* such concept: what the concept *is* consists of *both* the underlying rules of determination and its surface structure representation. That is, perceptual concepts are both deep *and* surface structures, and not either alone.

But there are other instances of concepts that are found in our knowledge which are not 'perceptual' the way 'triangularity' is and which are still contingent rather than a priori. How do we know, for example, that "all men are mortal" or that, as Newton proclaimed, "$F = ma$"? Predicates such as 'mortal', concepts such as force (or freedom), are experiential in the sense that our experiences are relevant to their determination, but they are not perceptual like 'triangular'. How do we recognize men, e.g., John Smith, as mortal? If recognition is restricted to sensory perception, to an iconic picturing notion, then it is obvious that we do *not* recognize men as mortal: mortality is not the sort of

[29] This was clear to philosophers such as Ernst Cassirer as far back as 1910: "The content of the concept cannot be dissolved into the elements of its extension, because the two do not lie on the same plane but belong in principle to different dimensions. The meaning of the *law* that connects the individual members is not to be exhausted by the enumeration of any number of instances of the law; for such enumeration lacks the generating *principle* that enables us to connect the individual members into a functional whole. If I know the relation according to which *a b c* . . . are ordered, I can deduce them by reflection and isolate them as objects of thought; it is impossible, on the other hand, to discover the special character of the connecting relation from the mere juxtaposition of *a*, *b*, *c* in presentation. . . . The unity of the conceptual content can thus be 'abstracted' out of the particular elements of its extension only in the sense that it is in connection with them that we become conscious of the specific rule, according to which they are related; but not in the sense that we construct this rule *out* of them through either bare summation or neglect of parts." E. Cassirer, *Substance and Function and Einstein's Theory of Relativity* (New York: Dover Publications, reprint, 1953), pp. 26, 17.

Cassirer's insight was lost to psychology during the fifty-odd year blight of behaviorism. Fortunately cognitive psychology is beginning to recover from behavioristic learning theory and its induction-by-enumeration model of knowledge acquisition; see U. Neisser, *Cognitive Psychology* (New York: Appleton-Century-Crofts, 1967).

thing that can be observed or photographed. Yet we do come to know that men are mortal, and that if John Smith is a man, he is mortal also. And the way in which we do this is not very different from the way in which we recognize a particular triangle as instantiating triangularity. In both cases a particular of a specific sort is classified as instantiating a generic relation that is not itself *any* experienced particular. Mortality as a concept is intimately linked, in our conceptual space, with the concept of man, and what we know in "John Smith will die" *is* that linkage. Our knowledge of such nonperceptual concepts is represented in some nonperceptual manner, specifically in whatever manner underlies our use of language. Whether this manner of representation is construed to be syntactic or semantic (there is considerable debate among linguists and psycholinguists here), it is clear that it is *abstract*, for no experiences that we can undergo are equivalent to it. We possess a conceptual schema that allows us to recognize (or perceive) instances of perceptual or linguistic formulations as relevant to that schema, but those latter formulations are *not* the schema itself. What we learn in learning such abstract concepts is to follow underlying rules of determination that can generate an indefinite number of surface structure manifestations.

Very similar considerations apply to the learning of consequential knowledge of nature, such as Newton's law $F = ma$. What we learn in doing science is how the basic law sketch $F = ma$ may be transformed and restructured in disparate instances of F's equaling ma's. We come to see (in the conceptual sense) that disparate surface structure manifestations share a common, underlying structural regularity that we may represent by the equation $F = ma$. The essence of concept formation is the abstraction of such rules of determination that generate surface structure instantiations. What the mind does is *construct* its concepts upon the basis of incomplete and often quite impoverished sensory experience. The psychology of concept formation requires a *constructive* theory of the mind rather than either an inductive or a deductive one.

Memory. In a sense, the heart of a constructive approach to concept formation and knowledge acquisition is a generative model of memory. Everyone knows that what constrains our inferences (or constructions) is, in some loose sense, our 'background knowledge', but virtually no philosophers have concerned themselves with characterizing the nature of human background knowledge. Thus a few words about the manner in which memory operates to provide that 'background knowledge' are appropriate. Once it is seen that memory does not "store" *particulars*, and that indeed the mind does not really "store" information *at all* (at least in anything like the customary sense), the appeal of an induction-from-particulars model of scientific concept formation vanishes.

The problem of prose recall is characteristic of the operation of memory in *meaningful* contexts (such as scientific concept formation, etc.). People just don't bother to memorize by rote in such situations: presented with an utterance pertaining to the "eminent arrival of the magistrate," they will remember that the gist of what was said was "Here come de judge." The latter utterance does contain the 'gist' of the former — it is indeed an adequate paraphrase — and yet there need not be a single 'surface structure' word shared in common by them. What the mind does with meaningful information is to represent it in terms of a semantically coherent schema, from which it can generate appropriate surface structure representations whenever the need arises. We do not store the given words of a sentence or even necessarily the paraphrase of a single sentence. Rather, what the head does with information is to *use* it, to construct an appropriate schema that can be used as the basis of future expectation. It is becoming quite clear from psychological research that sentences (or whatever is taken to be the appropriate unit in meaningful communication) do not 'carry' meanings in their surface structure manifestations — rather, they are 'triggers' to meanings that are already present in the (recipient) head. The structural representations that constitute human "understanding" are vastly richer than mere "words."[30]

[30] Cognitive psychology is beginning to acknowledge that comprehension involves more than the registration of words in sequence. For example, when presented with a series of "related" sentences, one usually "remembers" the 'gist' of the series as one syntactically complex but semantically integrated unit. Indeed confidence ratings indicate that we will be *more* confident that what was actually presented was the integrated unit than the actual simple sentences that were experienced. The semantic information is fitted into a plausible context by the hearer, and once the information is assimilated to such a schema the actual input is disregarded (see J. D. Bransford and J. J. Franks, "The Abstraction of Linguistic Ideas," *Cognitive Psychology*, 2 (1971), 331–50). If we can construct a meaningful context, virtually anything can be remembered, or judged comprehensible, etc. For example, paragraph passages that would be judged incomprehensible and all but impossible to recall by themselves can be recalled very well and are quite comprehensible when heard after exposure to a drawing which provides a context that is appropriate. J. D. Bransford and M. K. Johnson, Contextual Prerequisites for Understanding: Some Investigations of Comprehension and Recall," *Journal of Verbal Learning and Verbal Behavior*, 11 (1972),

To say that the head uses information to abstract "consequences" that are relevant to its expectations is not to say that the head stores information. The scientist's internal model of reality — his theory — is an abstract schema that is constantly being updated as the mind structures and restructures the information that sensory classification makes available. In modeling reality the mind functions as a generative concept forming device — it abstracts consequences from the input made available to it by relating particular surface structures to the sets of abstract rules of determination that constitute its "knowledge" or "memory." What human conceptual knowledge is, as Hayek said, is a set of patterns of action subsisting within the functioning CNS.

The upshot of this is clear: we should cease talking about science "learning by induction" and try to understand how the mind constructs its knowledge from its (tremendously impoverished) input. Experience per se, the perceived particulars determined by our initial sensory classifications, is not equivalent to our knowledge. *What we learn isn't facts* — instead we learn concepts which enable us to generate facts. Our concepts function as licenses to infer particulars, and those concepts are not just other particulars. What we remember is a conceptual schema, and experiences are only relevant to conceptualization insofar as they can be fitted into those conceptual schemata.

The role of experienced particulars: Learning from exemplars. The primary knowledge that we gain from the practice of science consists in the rules of determination that are the skeletal deep structure representation, the schema, of our concepts. Many concepts, such as energy, force, reinforcement, memory, evolution, etc., are purely conceptual rather than perceptual in the sense that, although sensory experience pertains to our usage of them, no sensory experience may be identified as a surface structure manifestation of the concept. What then is the role of experienced particulars in such abstract concept formation? The answer rein-

717–26. What we do with information is to use it, and that means to assimilate it to the knowledge structures we already possess and to the expectations we have for the future. If we can construct an appropriate context, we can understand or remember virtually anything; if we have no context, even information that is normally intelligible by itself is not understood or remembered. J. D. Bransford and N. S. McCarrell, "A Sketch of a Cognitive Approach to Comprehension: Some Thoughts about Understanding What It Means to Comprehend," in W. B. Weimer and D. S. Palermo, eds., *Cognition and the Symbolic Processes* (Washington, D.C.: Erlbaum Associates, 1974), pp. 189–229; J. J. Franks, "Toward Understanding Understanding," in *ibid.*, pp. 231–61.

forces the centrality of perception, as a *higher* mental process, in theory construction and knowledge acquisition.

What we learn from particular exemplars, or from exemplary puzzle solutions, has been well characterized by Kuhn: we *learn to see* disparate surface structure manifestations as similar in crucial respects to others. We learn to see particulars as instantiating the rules of determination that constitute the deep structural schema of a concept. By working with disparate surface structure configurations to the extent that we can see *their resemblance* to others, we tacitly become aware of the deep structural communality that underlies the particulars that we are familiar with and that can generate an indefinite range of further particulars. Scientific theories, as structural representations of reality, are ways of seeing that deep structure. Experienced particulars stand to that structure as instances stand to abstract rules of determination. What we learn in "doing" science are ways of perceiving. What we learn to see in "facts" are similarity relationships: we learn to see instances as instances of thing-kinds, and we learn how thing-kinds are structurally related. It is this perception of similarity relationships in the structural properties of reality that led Hanson to call theories *patterns*: they are patterns of constructions, according to which we come to conceive of reality. The interesting problems of inference and expectation that psychology must in the future address concern understanding the factors, such as our implicit background knowledge, that *constrain* the patterns of inference that we entertain.

To date, all that the historian of science or psychologist of knowledge acquisition can do is single out instances of "learning to see" similarity relationships in disparate structures. Thomas Kuhn has recently done this,[31] and it is not surprising that he has now held historical research in abeyance in order to study concept formation. And it is also not surprising that the psychology of perception is central to Kuhn's work: what the student of science learns in doing the exemplary problems of his discipline is a group-licensed way of seeing the interrelationships between various surface structures. Science is a matter of coming to recognize particular surface structure manifestations as being related to underlying deep structural rules of determination. And the purely structural knowledge of the underlying rules of determination constitutes the

[31] See "Second Thoughts on Paradigms," in F. Suppe, ed., *The Structure of Scientific Theories* (Urbana: University of Illinois Press, 1974), pp. 459–82.

knowledge by description which is our only knowledge of the nonmental realm.

Let us end this section with a slight digression. We are now in a position to understand a "fact of scientific life" that is all but inconceivable to the justificationist: that no scientist was ever persuaded to or dissuaded from a position by the "facts." It does not take much to see that Fries was correct, that science can neither prove nor disprove — contingent theories are as metaphysical in this regard as the wooliest of religious dogmas. But this has shocked the justificationist, because on his account *all that science can learn from experience is that a theoretical proposition is true or false*. Thus on justificationist criteria, science does not profit from, or make use of, experience.[32] This is but one more reason why science is irrational according to justificationist criteria.

But we can admit that science does not "induce" propositions from factual "experiences" without claiming that experience plays *no* role in scientific practice. Rather, we can relocate the role of experience in our learning from exemplars. What experienced particulars provide for the scientist is practice in perceiving — in perceiving instances as instances of thing-kind classifications specified by abstract rules of determination. In that sense experience remains the basis for our conceptual formulations, but the farther science progresses toward its quest for an adequate structural representation of reality, the sharper is the break between experience or knowledge by acquaintance and conceptual representation.[33] It is, as Hayek noted, precisely because all our knowledge is rooted

[32] Feyerabend loves to flaunt this in the face of the beleaguered justificationist: see his appendix "Science without Experience" in "Against Method," in M. Radner and S. Winokur, eds., *Minnesota Studies in the Philosophy of Science*, vol. 4 (Minneapolis: University of Minnesota Press, 1970), pp. 17–130.

[33] Cassirer put this beautifully when he wrote: "The goal of theoretical physics is and remains the universal laws of process. The particular cases, in so far as they are taken into account, serve only as paradigms, in which these laws are represented and illustrated. The further this scientific problem is followed, the sharper the separation becomes between the sytsem of our *concepts* and the system of the *real*. For all 'reality' is offered to us in individual shape and form, and thus in a vast manifold of particular features, while all *conception*, according to its function, turns aside from this concrete totality of particular features. . . . The direction of thought upon the 'concept,' and its direction upon the real, mutually exclude each other. For to the extent that the concept progressively fulfills its task, the field of perceptible facts recedes. . . . The final goal of the material sciences and of all other natural sciences is to remove empirical intuition from the content of their concepts. Science does not *bridge* the gap between "thoughts" and "facts," but it is science, which first creates this gap and constantly increases it" (1953, pp. 220, 221).

As Körner has indicated, the disconnection of scientific concepts from perceptual

in experience that our principles of conceptual determination must transcend experience. This being so, what we can *learn* from experience cannot be what the justificationist desired; we are instead forced to regard experience as an incomplete surface structure that is compatible with, but not sufficient to lead unambiguously to, its underlying purely conceptual rules of determination. No scientist ever learned "the truth" from "facts," any more than he accepted a theory on the evidence of "facts": yet this does not mean that experience plays no role in scientific concept formation.

But the point of this section remains: the role of experienced particulars in scientific concept formation is that of surface structure manifestations of abstract, underlying rules of determination. As such, *what* we learn from "factual experiences" is a way of seeing: a way of seeing certain experienced particulars as similar in some respect to others.[34] *How* we learn this from "facts" it remains for the psychology of concept formation to tell us.

Another Look at the Guidance of Inquiry

In unprofessional moments at least, virtually no one believes that scientists 'induce' theoretical formulations from observed instances, and although a few still believe in the rationality of concept formation according to the H-D methodology, no one has proposed an account of how one gets the 'theory' from which the 'deductions' occur *in the first*

experiences is total and complete. We never see mere particulars, only particulars as instances of thing-kinds. "To describe a group of phenomena, then, means not merely to record receptively the sensuous impressions received from it, but it means to transform them intellectually" (*ibid.*, p. 264).

[34] Once again it must be emphasized that science cannot deal with truly unique events: our classificatory schemata always assimilate particulars to abstract principles of determination, and the knowledge that results is purely structural. According to Cassirer, "No content of experience can ever appear as something absolutely strange; for even in making it a content of our thought, in setting it in spatial and temporal relations with other contents, we have thereby impressed it with the seal of our universal concepts of connection, in particular those of mathematical relations. The material of perception is not merely subsequently moulded into some conceptual form; but the thought of this form constitutes the necessary presupposition of being able to predicate any character of the matter itself, indeed, of being able to assert any concrete determination and predicates of it. Now it can no longer seem strange, that scientific physics, also, the further it seeks to penetrate into the "being" of its object only strikes new strata of numbers, as it were. It discovers no absolute metaphysical qualities; but it seeks to express the properties of the body or of the process it is investigating by taking up into its determination new 'parameters' " (*ibid.*, p. 150).

place. It is all very well to say that science is a matter of conjectures and refutations, as *both* Popper and his sophisticated justificationist antagonists agree upon, but that slogan is impotent without a psychology of theory formation.[35] 'Conjectures and refutations' is a fine methodology for a completed (i.e., a *dead*) science or for a mythical realm such as Plato's Third World, where discovery is *by definition* not a psychological issue. But since mortal human beings live (and practice science) in the real world rather than in the 'Third World', the recent attempts to preserve a purely *philosophically* rational (and logical) reconstruction of scientific growth in a logician's nirvana can be ignored. Logic is not the guide to scientific life, and there is no use killing science (and then mounting it in the Third World history museum) in order to make its growth seem purely logical.[36]

[35] For example, Salmon, *The Foundations of Scientific Inference*, attempting to revive a Reichenbachian pragmatic justification of induction and couple it with Bayes's theorem, argues for "plausibility considerations" as determinants of the prior probability of scientific hypotheses. His model of acceptance is basically induction *by elimination*, which he admits is impotent in the face of an unlimited supply of potential hypotheses: unless there is a principled way to eliminate a priori all but a few of them, elimination as a practical approach cannot get off the ground. "There are, as I have emphasized repeatedly, infinitely many possible hypotheses to handle any finite body of data, but it does not follow that there is any superabundance of *plausible* ones" (Salmon, p. 129). His next move is obvious: it is claimed that "plausibility arguments" suffice to weed out the indefinitely large class of potential hypotheses. "If we put plausibility arguments — perhaps I should say 'implausibility arguments' — to the purely negative task of disqualifying hypotheses with negligible prior probabilities, falsification or elimination becomes a practical approach. This is, it seems to me, the valid core of the time-honored method of induction by elimination" (*ibid*., p. 129).

But Salmon has no theory of 'plausibility' *at all*. To be sure, he lists, in sketchy fashion, three classes of characteristics that "may be used as a basis for plausibility judgments" (*ibid*., p. 125), but all such characteristics are impotent to determine plausibility in a given case except in an ad hoc manner. The prior probability of an hypothesis, regardless of whether it is equated with plausibility or not, is only determinable a posteriori. That is, the *true* prior probability of a scientific hypothesis is exactly the same as its posterior probability: as Fries indicated $P(h,e) = 0$ *always*. And precisely the same argument holds for prior and posterior plausibility considerations: all scientific hypotheses are as *im*plausible as they are *un*provable. If Salmon had any idea whatsoever how scientists determine the plausibility of hypotheses, he would ipso facto have a theory of scientific concept formation, and he would then have a genuine 'guide to life'. As it stands, he has no theory of guidance at all, only a relabeling of the problem as one of 'plausibility considerations'.

[36] This recent attempt to 'kill the patient' in order to preserve the appearance of the corpse, i.e., to render the growth of scientific concepts rational *from the logician's standpoint of rationality*, to implement a true *logic* of discovery, is due to Popper, *Objective Knowledge* (London: Oxford University Press, 1972); Lakatos, "Falsification and the Methodology of Scientific Research Programmes"; and A.

A number of philosophers and scientists have commented upon the futility of the traditional approaches to the 'guidance of life' problem. One persistent theme that emerges in their presentations is the centrality of what Polanyi has called *tacit knowledge*.[37] Polanyi's point is that we know far more than we can ever tell: our knowledge of reality is tacit in the sense that we could never 'formalize' or 'objectivize' it in the form of a scientific theory. "Tacit knowing is the way in which we are aware of neural processes in terms of perceived objects" (1966, p. x). His example of our ability to recognize physiognomic features is sufficient to make this point: we can *know as faces* (i.e., recognize) an indefinitely extended number of particular configurations, despite the fact that we have never seen them before, and no one can say how he can do this (indeed, there was no psychological theory that could address the issue until very recently). Our knowledge is in this sense vastly richer than our theories of our knowledge: both *what* and *how* we know are beyond the pale of traditional epistemologies. As Polanyi points out, our knowl-

Musgrave, "Impersonal Knowledge," Ph.D. thesis, University of London, 1969. The 'critical fallibilists', having spent three decades vanquishing the disillusioned justificationists by showing that the latter's theory of assessment (inductive logic) could not possibly explain science, are now facing the collapse of their own metatheory of scientific rationality. That is, the Popperians, having shown that 'instant' rational assessment was a chimerical quest, that the rationality of science does not lie in 'inductive logic' or 'confirmation theory', proposed instead that the rationality of science was to be found in its (logical) growth. But now it is obvious that deductive logic does not guide knowledge acquisition either — the guidance of life, and hence of scientific concept formation, is psychological rather than logical. The psychology of inference will (ultimately) explain scientific concept formation, but it will not reconstruct concept formation as a logic of discovery. There remain two alternatives for the concerned philosopher who wishes to understand science. One can, like Paul Feyerabend (see "Against Method" and "Consolations for the Specialist," in Lakatos and Musgrave, eds., *Criticism and the Growth of Knowledge*, pp. 197–230), maintain that science is irrational (since the critical fallibilist metatheory has collapsed) and let it go at that. Feyerabend is now 'gleefully' destroying the last vestiges of the Popperian rationality of science (with great success), but he appears to have no 'positive' program with which to replace it (other than an irrational appeal to 'love thy neighbor'). On the other hand, one can admit the failure of Popperian *and* justificationist rationality, and search for a psychology of inference and a sociology of knowledge transmission. This is the way of Thomas Kuhn in *The Structure of Scientific Revolutions*, and most especially in "Second Thoughts on Paradigms" and, from a slightly different perspective, Jean Piaget as a representative genetic epistemologist (see e.g., B. Kaplan "Genetic Psychology, Genetic Epistemology, and Theory of Knowledge," in T. Mischel, ed., *Cognitive Development and Epistemology* (New York: Academic Press, 1971, pp. 61–81). Needless to say, this essay is a beginning exercise in this latter approach, as is my *Psychology and the Conceptual Foundations of Science*.

[37] See especially his books *Personal Knowledge* (New York: Harper & Row, 1958) and *The Tacit Dimension* (New York: Doubleday, 1966).

edge of scientific problems as *problems* is an instance of this tacit knowledge — neither inductive nor deductive methodologies of science can tell us how to perceive a scientific problem or puzzle as *such*. But having raised this fundamental problem (and implied its converse: how do we recognize a solution as *such*) Polanyi lets us down, claiming that his *labeling* of the problem as an instance of 'tacit knowing' is its solution.[38] The question of guidance in scientific life comes down to this: can we provide a psychological account of concept formation that can explain the all-pervasive phenomenon of 'tacit knowledge' in science and its practice? Specifically, can we explain how problems are seen *as such*?

I think that the constructive theory of the higher mental processes sketched above, with its generative approach to concept formation, can be elaborated to the point where it at least begins to solve the problem of tacit knowledge. If our knowledge is a system of highly abstract, generative rules of action (in Hayek's sense), then the rules of determination (of concepts) which constitute our understanding are always 'tacit.' Our expectations of the future are determined by our constructive memory: hence our theory of memory is ipso facto our theory of 'expectation' which in turn is ipso facto what guides our concept formation, which in turn is what guides scientific life. Just as our (physiological) memory creates the richness of our sensory experience, so our constructive conceptual memory creates the richness of our inferential patterns. And the knowledge which is instantiated by these acts of construction is almost totally tacit in the sense that our conscious awareness is virtually never able to articulate the rules of determination which we are following in inferring our concepts. (But this tacit knowledge need not remain so forever: there is no reason why a completed psychology could not render explicit the rules of action according to which our knowledge is acquired and structured. Insofar as Polanyi claims that tacit knowledge is *inevitably* tacit, he is arguing from ignorance, and his position is to that extent obscurantist.)

The guidance of life is the psychology of 'seeing.' In normal science

[38] For this claim see Polanyi (1966) where he says: "We have here reached our main conclusions. Tacit knowing is shown to account (1) for a valid knowledge of a problem, (2) for the scientist's capacity to pursue it, guided by his sense of approaching its solution, and (3) for a valid anticipation of the yet indeterminate implications of the discovery arrived at in the end" (p. 24). But a description of a problem, even if correct, is not its explanation. What is required is a theory of tacit knowing and its interrelation to concept formation.

practice, when an enormous amount of background knowledge is stored away, and the practitioners of a science are free to explore nature to great depths, learning from exemplars (in Kuhn's sense) seems to constitute the group-licensed 'way of seeing.' By working with particulars, the practitioner gradually comes to see the abstract rule of determination *that makes the particular the kind of particular that it is.* The scientist's consequential knowledge of nature is tacitly generated by his exploration of exemplary particulars and the rules that determine their particularity. The growth of knowledge during periods of normal science consists in the search for new applications of known rules of determination and for new rules that bear a determinate (transformational) relation to the old ones. In forming new concepts of the structure of reality we take for granted a given conceptual-classificatory point of view and generate novel instances (applications) within that point of view. We 'see' the world as consisting of a specifiable group and the determinate transformations that the group may undergo: our 'inferences' consist in applying those transformations that we think adequately characterize the structure of reality, and what we 'test' is the match of expectation (generated by our model) and reality. As Kuhn emphasizes, we do not aim to uncover either fundamental novelties of fact or anomalies during such exploration.

But no set of concepts is adequate to the task of modeling reality exactly: we *do* turn up anomalies and novelties. Revolutionary reconceptualization can then provide a new way of seeing (i.e., a new group and set of transformations) that renders anomalies and novelties as expectancies, and the whole game of modeling reality may begin afresh. But during periods of revolutionary science, concept formation is *entirely* tacit and consists of finding new ways of seeing the entire domain in question. We *restructure* our representation of reality. During such reconceptualizations the research community involved very literally comes to see the domain from a different perspective, much as one can come to see the alternative perspective of an ambiguous figure that had not appeared to be ambiguous until the 'switch' occurred. Our perception of reality is inherently ambiguous in this same manner: since all our knowledge of reality is a function of the innate and learned modes of classification of our nervous systems, since we have no direct and unambiguous contact with external reality, our perception (and conceptualization) can shift whenever it is determined by a new rule of determination. All we can 'perceive' of external reality are disparate surface structure particulars that bear some (unknown) relationship to the actual deep structural nature of reality. That external reality determines (in the sense of causally influences) our conceptualization of it, but it cannot do so in an unambiguous manner: that is, we are free to conceptualize it according to one or another of the sets of rules of determination that our nervous systems are capable of constructing. This is not to say that reality per se is inherently ambiguous; rather, it is our perceptual and conceptual knowledge that is ambiguous in the sense that reality does not uniquely and unambiguously specify our classification of that reality. But the point remains: since reality does not unambiguously specify our conceptualization of it, the possibility of a scientific revolution that institutes a new 'group-licensed way of seeing' always faces *any* domain — no matter how thorough and certain our 'knowledge' of it may appear. And the only way we will be able to understand what is occurring in such reconceptualizations is to understand how we perceive and conceive, and how future knowledge acquisition is guided by present concept formation. As it is in normal science practice, the psychology of inference and expectation is indispensable to the understanding of revolutionary science.

There is a moral for the methodology of science in the psychological nature of guidance. The justificationist conception of 'inductive logic' as the guide to *rational* or 'good' scientific practice must be abandoned because it fuses guidance to assessment (by equating both to 'confirmation theory'). Confirmation theory is impotent with respect to scientific practice: even if a proposition could be proven true it would say nothing about where to look for the *next* 'truth'. But the Popperian methodology, which pictures science as conjectures and refutations *guided* by deductive logic (in the H-D explanatory framework), although a significant improvement upon the justificationist approach, likewise fails completely when it addresses the problems of guidance. Regardless of how much we may be able to learn from our mistakes (in the Popperian tradition), one thing we cannot learn is where to look next. And the pious claims of Lakatos (1970) that his methodology of scientific research programs reconstructed the rational growth of science (in his "improvement" of Popperian methodology) overlook the fact that Lakatos himself could not tell how a 'research program' would originate or what direction its course will take. All these 'rational reconstructions' of scientific growth are after the fact analyses that are totally incapable of providing even the weakest of methodological directives for actual research practice.[39] All the problems of guidance wind up in the 'wastebasket' of *external* history of science.

The practice of science is more a matter of matching to standards than of conjectures and refutations. What scientists *do* in gaining knowledge of reality is to search for patterns of regularity that will lead to rules of determination that may be imposed upon their initial sensory classifications. Our attempts to model reality by thought are attempts to determine invariances of structural properties that obtain between our theories and our sensory experiences. The brain creates a standard of expectation against which it attempts to match its sensory classifications. The process of matching to standards *is* the process of our knowledge acquisition. All our sophisticated hypothetico-inferential techniques are merely variations upon the fundamental theme of matching inferences to expectations. He who understands the psychology of inference and expectation understands the genesis of human knowledge.

Inference and the rationality of science. But what of the *rationality* of scientific practice? When *is* 'inference' rational? For the justificationist, if *scientific* inference cannot be *proven* to be a genuine source of proven or probable knowledge, then science-as-a-means-of-inference is no more rational than voodoo-as-a-means-of-inference (see Salmon, 1967, p. 55). According to justificationist intellectual honesty, one might then as well "do" voodoo as "do" science, since the one would be as *irrational* as the other. The tacit assumption is that inference cannot be rational unless it leads to knowledge (as the justificationist defines the term): since voodoo does not lead to (justificationist) knowledge, it is deemed irrational. Science, as our "good" means of inference, must be rescued from Hume's skepticism or it will be no better than voodoo and witchcraft.

[39] It is instructive to trace the evolution of practical methodological directives. The justificationist slogan was "aim at indubitable truths disclosed by indubitable method." The neojustificationist said, "Aim at probable truths . . ." with nothing else changed. Popper, after arguing that justificationist 'method' was chimerical, said, "Aim at bold, highly falsifiable theories and test them severely." Lakatos, attempting to bring Popperian methodology in line with actual scientific practice, exhorted us to "treat budding research programs leniently, and remember that criticism cannot kill" (he added 'quickly' after kill, but it is obvious that criticism *can never* kill a research program). Regardless of the hortatory appeal of these phrases, it is manifestly obvious that they provide no guidance to research at all: they are all directives for assessment, not acquisition. Beside them even Kuhn's modest proposal of "learn from exemplary puzzle solutions" is a revelation — for it is at least directed at the *acquisition* of knowledge, which, after all, *is* the concern of 'guidance'.

But if one abandons the justificationist framework the situation is quite different. All inference is equally 'inferential', equally conjectural and uncertain, and therefore there is nothing in the nature of the inference per se that separates science from voodoo. What renders science a rational source of knowledge (as opposed to voodoo) is not the use of a special type or pattern of inference: for both share inference in common. It is not that scientific inference is logical and justifiable, while voodoo inference is illogical and unjustifiable: both are unjustifiable and psychological rather than logical. If science is to be deemed rational and voodoo irrational, it cannot be on the grounds that their inferential procedures differ: the rationality of science must lie elsewhere.

The problems of demarcation (of science from its rivals) are as manifold as the problems of inference. Its alleged possession of a theory of instant rational assessment (the computational formula $P(h,e) = c$ of inductive logic is an 'instant' assessment procedure) is what the justificationist thought demarcated science from its rivals. But there is no instant assessment in science, as Popper and Kuhn have unceasingly pointed out. Hence science is irrational according to justificationist criteria of rationality. While denying any theory of instant assessment, Popper followed the justificationists in locating the rationality of science in its assessment (i.e., acceptance) of statements as warrantedly assertible. For Popper rationality lies in criticism of already inferred statements — and the 'critical approach' *un*fuses criticism from the attempt to justify (see Bartley, 1962). Within this nonjustificationist framework, inference is rational to the extent that it produces knowledge — to the extent that its end products are warrantedly assertible statements. Inference is thus unfused from rationality: rationality lies in the critical approach to the problems of assessment rather than in the 'logic' of induction. Voodoo *can* be as rational as science as far as its inferences are concerned: if it is to be deemed irrational, it must be on other grounds (for Popper, on the grounds that it is not critical of the warrant of its assertions).

The problem, of course, is that it is not at all obvious that Popperian criteria of demarcation (including Lakatos's (1970) progressive and degenerating problem shifts and research programs) are sufficient to capture the rationality of science. Kuhn (in the Postscript to *The Structure of Scientific Revolutions* and in "Reflections on My Critics," in Lakatos and Musgrave, eds., *Criticism and the Growth of Knowledge*) has argued convincingly (to my mind) that Popperian demarcation criteria are insufficient and that indeed there may be no satisfactory demarcation of science from its rivals *on methodological grounds*. But regardless of whether or not we can succeed in capturing the rationality of science, the thing to note is that inference must be unfused from rationality in order to see the true problems of either. Regardless of the adequacy of demarcation attempts, there is nothing in inference per se that renders science rational and its rivals irrational. One could think that science was rational because it possessed *the* true inference procedure only if the justificationist metatheory were presupposed. The nature of inference is one thing, the nature of rationality is another.[40]

Nevertheless, explaining how scientific knowledge grows (i.e., how scientific concepts are formed) ought to obviate most of the traditional problems of acceptance and its 'rationality'. Inference is rational when it produces knowledge. (Thus the Popperian phobia of commitment is not warranted — being committed to a way of 'seeing', as in embracing a normal science research tradition, is not inherently irrational. So long as commitment produces knowledge claims that *are* knowledge claims, it is perfectly rational; it is only when commitment ceases to produce knowledge that it is irrational.) The problem of rationality must be relocated into explaining the nature of factual truth and the nature of knowledge claims: when is an inferential claim warrantedly assertible, and what does warranted assertibility depend upon? What is the nature (and conditions) of factual truth? So far as rationality is related to inference, it has to do with the warrantability of inferences, i.e., with the *products* of inference, rather than with the processes per se. Discussing these issues is beyond our present scope.

The moral for the understanding of scientific *practice* that results from the study of knowledge acquisition comes down to this: no matter what philosophical reconstructions of the problems of assessment may

[40] This being so, defenses of the scientific 'game' *cannot connect methodology*, even synthetically through a contingent theory, *to epistemology*. This is because all sentient higher organisms, including all men (scientists and nonscientists alike), share the same inferential means of knowledge acquisition. Thus it is futile to connect verisimilitude to *science* (as did Lakatos in "Popper on Demarcation and Induction," in P. A. Schilpp, ed., *The Philosophy of Sir Karl Popper*, pp. 241–73) in a quasi-inductive justification of science over its alternatives. It is futile because all thought models reality, not just "scientific" thought (and because the empirical content of a theory cannot be equated with Popperian truth content). Such quasi-inductive "justifications" justify common sense, voodoo, etc., just as much as science, because they all share the same inferential processes. If science qua game is a better game to play than voodoo, it cannot be because of its method of inference.

yield, the rationality of scientific concept formulation is ultimately psychological (and to a certain extent sociological). The growth of scientific knowledge is rational in the sense that an ideally complete psychology of inference and expectation could explain the principles according to which human concept formation occurs. No purely philosophical methodology, which fails to base its account upon the psychological principles of knowledge acquisition, can be other than a reconstruction of the warrantability of knowledge claims *once they are put forward*. The philosopher's task is to worry about the assessment of theoretical formulations once they are at hand — he must surrender the acquisition of knowledge to the psychological sciences. This means, simply put, that there is no unitary problem of growth in science: there are at least two problems (assessment and acquisition), probably more. Purely philosophical approaches to a 'unitary' theory of growth, such as those of Popper and Lakatos, are bound to fail. And lest the philosopher feel complacent with retaining the problem of assessment in science, he need only be reminded that all extant methodologies are glaringly inadequate in one or another respect, and if he wants to preserve assessment as a philosophical problem he had better hurry, for the psychologist can say a few things about that problem, too. The philosopher had better come up with something more adequate than more volumes like this one on 'confirmation theory' or the scientists that he is supposed to be "educating" will cease listening entirely.

PETER CAWS

Mach's Principle and the Laws of Logic

In this paper I wish to raise a philosophical question about logic, namely, the question whether its laws can consistently be thought of as analogous to those of the empirical sciences, i.e., as subject in some sense or other to test and confirmation, or whether, as is more often maintained, they must be thought of as analytic and a priori if not as coventional. In order to float the question, some general idea of what kind of activity logic is must be presupposed. The *problem* of logic I take to be as follows: Given the truth (or probability) of sentences {*P*}, what can we say (with what degree of confidence, etc.) about the truth (or probability) of sentences {*Q*}? The *method* of logic I take to consist in performing operations on the sentences {*P*} or on supplementary sentences introduced for the purpose and in performing further operations on the sentences so generated, and so on until the sentences {*Q*} or some obviously related sentences are generated. According to the rules employed in these operations we may then say that the sentences {*Q*} are true or have a certain degree of probability in relation to the sentences {*P*}.

We thus arrive at a degree of confidence in the sentences {*Q*}. But what of our confidence in the whole procedure by which this degree of confidence is arrived at? Well, we can construct a second-order scheme for that and talk *about* the sentences {*P*} and the rules by which we operated on them. We thus arrive at a degree of confidence in the procedure. But what of our confidence in this second-order scheme? And so on.

It is tacitly agreed by almost everyone except Quine that this regressive problem presents itself in two distinct cases. The first covers deductive inference and gives us absolute confidence in the conclusion on the object level as well as in the rules at all subsequent levels. The second covers everything else; we can't even think of pursuing the regression

AUTHOR'S NOTE: The original version of this paper was presented at a conference on induction held at the University of Minnesota in July 1968.

more than one or two levels, and even there we have to cut off debate by shifting attention from truth, probability, etc., to acceptability, epistemic utility, and the like. Whenever bits of the problem under its second aspect can be so arranged as to yield to the techniques proper to its first, this is instantly done; we thus have deductive theories of probability, utility, acceptance, and so on — a veritable deductivizing of all tractable parts of inductive logic. It seems reassuring to be able to say with deductive certainty that the conclusion follows inductively, even if we can't assert the conclusion itself with anything more than inductive probability. This activity takes place mainly on the first meta-level, and represents a kind of sidestepping of second- and higher-order inductive issues. My admiration for the people who do it is great and sincere, but I have no contribution to make along these lines. Instead I wish to confront one of the issues they sidestep and to suggest in a slightly different way from Quine's that the separation of cases will not stand up under scrutiny.

I start with Hume and in particular with the distinction between matters of fact and relations of ideas. We tend on the whole to give Hume too much credit as a contemporary and to make too little allowance for his belonging to the eighteenth century. Given the tenor of immediately preceding discussions, especially Berkeley's, Hume may certainly be excused for believing in "ideas," but that does not excuse us for following him in this aberration, whether or not we call things by the same name. The main contention of this paper is that matters of fact are enough — are, indeed, all we have — and that the complex matters of fact which we *call* "relations of ideas" (or which we call "logic," deductive or inductive) are reflections of the inclusive matter of fact which we call "the world" and are as contingent (or as necessary) as it is.

This contention can be looked upon as a philosophical analogue of the generalized form of Mach's principle. What Einstein called Mach's principle was of course the restricted claim that the inertia of a given body is determined by the distribution of matter in the physical universe. The generalized claim is that the laws of nature are so determined, and the philosophical analogue is that the laws of logic are determined, not to be sure by the distribution of matter, but by some feature or features of the world as a whole, so that they would be different if it were different. This means among other things at least a reinterpretation of what we can mean by the expression "possible world," since as presently understood the limitations on possible worlds are precisely logical limitations, whereas if the world determines the logic (rather than the other way around) it would seem difficult to rule out a priori any world whatever, by any other argument, at least, than follows from the fact that this world already exists. *Given* logical rules (and the distinction between rules and laws is an important one in this context) we can of course explore the set of worlds in which they hold and the relative possibilities within this set; for that matter, we can devise systems with any rules we like, the only limitations being the scope of our imagination. (But that limitation is far from trivial.)

This, however, is running ahead somewhat. My proposal is really a double one, of which the analogical form of Mach's principle is only one element, the other element being the reconstruction of logic and all other relations of ideas as matters of fact. And even if both these elements were upheld, there would still remain the question of what light, if any, they threw on inductive logic and confirmation theory. The last question is the easiest. As I have already said, I claim to make no contribution to the technical discipline, but it may be that success in the philosophical enterprise would lead to a different view of what the technical discipline is all about and what can reasonably be expected of it. In particular it might help along the demystification of *deductive* logic, which as the unattainable paradigm has been responsible for driving so many inductive logicians to despair.

"Relations of ideas" are represented by Hume (for example, in the passage about consigning books to the flames) as the kinds of thing that form the object of reasoning about quantity or number. Matters of fact, on the other hand, are the kinds of thing that enter into *causal* relations. We can know some relations of ideas certainly, because (to use post-Humean language) they are analytic, definitional, and so on, and hence, as later came to be thought, empty of factual content. We can never know any matters of fact certainly, because in order to do so we would need certainty in causal relations, and these run into the notorious difficulty that future exceptions are always possible. (I include here of course future *knowledge* as well as future *events*.) I do not wish to rehearse all this, but to comment on the insertion of temporal considerations into the inductive case when they are absent from the deductive case. Suppose we were to ask, How do you know the deductive relations will hold in the future any more than the inductive ones?

After a bit of spluttering, and assuming that the question is taken seriously (as I mean it to be), the reply to this is likely to be a challenge: How could the deductive relations be different, since they are rules of procedure and not laws of nature? What would it be like for them to be different? What conceivable test could be proposed that would reveal exceptions to them? To which I readily answer that I have no idea, I can't imagine an exception, it doesn't even make sense to me — but nor do I conclude, from these limitations on my powers of fantasy, that the thing is impossible. I can readily think of exceptions in the case of induction; I can't conceive of them in the case of deduction — but there must surely be more to it than that, or if there isn't, then the debate is shifting to my territory. For imagining, conceiving, and the like aren't logical categories at all, but rather psychological and thus empirical ones.

But for counterinstances to arise in the deductive case would surely involve a self-contradiction. Here the deductive logician is likely to start writing things down, showing that the very rules of the language *L entail* the analyticity of this proposition or that. Now, however, I can afford to be generous. I have no objection whatever to analyticity within languages — for incorporating the law of contradiction as a rule in such languages; indeed, given the world we live in, it seems a sound move. But I can quite easily imagine a language in which the law of contradiction was not a rule, in which (let us say) truth-values were assigned at random; it might be spoken by some latter-day Epimenides, who would explain that Cretans aren't really liars, it's just that they don't care one way or the other, and if they contradict themselves then they contradict themselves and that's all there is to it. *We* care about truth — and that is a fact, not a principle of logic. And it is our caring about truth, not just about the rules of the language, that makes us choose the rules we do.

If during the argument the logician has written something down, that is particularly gratifying. For inscriptions have a factual presence, and as long as we hang on to them we can keep ideas (and propositions and a lot of other metaphysical baggage) at arm's length. The curious dependence of logicians on factual things like inscriptions (not to mention brains) ought to tip us off to a discrepancy between theory and practice. In theory logic reaches for immutable truth (most logicians are at heart Platonists); in practice the logician writes something down on paper or a blackboard, thinks for a bit, imagines, conceives, etc., and then writes something else down. But, you say, these are mere tokens — tokens stand for types, and the types refer to all those logical categories, truths, relations, and the like. That may be, but all I see are the inscriptions, the furrowed brow, the puzzled look. If I could give an adequate account of logic in terms of *them*, would we need the rest of the apparatus?

The kinds of matter of fact required for this reconstruction are two: first, obvious things like inscriptions and utterances, which can be located in the world easily enough alongside chairs and tables; second, a rather specialized kind of animal behavior (in this case the behavior of logicians) which can be located alongside eating and sitting, etc. I am prepared in the latter case to talk dispositionally about "abilities" provided it is understood that the possession of the ability in question is a factual matter to be judged in terms of behavior. What abilities does the logician need in order to ply his trade? They are not I think especially exotic abilities. The fundamental operation of logic is one that every functioning human being is capable of performing, indeed one that we all do perform all the time; I call it, borrowing from the classical English grammarians, "apposition," and it consists of taking two things — any two things — and putting (and holding) them together. I say "taking" and "putting" metaphorically; knowing a person's name is a case of apposition, so is knowing the French for an English word, so is using a metaphor, and so on. Apposition is a perfectly general binary operation, unconstrained by category considerations, and by means of it we build our world. The special behavior of logicians lies in the invention and following of special rules of apposition, which impose constraints under which we do not ordinarily work. (The laws of logic are not the laws of thought, any more than the laws of chess are the laws of moving objects on checkered surfaces. If Ayer or Berlin or Popper have black-and-white tiled bathrooms, that does not compel them to walk one square forward and one diagonally.)

The special rules of logic are formation-rules that preserve or at least safeguard type homogeneity and transformation-rules that preserve or at least keep track of truth. The chief talent that logic requires is an ability to stick to these rules; the looseness and redundancy, the ellipses and shortcuts of ordinary language give way to a more or less rigorous formalism (which has nothing to do with notation). This talent is rarer than might be supposed, which accounts for the fact that logic (as I have pointed out elsewhere)[1] is *too simple* for many people who look for subtlety in its elements. It rapidly gets complex, of course, but its complexities always break down into simple elements as the fuzzy complexities of everyday life do not. People are always putting more or less complicated objects and expressions in apposition with each other and one another; this activity is governed by no principles other than immediate utility or intelligibility and the conventions of ordinary language and behavior, and consequently the coherence and even relevance of any element of the resulting structure with or to any other element are not guaranteed, indeed it is normal for this structure to be incoherent and fragmented. These defects do not often appear because different parts of the structure are usually brought into play in fairly specific contexts which not not overlap; when incompatible parts of it are activated at the same time the various psychological phenomena of dissonance are observed, and there are also contexts which activate no parts because they are just not comprehended. (Note that these structures are specific to individuals.) *Logical* structures on the other hand are such that except under Russellian or Gödelian stress all their parts are mutually consistent and no parts are lacking. The very idea of such comprehensive austerity is well-nigh inconceivable to the ordinary talker or thinker in daily life.

But what the logician does is different only in degree from what the ordinary man does, and it too is governed in the end by utility and intelligibility and convention. What the *additional* constraints of logic make possible are just those logical properties that we think of as characteristic, namely, analytic precision and synthetic open-endedness. Everyday thought is at once grosser and more limited than logical inference: grosser because it works with unanalyzed complex wholes, more limited because these cohere imperfectly with one another and relate for the most part intransitively, so that inferential sequences are always short. Still it is adequate to the conditions of its world and survives because of this adequacy. Its world is an aspect of *the* world; the temptation we have to resist is that of supposing that because logical refinements enable us to transcend the limitations of everyday thought

[1] Peter Caws, "'. . . Quine/Is Just *Fine*,'" *Partisan Review*, 34, no. 2 (Spring 1967), 302.

they also enable us to transcend the actuality of the world. The logical operations we are capable of performing are just some of the things that our evolutionary development has equipped us to do, and like other features of our heredity they can be assumed to reflect a close adjustment to the facts of our environment. If these facts are contingent, then logical laws are also contingent, while if they are necessary the necessity of those laws is still extrinsic to logic and depends on an empirical relationship between it and the world.

In this light we need to look again at the doctrine that tautologies are empirically empty, perhaps to abandon it in favor of the doctrine that on the contrary they are empirically full, if I may so put it.[2] Identity and contradiction aren't just logical rules (although they are that), nor are they laws of thought; they are laws of nature, reflecting just those properties of the world which, according to the analogous form of Mach's principle, determine the logic we are bound to use in dealing with it. They aren't obviously falsifiable (as remarked earlier, our imaginations aren't up to envisaging counterinstances, and by now it should be obvious why not) but their nonfalsifiability is clearly of a different kind from the nonfalsifiability of pseudoscientific or metaphysical claims. For that matter, in an important sense the other inductively established laws of nature aren't falsifiable either. It is one thing to know (to be able to imagine, etc.) what a counterinstance would be *like*, another to be able to produce one. We can of course construct interesting but as yet useless systems incorporating alternative rules, but we can do that for logic too. Physical laws, like logical ones, function as rules within the theoretical systems to which they belong and acquire the status of laws only on the successful application of those systems to empirical problems.

What holds for the laws of deductive logic holds equally and under precisely the same conditions for the laws of inductive logic. In neither case are we really guaranteed success in advance; if in the deductive case it is claimed that consistency itself requires the outcome, it must be recognized that consistency in language is one thing, consistency in the world another, and that the former, again, reflects the latter. We count heavily on the world's consistency and are perpetually vindicated in

[2] See in this connection Charles Hartshorne, "Some Empty Though Important Truths," in Sidney Hook, ed., *American Philosophers at Work* (New York: Criterion Books, 1956), pp. 225ff.

doing so. We count heavily also on the world's continuity, on its regularity, and so on, and are vindicated in this too. The really serious difference between the cases lies in the information available in the premises, and from this point of view the current tendency to deductivize inductive problems seems to me entirely appropriate. But here a different distinction emerges, a form of the old distinction between theory confirmation and instance confirmation. In its traditional form the problem of induction focused mainly on scientific laws, which as conclusions of inductive inferences are regularly detached *without* modal qualifications and used as assertoric premises in making predictions. It would be absurd to say, every time we wanted to calculate quantities of chemical reagents or stresses in airframes, for example, "Well, the law is only confirmed to such and such a degree, so we can't really be sure how much we need," just as it would be absurd to place bets on the truth of special relativity theory. On the other hand in Bayesian estimation problems, statistical computations, etc., there isn't the same need for detachment; we can always go back to the original probabilities or the original figures and start again, and the probability estimate in the conclusion is of course the essential part of it. This difference in the *use* of inductive inference seems to me crucial. It is only in the former case that my version of Mach's principle can be thought of as applying, since in the latter our problem isn't with the behavior of the *world* exactly, but with the changing scope of our knowledge of it. The reason why the next raven should be black is *quite* different from the reason why the next ball drawn from the urn should be black.

I conclude by reiterating one or two points. Logic, like science, is a rule-governed human activity which consists in putting things (inscriptions, acts of judgment) in apposition with one another, in spatial juxtaposition, or in temporal sequence. Nobody can compel us to accept its conclusions, as the tortoise taught Achilles, but accepting (and detaching) them is one of the things it is useful to do in our dealings with the world. (Doing logic for no ulterior purpose is a form of having dealings with the world.) Within the systems we construct we can be as formal, as analytic, etc., as we like, but the choice among systems for use in dealing with the world rests, in the deductive as in the inductive case, on empirical (and in the long run on pragmatic) grounds. And there are limitations on the kinds of system we can construct, imposed by the finite scope of our intellect and its prior adaptation to the special circumstances in which we find ourselves. Success in theory, logical or scientific, consists in bringing it into parallel with the world so that it reflects essential features of the world. Some parallels are long-established and practically unquestioned, others are more recent and tentative. Also we can construct systems independently of questions of relevance, but their rules remain merely rules and are not to be confused with logical or empirical laws.

BRIAN SKYRMS

Physical Laws and the Nature of Philosophical Reduction

Introduction

This paper is addressed to a philosophical problem posed by physical laws. The problem is one of meaning and metaphysics. How are the meanings of physical laws to be distinguished from the meanings of accidental generalizations? Two prima facie alternatives are as follows:

(1) Physical necessity and possibility and causal relationships are *real relations* in the world. The claim made by a physical law is the claim made by the corresponding accidental generalization *plus* the claim that such relations obtain.

(2) The apparent difference in meaning between physical laws and accidental generalizations is an illusion. The meaning of the law "All *F*'s are *G*'s" is exhausted by "$(x)(Fx \supset Gx)$."

(1) and (2) are not without distinguished proponents. However, to many of us (1) is a metaphysical romance and (2) is unable to account for the role of laws and disposition terms in scientific (or for that matter nonscientific) thought. Attempts to avoid both (1) and (2) have centered on achieving a *philosophical reduction* of laws of nature and related suspect concepts to a *philosophically clear* language where they can be compared with extensional generalizations, and the difference between the two exhibited.

Despite an enormous expenditure of effort and intelligence, no reduction to date seems to work. Is it possible that no reduction can work? This question forces us to ask ourselves another. Just what constitutes an acceptable philosophical reduction? The answer to this question is by no means apparent. And this suggests the attractive speculation that at least some of the disheartening results have been the result of an overly narrow view of philosophical reduction.

In fact, I believe that this is the case; that progress has been blocked by an inadequate view of philosophical reduction; and that this view of reduction ultimately stems from an inadequate philosophy of language.

1. Ideal Language Games

The classical view of language goes something like this:[1]

In the study of language use, we distinguish three factors: (1) The language user together with the context in which he uses (speaks, hears, etc.) the elements of language; (2) those elements of the world that are referred to or designated by the elements of language; and (3) the elements of language themselves. The discipline which studies factor (1) (whether in connection with factors (2) and (3) or not) is called *pragmatics*. That which abstracts from (1) and deals only with the relation of (2) and (3) is called *semantics*. That which abstracts from (1) and (2) and deals with (3) alone is called *syntax*.

However, as long as we confine ourselves to the *description* of language use we can never really abstract completely from (1). We may get statements like "In 98 percent of the cases 'mond' was used to refer to the moon." Such statements may be called *descriptive semantic* statements because of their focus on designation relations. Nevertheless, they are part of pragmatics since they refer to incidents of language use.

The only way that reference to (1) can be really eliminated is by the construction of an idealized language system *L*. Then the contingent statement (A) is idealized to:

(A′) "Anyone operating in system *L* who uses 'mond' uses it to refer to the moon."

(A′) is taken to be definitive of L and thus necessary. (A′) licenses (or perhaps is equivalent to) the *semantical rule*:

(A″) 'mond' designates (in L) the moon.

(A″) is a statement of pure semantics (which, unlike descriptive semantics, is not a part of pragmatics).

Thus pragmatics corresponds with the descriptive study of language use, whereas pure semantics and syntax define an idealized language system.

If we identify pure semantics and syntax with the sorts of considerations which normally go under those names, then the preceding picture will lead us to identify an *idealized language system* with the familiar notion of a formal language. Such a conclusion, however, would be too hasty. If we look at the process of idealization closely, we will find that

[1] See Morris (1939), Carnap (1937a, 1937b, 1939), and appendix D of Carnap (1956).

it yields rules which do not fall within semantics or syntax as commonly conceived.

One example of idealization which led to an ordinary *semantical* rule has already been indicated. In order to get a syntactical rule of inference we need to idealize some notion of conditional acceptance or belief. Roughly, we start with

(B) Almost all instances in which an English user accepts a sentence of the form

α *and if* α *then* β

he is thereby disposed to accept β

and end up with modus ponens.

Here, I think, as in all other cases we are idealizing on certain attitudes, intentions, and dispositions that language users have toward the elements of language and toward each other. The same process yields the following:

(C) In almost all cases in which an English user uses the word "I" he uses it to refer to himself.

(C′) (x) Relative to a context where the sign user is x. "I" denotes (in *L*)x.

This sort of idealization leads to language systems with context-variable semantics as developed by Richard Montague and others[2] under the name "pragmatics."[3] This "pure pragmatics" exposes an illicit assumption of the classical view. That is, the assumption that *all* reference to the sign user and context of use must be eliminated in the process of idealization. What is important in the process is that empirical regularities are idealized into rules, and that following these rules is definitive of use of *L*.

Consider now the notion of conditional acceptance or belief utilized in (B) to deliver, under idealization, the syntactical notion of a rule of inference. If this notion is suitable for idealization, then why not a notion of partial conditional degree of acceptance or belief. Here acceptance of a probability measure is built into the definition of using a certain lan-

[2] See Montague, "Pragmatics" and "Pragmatics and Intensional Logic."

[3] It *is*, of course, pragmatics, as *defined* since it studies relations between signs and contexts of sign use. It is *pure* pragmatics since the process of idealization has taken place. But in terms of the purposes to which the contexts are put, it is closer to semantics than to other parts of pragmatics.

guage system *L*.[4] Likewise, "*x* takes *p* to be an adequate explanatory account for *q*" may be idealized to "*p* is an adequate explanatory account for *q*"[5] and "*x* regards *p* as relevant challenge to *q*" (where *q* might be a question or an imperative or a factual claim) can be idealized to build rules of rational dialectic[6] into the language system.[7]

Recalling that idealization need not eliminate reference to the sign user and context of use, we can go even further. One way is to build in practical reasoning — rules which connect belief or degree of belief with action. Another way is to build in connections between having certain experiences and changes in degree of belief of associated sentences.[8]

[4] Carnap's conception of probability usually seems quite different from this. But he appears to come close to this view in section 2, "Linguistic Frameworks," of "Empiricism, Semantics, and Ontology." Speaking of the thing-world framework, he says that questions of existence and reality ". . . are to be answered by empirical investigations. Results of observations are evaluated according to certain rules as confirming or disconfirming evidence for possible answers. (This evaluation is usually carried out, of course, as a matter of habit rather than a deliberate rational procedure. But it is possible, in a rational reconstruction to lay down explicit *rules* for the evaluation. This is one of the main tasks of a pure, as distinguished from a psychological, epistemology.) The concept of reality occurring in these internal questions is an empirical, scientific, non-metaphysical concept. To recognize something as a real thing an event means to succeed in incorporating it into the system of things at a particular space-time position so that it fits together with the other things recognized as real, according to the rules of the framework" (Carnap, 1950, p. 207).

This would make the justification of the confirmation rules an *external question*. It would be part of the question of justifying the adoption of *this linguistic framework* rather than another (and not necessarily a neatly separable part).

[5] q could have been either a sentence or an event. In the first place we end up with a syntactic relation (in the sense of being a relation between items of language); in the second a semantic relation (in the sense of being a relation between an item of language and an item in the world).

[6] See Skyrms (1967).

[7] I argued in Skyrms (1966) that these sorts of things could be built into the notion of an ideal language game. But I classified them as "pure pragmatics" on no better grounds than that they did not belong to semantics, conceived of as the theory of reference (and perhaps sense) or the theory of syntax, conceived of as the study of purely formal properties of signs. Of course the metalanguage defining the ideal language game can be divided up any way you please. But it need not be divided in such a way as to give semantics and syntax such restricted roles. See fn. 10.

[8] This sort of idea has been persuasively advanced by Richard Jeffrey. "[C]oming to have suitable degrees of belief is a matter of training — a *skill* which we begin acquiring in early childhood, and are never quite done polishing. The skill consists not only in coming to have appropriate degrees of belief in appropriate propositions in paradigmatically good conditions of observation, but also in coming to have appropriate degrees of belief between zero and one when conditions are less than ideal.

"Then learning to use a language properly is in large part like learning such skills as riding bicycles and flying aeroplanes. One must train oneself to have the right

Pure syntax, semantics, and pragmatics may now appear as rather motley categories.[9] Perhaps the traditional way is not the most illuminating way of dividing up the metalanguage. For the time being, I will simply refer to *the metalanguage* without worrying about how or whether to draw the syntax-semantics-pragmatics distinction.

The idea of a language system has already been stretched far beyond the customary notion of a formal language, to the point where I think that it is reasonable to begin using the term "ideal language game." Is there no end to this process?

When an ideal language game is intended as an *explication* of actual language use, the regularities which are idealized into rules should be those which can be taken as indicative of competence of language use. The limits on the process of idealization are then the limits of logical richness of actual language use. Of course *what is indicative of competence in language use* is rarely clear-cut. Too bad, but that's the way language use is, by and large.[10] For this reason, among others, empirical linguistic regularities will typically underdetermine the corresponding ideal language game, leading to a range of alternative explications.

sorts of responses to various sorts of experiences, where the responses are degrees of belief in propositions" (1968, pp. 179–80).

This view has the attractive feature of enabling empiricist epistemology to avoid the pretense of the myth of the given and the neoidealism of the coherence theory of knowledge (in this case, the coherence theory of rational belief).

[9] Actually, the formulation given is not precise enough to tell us where the boundaries are. Logical consequence is a relation between items of language, unlike the denotation relation. Is it therefore a syntactical relation rather than a semantical one? Or is the distinction an epistemological one; that in order to learn about logical consequence we must *first* study the denotation relation? Is semantics restricted to denotation and things definable in terms of it, or does it include all sorts of relations between items of language and items in the world? What about an idealized explanatory relation between explanatory accounts and events?

There are lots of alternative ways to answer such questions. One way is in terms of various functions of language. For instance, if we are interested in the language function of describing the world, we will need referential semantics which will include that part of traditional syntax *which is exclusive of inference rules.* If we are also interested in inferential functions of language more comes in. How much more depends on how broadly inference is conceived. Epistemological concerns cut across all categories. If we are interested in illocutionary force, etc., we bring in a whole new array of factors. The best way of sorting out all these factors will depend on what question is being investigated.

[10] As a situation where things are relatively clear-cut and which illustrates some of the richness of the notion of an ideal language game, Professor N. Belnap has suggested that we look at the legal procedures for a trial. For example, consider the concept of inadmissible evidence. If certain kinds of evidence are admitted in a criminal trial, an appellate court may declare a mistrial. In doing so, it is in effect saying the criminal trial game has not been played. Rules of admissibility of evi-

2. The Nature of Philosophical Reduction

Philosophical reduction is different in both aim and method from reduction in science.[11] In *scientific reduction*, the typical situation is that of an old theory being *replaced* by a new better theory. The new theory is *incompatible* with the old, but it is a consequence of the new theory that the old should give results which are in some sense an *approximation*[12] to the correct results when applied *within a limited domain* of phenomena. *Philosophical reduction* is not an attempt to *replace* a theory with a better one, but rather an attempt to *make sense* of language use which does not make sense on the face of it.[13]

Such an attempt is motivated by some set of criteria of prima facie philosophical intelligibility. These are often keyed to the type of objects referred to (sense data, physical objects, events, universals) or the way they are characterized (Are spatiotemporal relationships intelligible? Are causal relations?), but there are, no doubt, many other forms a philosophical conscience can assume.[14]

Traditionally, there are two kinds of reductions: those that are successful and those that aren't. Those that are successful establish some kind of translation[15] of prima facie unintelligible discourse into philosophically impeccable discourse. Good examples of such successful reductions are hard to come by (partly because of general disagreement on criteria of philosophical intelligibility and successful translation). An old favorite in Russell's Theory of Descriptions and despite recent criticisms it is still a powerful illustration. What I have in mind as an unsuccessful reduction is an instrumentalist treatment of a certain body of discourse.

dence are thus part of the metalanguage of the ideal language game explicating legal language use in the criminal courts.

[11] See Bohm (1957) and Feyerabend (1962). (This is not a blanket endorsement.)

[12] There are lots of different kinds of approximations. A close look at statistical mechanics and thermodynamics is a good antidote to overly simplistic views of what approximations must be like.

[13] This fundamental distinction is often lost sight of in discussions of Craig's theorem. No phenomonalist in his right mind wants to say that scientific theories should be *replaced* by their Craigian counterparts. The phenomonalist wants to make sense of scientific theories in phenomonalistic terms.

[14] For example, consider the intuitionist criticism of classical logic.

[15] Criteria for a correct translation have come in many shapes and sizes; some stringent, some easygoing, some philosophically unintelligible from other viewpoints. If the criterion for having a correct translation is represented as *having the same meaning*, then the paradox of analysis rears its head. In the general framework given here, however, it is hard to motivate the paradox of analysis at all.

It is the sort of attempt to make sense of a body of discourse that is grasped at when all else fails, but the discourse seems too important to just dismiss as nonsense. The prima facie unintelligible discourse becomes intelligible as a machine which processes intelligible inputs and produces intelligible outputs. The classical example of this sort of move is the treatment by finitists, nominalists, and some constructivists of those parts of mathematics which do not satisfy their respective criteria of philosophical intelligibility.

I wish to consider a conception of philosophical reduction which has these two classical forms as special cases, but which covers a rich spectrum of other cases as well.

Suppose we start with a body of language usage some parts of which are philosophically intelligible and some parts of which are not. A *philosophical reduction* of the prima facie unintelligible parts will consist of (1) an *ideal language game* which is an explication of that body of usage such that the metalanguage used to define the language game is itself prima facie intelligible; and (2) a *justification* of the ideal language game.

The notion of an ideal language game was introduced in the last section in a rather elastic way. The metalanguage defining such a game can have quite a lot or quite a little built into it. Consequently, the concept of a philosophical reduction here introduced partakes of the same character. Degenerate cases of philosophical reduction are almost always achievable. Thus, when approaching some prima facie unintelligible usage we should not ask, "Is a philosophical reduction possible?" but rather, "How rich a philosophical reduction is possible?" "How much can be built into the metalanguage?" "How extensive are the connections set up between the prima facie intelligible and unintelligible usages?" The *richness* of the reduction determines the *degree* to which the prima facie unintelligible usage is rendered intelligible.

It is time now for this general conception to be tied down a little by looking at particular cases. Let us first note that the body of language usage that we start out with might (in whole or in part) consist of the use of some ideal language game.[16] Such a fortunate special case eliminates some of the looseness introduced into a philosophical reduction by

[16] For example, the problem might be in relating an unintelligible language game to an intelligible one, or an intelligible game to unintelligible usage.

the explication relation. Typically, however, the demand for reduction arises out of language usage which does not hew to the lines of the idealized structure of an ideal language game.[17]

In the special case where all that the metalanguage tells us about the prima facie unintelligible portion of discourse consists of formal rules of formation and of inference both within that part and between it and the intelligible part, we have the substance of a classical *instrumentalist* picture.

As an illustration, consider Boolean algebra as originally formulated by Boole.[18] Boole gives a logical interpretation to algebraic symbols as follows: x, y, z are *elective* symbols which represent an election of objects from a universal set. The *product* $x \cdot y$ represents intersection and the *sum* $x + y$ (which is only defined for disjoint sets) represents union. *Difference* $x - y$ represents the class resulting from removing the members of y from the members of x. The symbol 1 represents the universal class and the symbol 0 the empty class. As rules of inference Boole adopts all the procedures of ordinary numerical algebra. This allows the derivation of equations which have no logical interpretation (e.g., equations with numerical coefficients different from 0 or 1). Further derivation from such equations can lead back to logically interpretable equations. Taking the logically uninterpretable equations as the philosophically unintelligible ones, Boole took the classical instrumentalist position. The rules of inference are the rules of ordinary algebra together with the rule $x^2 = x$ for elective symbols. (This rule does not apply to numerical coefficients.) They are clearly sound for the intelligible sentences. The reduction is completed by a *justification* which, in this case, is a proof that chains of inference leading from intelligible sentences through unintelligible sentences back to intelligible sentences are truth-preserving.[19]

As has been already mentioned, a similar instrumentalist position toward large parts of classical mathematics has been adopted by various philosophical schools. Here, however, reduction has failed for lack of an appropriate *justification*.

[17] However, this matter, like so many others, is really a matter of degree. As soon as we talk of *rules* of a language, no matter how few, we have begun the process of idealization. To this extent, English, French, and German are ideal language games.

[18] See C. I. Lewis (1960), pp. 51–72.

[19] See E. W. Beth (1966), secs. 23 and 25.

It may be worthwhile at this point to take a quick look at the relevance of Craig's theorem to the instrumental view of theories. Assume that the vocabulary of a language is divided into theoretical terms and observation terms and that sentences containing only observation terms are prima facie intelligible while others are not. Formal rules of formation and inference are specified for the whole language so that its theorems are recursively enumerable. The instrumentalist wants to regard the theoretical part of the language as a machine for deriving theorems which contain only observation terms. This it certainly is, whatever else it may be. An intelligible metalanguage appropriate to an instrumentalist has two major problems left. The first is to find an appropriate justification.[20] The second is that the instrumentalistic position *assumes* that an instrumental reduction is the strongest reduction possible. If the theoretical part of the language does other things which can be rendered philosophically intelligible, then the instrumentalist must move from his cramped perspective to a richer appreciation of theories.[21] Having gotten thus far without Craig's theorem, is the instrumentalist helped by Craig's theorem with either of his two outstanding problems?

Craig's theorem tells us that there is a recursively axiomatizable theory, couched only in observation terms, which has as its theorems exactly the theorems of the original theory which contain only observation terms. So what? Given the instrumentalist characterization of the theoretical machine, Craig's theorem tells us that, with its aid, we can build another machine, an observational machine, that will do the same job. This neither helps us justify what these machines do nor gives us any reason to think that we have not neglected other vital functions of the theoretical machine.

Another type of reduction arises when the metalanguage establishes an equivalent intelligible sentence for every prima facie unintelligible sentence. Then we have a classical *translational* reduction (with "equivalent" mirroring the vagueness of "translation"). To be more specific, consider the case in which equivalence is logical equivalence in the fol-

[20] This of course is a problem which, in different guise, plagues realists as well. But we are not here assigning points in metaphysical debate. It is still correct that the reduction cannot be called complete without some such justification.

[21] I find Hempel's point about inductive systematization relevant in just this way. See Hempel (1958), pp. 37–93.

lowing sense: The metalanguage defines, in intelligible terms,[22] a sense of "interpretation of *L*" "true in an interpretation of *L*" for the language game *L*. (These definitions cover both intelligible and unintelligible sentences of *L*.) Two sentences are equivalent if they have the same truth-value under all interpretations. As an example, let us consider again Russell's Theory of Descriptions. The ideal language game which we are considering contains formulas of the first-order functional calculus with identity (intelligible) together with formulas containing '$(\iota\, x)$', '$(E!)$', etc., in specified ways (unintelligible). Here the sense of interpretation is the standard one for the intelligible part, and the task of finding a translation is trivial once the semantics of the prima facie unintelligible part is set up. Neither of these features holds in general for translational reductions. Another example of a successful translational reduction, for those who don't like Russell's Theory of Descriptions, is the reduction of numerical quantifiers (There are at least *n*; there are at most *n*; there are exactly *n*; where *n* is a positive integer) to first-order logic with identity.

The type of translational reduction just discussed is a special case of a semantical reduction. In a *semantical* reduction the metalanguage establishes, *in intelligible terms*, a semantics for each sentence of the language game *L*. This sort of reduction has a special significance because of its connection with metaphysical questions. If we have an intelligible semantics for a sentence, then it makes sense to say that sentence is *true* or *false*. If we have an intelligible sense of denotation for a class of terms, then it makes sense to say that such things *exist*. If we have an intelligible sense of satisfaction for a property or relational predicate, then it makes sense to say that the corresponding property or relation is *real*.

It is of some importance to notice that a semantical reduction may be possible where a translational one is not, since metaphysical discussions

[22] This requirement may seem too strong for certain kinds of translational reduction. For instance the virtual theory of classes establishes translations of a certain (small) portion of set theoretical discourse into set-free discourse. If set talk were taken as prima facie unintelligible and first-order logic intelligible, this would appear to be an acceptable reduction. But the metalanguage would appear to require set theory (or something which might be considered just as bad) to accomplish its task. Perhaps the best way to look at it is this: Separate the criteria of philosophical intelligibility of the object level C_o and the meta-level C_{2m}. Speak of a reduction having been achieved *from the standpoint m*. Speak of a thorough reduction having been achieved just when $m = o$.

often seem to focus exclusively on translational reductions.

An example which is interesting for just such reasons is the comparison of tense and date languages for dealing with time. Suppose we take the view that date languages are intelligible while tense languages are prima facie unintelligible. Consider the following simplified ideal language:[23]

Predicates:

P (a monadic predicate of times).
= (identity).

Terms:

0 is a term.
Now is a term.
If **a** is a term, **Aa** (one unit of time after α) and **Ba** (one unit of time before α) are terms.

Sentences:

If x and y are terms, **P**x, **P**y, x = y, y = x are sentences.

Sentences containing **Now** are here regarded as unintelligible, all others as intelligible. A semantical reduction is accomplished by the following. A world, *W*, consists of (1) *D* = the set containing 0 and the positive and negative integers (the set of times).[24]
(2) Sw = a subset of *I*w (the extension of ϕ).

Denotation:

0 denotes 0 at *t*.
Now denotes *t* at *t*.
If x denotes y at *t* then **A**x denotes y + 1 at *t* and **B**x denotes y − 1 at *t*.

Truth:

Px is true in *W* at *t* just in case $(\exists\eta)$ x denotes η at *t* and $\eta \epsilon$ Sw.
x = y is true in *W* at *t* just in case $(\exists n)(\exists m)$ x denotes *n* at *t* and

[23] Here a name of a symbol is formed by printing it in boldface. The name of a concatenation is formed by concatenating the names of its parts. Boldface type was not available for the identity sign; instances of it should be considered boldface as appropriate.

[24] We represent time as a framework concept by choosing D_w to be the same for all worlds.

y denotes m at t and $n = m$.

Logical Equivalence:

x is logically equivalent to y just in case for all worlds w and times t x is true in w at t if y is true in w at t.

The foregoing semantical reduction seems trivial. Yet it establishes that no translational reduction is possible. Consider the sentence "Now = 0." It is true at times 0 and false otherwise. No intelligible sentence has this character, and thus no intelligible sentence is an adequate translation of it. The same argument holds good if we enrich the language by adding truth functions, temporal variables, and quantifiers, earlier than and later than relations and other predicates (allowing the recovery of a full-tensed language by definition). The argument remains the same if we let "0" denote different times in different worlds. A slightly different argument establishes the same result if we make sentence tokens rather than sentences the bearers of truth.[25]

[25] For every predicate, P, and for every term, T, in the foregoing type language there is an infinite family of tokens $\{Pu\}$, $\{Tu\}$ indexed by the positive integers. If $T(u)$ is a term token then $\mathbf{P}(s)\ T(u)$ is a sentence token provided that $(u = s)$. If $T(u)$ and $T(v)$ are term tokens then $T(u) = (s)\ T(v)$ and $T(v) = (s)\ T(u)$ are sentence tokens provided that $u = s = v$. In the token language a world, w, must also specify a function fw from the positive integers into Dw [specifying the time of production $fw(v)$ of a token, e.g., $\phi(u)$ **Now**(u), in w.] The semantics must be changed as follows:

Denotation:

0(u) denotes 0 in w.

Now(u) denotes $fw(u)$ in w.

"at t" is replaced by "in w" in the other definitions.

Truth:

P(u)x is true in w just in case $(\exists n)$x denotes n in w and $n \epsilon Sw$.

x = (**u**)y is true in w just in case $(\exists n)$ $(\exists m)$ x denotes m in w and y denotes n in w and $n = m$.

Logical equivalence:

Tokens x and y are logically equivalent if their truth-value agrees in all worlds. Logical necessity and impossibility for tokens are defined as truth in all or no worlds.

That there is still no translational reduction possible can be seen from the following: Consider the sentence token **0**(u) = (u) **Now**(u). It is true in a world, w, just in case $f_w(u) = 0$. Thus there are two worlds differing only in the function f, such that in one the chosen sentence token is true; in the other it is false. Since the truth-value of sentence tokens not containing Now-tokens does not depend on choice of f, there can be no sentence token not containing a Now-token logically equivalent to the token in question.

This result, of course, comes directly from allowing a token to be produced at different times in different worlds. This is not a trick, however, but rather is necessary for an adequate explication. For if a token had to be produced at the same time

The miniature temporal language game that I used to illustrate semantical reduction did not incorporate any formal rules of inference. In other cases (modal logic, intuitionistic logic) inference rules will play a prominent role in the language game accomplishing the reduction.

In these cases a justification of the inference rules is called for. This may take the form of a proof of soundness or even completeness. The exact role of the inference rules and nature of the justification required will depend on the character of the language use being explicated.[26]

The types of reduction discussed fall within the purview of explications accomplished by formal languages as traditionally conceived. The generalization to reductions accomplishable by an ideal language game multiplies the types of reduction possible. For example, consider an ideal language game where degrees of belief are required to obey the rules of the probability calculus and where players are required to coordinate them with choices by way of a rule of rational decision — say the rule to maximize expected utility. Suppose that statements of fact and attributions of value to world states are taken to be intelligible, but that, because of complementarities of goods, attribution of values to less specific goods are held to be unintelligible (G. E. Moore's principle of organic unities). These prima facie unintelligible values may be reduced by defining them as probability weighted averages of the true values of world states.[27] The corresponding value attribution statements are not now to be regarded as true or false, since they depend both on the values of the world states and on the subjects state of belief. But they have a pragmatic significance by virtue of their relation to rational decision.

Any of the other possible facets of an ideal language game may figure in a philosophical reduction. The more closely the unintelligible usage is knitted to the intelligible usage, the better the reduction. Thus, the logic of explanation, the logic of practical reason, and other varieties of what is now called "philosophical logic" can join with the logic of partial belief in effecting a philosophical reduction. It is hard to specify precisely in general what sort of justifications are wanted for philosophical reductions. Roughly, however, justifications always seem to be proofs of

in each world, tokens such as **O**(u) = (u) **Now**(u) would have to be either logically necessary or logically impossible.

[26] For instance, given the historical context in which the intuitionist propositional calculus was advanced, a proof of completeness rather than merely soundness is required for a successful philosophical reduction.

[27] See Jeffrey (1965).

some type of coherence between the various strands of the ideal language game effecting the reduction. The proofs forming the basis of Bayesian decision theory[28] constitute just such a coherence proof linking a type of logic of partial belief to a type of logic of practical reason.

A philosophical reduction bestows *meaning* on prima facie unintelligible usage. Due appreciation of the various forms of philosophical reduction leads to the conclusions:

(1) There is a broad spectrum of different *kinds* of meaning.

(2) The meaning conferred by a reduction is a matter of *degree*.

(3) The meaning is conferred to a body of usage. It is only derivatively or in special cases that we can speak of the meaning of a term or a sentence.

The entire discussion so far has been relativized to a set of criteria of philosophical intelligibility. But how do we find the correct criteria of philosophical intelligibility? Certainly there is no consensus (and if there were one it would probably only be of sociological interest). I am not sure that we ought to think that there is an answer to this question. Differing philosophical consciences may just be what we are stuck with. And I do not think that this is necessarily a philosophical disaster. Suppose that from perspective A a certain body of usage is prima facie unintelligible and from perspective B it is intelligible. A philosophical reduction, achieved from perspective A, is not *worthless* from perspective B. From perspective B *conceptual connections* are still being uncovered, although within a body of discourse which is *already* intelligible. Is it then *more* intelligible from perspective B? It is certainly better understood. It is reasonable to attempt a philosophical reduction of prima facie intelligible usage. In general, then, it might be better to switch to more neutral labels: the *reducing* usage and the *reduced* usage. Philosophical reduction, then, has a rationale independent of philosophical perspective — the rationale of charting conceptual interconnections. What philosophical conscience brings to a particular project is a *sense* of urgency.

3. Laws and Extensional Generalizations

The philosophical perspective operative for the rest of this paper is that laws are prima facie unintelligible and that sentences of extensional

[28] See Ramsey (1960); Jeffrey (1965).

logic are prima facie intelligible. The simplest reduction imaginable would be at hand if a law, "All *F*'s are *G*'s" could be adequately represented by its extensional counterpart "(x) $(Fx \supset Gx)$"; and the acceptability of the law would be construed to be just the probability of that extensional counterpart. Such an identification, however, gives rise to a cornucopia of puzzles and paradoxes. These difficulties are grounds for giving serious attention to radical alternatives.

The radical alternative with the most distinguished pedigree is *qualified instance confirmation*. The qualified instance confirmation of the law "All *F*'s are *G*'s" is the conditional probability Pr (Ga given Fa) for some new individual constant a not mentioned in our evidence. (The instance confirmation is Pr $(Fa \supset Ga)$.) Carnap, like Ramsey before him, suggested that it is this quantity — rather than the probability of the extensional counterpart of the law $Pr(x)$ $(Fx \supset Gx)$ — which is relevant to grading the acceptability of laws.

There are powerful considerations pressing us in both the "instance" direction and the "qualified" direction. But there are also strong arguments that appear to block any such treatment of laws and throw us back to the identification of laws and extensional generalizations. We shall see, however, that it is possible in large measure to combine the advantages of both views; and that the means of effecting this combination illuminate the nature of nomic necessity.

4. The Paradox of Provisional Acceptance

The argument pushing us in the *instance* direction is quite simple. When we accept a law, we do so with the expectation that it will eventually be superseded and replaced by a more adequate law. This expectation is inductively well founded in the history of science.[29] Thus, when we accept a law we do so with the expectation that someday a counterinstance will be found. But if there *is* a counterinstance, whether it has been found yet or not, the factual claim made by the extensional counterpart of the law is *false*. So it appears that when we accept a law, we take it as making a certain factual claim and we believe both that that claim is true and that it is false.

To say that we accept laws only provisionally is not very helpful without an explanation of why we should provisionally accept something

[29] This point is developed in great detail in Bohm (1957).

we believe to be false. To say that laws acquire high probability rather than certainty leaves us with the embarrassing question: How can a statement and its denial both have high probability?

In short, if, when we hold a law in high regard, we believe it probable that we will eventually find a counterinstance to that law, then that holding in high regard cannot consistently be construed as attributing high probability to the relevant extensional generalization.

That regard can consistently be interpreted as attributing high instance confirmation to the law, since one can quite consistently hold (I) that it is probable that some item constitutes a counterexample and (II) that there is no item such that it is probable that *that* is a counterexample.

And high instance confirmation gives some reason for holding in high regard some statements we believe to be false, since each application of such a statement would be a low risk operation.

5. Predictive Force

The qualified instance confirmation of a law measures a certain kind of predictive force. It is this kind of predictive force which is relevant when we come to know Fa with certainty or near certainty and predict Ga on the basis of this knowledge together with the law.

The classical account of the predictive function of laws goes something like this: A law makes an assertion about the world; the assertion expressed by the associated extensional generalization. The probability of a law is just the probability of this assertion being correct. Thus, if the probability of the law "All *F*'s are *G*'s" is high, we run little risk in taking the generalization $(x)(Fx \supset Gx)$ as true and proceeding deductively from there.

One of the assumptions of the classical view — that what is at stake with a law is the probability of its extensional counterpart — led to the paradox of provisional acceptance. But even if we modify the classical view by substituting high instance confirmation of a law for high probability of its extensional counterpart, it still retains the following assumption:

> If the instance confirmation of a law L_1 is high and the extensional counterpart of another law L_2 is a logical consequence of that of L_1, then the predictive force of L_2 must also be high.

This assumption is false. The instance confirmation of a law may be as close to 1 as you please, and yet the predictive force of some consequence of it, equivalent of it, or of it itself, may be as close to 0 as you please. The classical account of the predictive function of laws rests on probabilistic fallacies!

In short, if, when we hold a law "All *F*'s are *G*'s" in high regard, it is to follow that upon learning Fa we are entitled to predict with high probability Ga, then that holding in regard cannot be consistently construed as attributing high instance confirmation to the law.

It can consistently be construed as attributing high qualified instance confirmation to the law. In fact, that construal yields just what is here required and nothing more.

The way in which the classical account of the predictive function of laws fails casts some light on a number of odd cases. Some of them are given below.

Necessarily vacuous antecedents and necessarily universal consequents. If a law saying "There aren't any *F*'s" has high instance confirmation, then so does any law saying "All *F*'s are *G*'s" since $Pr(Fa \supset Ga) \geqq Pr(\sim Fa)$. Science doesn't work this way. For example, let Fx be:

> x is an event consisting of a closed system consisting of a reservoir and a particle undergoing a process whose net effect is to take heat from the reservoir and do work on the particle;

and let Gx be:

> x is an event consisting of a closed system undergoing a change in total energy.

Then, "All *F*'s and *G*'s" is the unlikely law that if an event were to violate the second law of phenomonological thermodynamics, it would violate conservation of energy. No scientist would have accepted this as a law, but would have chosen, if anything, its contrary, "All *F*'s are non-*G*'s" since the law of conservation of energy was supported by a mass of evidence independent of the evidence for the second law.

Examples of this kind can be multiplied. The antecedent of Newton's first law (Every body not acted on by any external force . . .) is rendered necessarily vacuous by his law of universal gravitation. Ideal gases are not only nowhere to be found, but are regarded as physically impossible, yet the ideal gas laws are not regarded as trivial consequences of this fact.

It is interesting to note that necessary universality of the consequent is *not* on a par with necessary vacuousness of the antecedent. In most cases (though not all) the inference from the law "Everything is a G" to "All F's are G's" appears to be quite in order. (Note that contraposition will transform a necessarily vacuous antecedent into a necessarily universal consequent, and conversely.)

How does qualified instance confirmation (QIC) explain what's going on? Let $Pr(Fa \,\&\, Ga) = a$; $Pr(Fa \,\&\, {\sim}Ga) = b$; $Pr({\sim}Fa \,\&\, Ga) = c$; $Pr({\sim}Fa \,\&\, {\sim}Ga) = d$. Then QIC "All F's are G's" is $a/(a+b)$. And interpreting "Nothing has F" as "All U's have non-F" where U is a universal property, its qualified instance confirmation, its QIC, is $c+d$. The quantity $c+d$ may be as high as you please (short of 1) and yet $a/(a+b)$ may take on any probability value whatsoever. In our example, the remark that "the law of conservation of energy is supported by a mass of evidence independent of the evidence for the first law" is represented as $b/(a+b)$ being large. In terms of QIC, this explains the failure of "All F's are G's" and the success of "All F's are non-G's."

When we come to necessarily universal consequents things change. QIC "Everything is G" is $a+c$, which certainly has a more favorable connection to $a/(a+b)$ than did $c+d$. In fact if "Fa" and "Ga" are *independent*, then $a/(a+b) = a+c$. Things *can* still go wrong, however, if F is *unfavorable* to G. Let Bx be:

> x is an event consisting of a closed system consisting of a reservoir and a (Brownian) particle undergoing an acceleration.

When it was still reasonable to accord high QIC to "Everything is a non-F" it would not have been reasonable to accord high QIC to the (thought experiment) law "Every B is a non-F."

A variant of Hempel's paradox.[30] Consider the law "All non-white things are non-swans." It is plausible to assign it high instance confirmation on a sufficiently long string of observations of non-white non-swans (red shoes, yellow butterflies, etc.). Its contrapositive "All swans are white" must have equally high instance confirmation on this evidence, although the evidence appears to have no *relevance* to this law.

Viewing the situation in terms of qualified instance confirmation brings some interesting facts to light. If our evidence consisted solely of observation of an enormously long string of non-white non-swans, it

[30] Compare Hempel (1948).

seems reasonable that our probabilities regarding an unexamined individual, a, might look like this:

$Sa \& Wa$	ϵ
$Sa \& \sim Wa$	ϵ
$\sim Sa \& Wa$	ϵ
$\sim Sa \& \sim Wa$	$1 - 3\epsilon$

Such a probability assignment would indeed yield high QIC for "All non-swans are non-white" but would make QIC "All swans are white" only 1/2. This fact may well make us queasy about saying that "All swans are white" is well confirmed, for it lacks the predictive force that we expect from a well-confirmed law.

Indeed, suppose that under such circumstances someone finally sighted a swan (at dusk) but was unable to determine its color. Our body of evidence now looks like this:

$$\sim Wa \& \sim Sa \& \sim Wb \& \sim Sb \& \ldots \& Sz.$$

If we change our degrees of belief by conditionalization, both $Sz \& Wz$ and $Sz \& \sim Wz$ which previously had probability ϵ now receive probability 1/2. *But $Sz \& \sim Wz$ is incompatible with the law in question! Spotting that swan disconfirmed the law even though we couldn't see its color.* It is a paradoxical sort of confirmation for a law that yields the result that the sort of conditions which would normally set up the predictive inference automatically disconfirm the law. Such is, however, the case whenever the QIC of a law is low.

A variant of Goodman's paradox:[31] Consider the law (L_1) "All emerald stages[32] are either green and before 2000 A.D. or blue and at or after 2000 A.D." Observation of a long enough string of emerald stages before the year 2000 might be thought to give high instance confirmation to (L_0) "All emerald stages are green and before 2000" and thus to (L_1) and thus to (L_2) "All emerald stages after the year 2000 are blue."

Now L_0's positive instances support neither it nor L_1 in the way assumed. Goodman's tough question is "Why don't they?" and I have no answer for it here. What I would like to point out is that even if L_0 had high *qualified* instance confirmation, L_2 need not have high

[31] Compare Goodman (1955).
[32] Time slices of emeralds.

qualified instance confirmation. The move from L_0 to L_1 does preserve QIC, but the move from L_1 to L_2 need not.

6. The Claims of Deduction

Given the facts discussed in the last section, why not take QIC itself as the prime measure of the acceptability of scientific laws? The main reason against such a radical revision was clearly stated in Hempel's "Studies in the Logic of Confirmation."[33] However essential high predictive force may be, it is not in itself sufficient to account for the role that laws play in science. In scientific practice, laws are not distinguished from their extensional counterparts. Inferences from law to law are freely made in accordance with rules of deduction for their extensional counterparts. Short of challenging fundamental scientific practice, any acceptable account of laws must provide a justification for such inferences.

Doesn't the failure of these inferences put qualified instance confirmation on a collision course with scientific practice? Only if we take high QIC as both necessary and *sufficient* for holding a law in high regard. Since the type of prediction to which QIC is relevant is *important* in science, high QIC should be necessary. Since inferences which need not preserve high QIC are important to science, high QIC is not sufficient to qualify a statement to function as a premise for such an inference.

We now have some bounds on reasonable requirements for a law to be held in high regard. They should not be as stringent as high probability of its extensional counterpart, on pain of the paradox of provisional acceptance. They should guarantee high qualified instance confirmation. They should be preserved by all or most customary scientific inferences. Are there sets of requirements that fill this bill?

7. Prediction and Deduction Reconciled

For any set of elementary monadic predicates, a complex predicate which is a conjunction containing for each elementary predicate either it or its negation (but not both) and nothing else is called a *Q-predicate* for that set. (If the set is the set of predicates for a language then it is a Q-predicate for the language.) Thus, let the set be:

[33] Hempel (1948).

Sx: x is a swan.
Wx: x is white.

The *Q*-predicates are as follows:

Q_1: Sx & Wx
Q_2: Sx & ~Wx
Q_3: ~Sx & Wx
Q_4: ~Sx & ~Wx.

Now consider the law:

L: (x) Sx → Wx.

Let us assume that substitution of *L*-equivalents *within* the antecedent a consequent leaves the law essentially the same. (It can be shown that such substitutions leave the predictive force of the law unchanged.) Then the laws that can be written *in terms of S and W* and whose extensional counterparts are equivalent to that of *L* are limited essentially to the following, written in terms of their *Q*-predicates:

*L_1: $(x)[Q_1x \vee Q_2x \vee Q_3x \vee Q_4x \rightarrow Q_1x \vee Q_3x \vee Q_4x]$
*L_2: $(x)[Q_1x \vee Q_2x \vee Q_3x \rightarrow Q_1x \vee Q_3x]$
L_3: $(x)[Q_1x \vee Q_2x \vee Q_3x \rightarrow Q_1x \vee Q_3x \vee Q_4x]$
*L_4: $(x)[Q_1x \vee Q_2x \vee Q_4x \rightarrow Q_1x \vee Q_4x]$
L_5: $(x)[Q_1x \vee Q_2x \vee Q_4x \rightarrow Q_1x \vee Q_3x \vee Q_4x]$
*L_6: $(x)[Q_2x \vee Q_3x \vee Q_4x \rightarrow Q_3x \vee Q_4x]$
L_7: $(x)[Q_2x \vee Q_3x \vee Q_4x \rightarrow Q_3x \vee Q_4x \vee Q_1x]$
*L_8: $(x)[Q_1x \vee Q_2x \rightarrow Q_1x]$
L_9: $(x)[Q_1x \vee Q_2x \rightarrow Q_1x \vee Q_3x]$
L_{10}: $(x)[Q_1x \vee Q_2x \rightarrow Q_1x \vee Q_4x]$
L_{11}: $(x)[Q_1x \vee Q_2x \rightarrow Q_1x \vee Q_3x \vee Q_4x]$
*L_{12}: $(x)[Q_2x \vee Q_3x \rightarrow Q_3x]$
L_{13}: $(x)[Q_2x \vee Q_3x \rightarrow Q_3x \vee Q_1x]$
L_{14}: $(x)[Q_2x \vee Q_3x \rightarrow Q_3x \vee Q_4x]$
L_{15}: $(x)[Q_2x \vee Q_3x \rightarrow Q_3x \vee Q_1x \vee Q_4x]$
*L_{16}: $(x)[Q_2x \vee Q_4x \rightarrow Q_4x]$
L_{17}: $(x)[Q_2x \vee Q_4x \rightarrow Q_4x \vee Q_1x]$
L_{18}: $(x)[Q_2x \vee Q_4x \rightarrow Q_4x \vee Q_3x]$
L_{19}: $(x)[Q_2x \vee Q_4x \rightarrow Q_4x \vee Q_1x \vee Q_3x]$
*L_{20}: $(x)[Q_2x \rightarrow Q_1x]$

L_{21}: $(x)[Q_2x \rightarrow Q_1x \vee Q_3x]$
L_{22}: $(x)[Q_2x \rightarrow Q_1x \vee Q_4x]$
L_{23}: $(x)[Q_2x \rightarrow Q_1x \vee Q_3x \vee Q_4x]$
L_{24}: $(x)[Q_2x \rightarrow Q_3x]$
L_{25}: $(x)[Q_2x \rightarrow Q_3x \vee Q_4x]$
L_{26}: $(x)[Q_2x \rightarrow Q_4x]$
L_{27}: $(x)[Q_2x \rightarrow Q_1x \,\&\, Q_2x]$.

Now, note that the predictive force of the following groups of laws must be the same: (L_2L_3) (L_4L_5) (L_6L_7) $(L_8L_9L_{10}L_{11})$ $(L_{12}L_{13}L_{14}L_{15})$ $(L_{16}L_{17}L_{18}L_{19})$ $(L_{20\text{-}27})$. This is because their antecedents are *L*-equivalent and, although their consequents are not *L*-equivalent, the conjunctions of consequent and antecedent are *L*-equivalent. That is, their reference classes do not coincide absolutely, but coincide within their attribute classes. Let us treat such laws as equivalent. Then the number of classes of equivalent laws reduces to eight, those whose simplest members are starred. "All swans are white" appears as L_8 and "All non-white things are non-swans" appears as L_{16}. L_{20} and its equivalents are in a rather strange form for predictive instruments, their consequents being inconsistent with their antecedents, giving a QIC of absolutely zero. What is required to assure high QIC to L_8, L_{12}, and L_{16} is that the probability that an unexamined individual is a falsifier $[Pr\ Q_2\ (a)]$ is small compared with the probabilities of each of the other *Q*-properties. Happily this condition assures that the QIC of every other member of the list (excepting L_{20} and its equivalents) is also high! Thus, any quantity which must be less than or equal to QIC (L_8), QIC (L_{12}), QIC (L_{16}) is such that the predictive force of any member of the group whose consequent is consistent with its antecedent must be at least as high as it.

Unfortunately, this no longer holds good when we consider laws with equivalent extensional generalizations that we can write with the addition of new primitive predicates. For instance assume that QIC $(L_8) = [Pr(Q_1a)]/[Pr(Q_1a) + Pr(Q_2a)]$ is high and that H is another primitive predicate. Expending L_5 in terms of it we get $(x)\ [(Q_1x \,\&\, Hx) \vee (Q_1x \,\&\, {\sim} Hx) \vee (Q_2x \,\&\, Hx) \vee (Q_2x \,\&\, {\sim}Hx) \rightarrow (Q_1x \,\&\, Hx) \vee (Q_1x \,\&\, {\sim}Hx)]$ and

$$\text{QIC}\ (L_8) = \frac{Pr(Q_1a \,\&\, Ha) + Pr(Q_1a \,\&\, {\sim}Ha)}{Pr(Q_1a \,\&\, Ha + Pr(Q_1a \,\&\, {\sim}Ha) + Pr(Q_2a \,\&\, Ha) + Pr(Q_2a \,\&\, {\sim}Ha)}$$

but that this quantity is high in no way guarantees that Pre $(x)[(Q_1x \,\&\, Hx) \vee Q_2x \rightarrow Q_1x \,\&\, Hx]$ will be high, for $Pr(Q_1a \,\&\, Ha)$ might be as small as you please with the difference being made up by $Pr(Q_1a \,\&\, {\sim}Ha)$.

The measure of "convertibility" of laws that we have gained must then be relativized to some finite set of predicates.

The situation, in general, can be summed up as follows:

Let L be a law, S be a finite set of predicates, $\{Q_i\}$ be the Q-predicates constructed out of S, and $Q_j \ldots Q_{j+n}$ be the Q-predicates the possession of which would falsify L. A measure of *convertible confirmation* of L with respect to S $\{CC_s(L)\}$ is a quantity which is always less than or equal to the minimum of:

$$\frac{Pr(Q_i(a))}{Pr(Q_i(a)) + Pr(Q_j(a)) + \ldots + Pr(Q_{j+n}(a))}$$

for all i different from $j \ldots j+n$.[34] Call any law an instance of whose consequent is logically incompatible with the corresponding instance of its antecedent *Pickwickian*. Call any law whose extensional counterpart is logically equivalent to that of L *a quasi equivalent of L.*

Then: If L_1 and L_2 are quasi equivalents containing only predicates in S then $CC_s(L_1) = CC_s(L_2)$ and if L_2 is non-Pickwickian $QIC\ (L_2) \geqq CC_s(L_1)$.

Looking back at the familiar cases of swans and ravens, we find that if we do *not* suppress background evidence, typical cases where we regard the law as well confirmed are cases where its convertible confirmation is high over a fairly rich set of predicates. And this explains why we do not typically fall into paradox when we treat quasi equivalents of a law as genuine equivalents. Put in the following way, the requirement that a well-confirmed law have high convertible confirmation seems almost trivially correct. For "All swans are white" to be well confirmed the probability that the something is not a non-white swan must be low not only given that it is something or other; but also given that it is a swan; given that it is a non-white thing; given that it is a fat swan; given that it is a non-white bird; etc.

Our attempts to recapture the equivalents have met with a certain degree of success. What about consequences? Let us call L_2 a *quasi*

[34] Thus the minimum over i of these quotients is itself a measure of convertible confirmation. So is the product over i.

consequence of L_1 if its extensional counterpart is a consequence of that of L_1. A law, with an antecedent and consequent written as disjuncts of Q-properties, would look like this:

$$(x)[Q_ix \vee \ldots \vee Q_{i+m}x \vee \ldots \vee Q_jx \vee \ldots \vee Q_{j+n}x \rightarrow Q_ix \vee \ldots \vee Q_{i+m}x \vee \ldots \vee Q_kx \vee \ldots \vee Q_{k+o}x]$$

where $Q_j \ldots Q_{j+n}$ are the falsifier properties. $\{Q_i \ldots Q_{i+m}\}$ must contain at least one property if the law is to be non-Pickwickian. $\{Q_k \ldots Q_{k+o}\}$ may be empty and if it is not the law it is equivalent to that which results when it is.

L_2 is a quasi consequence of L_1 if L_2's falsifier properties are a subset of those of L_1. It follows that if L_1 has high convertible confirmation L_2's falsifier properties must have low probability relative to all the Q's which are not falsifiers of L_1. But it does *not* follow that L_2's falsifiers must have high probability relative to all the Q's which are not falsifiers of L_2. Thus quasi consequence need not preserve convertible confirmation. For example, let there be two primitive predicates in S and thus four Q-predicates.

$$L_1\colon (x)[Q_1x \vee Q_2x \vee Q_3x \rightarrow Q_1x]$$
$$L_2\colon (x)[Q_2x \vee Q_3x \rightarrow Q_2x]$$
$$Pr[Q_1a] = .5 - \epsilon$$
$$Pr[Q_2a] = \epsilon$$
$$Pr[Q_3a] = \epsilon$$
$$Pr[Q_4a] = .5 - \epsilon.$$

$CC_s(L_1)$ is high but $CC_s(L_2)$ and $QIC\ (L_2)$ are low. This sort of failure, however, does not contravene the actual use of deduction in scientific practice. The antecedent of L_2 cannot be instantiated unless L_1 is falsified. Thus, the only conditions under which L_2 would be of any use as a predictive instrument are conditions under which L_1 would no longer be available to serve as a premise.

This suggests that we define a relative sense of convertible confirmation as follows: Let F be a set of Q-predicates that we treat as immune to instantiation. Let $Q_j \ldots Q_{j+n}$ be the falsifier predicates for L. Then $CC_{sf}(L)$ is a quantity which is always less than or equal to the minimum of:

$$\frac{Pr(Q_i(a))}{Pr(Q_i(a)) + Pr(Q_j(a)) + \ldots + Pr(Q_{j+n}(a))}$$

for all Q_i which are neither falsifier predicates for L nor members of F. Now we can say that if L_2 is a quasi consequence of L_1 and F is the set of falsifier predicates of L_1, then $CC_{sf}(L_2) \geqq CC_s(L_1)$.

But what is really important is the upshot of this for the predictive force of L_2 and its equivalents. Let us say that a law is *Pickwickian with respect to F* if the instantiation of both its antecedent and consequent is impossible without instantiating some member of F. And a law, L_1, is Pickwickian with respect to another L_1 just in case it is Pickwickian with respect to the set of falsifiers of L_2. Then if $CC_{sf}(L_2)$ is high and L_3 is a quasi equivalent of L_2 which is not Pickwickian with respect to F, then $QIC(L_3)$ is at least as high. Consequently, *if L_2 is a quasi consequence of L_1 written in terms of S which is non-Pickwickian with respect to L_1, then $QIC(L_2) \geqq CC_s(L_1)$*. Finally, we want to talk about consequences of sets of laws rather than just single laws. Let us define a measure of convertible confirmation of a *network of laws* as a quantity always less than or equal to

$$\mathrm{Min}_{i} \quad \frac{Pr(Q_i(a))}{Pr(Q_i(a)) + Pr(Q_j(a)) + \ldots + Pr(Q_{j+n}(a))} \qquad i \neq j \ldots i \neq j+n$$

where $Q_j \ldots Q_{j+n}$ here represent the Q-predicates the instantiation of which would falsify at least one member of the network. The term *network* is used to emphasize the interdependent nature of the convertible confirmation. For the convertible confirmation of a network of laws may be high while the convertible confirmation of its members is not. For example:

L_1: $(x)[Q_1s \vee Q_2x \rightarrow Q_1x]$
L_2: $(x)[Q_1x \vee Q_3x \rightarrow Q_1x]$
$Pr\, Q_1(a) = .5 - \epsilon$
$Pr\, Q_2(a) = \epsilon$
$Pr\, Q_3(a) = \epsilon$
$Pr\, Q_4(a) = .5 - \epsilon.$

The convertible confirmation of the network $\{L_1,L_2\}$ is high. $CC_s \{L_1L_2\} = [(.5 - \epsilon)/(.5 - \epsilon)]^2$. But $CC_s\,(L_1) = CC_s\,(L_2) = 1/2\,[(.5 - \epsilon)/(.5)]^2$. The reason is, of course, that the only conditional probabilities that we required to be high for CC of the network to be high are conditions which do not falsify any law in the network. Such exemptions are not available for laws standing alone. So if F is the class of falsifiers for a network, whenever CC_s of the network is high, CC_{sf} of each member of the network is high. Notice that networks of laws may, for this reason, exhibit a hierarchical organization. If CC of a subnetwork is as high as CC of the network, this means that the predictive force of the subnetwork will survive intact upon the discovery of instances which falsify other members of the network. It is interesting that in this context convertible confirmation becomes a measure of *independence.*

The situation with respect to quasi consequences of a network of laws and the relation of being Pickwickian with respect to a network of laws is a straightforward generalization of the foregoing. If CC_s of a network is high and F is the set of falsifiers for the network then CC_{sf} of each quasi consequence of the network written in terms of predicates in S is at least as high. And each of those quasi consequences which is not Pickwickian with respect to the network has high qualified instance confirmation.

The concept of a network of laws also allows us to look at the question of quasi consequences in the following rather attractive way:

I: Let N be a network of laws; L be a quasi consequence of N written in terms of predicates of S; and N_L the network formed by adding L to N. Then

$$CC_s\,(N_L) = CC_s(N).$$

II. Let N be a network of laws and L be a member of N (written in terms of predicates of S) which is not Pickwickian with respect to N. Then

$$QIC(L) \geqq CC_S(N).$$

We traced many of the paradoxes of section three to a failure of predictive force. But the quasi consequence relation proved so poor at preserving predictive force that it seemed quite hard to make any sense at all of the use of deductive technique in science. I and II above reestablish a broad scope for nonparadoxical deductive inference. Paradoxical inferences arise when CC of the premises is not high (though some other supposedly relevant measure may be) or when the conclusion is Pickwickian with respect to the relevant network.

8. Law, Accident, and Observation

Any quantity which is a measure of convertible confirmation has the virtues exhibited in the last two sections. But these virtues, by themselves, are not enough. Laws pose additional problems with which we must deal. A fundamental one is the distinction between law and accident. I wish to introduce it here, because the solution seems almost within our grasp with methods already developed.

The paradox of provisional acceptance showed us that the law "All F's are G's" can be held in high regard while $Pr\ (x)(Fx \supset Gx)$ is low. The opposite situation is also possible. Consider Goodman's famous example: "All coins in my pocket on V.E. day 1947 are silver." We can imagine events which would lend credence to the law (i.e., dropping copper pennies into the pocket and finding upon removal that they had been transmuted to silver, etc.) but we can also imagine less bizarre events which would lend credence to the extensional generalization but not the law (looking in my pocket on V.E. day, etc.). Thus, the nomic force of a generalization is determined not by its syntax or semantics, *but by the character of its evidential support.*

It is easy to think of accidental generalizations of spatiotemporally limited scope and tempting to think that there might be no others. In order to keep ourselves honest, we should also bear in mind Hempel's example: "All bodies of pure gold have a mass of less than 100,000 kilograms." (A little imagination will show that here too there are logically possible evidential situations which would lead us to regard this generalization as a law rather than as a global accident.) — So much for the problem.

Now remember that we defined QIC of "All F's are G's" as $Pr(Ga$ given $Fa)$ where 'a' is a "new" individual constant *not mentioned in our evidence.* But it might be difficult ultimately to maintain an epistemological picture where evidence comes as a neat list of sentences. This suggests that we might consider redefining QIC as the minimum value (or greatest lower bound) of $Pr(Ga$ given $Fa)$ for all individual constants, a, in the language, and making the coordinate redefinition of a measure of convertible confirmation. Such a redefinition has two pleasant consequences. First, the new qualified instance confirmation (QIC*) of a law is less than or equal to the probability that a is a counterinstance to that law, for each a:

$$\text{QIC* "All } F\text{'s are } G\text{'s"} \leq \text{Min}_{\alpha}\, 1 - Pr(F\alpha \,\&\, {\sim}G\alpha).$$

Second, it appears to weed out just those accidental generalizations which have been bothering us.

Let us look at our two examples. It may be very probable that every coin in my pocket on V.E. day was silver. I remember turning my pocket inside out and looking very carefully. Yet consider the coin I got in change for my ninety-nine–cent lunch. I looked at it carefully too, and a coppery A. Lincoln looked back. It is clearly quite unlikely that this coin was in my pocket on V.E. day. But that small unlikelihood is the sum of the probabilities that it was in my pocket and silver and that it was in my pocket and not silver. Their *relative* size is what is important in determining the conditional probability that the coin in question was silver *given* that it was in my pocket on V.E. day. It seems clear that this probability must be low; that should I suddenly become convinced that I was mistaken and that this coin really was in my pocket I should revise my belief about the composition of coins in my pocket rather more than my belief about the composition of this coin. On the other hand, in the bizarre "transmutation" case, where Goodman's generalization *does* have nomic force, it seems clear that this conditional probability *is* high.

In general, it seems that this will work for any accidental conditional about a finite class. Assume that our observations have rendered it likely that there are only a finite number of F's, $a_1, a_2, \ldots, a_n$ and that each of these has G. What is $Pr(Ga_{n+1}$ given $Fa_{n+1})$ where a_{n+1} is known to be distinct from $a_1 \ldots a_n$? Thus we have as mutually exclusive and exhaustive categories:

a. $(x)(Fx \supset x = a_1 \vee \ldots \vee x = a_n) \,\&\, Ga \,\&\, \ldots \,\&\, Ga_n$.
b. $(x)(Fx \supset x = a_1 \vee \ldots \vee x = a_n) \,\&\, (\exists x)({\sim}Gx)$.
c. $Fa_{n+1} \,\&\, Ga_{n+1} \,\&\, (x)(Fx \supset Gx)$.
d. $Fa_{n+1} \,\&\, Ga_{n+1} \,\&\, (\exists x)(Fx \,\&\, {\sim}Gx)$.
e. $Fa_{n+1} \,\&\, {\sim}Ga_{n+1}$.
f. None of the above and $(x)(Fx \supset Gx)$.
g. None of the above.

$$Pr\,[(x)(Fx \supset Gx)] = Pr\,(a) + Pr\,(c) + Pr\,(f).$$

$$Pr(Ga_{n+1} \text{ given } Fa_{n+1}) = \frac{Pr\,(c) + Pr\,(d)}{Pr\,(c) + Pr\,(d) + Pr\,(e)}.$$

If the generalization is "accidental" its probability will be high by virtue of *Pr* (a) being high and the conditional probability may fail to be high. If it is "lawlike" *Pr* (c) should be high *relative* to *Pr* (d) and *Pr* (e), and *Pr* (f) should be high relative to *Pr* (e).

Indeed, we seem to have equally good luck with universal accidents. We may assume that the old instance and qualified instance confirmations of "All bodies of pure gold have a mass of less than 100,000 kilograms" are high, but the probability that Ganymede has a mass of less than 100,000 kilos, *given* that Ganymede is pure gold, is not high. Thus, QIC* and coordinate measures of convertible confirmation will be appropriately low.

Unfortunately, in one peculiar sort of case, QIC* appears to disqualify perfectly decent laws as well. Suppose that we are in a position to accord high CC* to the law "All swans are white" and that we then observe a black bird without being able to determine whether it is a swan or not. If we believe the law, we shall of course *believe* that this bird is not a swan, but what we are interested in here is the probability that it is white *given* that it is a swan, and there is good reason to believe that now *that* conditional probability must be low. For if we subsequently determined that the bird *was* a swan, we would typically assign high probability to the proposition that it was a *black* swan and a disconfirmer of the law, rather than taking this new observation as grounds for repudiating our previous conviction that it was a black bird. This means that if we change our degrees of belief by conditionalization, *Pr*(White a given Swan a) *must have been low*. Thus, an observation that determines that a bird is black without determining that it is a swan or not renders QIC* and CC* of "All swans are white" low, although the law should certainly retain its viability in these circumstances.

Example:

		a is black	*a is a swan*
	Pr_1	Pr_2	Pr_3
Sa & Wa	.33	.01	.36
Sa & ∼Wa	.01	.02	.66
∼Sa & Wa	.33	.01	.0002
∼Sa & ∼Wa	.33	.96	.0198

If we choose more powerful laws as examples there is more of a tendency to retain the law at the expense of the earlier observation *but this tend-*

ency cannot be overriding lest it render the law immune to instantial disconfirmation.

It seems that we need to exempt cases where a has been examined for W but not for S or for S but not for W from those over which we take CC*. But to state such an exemption we need to presuppose that neat distinction between knowledge by observation and knowledge by inference which is so much in doubt.

9. Resiliency

In connection with various problems, the distinction has been made between intensional and extensional evidence for a disjunction. The idea is that evidence for one of the disjuncts will automatically count as *extensional* evidence for the disjunction but not necessarily as intensional evidence. Intensional evidence should be evidence for the disjunction as a unified, connected whole — so that if one disjunct were cast into doubt, the other would tend to take up the slack.

This image naturally suggests:

$$I(p \vee q) = \text{Min}\{Pr(p \text{ given } {\sim}q), Pr(q \text{ given } {\sim}p)\}$$

as a measure of the strength of the intensional evidence for $p \vee q$.

Since any truth function can be put in disjunctive normal form, we have a natural extension of this idea for all truth functions. Let $S = \{p_1 \ldots p_n\}$ a set of atomic sentences including the atomic sentences of q and let $\{S_i\}$ be the n disjuncts when q is written in disjunctive normal form in terms of the members of S. Then the *resiliency* of q over S is as follows:

$$\begin{aligned} Rs(q) = \text{Min}\{&Pr\,[S_1 \text{ given } {\sim}(S_2 \vee S_3 \vee \ldots \vee S_n)] \\ &Pr\,[S_2 \text{ given } {\sim}(S_1 \vee S_3 \vee \ldots \vee S_n)] \\ &\cdot \\ &\cdot \\ &\cdot \\ &Pr\,[S_n \text{ given } {\sim}(S_1 \vee S_2 \vee \ldots S_{n-1})]\}.\end{aligned}$$ [35]

Since $Pr\,[S_1 \text{ given } {\sim}(S_2 \vee S_3 \vee \ldots \vee S_n)]$ is just $Pr\,[q \text{ given } {\sim}(S_2 \vee S_3 \vee \ldots \vee S_n)]$ and since the members of $\{S_i\}$ are just the state-descrip-

[35] If we consider only belief states where just tautologies get probability 1 and just contradictions get probability zero, then these conditional probabilities will always be well defined. Such a restriction can be independently motivated.

tions over S which are compatible with q we might further generalize the concept of resiliency to any set of atomic sentences.

Let S be any set of atomic sentences, $\{S_i\}$ be the set of n state-descriptions in terms of members of S compatible with q, and let S_{-j} be the disjunction of all members of $\{S_i\}$ except S_j. Then

$$R_s\,(q) = \underset{j=1,\,n}{\text{Min}}\; Pr\;[q \text{ given } S_{-j}].$$

If S is disjoint from the set of atomic sentences of q, then $Rs(q)$ is just the minimum of the conditional probabilities given each state-description over S.

The generalization is in a sense justified by the following fact: If r is compatible with q and S contains all the atomic sentences occurring in r, then $Pr\;(q \text{ given } r) \leqq R_s\;(q)$.

There appear to be just two legitimate sources of resiliency of belief in a contingent statement; the first being habits stemming from recurrent patterns in our experience, the second being from the vividness of current observation.

With regard to the first source, I wish to make two connections between resiliency and the preceding discussions. The first is the obvious connection with convertible confirmation. The resiliency of the *material* conditional for an unexamined individual is a measure of convertible confirmation:

$$R_T\;(Fa \supset Ga) = CC_S \text{ “All } F\text{'s are } G\text{'s”}$$

(Where $\lambda \in S$ iff $\lambda\, a \in T$).

The second is the connection with lawlikeness itself. Suppose we expand our language by the addition of one new individual constant, δ. Then the discussion of the last section suggested that lawlikeness has to do with the resiliency of “$F\delta \supset G\delta$” over (‘$\delta = a_1$’, ‘$\delta = a_2$’ . . .) for each individual constant in the language — barring certain exceptional cases.

This brings us back to our outstanding problem, which consists of just those exceptional cases. It may now be viewed as a collision between resiliency owing to the first source (patterns in experience) and resiliency owing to the second source (current observation).

High resiliency is a strong requirement, and it is easy for conflicts in resiliency to arise. For instance, if p and q are atomic sentences it is impossible for $R_{\{p,q\}}$ (p) and $R_{\{p,q\}}$ $(p \supset q)$ both to be high. For the first to be high, Pr (p given $\sim q$) must be high. For the second to be high, Pr ($\sim p$ given $\sim q$) must be high. Whichever statement wins in the crunch (autonomous increase in probability of $\sim q$) gets the high resiliency.

We have already seen that where the contest is between resiliency of $p \supset q$ deriving from a law and resiliency of p (or of $\sim q$) deriving from observation, the resiliency deriving from observation must in some cases (good clear current observation) be allowed to win if we are to account for instantial disconfirmation of laws (although it is worth noticing that a highly confirmed physical law may be so resilient that *repeated* counterobservations are required to disconfirm it).

But although p and $p \supset q$ cannot both be resilient (over $\{p,q\}$) their conjunction can, since R_s $(p \,\&\, (p \supset q)) = R_s$ $(p \,\&\, q)$ and $R_{\{p,q\}}$ $(p \,\&\, q) = Pr$ $(p \,\&\, q)$. Thus $R_{\{Fa,\, Ga\}}$ $[(Fa \supset Ga) \,\&\, Fa]$ can be as high as can $R_{\{Fa, Ga\}}$ $[(Fa \supset Ga) \,\&\, {\sim}Ga]$ but $R_{\{Fa, Ga\}}$ $[(Fa \supset Ga) \,\&\, Fa \,\&\, {\sim}Ga]$ is, of course, zero. The preceding statement holds good also if resiliency is taken over a broader basis.

This suggests that the problem has been put wrong, that what receives high confirmation is not a law nor a network of laws but a network of both laws and observation statements. The smaller the subnetworks containing a law which retain resiliency, the more central the law is to our system of beliefs. The number of statements in such a system which owe their resiliency directly to observation will be small, for only current observation in favorable conditions will provide the required resiliency.

The idea of a *network* which emerged in the context of convertible confirmation and quasi consequence now extended to include nonlaws through resiliency thus solves our outstanding problem. It is interesting that the logic of laws appears to *force* this wholistic viewpoint upon us.

10. Conclusion

This paper has taken only the first steps toward a philosophical reduction of laws of nature. But if these steps are in the right direction, we can already draw some nontrivial philosophical conclusions. The logic of laws is, in the most general case, a logic of networks. Our most useful measures of virtue for laws and networks of laws, *resiliencies*, are functions of our probability assignments on an extensional language although they are not themselves probabilities.[36] The status of a law within a network is determined by the resiliency of *sub*networks which contain it. Resiliency accounts both for the probabilistic presuppositions of deductive science and for the nomic force of laws. Resiliency is a kind of modal notion. If a sentence is resilient over S in a belief state, it will retain high probability in any belief state accessible from the first by conditionalization[37] on any sentence which is consistent with it and written in terms of S, and conversely. Thus the resiliency of S in a given belief state may be thought of in two ways — as a function of the probabilities that S has in certain related belief states, or as a function of the probabilities of other statements in the original belief state. It is thus a sort of modality without metaphysics, reducing nomic necessity from an ontological to an inferential status.

[36] Resiliency is not additive.

[37] Or by a weighted average of conditionalizations on p and $\sim p$ à la Jeffrey. See Jeffrey (1965), ch. 11. This sort of transformation is to be preferred since it avoids introducing zeros and ones into the probability distribution.

REFERENCES

Adams, E. W. (1965). "The Logic of Conditionals," *Inquiry*, vol. 8, pp. 166–97.

Adams, E. W. (1966). "Probability and the Logic of Conditionals," in J. Hintikka and P. Suppes, eds., *Aspects of Inductive Logic*. Amsterdam: North Holland. Pp. 265–316.

Beth, E. W. (1966). *The Foundations of Mathematics*. New York: Harper Torchbooks.

Bohm, D. (1957). *Causality and Chance in Modern Physics*. New York: Van Nostrand.

Carnap, R. (1937a). *Logical Syntax at Language*. New York: Harcourt, Brace.

Carnap, R. (1937b). *Introduction to Semantics*. Cambridge, Mass.: Harvard University Press.

Carnap, R. (1939). "Foundations of Logic and Mathematics," in *International Encyclopedia of Unified Science*, vol. 1. Chicago: University of Chicago Press.

Carnap, R. (1950). *Logical Foundations of Probability*. Chicago: University of Chicago Press.

Carnap, R. (1956). *Meaning and Necessity*, enlarged ed. Chicago: University of Chicago Press.

Chisholm, R. M. (1946). "The Contrary-to-Fact Conditional," *Mind*, vol. 55, pp. 289–307.

Ellis, B. (1968). "Probability Logic," mimeo.

Feyerabend, P. K. (1962). "Explanation, Reduction, and Empiricism," in H. Feigl and G. Maxwell, eds., *Minnesota Studies in the Philosophy of Science*, vol. 3. Minneapolis: University of Minnesota Press.

Goodman, N. (1955). *Fact, Fiction and Forecast*. Cambridge, Mass.: Harvard University Press.

Hempel, C. G. (1948). "Studies in the Logic of Confirmation," *Mind*, vol. 54, pp. 1–29, 97–121.

Hempel, C. G. (1958). "The Theoretician's Dilemma," in H. Feigl, M. Scriven, and G. Maxwell, eds., *Minnesota Studies in the Philosophy of Science*, vol. 2. Minneapolis: University of Minnesota Press.

Hempel, C. G. (1966). *Philosophy of Natural Science*. Englewood Cliffs, N.J.: Prentice-Hall.

Hempel, C. G. (1968). "On a Claim by Skyrms Concerning Lawlikeness and Confirmation," *Philosophy of Science*, vol. 35, pp. 274–78.

Jeffrey, R. C. (1964a). "If," abstract in *Journal of Philosophy*, vol. 61, pp. 702–3.

Jeffrey, R. C. (1964b). "Review of *Logic Methodology and Philosophy of Science*," *Journal of Philosophy*, vol. 61, pp. 79–88.

Jeffrey, R. C. (1965). *The Logic of Decision*. New York: McGraw-Hill.

Jeffrey, R. C. (1966). "Goodman's Query," *Journal of Philosophy*, vol. 62, pp. 281–83.

Jeffrey, R. C. (1968). "Probable Knowledge," in I. Lakatos, ed., *Inductive Logic*. Amsterdam: North Holland. Pp. 166–80.

Lewis, C. I. (1960). *A Survey of Symbolic Logic*. New York: Dover. Pp. 51–72.

Montague, R. (1968). "Pragmatics," in R. Klibansky, ed., *Contemporary Philosophy*. Florence: La Nuova Italia Editrice.

Montague, R. (1972). "Pragmatics and Intensional Logic," in D. Davidson and G. Harman, eds., *Semantics of Natural Language*. Dordrecht: Reidel.

Morris, C. W. (1939). "Foundations of the Theory of Signs," in *International Encyclopedia of Unified Science*, vol. 1. Chicago: University of Chicago Press.

Pap, A. (1958). "Disposition Concepts and Extensional Logic," in H. Feigl, M. Scriven, and G. Maxwell, eds., *Minnesota Studies in the Philosophy of Science*, vol. 2. Minneapolis: University of Minnesota Press. Pp. 196–224.

Ramsey, F. P. (1960). *The Foundations of Mathematics*. New York: Humanities Press.

Rescher, N. (1964). *Hypothetical Reasoning*. Amsterdam: North Holland. (See appendix 3.)

Sellars, W. (1958). "Counterfactuals, Dispositions, and Causal Modalities," in H. Feigl, M. Scriven, and G. Maxwell, eds., *Minnesota Studies in the Philosophy of Science*, vol. 2. Minneapolis: University of Minnesota Press. Pp. 225–308.

Sellars, W. (1962). "Time and the World Order," in H. Feigl and G. Maxwell, eds., *Minnesota Studies in the Philosophy of Science*, vol. 3. Minneapolis: University of Minnesota Press. Pp, 527–616.

Skyrms, B. (1964). "On Failing to Vindicate Induction," *Philosophy of Science*, vol. 31, no. 1, pp. 62–64.

Skyrms, B. (1966). "Nomological Necessity and the Paradoxes of Confirmation," *Philosophy of Science*, vol. 33, pp. 230–49.

Skyrms, B. (1967). "The Explication of 'x knows that p,'" *Journal of Philosophy*, vol. 64, pp. 373–89.

Stalnaker, R. C. (1968). "A Theory of Conditionals," in N. Rescher, ed., *Studies in Logical Theory*. Oxford: Oxford University Press. Pp. 98–112.

Stalnaker, R. C. (1970). "Probability and Conditionals," *Philosophy of Science*. vol. 37, pp. 64–80.

Stevenson, C. L. (1970). "If-ICULTIES," *Philosophy of Science*, vol. 37, pp. 27–49.

ROBERT M. ANDERSON, JR.

Paradoxes of Cosmological Self-Reference

> Ken-O and his disciple Menzan (1683–1769, Soto)
> were eating a melon together. Suddenly the master
> asked, "Tell me, where does all this sweetness
> come from?"
> "Why," Menzan quickly swallowed and answered,
> "it's a product of cause and effect."
> "Bah! That's cold logic!"
> "Well," Menzan said, "from where then?"
> "From the very 'where' itself, that's where."
>
> A Zen Anecdote[1]

When I first considered writing this paper, I thought that perhaps one way to do it would be to have one page contain the title and then devote a few blank pages to what I want to say. Instead I have decided to put some marks on the pages. I will relate some parables and poems of science, some tales of science fiction-fact, which will (like the koans of the Zen master) lead you into paradoxes. In this way I will show you the paradoxes rather than tell them to you. Sometimes I will try to tell you them. But then:

My propositions serve as elucidations in the following way: anyone who understands me eventually recognizes them as nonsensical, when he has used them — as steps — to climb up beyond them. (He must, so to speak, throw away the ladder after he has climbed up it.)

AUTHOR'S NOTE: The ideas for this paper came into my head while I was being supported by a grant from the Carnegie Corporation as a research assistant at the Minnesota Center for the Philosophy of Science. It has been written while receiving support as a biological sciences training fellow on a grant from the National Institute of Mental Health: MH8304-09. I have profited from discussions with Professor Grover Maxwell and many others, but they are not in the least responsible for the absurdities that follow.

[1] L. Stryk and T. Ikemoto, eds., *Zen: Poems, Prayers, Sermons, Anecdotes, Interviews* (Garden City, N.Y.: Doubleday, 1963), p. 107.

He must transcend these propositions, and then he will see the world right.[2]

I begin with an innocent looking criticism of the use of meaning by contemporary philosophers of science in their analyses of theories. Although meaning has been used by philosophers for many years as an analytical tool in their accounts of theories, I am concerned here more particularly with it as it has been used by the "new wave" of philosophers of science: Norwood Russell Hanson, Stephen Toulmin, Thomas Kuhn, and Paul K. Feyerabend. All of these philosophers have attacked the more traditional ideas on the meaning of terms in theories. According to more traditional philosophers of science (logical empiricists), there are two importantly different kinds of terms in a theory — observation terms and theoretical terms. Observation terms refer directly to experience (or to middle-sized physical objects). Their meaning does not vary according to the theory with which they are associated by way of correspondence rules. Theoretical terms are interpreted by reference to observation terms and derive their meaning, at least in part, from such terms. As Feigl puts it, "There is an 'upward seepage' of meaning from the observational terms to the theoretical concepts."[3]

The "new wave" of philosophers of science (or "radical meaning variance theorists"[4] as they have been called) reject in its entirety this account of scientific theories. In particular, for theories of general scope (i.e., theories that say something about everything) they deny the possibility of theory neutral observation terms. When you have such theories, they claim, even the observations cannot escape the theories' nomological nets. Observations cannot be described in language that is not theoretically contaminated. The observation terms of any theory are theory laden — they obtain part of their meaning from the theoretical context in which they are embedded.

Paul K. Feyerabend makes this point in the course of his more general attacks on logical empiricist philosophy of science.[5] He rejects the

[2] L. Wittgenstein, *Tractatus Logico-Philosophicus* (London: Routledge & Kegan Paul, 1961), 6.54.

[3] H. Feigl, "The 'Orthodox' View of Theories: Remarks in Defense as Well as Critique," in M. Radner and S. Winokur, eds., *Minnesota Studies in the Philosophy of Science*, vol. 4 (Minneapolis: University of Minnesota Press, 1970), p. 7.

[4] C. R. Kordig, *The Justification of Scientific Change* (Dordrecht: Reidel, 1971).

[5] In what follows I wish to concentrate on Feyerabend's philosophy of science. I hope to unravel a coherent strand from the seemingly inconsistent tangle of Feyer-

logical empiricist principle that the meanings of the terms of theories must be invariant with respect to scientific progress; "that is, all future theories will have to be formed in such a manner that their use in explanations does not affect what is said by theories, or factual reports to be explained"[6] (condition of meaning invariance). According to Feyerabend, science does not and should not follow this principle. Knowledge grows by changes taking place that are more akin to the metamorphosis of a caterpillar into a butterfly than to the development of a chick into a hen. When this sort of change takes place, the meanings of the terms of our theory of the world undergo a radical transformation. A common example of this is the Newtonian world view compared with the Einsteinian. Length, according to each of these theories, is something quite different. According to the Newtonian world view length is independent of the frame of reference of the observer. In the Einsteinian view, it is dependent. Since observation terms acquire part of their meaning from their theoretical context, it follows that they also differ radically in their meanings from theory to theory. According to Feyerabend, "Each theory will have its own experience."[7] It is important to note here that the controversy about whether a scientist's belief in a particular theory affects how he sees the world or brings about phenomenological differences in his experience is irrelevant to this point. The point is not that theories change experience (e.g., make the interferometer bands appear to dance, whereas before they just floated by), but that theories characterize experience, tell us how it fits in with everything else. "Experience is one of the processes occurring in the world. It is up to detailed research to tell us what its nature is, for surely we cannot be allowed to decide about the most fundamental thing without careful research."[8]

For Feyerabend, each theory has its own set of facts. It is this consequence of Feyerabend's account of scientific theories (and of those of the other "radical meaning variance theorists") that has prompted the question, "How is it possible to compare theories if they have no

abend's philosophy. For some of Feyerabend's inconsistencies, see D. Shapere, "Meaning and Scientific Change," in R. G. Colodny, ed., *Mind and Cosmos* (Pittsburgh: University of Pittsburgh Press, 1966).

[6] P. K. Feyerabend, "Problems of Empiricism," in R. G. Colodny, ed., *Beyond the Edge of Certainty* (Englewood Cliffs, N.J.: Prentice-Hall, 1965), p. 164.

[7] *Ibid.*, p. 214.

[8] *Ibid.*, p. 151.

domain of overlap?" A critical experiment that would simultaneously refute one of two competing theories and confirm the other is clearly impossible. For example, an experiment that would decide between Newtonian and Einsteinian theory is impossible (among other reasons) because the instruments and other entities involved in the crucial observations are, in the Newtonian theory, three-dimensional spatial solids which endure through time, whereas in the Einsteinian account, they are four-dimensional invariants in space-time.

Although I am in agreement generally with Feyerabend's approach to the philosophy of science, it seems that in his presentation of his account of theories, he has made an important error. This is his use of "meaning" as a metatheoretical term. He is not alone in making this mistake, however, since the other "radical meaning variance theorists" also make it. And it is further magnified by the critics[9] of the view fastening on the arguments which turn on how meaning is construed. This is no surprise, considering present-day fashion of assimilating philosophy to linguistic analysis and the philosophy of language.

The point I wish to make is that meaning is as much a process occurring in the world as is experience. Meaning has to do with the operation of living brains in their biological and extrabiological environments. A highly simplified kind of meaning is present in coding operations. For example, some general purpose computers must be programmed through a set of twelve binary switches which may be in either the up or the down position.[10] This makes it necessary to communicate with the computer with sequences of

DUUDDDUDUDDU
UUDDDUDDDUDD

To avoid having to memorize long lists of binary numbers, programmers developed a code. They divided the twelve switches into sets of three and labeled the up position with an integer in the order 4-2-1. When two or three switches are up, the sum is represented. Thus

0 means DDD
1 means DDU

[9] The critics I have specifically in mind are Shapere, Kordig, and P. Achinstein, "On the Meaning of Scientific Terms," *Journal of Philosophy*, 61 (1964), 497–509.

[10] K. H. Pribram, *Languages of the Brain* (Englewood Cliffs, N.J.: Prentice-Hall, 1971), p. 67.

2 means DUD
3 means DUU
etc.

and any twelve-number binary sequence can be represented by four numbers. This code can be instantiated as an electronic circuit. It is plausible that similar codes are "wired in" to our brains and are important for memory and meaning. I must add, however, that almost nothing is known about memory and even less is understood about meaning (this example is misleading since it ignores the core problem of meaning — context).[11]

Whether or not you accept a biological approach to the problem of meaning, if you grant that meaning is part of the world, then Feyerabend must draw the same conclusions about meaning that he draws about experience. That is, that every theory will have its own meaning. Just as experience is characterized independently by each theory, so is meaning. What meaning is, is dependent on the particular theory being entertained. Observation is not theory neutral and neither is meaning. Comparison of theories in terms of meaning must be rejected for the same reasons that comparison of theories in terms of experience is. Neither realm of existence is safe from the infinitely thorough purge of the universe that accompanies a new theory. Neither can bridge the gap of indefiniteness that separates two theories.

Feyerabend doesn't always make the mistake of using "meaning" as a metatheoretical term for stating his view of theories. He sometimes, instead, puts it in terms of the complete replacement of the ontology of one theory with that of another, a complete ontological housecleaning.[12] In fact, when Feyerabend is pressed about what he means by two theories being radically different, he explicates himself by saying that they are different if they postulate different *kinds of entities*.[13] In the same

[11] For an interesting paper on the problem of context, see J. D. Bransford and N. S. McCarrell, "Some Thoughts about Understanding What It Means to Comprehend," in W. B. Weimer and D. S. Palermo, eds., *Cognition and the Symbolic Processes*, in press.

[12] P. K. Feyerabend, "Explanation, Reduction, and Empiricism," in H. Feigl and G. Maxwell, eds., *Minnesota Studies in the Philosophy of Science*, vol. 3 (Minneapolis: University of Minnesota Press, 1962), p. 29.

[13] P. K. Feyerabend, "On the 'Meaning' of Scientific Terms," *Journal of Philosophy*, 62 (1965), 267.

paper where he admit this, he also says that meaning isn't very important.[14]

The "radical meaning variance theorists," and especially Feyerabend, have been taken to task by some philosophers because of the broad meaning they have given to the word "theory." Consider the following definition given by Feyerabend: "The term 'theory' will be used in a wide sense, including ordinary beliefs (e.g., the belief in the existence of material objects), myths (e.g., the myth of eternal recurrence), religious beliefs, etc. In short, any sufficiently general point of view concerning matters of fact will be termed a 'theory.' "[15] Rather than hassle endless debates about semantics, I will follow Kuhn and use a word different from "theory." A *cosmology* will be pretty much what for Feyerabend is a theory. (Feyerabend, at times, talks about cosmologies instead of theories.) Cosmologies are attempts to give an accounting of what there is in the world and of the relations (if any) that these things bear to one another. Examples of cosmologies are the Elizabethan world view (which ordered the things in the world — God, angels, people, animals, the elements — into a chain of being),[16] the scientific world view as expressed in general relativity theory, and even the washed-out world picture of the suburban "nowhere man."

The task of first importance for the philosopher is the construction of a cosmology. This is one of the lessons that is to be found in Feyerabend's (and Russell's) philosophy of science. Since Descartes, philosophy has often been done backward. Philosophers have asked the question, "How can I know anything?" and have tried to answer their question ex nihilo without reference to any theoretical frame. They tried to do this by developing a logic that would certify inferences from an incorrigible factual core of experience. The hopelessness of trying to develop such a logic has been shown by many philosophers.[17] (It appears, however, that some of the contributors to this volume have not yet given up this quest.) Feyerabend's philosophy of science shows that experience itself is open to theoretical interpretation. What experience is depends on the cosmology that one adopts.

The general lesson to be learned is that cosmology precedes episte-

[14] *Ibid.*, p. 273.

[15] Feyerabend, "Problems of Empiricism," p. 219.

[16] E. M. W. Tillyard, *The Elizabethan World Picture* (New York: Random House, 1944).

[17] See numerous articles by Popper and his students.

mology rather than the other way around. If one wishes to understand what we can know about the world and how we can know it, one must (as Russell did) first have a cosmology (for Russell it was the description of the world given by the physics of his day).[18] Given your account of the world, you can then ask and answer the questions that interest you. If you are interested in problems of perception and knowledge, the knower-known relation is worked out in terms of the cosmology. For example, your cosmology might be that of a Newtonian materialist. You believe that there are atoms and the void and that the atoms obey Newton's laws. You might, given this cosmology, try to make out the knower-known relationship in terms of the relation of the neural activity of brains (interpreted as movements of atoms) and clusters of atoms outside these brains. Intentionality and meaning might be made out in terms of brain states. Truth could be the existence of a structural similarity between the configuration of atoms in the brain and the clusters of atoms outside.

Problems of justification can arise with respect to local theories, but they must be solved within the cosmology. The question of justification does not arise with respect to the cosmology. The cosmology is not justified by reference to other contingent theories, nor does it stand in need of justification. All justification takes place within the cosmology.

Just as experience and meaning are part of the world, so is cosmology. Just as experience and meaning are defined by the cosmology, the cosmology must be also. The cosmology must say what itself is. "Every cosmology must have its own cosmology," to substitute new values for the variables in Feyerabend's proposition. It is clear that one cannot appeal to a metacosmology to define the cosmology. The cosmology, in giving an account of what is in the world, exhausts the field. Feyerabend seems often to not see this (if he agrees with it at all). He often calls for the proliferation of theories and for scientists to use their imaginations in the name of scientific progress. Ignoring the problem that Feyerabend has with maintaining his philosophy of science consistently and also defining scientific progress (how can two different comprehensive theories of the world be compared, and one be meaningfully labeled "better" than the other?), why does Feyerabend believe that a proliferation of theories will bring about scientific progress? One explanation might be as follows: The scientific community might be thought of as

[18] B. Russell, *Human Knowledge* (New York: Simon and Schuster, 1967).

an Ashby set of homeostats interacting with the environment.[19] Such a set of homeostats is highly adaptable not only because it is a stable system but also because random processes can take command and the entire system realign itself. Scientists should use their imaginations to keep the random element functioning so that the scientific community can realign itself. This reason for scientists using their imaginations could be put in terms of the Newtonian materialist cosmology augmented by a local theory of homeostats. But Feyerabend supplies no such underpinning for his exhortations. His maxim that one should proliferate theories has the status of a metacosmological precept independent of any particular world view.

It is because of the impossibility of a metacosmology that I said at the beginning of this paper that I could show you what I want to say but couldn't really tell it to you. To tell it to you I would have to make general statements about cosmologies, do metacosmology. But this I can't do. Nevertheless, I have written many propositions which you must transcend if you wish to attain truth.

It is also due to the impossibility of metacosmology that if an account is to be given of the cosmology, it must be done in terms of the cosmology. But is this possible? Is it possible for the cosmology to define itself?

The cosmology gives an account of the world.

Only what the cosmology says goes. (Contra Feyerabend.)

The cosmology says what meaning is.

The cosmology says what it is to communicate.

The cosmology says whether anything is contingent,
whether anything is necessary.

And what it is for anything to be
contingent or necessary.

The cosmology says how things can be known
and therefore what is known.

It therefore says how itself is known.

The cosmology says what it is for the
cosmology to say something.

The cosmology says what the cosmology is.

[19] W. R. Ashby, *Design for a Brain* (New York: John Wiley, 1960).

Is there a nest of paradoxes concealed in this aphoristic? Consider a second set of statements set more in the "material" mode.

What is, is.

What is known and what knows is part of what is.

What is, determines what is known and what knows.

When a person says what is, his saying of what is, is part of what is.

What is, determines what it is to say what is.

What is, says what is.

I am using these chants to emphasize the bizarreness of the idea of something defining its own nature.

Let's take a world view and see how these maxims work within it. For simplicity, let's take the Newtonian materialist cosmology. According to this view, there are atoms and the void and the atoms obey Newton's laws. I have already said how truth, meaning, and knowledge might be worked out within this cosmology, but what of the cosmology itself? The cosmology must be certain collections of atoms. It must consist of configurations of atoms in people's brains, of certain other configurations similar in structure on paper, and of others in the memory banks of computers. If this is the cosmology, how does it say what there is in the universe? It does so by bearing a structural resemblance to the rest of the universe.[20] It postulates structures similar to its own to exist in the universe. The cosmology is a part of the universe miming the universe. It mimes the universe by following and matching the movements of the universe. Its structure unfolds as similar to the structure of the universe. But if this is how the cosmology can say what exists in the universe, how does it say something about what itself is? How can it bear a structural resemblance to itself? (Perhaps the cosmology's macrostructure, e.g., marks on paper, can bear a structural resemblance to its microstructure, atomic composition.) How can it mime itself? It simply is itself so how can it bear a resemblance to itself or mime itself? How then can it define its own nature?

We are confronted by a dilemma. Either we try to understand the world (i.e., do philosophy) without reference to a presuppositional frame or cosmology (i.e., try to develop an epistemology without refer-

[20] G. Maxwell, "Theories, Perception, and Structural Realism," in R. G. Colodny, ed., *The Nature and Function of Scientific Theories* (Pittsburgh: University of Pittsburgh Press, 1970), p. 17.

ence to an ontology) or we adopt a cosmology and confront the paradox that the cosmology must define itself. Philosophy is either barren or paradoxical.

In one koan, the Zen master brandishes a stick over the pupil's head, and says fiercely, "If you say this stick is not real, I will strike you with it. If you say this stick is real, I will strike you with it. If you don't say anything, I will strike you with it."[21]

Can the cosmos or world as a whole say what itself is? To say something about itself it must be able to refer to itself. But how can it refer to itself if there is nothing outside itself? The famous sentence in Gödel's proof that says of itself that it can't be proved can do this because its self-reference is mediated by things (people and institutions) outside of itself. But the cosmos can't transcend itself. Neither can it compare itself relative to some other form of being and define itself relative to this other form. Within the cosmos, however, things may be able to define themselves with respect to each other. The cosmos may, on the other hand, be the most perfect theory of itself. Since it is self-identical, it bears a perfect resemblance to itself. The cosmos represents itself by being itself.

Can I say what I am? I am a collection of atoms with such and such properties. But what has been accomplished with such a phrase? Some atoms have arranged themselves in a particular pattern within me. And what has been accomplished in this answer? Some atoms have arranged themselves in a pattern within me. My atoms say that they are atoms. (I am still using the Newtonian materialist cosmology.)[22]

[21] R. E. Ornstein, *The Psychology of Consciousness* (San Francisco: W. H. Freeman, 1972), p. 156.

[22] Bertrand Russell's philosophy, especially as currently modified and augmented by Grover Maxwell, may escape this paradox. Experience is not interpreted for Russell and Maxwell but is direct. Nothing is hidden. A "person" is certain events in the universe, and in being these events he directly experiences them. He can know *what* these events are, know their natures. This is similar to the world's being the best theory of itself by being the same as itself. (A quotation from Schopenhauer feels appropriate here: ". . . a consciousness when directed inwardly becomes *self-consciousness*. Then that inner being presents itself to this self-consciousness as that which is so familiar and so mysterious, and is denoted by the word *will*." ". . . by this word nothing is further from our intention than to denote an unknown *x*; but, on the contrary, we denote that which at least on one side is infinitely better known and more intimate than anything else." R. Taylor, ed., *The Will to Live: Selected Writings of Arthur Schopenhauer* (Garden City, N.Y.: Doubleday, 1962), p. 33. Except for the implicit reference to an act-object distinction, I think this is how Maxwell and Russell feel.

When your practice is not good, you are poor Buddha. When your practice is good, you are good Buddha. And *poor* and *good* are Buddhas themselves. *Poor* is Buddha and *good* is Buddha and *you* are Buddha also. Whatever you think, say, every word becomes Buddha. I am Buddha. *I* is Buddha, and *am* is Buddha, and *Buddha* is Buddha. Buddha. Buddha. Buddha. Buddha.[23]

Perhaps when Western analysis, the scientific method, is pushed to extremes it leads to Eastern mysticism. I have taken the philosophy of science (Feyerabend's) that seems to me the strongest of all present-day contenders. And I have tried to develop it consistently with its precept that everything is to be interpreted within a presupposed cosmology. This led to cosmological self-references which seemed to be paradoxical and to our consequent puzzlement. But perhaps it also leads to enlightenment?[24]

[23] From a lecture by Shrinya Suzaki Roshi at Zen Mountain Center in 1968. C. Naranjo and R. E. Ornstein, *On the Psychology of Meditation* (New York: Viking Press, 1971), p. 84.

[24] If you have had a great deal of trouble with this essay, and especially with the last part, I suggest you indulge yourself in some wine or other intoxicant. It may be more fun the second time through. Once I was telling these ideas to some friends over wine. One of them suddenly said, "Yes, I see what the paradox is. It's . . ." She stopped in midsentence with her mouth hanging open.

INDEXES

Name Index

Subject Index